20

6

7

19

7

6

8

12

9

9

10

11

**REGIONS OF
THE UNITED STATES
AND CANADA**

REGIONAL GEOGRAPHY
OF THE UNITED STATES
AND CANADA

REGIONAL GEOGRAPHY OF THE UNITED STATES AND CANADA

Second Edition

Tom L. McKnight

Professor Emeritus of Geography
University of California, Los Angeles

Prentice Hall, Upper Saddle River, New Jersey 07458

Library of Congress Cataloging-in-Publication Data

McKnight, Tom L. (Tom Lee)
    Regional geography of the United States and Canada / Tom L.
  McKnight. -- 2nd ed.
        p.   cm.
    Includes bibliographical references and index.
    ISBN 0-13-456484-7
    1. United States--Geography.   2. Canada--Geography.   I. Title.
  E161.3.M35   1997
  917.3--dc20   96-46115
                                              CIP

Acquisitions Editor: Daniel Kaveney
Editor-in-Chief: Paul F. Corey
Editorial Director: Tim Bozik
Assistant Vice President of Production and Manufacturing: David W. Riccardi
Executive Managing Editor: Kathleen Schiaparelli
Assistant Managing Editor, Production: Shari Toron
Production Editor: Alison Aquino
Cover Designer: Bruce Kenselaar
Cover Photo: Mount Adams, Washington from Hood River Valley, Oregon. Courtesy of Superstock, Inc.
Manufacturing Manager: Trudy Pisciotti
Director, Image Resource Center: Lori Morris-Nantz
Photo Research Supervisor: Melinda Lee Reo
Image Permission Supervisor: Kay Dellosa
Photo Researcher: Beaura Katherine Ringrose
Image Coordinator: Reynold Reiger
Copy Editor: Barbara Liguori
Composition: Preparé Inc. / Emilcomp srl
Photo Credits: p. xv—Joan Clemons; pp. 1, 8, 29, 52, 82, 93, 117, 163, 200, 229, 258, 290, 316, 351, 377, 394, 427—Tom McKnight;
    p. 134—Port Authority of New York and New Jersey; p. 180—Aerial Innovations/Atlanta Convention & Visitors Bureau;
    p. 452—BP America, Inc.

© 1997, 1992 by Prentice-Hall, Inc.
Simon & Schuster / A Viacom Company
Upper Saddle River, New Jersey 07458

Printed in the United States of America
10   9   8   7   6   5   4   3

ISBN 0-13-456484-7

Prentice-Hall International (UK) Limited, *London*
Prentice-Hall of Australia Pty. Limited, *Sydney*
Prentice-Hall Canada Inc., *Toronto*
Prentice-Hall Hispanoamericana, S.A., *Mexico*
Prentice-Hall of India Private Limited, *New Delhi*
Prentice-Hall of Japan, Inc., *Tokyo*
Simon & Schuster Asia Pte. Ltd., *Singapore*
Editora Prentice-Hall do Brasil, Ltda., *Rio de Janeiro*

To my life-long friends, with great affection…

Diana and David
Herb
Jean and Jimmy
Lydia and Gene
Nancy Wacil
Pat
Phyllis and John
Shirley and Bill
Sidna
Toogie and Elmer

CONTENTS

8
MEGALOPOLIS 134

9
THE APPALACHIANS AND THE OZARKS 163

10
THE INLAND SOUTH 180

11
THE SOUTHEASTERN COAST 200

12
THE HEARTLAND 229

13
THE GREAT PLAINS AND PRAIRIES 258

LIST OF VIGNETTES

PREFACE

Regional geography has recently passed through a period of relative disfavor. Its critics, encouraged by the quantitative revolution in many fields of learning, including geography, have found fault with its alleged lack of precision and methological rigor. This is not the place to debate such charges, but the author of this book is a strong adherent of the conviction that regional geography meets a basic need in furthering an understanding of Earth's surface and that no surrogate has yet been devised to replace it.

The complexities of human life on Earth are much too vast and intricate to be explained by multivariate analysis and related mathematical and model-building techniques. Such models may be useful, but any real understanding of people and their earthly habitat requires that words, phrases, photos, diagrams, graphs, and especially maps be used in meaningful combination. Conceptualization and delimitation of regions are critical exercises in geographic thinking, and the description and analysis of such regions continue as a central theme in the discipline of geography; some would call it the essence of the field.

It is a fundamental belief of the author that a basic goal of geography is *landscape appreciation* in the broad sense of both words, that is, an understanding of everything that one can see, hear, and smell—both actually and vicariously—in humankind's zone of living on Earth. In this book, then, there is heavy emphasis on landscape description and interpretation, including its sequential development.

The flowering of civilization in North America is partly a reflection of the degree to which people have levied tribute against natural resources in particular and the environment in general. From the Atlantic to the Pacific and from the Arctic to the Gulf, people have been destroyers of nature, even as they have been builders of civilizations. Overcrowding of population and overconsumption of material goods have now become so pervasive that reassessment of goals and priorities—by institutions as well as by individuals—is widespread. Reaction to environmental despoliation is strongly developed, and ecological concerns are beginning to override economic considerations in many cases. In keeping with such reaction, environmental and ecologic issues are frequently discussed in this book.

Canada and the United States have a common heritage and have been moving toward similar goals. These factors, along with geographical contiguity and the binational influence of mass media, have produced both commonality of culture and mutual interdependence. Nevertheless, there are clear-cut distinctions between the two countries, and there is a particular concern among many Canadians to define a national character that is separate from both psychological and economic domination of the United States. These national interests and concerns, however, do not mask the fact that the geographical "grain" of the continent often trends north-south rather than east-west; thus, several of the regions delimited in this book cross the international border to encompass parts of both countries.

A renewal of interest in regional geography has become apparent, and nowhere is this more clearly shown than in the increase in publications dealing with aspects of the regional geography of the United States and Canada. There are more journals and more journal articles, and special publications of great significance have appeared. Perhaps most important, the output of state, provincial, regional, and city atlases continues unabated. The chapter-end bibliographies of this book are overflowing with references to useful recent publications; indeed, the bibliographies could easily have been expanded severalfold.

This volume is based significantly on its long-established predecessor, *Regional Geography of Anglo-America*, which flourished six editions, beginning in 1943, with C. Langdon White and Edwin J. Foscue as senior authors. One of the more popular features of the previous book was the use of boxed vignettes that permitted a more detailed discussion of specific issues and topics. The use of boxes, or vignettes (titled "A Closer Look"), has been expanded in this, the second edition of *Regional Geography of the United States and Canada*, and about half of them are written by guest authors on an invited basis.

ACKNOWLEDGMENTS

The author has leaned heavily on many geographers and colleagues in associated disciplines. A great many people, including several students, have contributed thoughts, ideas, information, and critiques of value. Accordingly, the number to whom I am grateful is so large that only a general acknowledgment is possible. To three groups, however, my special thanks are due.

Twenty-one noted geographers contributed sprightly and insightful vignettes. Their names are listed following the Contents.

Several geographers went beyond the norms of collegiality in providing helpful suggestions and comments. Chief among these were:

John Hudson of Northwestern University
J. Lewis Robinson of the University of British Columbia

Other colleagues commented critically upon specific chapters or issues of importance. These include:

John Alwin, Central Washington University
Joseph P. Beaton, California State Polytechnic University, Pomona
Marshall E. Bowen, Mary Washington College
John Dietz, University of Northern Colorado

P.D. Herrem, University of Calgary
John Hudson, Northwestern University
Richard H. Jackson, Brigham Young University
Elwood J.C. Kureth, Eastern Michigan University
David W. Lantis, California State University, Chico
A.J. Larson, University of Illinois at Chicago
Jonathan Leib, Georgia Southern University
Kenneth C. Martis, West Virginia University
Richard H. Schein, University of Kentucky
Barney Warf, Florida State University
Wilbur Zelinsky, Pennsylvania State University

As always, the editorial and production people at Prentice Hall were efficient, friendly, and effective. I am particularly grateful to *Dan Kaveney*, *Alison Aquino*, and *Barbara Liguori*.

Tom L. McKnight
Los Angeles

ABOUT THE AUTHOR

Tom McKnight was born and raised in Dallas, Texas. He started out to be a geologist, but discovered geography through the tutelage of Edwin J. Foscue, and soon shifted his interest to the "mother science." His training includes a B.A. degree (geology major, geography minor) from Southern Methodist University, an M.A. degree (geography major, geology minor) from the University of Colorado, and a Ph.D. degree (geography major, meteorology minor) from the University of Wisconsin. He has been lucky enough to live in Australia for extended periods on eight different occasions. Most of his professional life has been based at U.C.L.A., but he has also taught temporarily at nine American, three Canadian, and three Australian universities. He served as Chair of the U.C.L.A. Geography Department from 1978 to 1983. His favorite places are Australia, Colorado, Yellowstone Park, and Dallas.

The North American Continent

Two of the world's largest countries—the Dominion of Canada and the United States of America—occupy the northern third of the Western Hemisphere. Their burgeoning populations, rapidly diversifying societies, and dynamic economies present a fascinating human tableau on an expansive and diverse continental landscape.

CONTINENT, SUBCONTINENT, OR CULTURE REALM?

The Western Hemisphere can be grossly subdivided in various ways (Fig. 1-1):

1. Two continents—North America and South America;
2. Three subcontinents—North America, Middle America, and South America;
3. Two culture realms—Anglo-America and Latin America.

Commonly in the past Canada and the United States were referred to jointly as *Anglo-America*, for their predomi-

nant ancestry was English, and most of the economic and political institutions of both countries derived from this heritage. More recently, however, the term Anglo-America has become increasingly anachronistic due to the rapidly growing components of both population and culture that are non-Anglicized.

For simplicity's sake, in this book we will refer to the United States and Canada jointly as the *North American subcontinent*.

CONTINENTAL PARAMETERS

North America, as defined here, encompasses an area of nearly 7.5 million square miles (19.4 million km^2), which is larger than each of the seven recognized continents except Asia and Africa. It sprawls across 136 degrees of longitude, from 52° west longitude at Cape Spear in Newfoundland to 172° east longitude at Attu Island in the western extremity of the Aleutians. Its latitudinal extent is 64 degrees, from 83° north latitude at Ellesmere Island's Cape Columbia to 19° north latitude on the southern coast of the Big Island of Hawaii.

(a)

(b)

(c)

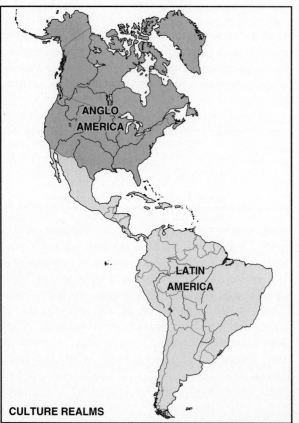

FIGURE 1-1 Major subdivisions of the Western Hemisphere: (a) continents (b) subcontinents (c) culture realms.

A VIEW FROM SPACE

The subcontinent is roughly wedge shaped, with its broadest expanse toward the north. The great bulk of North America is thus in the middle latitudes, with a considerable northern extension into the high latitudes and only Hawaii reaching into the tropics.

If we were to view the entire subcontinent from an orbiting space station on a clear day, certain gross features would appear prominently. Perhaps the most conspicuous configuration is the irregular continental outline; some extensive coastal reaches are relatively smooth, but by and large the margin of the continent is irregular and embayed and there are numerous prominent offshore islands (Fig. 1-2). [1]

The most notable indentation in the continental outline is Hudson Bay, which protrudes southward for some 800 miles (1280 km) from Canada's north coast. Despite its vastness, however, Hudson Bay is relatively insignificant in its influence on the geography of North America. Its surface is frozen for many months, and even the open water of summer supports only one sea route of importance and has minimal climatic effects on the surrounding lands.

Much more significant geographically are two extensive oceanic areas whose margins impinge less abruptly on the continent; both the Gulf of Mexico to the southeast and the Gulf of Alaska to the northwest constitute gross irregularities in the continental configuration that have major climatic influences and considerable economic importance. Other coastal embayments that might be conspicuous from the viewpoint of a satellite include the Gulf of St. Lawrence, the Bay of Fundy, and Chesapeake Bay on the east coast; and Puget Sound, Cook Inlet, Bristol Bay, Norton Sound, and Kotzebue Sound on the west.

More than 70,000 islands are another feature that commands attention in the gross outline of the subcontinent. Easily the most prominent island group is the Canadian Arctic Archipelago, an expansive series of large islands to the north of the Canadian mainland that constitutes more than 14 percent of the total area of Canada. The largest islands of the archipelago are Baffin and Ellesmere, and 9 of the 10 largest islands of the subcontinent are in the group (Fig. 1-3).

Four other sizable islands are clustered around the Gulf of St. Lawrence. Prince Edward Island is a Canadian province, the island of Newfoundland is part of the province of the same name, Cape Breton Island is part of Nova Scotia, and Anticosti Island is part of Quebec.

---

[1] The coastline of the high-latitude portions of the subcontinent is much more uneven than that of other sections. For example, Alaska has more coastline mileage than the other 24 coastal states combined.

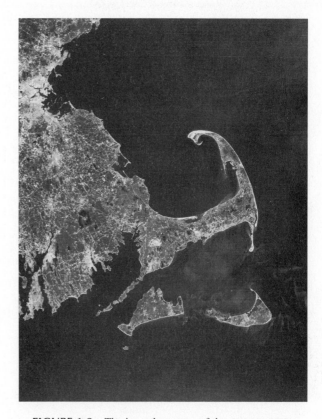

FIGURE 1-2   The irregular nature of the eastern coastline of North America is evident in southeastern Massachusetts and eastern Rhode Island. Cape Cod is composed of a complex of glacial moraines and sandy, current-built bars, spits, and hooks. From east to west the conspicuous islands are Nantucket and Martha's Vineyard. Cape Cod actually is separated from the mainland by the smooth curve of Cape Cod Canal, which extends from Cape Cod Bay to Buzzards Bay. Lighter tones distinguish the Boston urbanized area in the upper left of the photo and a portion of metropolitan Providence at the head of Narragansett Bay (*Landsat* image).

The islands off the east coast of the United States are small and sparsely populated, with one major exception, namely, Long Island, whose 1400 square miles (3600 km[2]) support a population in excess of 7 million. The other coastal islands of the Atlantic Ocean and Gulf of Mexico are nearly all long, narrow, low-lying sand ridges. Although there are a great many such islands, most are so close to shore and so narrow as to be indistinguishable from our theoretical high-altitude viewpoint.

The island pattern off the Pacific Coast is quite uneven. Only a few islands are found off the southern portion of the coast and of these only the Channel Islands of California en-

FIGURE 1-3   Some important place names in North America.

compass much acreage. The coast of British Columbia and southern Alaska, on the other hand, is extensively bordered by islands, many of which are large. Most notable is Vancouver Island, which constitutes the extreme southwestern corner of Canada. Other major islands on this coast include the Queen Charlotte Islands of British Columbia, the Alexander Archipelago of southeastern Alaska, Kodiak Island of southern Alaska, and the far-flung Aleutian group.

THE COUNTRIES

The material in most of this book is presented by regions. As a prelude, it is well to look briefly at the nations as entities,

noting a few general facts to serve as a context for regional analysis.

The United States of America

The total area of the United States is 3,615,200 square miles (9,363,400 km²), a figure exceeded by only three other countries—Russia, Canada, and China. The 1996 population was about 266,000,000, which ranked third among the nations of the world, after China and India.

The United States is a federal republic with a division of power between federal and state governments. Both levels of government have a threefold administration: executive, legislative, and judicial branches. There are 50 states, which vary greatly in size and population. The 48 "old" states, excluding Alaska and Hawaii, are collectively referred to as the *conterminous states*.

The states are subdivided into local governmental units called counties,[2] with the following exceptions: in Louisiana the units are called parishes; in Maryland, Missouri, Nevada, and Virginia there are cities that are independent of any county organization and thus constitute, along with counties, primary subdivisions of these states; and in Alaska the populated parts are subdivided into boroughs. Altogether, there are about 3100 counties and county equivalents in the United States.[3]

The national capital is the District of Columbia, which is territory on the northeastern side of the Potomac River that was ceded to the nation by the state of Maryland.[4] In addition to the 50 states and the District of Columbia, the United States governs a number of small islands in the Caribbean Sea, the most important of which is Puerto Rico, as well as Guam, American Samoa, Midway, and Wake Island in the Pacific Ocean.

The Dominion of Canada

Canada is the world's second largest country, with an area of 3,851,800 square miles (9,976,200 km[2]). It is not densely populated, however, and its 1996 population of about 30,000,000 ranked only 31st among the nations of the world.

The governmental organization of Canada is a confederation with parliamentary democracy that combines the federal form of the United States with the cabinet system of Great Britain. The cabinet system partially unites the executive and legislative branches of government; the prime minister and all, or nearly all, the cabinet are members of the House of Commons. The reigning monarch of Great Britain is also the head of the Canadian state and is represented by a governor-general whose duties are formal and rather perfunctory.[5] The prime minister is the active head of the govern-

ment. Members of the House of Commons are elected by the people of Canada. Members of the Senate, on the other hand, are appointed for life by the cabinet. The House of Commons is the dominant legislative body, with many more powers than the Senate.

The Canadian confederation contains 10 provinces and two territories. The easternmost provinces of Newfoundland, New Brunswick, Nova Scotia, and Prince Edward Island are often referred to as the Atlantic Provinces; the latter three are collectively called the Maritime Provinces. Alberta, Saskatchewan, and Manitoba, the three provinces of the western interior, are known as the Prairie Provinces. The other three provinces, Quebec, Ontario, and British Columbia, are not normally considered members of groups. Most of northern Canada is encompassed within the Yukon Territory and the Northwest Territories. The various provinces and territories have different systems for administering local government; each is usually subdivided into counties or districts, which may be further fragmented into minor civil divisions.

The national capital of Ottawa does not occupy a special territory but is within the province of Ontario, adjacent to its border with Quebec. The creation of a federal district (analogous to the District of Columbia) was first officially proposed in 1915 and has been discussed in varying degrees of seriousness ever since, but with no formal action. Nevertheless, a planning district, called the "National Capital Region," has been designated. It comprises some 900 square miles (2300 km[2]) and is divided about equally between Ontario and Quebec. Indeed, in some quarters the capital is now referred to as "Ottawa-Hull," since Hull is Ottawa's principal suburb on the Quebec side of the boundary and contains the offices of numerous government agencies.

AMICABLE NEIGHBORS

In the early years of their separate political existence, the United States and Canada battled against each other five times, most seriously in the War of 1812. For well over a century, however, neighborliness has prevailed, and their common 5525-mile (8840-km) boundary is rightly referred to as "the longest undefended border in the world."

Americans and Canadians are alike in many ways, both as a people and as a society, although the differences are sometimes notable and often are emphasized by commentators. Life in the two countries is so similar that citizens of nei-

---

[2] Counties vary greatly in size and population. Delaware has the fewest, with 3, whereas Texas has the most, with 254.

[3] The total number of counties is not static, for new ones are added from time to time. In 1981 a new county—Cibola—was created in northwestern New Mexico, and in 1983 La Paz County was carved out of Yuma County in southwestern Arizona.

[4] The District of Columbia was created in 1790 when Maryland and Virginia both ceded territory for its establishment. Virginia reannexed its ceded land in 1847, however, so the present District of Columbia consists entirely of land that was originally part of the state of Maryland.

[5] Despite the "patriation" of the Canadian Constitution in 1982, Canada is still a constitutional monarchy and the roles of the queen and the governor-general are unchanged. What is changed is that the Constitution (previously

called the British North America Act) has been expanded, principally by the addition of a Charter of Rights and Freedoms, and removed from the legal control of the British Parliament and placed wholly in Canadian hands.

ther country experience "culture shock" when visiting the other. Indeed, the movement of people through the 130 border-crossing stations is of enormous magnitude: In the mid-1990s Canadians crossed into the United States on about 40 million occasions annually, and the reverse flow amounted to about 33 million annual excursions.

Of at least equal significance as an indicator of an amicable relationship is the amount of commerce between the two countries, which is the most voluminous two-way trade in the world. Each is the other's best customer. In recent years more than $170 billion worth of goods and services were exchanged annually. This total undoubtedly will continue to rise as a result of the U.S.-Canadian Free Trade Agreement, promulgated in 1989, which will eliminate all tariffs between the two countries within a decade.

This is not to say that there are no problems between these neighbors. Significant contentious issues—such as acid rain, Great Lakes pollution, and commercial fishery allotments—continue to strain the relationship.

Despite the similarities of the people and their institutions, it probably is inevitable that tension between these continental neighbors will persist, simply because they are unequal partners; one is 10 times as large (in population and economy) as the other. Canada has always had to face the problem of building a nation in the shadow of a giant. Or, as ex-Prime Minister Pierre Trudeau is purported to have said, "Occupying a continent with the United States is like sleeping with an elephant; no matter how benign the beast is, its slightest wiggle shakes the bed."

This inequality certainly influences attitude. According to a major 1989 public opinion poll, about half the Canadian respondents characterized Americans in negative terms, whereas less than 5 percent of the American respondents felt negatively about Canadians.

Simply stated, it is probably fair to observe that many Canadians view the United States with concern, but most Americans are indifferent to Canada. Canadian novelist Margaret Atwood has noted that what separates the two countries is the world's longest one-way mirror. Canadians gaze south, obsessed with fascination and sometimes fear of the American colossus, but Americans rarely bother to look north. The comment attributed to Chicago mobster Al Capone encapsulates the American viewpoint: "Canada? What street is that on?" Such a statement is not snobbish; it simply represents the inconsistent interest in, and knowledge of, world affairs that is so prevalent in the United States.

SELECTED GENERAL BIBLIOGRAPHY ON CANADA AND THE UNITED STATES

AGNEW, JOHN, *The United States in the World Economy: A Regional Geography.* New York: Cambridge University Press, 1987.

BIRDSALL, STEPHAN, AND JOHN FLORIN, *Regional Landscapes of the United States and Canada* (2d ed.). New York: John Wiley & Sons, 1981.

BROWN, RALPH H., *Historical Geography of the United States.* New York: Harcourt, Brace and Company, 1948.

CHAPMAN, JOHN D., AND JOHN C. SHERMAN, EDS., *Oxford Regional Economic Atlas: United States and Canada.* London: Oxford University Press, 1975.

CONZEN, MICHAEL P., ED., *The Making of the American Landscape.* Boston: Unwin Hyman, 1990.

DUNBAR, GARY S., "Illustrations of the American Earth: An Essay in Cultural Geography," *American Studies*, 12 (Autumn 1973), 3–15.

ELLIOTT, J. L., *Two Nations, Many Cultures* (2d ed.). Scarborough, Ont.: Prentice-Hall of Canada, Ltd., 1983.

GENTILCORE, R. LOUIS, ED., *Historical Atlas of Canada.* Vol. 2, *The Land Transformed 1800–1891.* Toronto: University of Toronto Press, 1993.

GERLACH, ARCH C., ED., *The National Atlas of the United States of America.* Washington, D.C.: United States Department of the Interior, Geological Survey, 1970.

GRANATSTEIN, J. L., AND NORMAN HILLMER, *For Better or for Worse: Canada and the United States to the 1990s.* Toronto: Copp Clark Pitman Ltd., 1991.

HAMELIN, LOUIS-EDMOND, *Canada: A Geographical Perspective.* Toronto: Wiley Publishers of Canada Limited, 1973.

———, *Canadian Nordicity: It's Your North Too.* Montreal: Harvest House, 1979.

HARRIES, KEITH D., *The Geography of Crime and Justice.* New York: McGraw-Hill Book Co., 1974.

———, AND STANLEY D. BRUNN, *The Geography of Laws and Justice: Spatial Perspectives on the Criminal Justice System.* New York: Praeger Publishers, 1978.

HARRIS, R. COLE, ED., *Historical Atlas of Canada.* Vol. 1, *From the Beginning to 1800.* Toronto: University of Toronto Press, 1987.

———, AND JOHN WARKENTIN, *Canada Before Confederation: A Study in Historical Geography.* New York: Oxford University Press, 1974.

HART, JOHN FRASER, *The Look of the Land.* Englewood Cliffs, NJ: Prentice-Hall, 1975.

JACKSON, RICHARD H., *Land Use in America,* New York: John Wiley & Sons, 1981.

KERR, DONALD AND DERYCK W. HOLDSWORTH, EDS., *Historical Atlas of Canada*, Vol. 3, *Addressing the Twentieth Century 1891–1961.* Toronto: University of Toronto Press, 1990.

MASON, ROBERT J., AND MARK T. MATTSON, *Atlas of United States Environmental Issues*. New York: Macmillan, 1990.

MATTHEWS, GEOFFREY J., and ROBERT MORROW, JR., *Canada and the World: An Atlas Resource*. Scarborough, Ont.: Prentice-Hall of Canada, 1985.

MATTSON, CATHERINE, AND MARK T. MATTSON, *Contemporary Atlas of the United States*. New York: Macmillan, 1990.

McCANN, LARRY D., ED., *Heartland and Hinterland: A Geography of Canada* (2d ed.). Scarborough, Ont.: Prentice-Hall of Canada, Ltd., 1987.

MEINIG, D. W., *The Shaping of America: A Geographical Perspective on 500 Years of History*. Vol. 1, *Atlantic America, 1492–1800*. New Haven: Yale University Press, 1989.

———, *The Shaping of America: A Geographical Perspective on 500 Years of History*. Vol. 2, *Continental America, 1800–1867*. New Haven: Yale University Press, 1993.

MITCHELL, ROBERT D., AND PAUL A. GROVES, EDS., *North America: The Historical Geography of a Changing Continent*. New York: Rowman & Littlefield, 1987.

PATERSON, J. H., *North America* (9th ed.). New York: Oxford University Press, 1994.

ROBINSON, J. LEWIS, *Concepts and Themes in the Regional Geography of Canada*. Vancouver: Talonbooks, 1983.

ROONEY, JOHN F., JR., *A Geography of American Sport*. Reading, MA: Addison-Wesley Publishing Co., 1974.

ROONEY, JOHN F., JR., WILBUR ZELINSKY, and DEAN R. LOUDER, EDS., *This Remarkable Continent: An Atlas of United States and Canadian Society and Cultures*. College Station: Texas A&M University Press, 1982.

SAUER, CARL O., *Sixteenth Century North America*. Berkeley: University of California Press, 1975.

SHORTRIDGE, BARBARA, *Atlas of American Women*. New York: Macmillan, 1987.

Statistics Canada, *Canada*. Ottawa: Queen's Printer, annual.

———, *Canada Year Book*. Ottawa: Queen's Printer, annual.

Surveys and Mapping Branch, *The National Atlas of Canada*. Ottawa: Surveys and Mapping Branch, Department of Energy, Mines and Resources, 1973.

United States Bureau of the Census, *Statistical Abstract of the United States*. Washington, D.C.: U.S. Government Printing Office, annual.

WARD, DAVID, ED., *Geographic Perspectives on America's Past*. New York: Oxford University Press, 1979.

WARKENTIN, JOHN, ED., *Canada: A Geographical Interpretation*. Toronto: Methuen Publications, 1968.

ZELINSKY, WILBUR, *The Cultural Geography of the United States* (2d ed.). Englewood Cliffs, NJ: Prentice Hall, 1994.

The Physical Environment

2

North America is largely a midlatitude subcontinent, lying entirely north of the Tropic of Cancer, except for Hawaii, and mostly south of the Arctic Circle. It spreads broadly in these latitudes, fronting extensively on all three Northern Hemisphere oceans—Atlantic, Pacific, and Arctic. Physical features vary widely, partly as a result of the great size of the subcontinent; indeed, this diversity of environmental conditions is the keystone to understanding its physical geography.

THE PATTERN OF LANDFORMS

The basic pattern of physiographic features is a fourfold division roughly oriented north–south across the subcontinent (Fig. 2-1). In the west is a complex series of mountain ranges and lengthy valleys interspersed with numerous desert basins and plateaus; in the center is an extensive lowland area that widens toward the north; in the east is a broad cordillera of mountains and hills; and along part of the east coast is a coastal plain that swings westward to join the central lowland along the Gulf Coast.

In the conterminous states the western mountain complex consists of two major prongs. In the coastal states a number of steep-sided ranges more or less parallel the coast

from Mexico to Canada; they vary in height from only a few hundred feet in parts of the Coast Ranges of Oregon and Washington to more than 14,000 feet (4200 m) in the Sierra Nevada of California and the Cascades of Washington.

The Rocky Mountain cordillera consists of a series of southeast–northwest trending ranges that rise abruptly from the central plains and extend with only one significant interruption from north-central New Mexico to the Yukon Territory (Fig. 2-2). Between the Rockies on the east and the Sierra Nevada–Cascade ranges on the west is an extensive area of dry lands where plateaus, mesas, desert basins, and short but rugged mountain ranges intermingle.

In western Canada the coastal mountains become more rugged and complex and are separated from the Canadian Rockies by mostly forestcovered, plateaulike uplands. To the northwest the highland orientation changes from south–north to east–west, with the broad lowland of the Yukon–Kuskokwim basins separating the wilderness of the Brooks Range to the north from the massive ranges of southern Alaska. These mountains are the highest and most heavily glaciated of the continent.

The central lowland should not be considered uninterrupted flatland. Much of the terrain is undulating or rolling, and there are many extensive areas of low hills, some quite

FIGURE 2-1   Physiographic diagram of the subcontinent of North America (original map by
A. K. Lobeck: reprinted by permission of Hammond, Inc.).

steep-sided. By and large, however, most of the area between the Rockies on the west and the Appalachians on the east is a lowland of gentle relief. Its narrowest extent is toward the south, from which there is notable widening until the longitu- dinal extent in the north encompasses almost the entire width of the continent. The major exceptions to this pattern are in the rugged and glaciated eastern islands of the Canadian Arctic Archipelago: Baffin, Devon, Ellesmere, and Axel Heiberg.

FIGURE 2-2 The mountains of western North America generally are high, steep, rocky, and rugged. This scene shows Mt. Evans towering above Echo Lake in the Colorado Rockies a few miles west of Denver (courtesy Colorado Department of Public Relations).

The mountains of the east are not as high, rough, or obviously glaciated as those of the west, and the eastern cordillera is only half as long (Fig. 2-3). Nevertheless, the Appalachian Mountain system extends almost without interruption from Alabama to the Gulf of St. Lawrence, and related ranges carry the trend through most of eastern Quebec and Labrador. The highest peaks are less than 8000 feet (2400 m), and most crests are less than half that height. The Appalachian system, however, is a broad one and over much of its extent the slopes are steep and heavily wooded. An important outlier of the eastern highlands is found in the tristate area of Arkansas, Missouri, and Oklahoma where the extensive hills of the Ozarks and Ouachitas are found.

The east-coastal lowland is a feature of lesser magnitude than the other three divisions mentioned. It is a classic example of an embayed coastal plain, with many estuaries, bays, and lagoons. From its narrow northern extremity in New England it slowly widens southward and then swings west in Georgia and Florida to link indistinguishably with the southern margin of the central lowland.

Hydrography

Two prominent drainage systems, the Great Lakes–St. Lawrence and the Mississippi–Missouri, dominate the hydrography of North America (Fig. 2-4). The five Great Lakes (Superior, Huron, Erie, Ontario, and Michigan, the first four shared by the United States and Canada) drain northeastward to the Atlantic via the relatively short St. Lawrence River. Many rivers flow into the Great Lakes, but all are short. The drainage basin of the Great Lakes watershed is remarkably small; it is, for example, only one-sixth the size of the Mississippi–Missouri drainage area (Fig. 2-5).

Most of the central part of the United States and a little of southern Canada are drained by the Mississippi River and its many tributaries. The Mississippi itself flows almost due south from central Minnesota to the Gulf of Mexico below New Orleans. Its principal left-bank tributary is the Ohio River. The Ohio River drains much of the northern Appalachians and the Midwest before being joined by the Tennessee River, which drains much of the southern Appalachians and adjacent areas. The far-reaching Missouri River, emanating from the northern Rockies of Montana and gathering tributaries all across the north-central part of the country, is the major right-bank tributary of the Mississippi.

Between the mouths of the St. Lawrence and Mississippi Rivers the well-watered east coast of North America is drained by a host of rivers, most of moderate length but carrying much water. West of the Mississippi the rivers that enter the Gulf are longer but do not have a large volume of flow.

FIGURE 2-3   The mountains of the eastern part of the continent are not so high and rugged, and most are completely forested. This scene is from central West Virginia (TLM photo).

In the western United States there is great variety in the rivers that reach the Pacific Ocean. The Colorado River is a lengthy desert stream that drains much of the arid Southwest and eventually debouches into the sea in Mexico. The complex Sacramento–San Joaquin system drains much of California. The Columbia, principal river of the Pacific Northwest, originates in Canada and traverses some 465 miles (745 km) before crossing into the United States, where it is joined by its main tributary, the Snake, and finally flows for another 300 miles (480 km) as the boundary between Oregon and Washington.

The Pacific drainages of Canada and Alaska encompass many short rivers and a few long ones, but heavy precipitation ensures that the streams have a large volume of flow. The most notable rivers are the Fraser in southern British Columbia and the Yukon in the Yukon Territory and Alaska.

Most of the Arctic drainage of North America is accomplished by the Mackenzie system or the myriad streams that flow centripetally into Hudson Bay. The Mackenzie system is an unusually complex one; its major water sources are

the Liard River and Great Slave Lake, the latter being fed by the extensive watersheds of the Hay, Peace, Athabasca, and Slave Rivers.

It is noteworthy that thousands of square miles in the western United States are not served by external drainage. Particularly in Nevada, Utah, and California, basins of internal drainage abound. Most of these basins contain either shallow or dry lakes in their center (of which Utah's Great Salt Lake is the most conspicuous) and are fed by streams that usually flow only intermittently.

In some parts of the subcontinent, lakes are a significant element in the landscape (Fig. 2-6). As a result of more extensive glaciation in the past, Canada has a much larger proportion of its surface area in lakes (as well as marshes and swamps) than does the United States. (There is some truth to Minnesota's claim to be the "land of 10,000 lakes" and Ontario's assertion to be the "land of 100,000 lakes" is equally valid.) Furthermore, many of Canada's lakes are large ones; Great Bear, Great Slave, and Winnipeg, for example, are larger than Lake Ontario, and Canada contains no fewer than eight lakes that are larger than any wholly United States lake

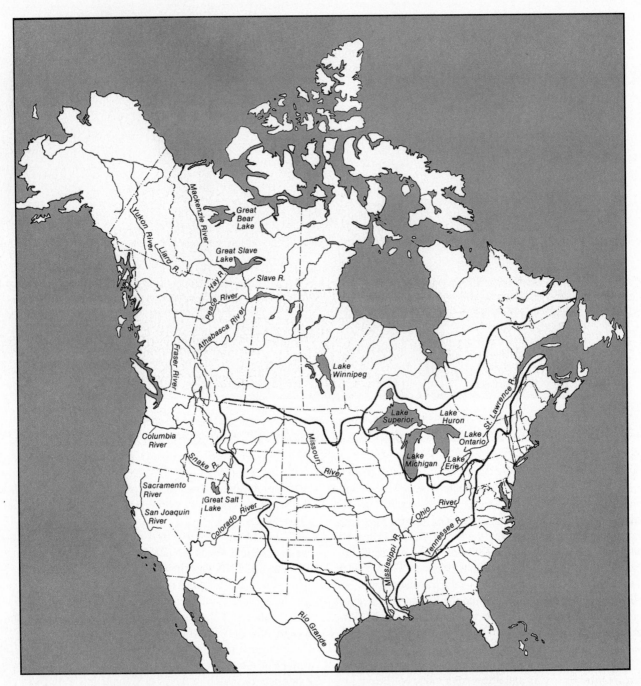

FIGURE 2-4   The major hydrographic features of North America. The drainage basins of the Mississippi River watershed and of the Great Lakes–St. Lawrence River systems are bounded by a heavy black line.

except Lake Michigan.[1] The United States has relatively few natural lakes except in the glaciated sections (primarily New England, New York, and the upper Lakes states), in interior basins of the arid West, and in the flat limestone country of central and southern Florida.

There has been a great proliferation of artificial lakes (reservoirs), particularly in the southeastern, south-central,

[1] The freshwater lakes in Canada total more than 290,000 square miles (755,000 km^2), which is an area larger than the state of Texas.

FIGURE 2-5   The most famous waterfall on the continent is Niagara. The Niagara River bifurcates to flow around Goat Island, with the American Falls on the left of the photo and the more voluminous Canadian (Horseshoe) Falls on the right, immediately beyond the Oneida Observation Tower (courtesy Ontario Ministry of Industry and Tourism).

and southwestern states, where natural lakes are rare. Most such reservoirs are formed by the simple damming of rivers, producing fingerlike bodies of water that extend for long distances up former stream valleys and are now prominent features on the large-scale maps of almost any area from Carolina to California.

Glaciation

An understanding of the subcontinent's physical geography requires that attention be paid to the role of glaciation in cre-

ating the contemporary landscape, particularly in Canada. During the most recent "Ice Age" (the Pleistocene epoch), which began 1 million to 2 million years ago and may have ended less than 8000 years ago, extensive continental ice sheets formed in northern and central Canada and made at least four major advances southward (Fig. 2-7). At the height of Pleistocene glaciation the ice sheets extended as far south as Long Island, the Ohio River, the Missouri River, and the middle Columbia River. At the same time, mountain glaciers of considerable size developed throughout the ranges of the

FIGURE 2-6   A satellite image of the Great Lakes (NASA/Johnson Space Center).

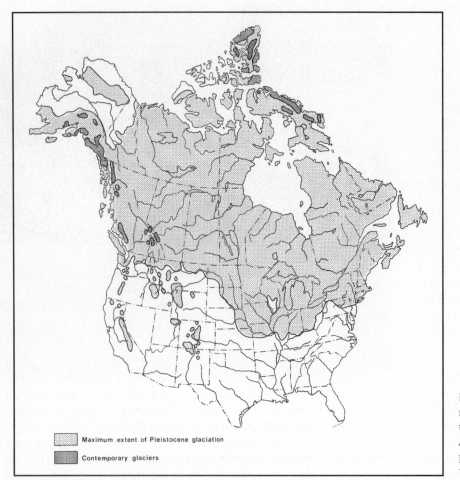

FIGURE 2-7 Recent and contemporary glacial extent in North America (after *National Atlas of the United States of America*, Washington, D.C.: U.S. Department of the Interior, Geological Survey, 1970, p. 76).

Maximum extent of Pleistocene glaciation

Contemporary glaciers

West; evidence of Pleistocene glaciation has been discovered as far south as central New Mexico and the San Bernardino Mountains of southern California.

Glacial erosion and deposition completely reshaped the terrain in all areas covered by Pleistocene ice. From the standpoint of geologic time the retreat of the ice is so recent that the critical factors of slope, drainage, and surficial material are more directly the result of the action of ice, and meltwater, than of any other landscape-shaping element. This fact is clearly shown by the deranged drainage patterns and the large amount of standing water (lakes, swamps, and marshes) now found in areas once covered by Pleistocene ice sheets.

Except in a few locations, ice is but a minor feature in the contemporary topography. The only sizable ice sheets, although much smaller than those of Antarctica or Greenland, occur on the four large eastern islands of the Canadian Arctic Archipelago.

Mountain glaciers are also still found, but they are slight remnants of their past extent. Small living glaciers, on-

ly a few acres in size, occur as far south as central Colorado in the Rockies and the Sierra Nevada of California. Mountain glaciers appear with increasing frequency farther north in the western cordilleras, but the only place in the conterminous states where their length is reckoned in miles is in the northern part of the Cascade Range.

Contemporary glaciation is much more extensive in western Canada, and mountain icefields occur in some parts of the Canadian Rockies and British Columbia's Coast Mountains (Fig. 2-8). Mountain glaciation reaches its greatest extent in the ranges of southern Alaska and the southwestern corner of the Yukon Territory, where the length of many valley glaciers is measured in tens of miles, and the areal extent of one piedmont glacier (the Malaspina) is greater than that of the state of Rhode Island.

Major Landform Regions of North America

Several geographers and geomorphologists have subdivided the United States and Canada into physiographic, or landform, regions based on various criteria but primarily on the gross distribution of terrain features. Figure 2-9 is a combination and modification of the work of a number of these scholars.

The *Southeastern Coastal Plain* (1) is one of the flattest portions of North America. Most of the region slopes gently seaward, at the rate of only a few feet or even a few inches per mile, with the gentle slope continuing underwater for dozens of miles as a continental shelf. The uninterrupted flatness of the landscape is relieved only sporadically by low (10 to 50 feet [3 to 15 m]) riverine bluffs alongside the broad river valleys or, particularly west of the Mississippi River, by low linear ridges (called *cuestas*) that parallel the coast.

Most of the region was submerged in relatively recent geologic time, and its surface layers consist of loosely consolidated sands, gravels, marls, and clays. The coastal margin is exceedingly irregular as a result of recent submergence and embayment; it is characterized by extensive drowned valleys (estuaries) and many bays, swamps, lagoons, and low-lying sandbar islands. The portion abutting the Appalachians encompasses the Piedmont, where resistant crystalline bedrock and rising elevation produced a more undulating surface.

The *Appalachian Uplands* (2) extend from Alabama to the Gulf of St. Lawrence and the physiographic trend is continued in the island of Newfoundland. The region includes the "mountainous" part of eastern North America, although much of the terrain consists of low, forested hills. From Pennsylvania southward most of the surface is underlain with sedimentary rocks. These sediments have been tightly folded in the eastern portion of the region to produce a remarkable sequence of parallel valleys and long, steep-sided ridges, although the easternmost ridge (the Blue Ridge), which contains the highest peaks in the eastern part of the subcontinent consists of ancient crystalline rocks.

To the west of the ridge-and-valley section the sediments are more horizontal; here the so-called Allegheny and Cumberland plateaus are thoroughly dissected by streams and give the appearance of an endless region of low hills. Most of the hilltops are composed of sandstone cut through by the major streams to form relatively deep and narrow valleys. North of Pennsylvania, the underlying rock is mostly crystalline (igneous or metamorphic) and the surface form varies greatly, although hills and low mountains dominate the landscape. The most conspicuous ranges are the Adirondacks in New York, Green Mountains in Vermont, White Mountains in New Hampshire, Notre Dame and Shickshock ranges in Quebec, and Long Range in Newfoundland.

The *Interior Uplands* (3) bear considerable physiographic resemblance to the Appalachians, although they have less altitude and local relief. The Ozark section, mostly in southern Missouri and northern Arkansas, is a dissected plateau that consists of an amorphous pattern of low hills.

FIGURE 2-8 Athabaska Glacier is one of the longest in the Canadian Rockies. The distance from the toe to the highest icefall, where the glacier issues from the Columbia Ice Field, is about 8 miles [13 km] (TLM photo).

Separated from the Ozarks by the transverse valley of the Arkansas River is the Ouachita section in western Arkansas and eastern Oklahoma. This is an area of east–west trending, linear ridges and valleys; it is markedly similar to the ridge-and-valley section of the Appalachians.

The *Interior Plains* (4) are a vast area of gentle relief that occupies much of the central portion of the subcontinent. For the most part the landscape is that of a typical stream-eroded region of low ridges and shallow valleys that reflects the differences in the rock on which it is carved. Some portions are remarkably flat, such as the High Plains of west Texas or the prairies of central Illinois; however, most of the region is characterized by undulating terrain or low, even-topped hills. Relatively flat-lying sedimentary rocks underlie the surface in most places. The terrain was conspicuously modified by the action of Pleistocene ice sheets north of the Ohio and Missouri Rivers. Numerous lakes, marshes, and ponds occur in the glaciated sections of the region, as well as many long rivers.

The *Rocky Mountains* (5), located just west of the flattish Great Plains portion of the Interior Lowlands, constitute a very abrupt physiographic change when approached from the east. They are characterized by high elevations—more than 50 peaks in Colorado reach above 14,000 feet (4,200 m)—great local relief, rocky ruggedness, and spectacular scenery. Only in the Wyoming Basin area of southern and central Wyoming is there any significant section that is not dominated by mountainous terrain. There was great variety in the mountain-building processes that produced the Rockies.

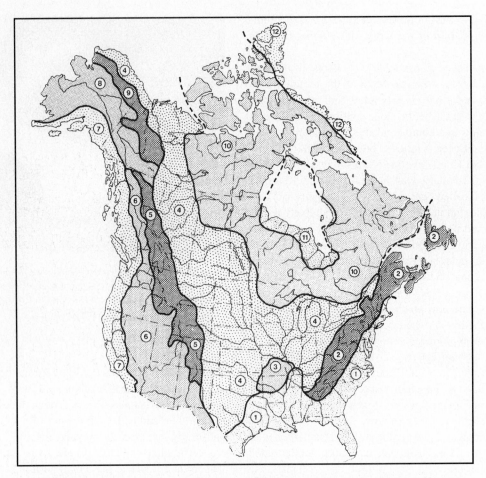

FIGURE 2-9   Major landform regions of North America are the (1) Southeastern Coastal Plain, (2) Appalachian Uplands, (3) Interior Uplands, (4) Interior Plains, (5) Rocky Mountains, (6) Intermontane Region, (7) Pacific Coast Region, (8) Yukon Basin and Plateaus, (9) Northwestern Highlands, (10) Canadian Shield, (11) Hudson Bay Lowland, and (12) High Arctic Mountains (N. M. Fenneman, "Physiographic Divisions of the United States," *Annals*, Association of American Geographers, 18 [1928], 261–353; W. W. Atwood, *The Physiographic Provinces of North America* [Boston: Ginn and Co., 1940]; A. K. Lobeck, "Physiographic Diagram of the United States" [Madison: Wisconsin Geographic Press, 1922]; *The National Atlas of the United States of America* [Washington, D.C.: U.S. Department of the Interior, 1970], pp. 61–64; *Atlas of Canada* [Ottawa: Department of Mines and Technical Surveys, 1957], plate 13).

In the southern portions extensive granitic intrusions were thrust up many thousands of feet, whereas farther north sedimentary rocks were drastically folded, and thrust and block faulting occurred on a large scale. Throughout the region the recent action of mountain glaciers has deepened the valleys, steepened the slopes, and sharpened the peaks.

The *Intermontane Region* (6) encompasses a bewildering variety of terrain formed in many different ways. The southern and southwestern portions are basin-and-range

country where numerous discrete short rugged mountain ranges are interspersed with flat alluvial-filled valleys. In the "Four Corners" country, plateaus, mesas, cliffs, and buttes dominate the landscape, for weakly consolidated horizontal sedimentary rocks were stripped and fretted by arid-land erosion processes (Fig. 2-10). The Columbia and Snake lava plateaus in Idaho, Oregon, and Washington have been deeply incised by major rivers and eroded into rolling hills in some areas. The various plateaus of the Canadian portion of the re-

FIGURE 2-10   The spectacular starkness of the Intermontane West is demonstrated dramatically in this view of the Mitten Buttes in Utah's Monument Valley (TLM photo).

gion (Fraser, Nechako, and Stikine) were severely dissected and in many places appear as hills or mountains.

The *Pacific Coast Region* (7) is largely mountainous, with the trend of the ranges generally paralleling the coast. High and rugged mountains extend the entire length of the region, from the Sierra Nevada of California to the Alaska Range in the north. Several major valleys, particularly the Central Valley of California and the Willamette Valley of Oregon, are sandwiched between massive interior ranges and numerous smaller coastal ranges. The coastline itself is quite regular in the south, where the steep coastal ranges plunge abruptly into the sea; northward the coastline becomes exceedingly irregular, with lengthy bays and fiords interspersed with sinuous peninsulas and numerous islands.

The *Yukon Basin and Plateaus* (8) section occupies most of central Alaska and the southern part of the Yukon Territory, primarily within the drainage basins of the Yukon and Kuskokwin rivers. Hill land predominates in the upstream areas, but much of central Alaska is a broad, flat-floored, poorly drained lowland. The lower courses of the two major rivers have built complex deltas.

The *Northwestern Highlands* (9) section includes the massive barren slopes of the Brooks Range in northern Alaska and a series of rugged mountains in northwestern Canada.

The *Canadian Shield* (10) is an extensive, ancient, stable region floored with some of the world's oldest known crystalline rocks. Except for some relatively rugged hills in eastern Quebec and Labrador, this is a gently rolling landscape typified by many outcrops of bare rock and an extraordinary amount of surface water in summer. There are hundreds of thousands of water bodies, ranging in size from gigantic to minute, connected by tens of thousands of rivers and streams.

The *Hudson Bay Lowland* (11) is a flat coastal plain that slopes imperceptibly toward the sea. It is underlain by recent sedimentary deposits that distinguish it from the surrounding, and underlying, Laurentian Shield.

High Arctic Mountains (12) occupy much of the four large eastern islands of the Canadian Arctic Archipelago. The region is typified by rugged and rocky slopes, large glaciers and ice sheets—several of which are larger than the province of Prince Edward Island—lack of vegetation and soil, and deep permafrost.

The Hawaiian Islands (13) represent isolated and severely eroded tops of a submarine mountain range that was created by volcanic activity as a tectonic plate slowly passed over a "hot spot" (a mantle plume) in the crust. The larger islands are rugged and steep; the smaller ones have a gently undulating surface.

THE PATTERN OF CLIMATE

Most of the North American subcontinent experiences a climate in which seasonal changes are marked, as is characteristic of middle- and high-latitude portions of the world. Summer is generally hot and is the season of maximum precipitation; winter is cold and somewhat drier; and spring and fall are stimulating seasons of transition. There are many exceptions to these generalizations, particularly on the Pacific littoral, where winter is the wet season and summer is quite dry.

For the most part the climate is dominated by weather systems that move across the continent from west to east, in association with the westerly air flow that prevails over the mid-latitudes. Some of these migratory systems are focused around low-pressure cells (extratropical cyclones) that attract unlike air masses, producing frontal disturbances, stormy or unsettled weather, and precipitation that is sometimes abundant (Fig. 2-11). Other systems are dominated by high-pressure cells (anticyclones) that bring clear skies and extreme temperature conditions (usually cold, sometimes hot).

These air circulation patterns (mostly westerly winds) combine with the topographic patterns (north–south trending mountain ranges) and land/sea relationships (oceanic moisture sources to the west, east, and south) to produce plentiful precipitation along the northern portion of the west coast, the windward sides of western mountain ranges, and the southeastern part of the continent. Drier conditions prevail in the Arctic, over much of the central plains area, and particularly in the interior West. This simplified pattern is modified and interrupted by a variety of factors, particularly cold air invasions from the north in winter and onshore flows of warm, moist air from the Gulf of Mexico in summer.

Canada has a climate that is dominated by winter. Indeed, Canada is the world's coldest country, with an average overall temperature of 20°F (−6°C). Ottawa is the second coldest national capital in the world, after Ulaanbaatar, Mongolia.

Major Climatic Regions

There are almost as many classifications of climate as there are climatologists. Although most classifications are fundamentally the same, their minor variations and specialized nomenclature may confuse the reader. No standard classification has been adopted for climates; the map used in this book shows one of the most widely accepted schemes (Fig. 2-12).

The basic pattern that emerges is one of east–west trending climatic zones in the eastern and northern portions of the continent, with north–south trending zones in the west. Such a pattern reflects the fundamental significance of latitude, with topographic modification in the west. Generally similar patterns are evident in the succeeding maps of vegetation and soils.

In the southeastern quarter of the United States the climatic type is classified as *humid subtropical* (1). Summer is the most prominent season. Humid tropical air usually pervades most of the region and high temperatures frequently

FIGURE 2-11 Winter is a prominent fact of life over most of the subcontinent. This January scene is near Racine in southeastern Wisconsin (TLM photo).

FIGURE 2-12 Major climatic regions of North America are (1) humid subtropical; (2) humid continental with warm summers; (3) humid continental with cool summers; (4) steppe; (5) desert; (6) mediterranean, or dry summer subtropical; (7) marine west coast; (8) subarctic; (9) tundra; (10) ice cap; and (11) undifferentiated highlands (after Glenn T. Trewartha, Arthur H. Robinson, and Edwin H. Hammond, *Elements of Geography*, 5th ed. [New York: McGraw-Hill Book Company, 1967]).

combine with high humidity. Winter is short and relatively mild, although significant cold spells occur several times each year. Precipitation is spread throughout the year, with a tendency toward a maximum in summer, except in Florida, where rains associated with hurricanes bring a fall maximum.

In east-central United States there is a large area of *humid continental with warm summer* (2) climate. This is a zone of interaction between warm tropical air masses from the south and cold polar air masses from the north, which results in frequent, stimulating weather changes throughout the year. Summers are warm to hot and are the time of precipitation maxima. Winters are cold and there is considerable snowfall.

There is a broad east–west band of *humid continental with cool summer* (3) climate along the international border in eastern North America. It is distinguished from the previous climatic type by its shorter and milder summer and its longer and more rigorous winter.

The *steppe* (4) climate of the Great Plains and Intermontane areas is basically a semiarid climate with marked seasonal temperature extremes. Summers are hot, dry, and

windy and are punctuated by occasional abrupt thunderstorms; winters are cold, dry, and windy, with occasional blizzards. Intense and dramatic weather—hail, heat waves, tornadoes, windstorms, and the like—are typical of this region.

The *desert* (5) climate of the southwestern interior is characterized by clear skies, brilliant sunshine, and long periods without rain. Aridity is universal except at higher elevations. Summer is long and scorchingly hot; winter is brief, mild, and delightful.

Central and coastal California is a region of *mediterranean* (6) climate, with its anomalous precipitation regime in which sequential winter frontal storms move in from the Pacific, bringing alternating periods of rain and sunshine. Summer is virtually rainless, for subtropical stable high-pressure conditions dominate. Coastal sections have marine-induced mild temperatures throughout the year, whereas inland locations experience hot summers and cool winters.

The *marine west coast* (7) climate of the Pacific littoral is characterized by long, relatively mild, very wet winters and short, pleasant, relatively dry summers. The frequent movement over the region of maritime polar air from the Pacific, and its associated fronts, accounts for the prominence of winter conditions. Exposed slopes at higher elevations receive some of the greatest precipitation totals (both rain and snow) in the world.

The climate of most of central Canada and Alaska is classified as *subarctic* (8). Winter, the dominant season, is long, dark, relatively dry, and bitterly cold. Summer occupies only a brief period, but the succession of long hours of daylight produces several weeks of warm-to-hot weather. Summer rainfall is scanty, but evaporation rates are low and so moisture effectiveness is high. Except in Quebec and Labrador, winter snowfall is also scanty, but there is little melting from October to May.

The *tundra* (9) climate of arctic Canada and Alaska is virtually a cold desert. There is little precipitation at any season; however, evaporation rates are also quite low. Winter is very long, although not as cold as in the subarctic areas. Summer is short and cool.

The *icecap* (10) climate of the High Arctic is rigorous in the extreme. Low temperatures and strong winds make these areas unendurable for humans and animals, except briefly.

The major mountain areas are classified as having a *highland* (11) climate. Generalizations concerning weather and climate in such areas can be made only with reference to particular elevations or exposures. On the average, temperature and pressure decrease with altitude, whereas precipitation and windiness increase. Thus vertical zonation is the key to understanding highland climates.

THE PATTERN OF NATURAL VEGETATION

Unlike landforms and climate, the natural vegetation has undergone major changes caused by humankind. The native flora has been so thoroughly removed, rearranged, and replaced that a discussion of natural vegetation in many areas is largely a theoretical or historical exercise. Still, from a broad standpoint it is possible to reconstruct the major vegetation zones and make some meaningful generalizations about them (Fig. 2-13). The reader should keep in mind that in many parts of the continent introduced plants—particularly crops, pasture grasses, weeds, and ornamentals—are much more conspicuous than native flora.

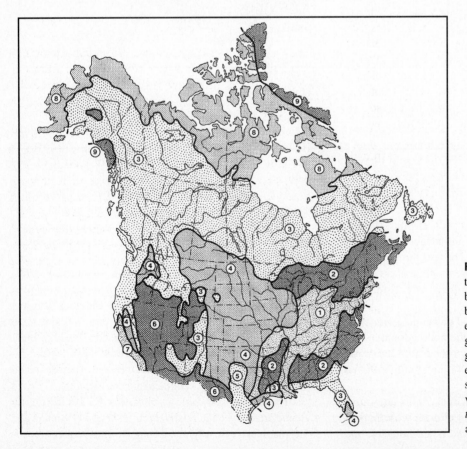

FIGURE 2-13   Major natural vegetation regions of North America are (1) broadleaf deciduous forest, (2) mixed broadleaf deciduous and needleleaf evergreen forest, (3) needleleaf evergreen forest, (4) grassland, (5) mixed grassland and mesquite, (6) broadleaf evergreen shrubland, (7) mediterranean shrubland, (8) tundra, and (9) little or no vegetation (after *The National Atlas of the United States of America*, pp. 90–92; and *Atlas of Canada*, plate 38).

The continent's natural vegetation associations can be divided into three broad classes: forests, grasslands, and shrublands.

Forests

The forests occur in six zones:

1. Northern coniferous forests, or taiga
2. Eastern hardwood zone [2]
3. Eastern mixed forests
4. Southern pineries
5. Rocky Mountain forests
6. Pacific Coast forests

The taiga, called *boreal forest* in Canada, is composed mostly of coniferous trees growing under conditions of long, bitterly cold winters. It comprises about three-fourths of Canada's productive forest land, is widespread in central and southern Alaska, and dips slightly into the states of Minnesota and Wisconsin. For the most part, the trees grow in relatively pure stands, with white and black spruce being the most widespread. Also common are tamarack, balsam, and alpine fir, and jack and lodgepole pine. Sometimes non-conifers—particularly aspen, white birch, and balsam poplar—cover extensive areas. Although the boreal forest covers nearly half of the total area of Canada and makes up nearly one-tenth of all the forests in the world, it is composed of fewer than 40 tree species. The short growing season and relatively poor drainage of the region inhibit rapid growth, so that even old trees are not very tall.

Three belts of forested land extend southward from the taiga. One prong follows the Appalachian Mountains and adjacent lowlands, another the Rocky Mountains, and a third the ranges of the Pacific coastal states. The easternmost prong grades southward from the relatively pure softwood stands of the taiga through a zone of mixed broadleaf and needleleaf trees to an expansive region where hardwoods constituted the original vegetation. South of the hardwood region is another area of mixed forest that gives way to extensive forests dominated by southern yellow pine in the Southeast.

The Rocky Mountain prong is mostly mountain forest, with various species occupying altitudinal zones vertically arranged on the mountainsides. In the southern Rockies trees grow only in the uplands, but in the northern Rockies many of the lowlands between the ranges are also forested. Most trees are conifers, with spruce and fir occupying the wetter areas; pine and juniper grow in drier localities. Aspen, the first species to occupy an area after a fire, is one of the few deciduous trees and is quite widespread in some places.

The Pacific Coast prong also consists primarily of softwoods. Although the forested zone follows the mountain trend, most of the lowlands north of the Sacramento Valley (except the Willamette Valley) were also originally forested. This is a region of huge trees, generally the largest to be found on Earth (Fig. 2-14). Fir, spruce, and hemlock dominate in the north, whereas redwoods become conspicuous in northern California. The drier slopes are almost invariably pine covered.

Grasslands

Grasslands are usually found in areas where the rainfall is insufficient to support trees, and most have never had any other type of vegetation. Where grass is tall, it is often called *prairie*; where short, *steppe*. Toward the drier margin the short-grass association grades into bunchgrass.

The eastern portion of the Great Plains was originally clothed in prairie grasses, with a significant eastward extension in the so-called Prairie Wedge of Illinois. The western part of the Plains had a steppe association, and there were other significant areas of grassland in the Central Valley of California and in the northern part of the Intermontane Region.

The tundra of the far north can be viewed as a pseudo-grassland. It is characterized by a low-growing mixture of sedges, grasses, mosses, and lichens, occurring in an amazing variety of species.

Shrublands

Shrublands vary considerably in their characteristics but usually develop under an arid or semiarid climate. The typical plant association is a combination of bushes or stunted trees and sparse grasses. In southern and central Texas the shrubby mesquite tree is the dominant plant. In the area from Wyoming to Nevada sagebrush provides the most common ground cover. In the Southwest cacti and other succulents are

---

[2] The terms *hardwoods* and *softwoods* are the most generally accepted popular names for the two classes of trees, the *Angiosperms* and the *Gymnosperms*. Most Angiosperms, such as oak, hickory, sugar maple, and blacklocust, are notably hard woods and many Gymnosperms, such as pines and spruces, are rather soft woods. But there are a number of outstanding exceptions. Basswood, poplar, aspen, and cottonwood, all classified as hardwoods, are actually among the softest of woods. Longleaf pine, on the other hand, is about as hard as the average hardwood, although classified as a softwood. The most accurate popular descriptions for the two groups are *trees with broad leaves* for the Angiosperms and *trees with needles and scalelike leaves* for the Gymnosperms (from Forest Products Laboratory, *Technical Note 187*, Madison, Wisconsin).

FIGURE 2-14   The evergreen forests of the Pacific Northwest are dominated by big trees, as shown here in western Washington (courtesy Weyerhaeuser Company).

characteristic (Fig. 2-15), although some large expanses are dominated by a sparse covering of spindly creosote bush and burroweed. The chaparral-manzanita association of southern and central California is interspersed with some broad reaches of grassland and open oak woodland.

THE PATTERN OF SOIL

The most complex feature of the environment is the soil. This complexity is largely attributable to complicated chemical factors and reactions in soil development that not only are imperfectly understood but also leave little or no mark on the landscape and so are difficult to ascertain without detailed microstudies. Also, minor differences in such related features

as slope or drainage can be much more important than broader environmental parameters in determining significant soil characteristics; thus soil variations over short distances are often much more significant than broader regional variations.

The identification and distribution pattern of meaningful soil categories is consequently difficult to determine except on a very large scale. Most maps of soil categories are hopelessly complex for purposes of macrostudy. Moreover, the past few years have been a period of fundamental change in the way that soils are classified and categorized by pedologists and other scholars interested in soil distribution. Thus an entirely new classification scheme was adopted, and many changes in nomenclature and terminology were accepted.

With these caveats in mind, we can make only broad generalizations about the distribution of soil categories. Figure 2-16 is based on the United States Comprehensive Soil Classification System—now officially called *Soil Taxonomy*—that was developed slowly and meticulously during the 1950s and 1960s by the Soil Survey Staff of the U.S. Department of Agriculture.[3] The principal *orders* (the major categories in the hierarchical classification) are briefly described here.

Alfisols (A) are soils with mature profile development that occur in widely diverse climatic and vegetation environments. They have gray-to-brown surface horizons and a clay accumulation in subsurface horizons. They are most widespread in the Great Lakes area, the Midwest in general, and the northern part of the Prairie provinces.

Andisols (N) are developed from volcanic ash and have been deposited in relatively recent geologic time. They are not highly weathered and have minimum profile development but relatively high inherent fertility. They occur sparsely primarily in the Pacific Northwest of the United States and the southwestern corner of Canada.

Aridisols (D) are mineral soils that are low in organic matter and dry in all horizons most of the time. Associated primarily with arid climatic regimes, they are most extensively found in the western interior of the United States, particularly the Southwest.

Entisols (E) are primarily of immature development, with a low degree of horizonation. Characteristically they are

[3] Soil Survey Staff, *Soil Classification: A Comprehensive System—7th Approximation* (Washington, DC: U.S. Department of Agriculture, Soil Conservation Service, 1960); Soil Survey Staff, *Supplement to Soil Classification System—7th Approximation* (Washington DC: U.S. Department of Agriculture, Soil Conservation Service, 1967); Soil Survey Staff, *Soil Taxonomy: A Basic System of Soil Classification for Making and Interpreting Soil Surveys* (Washington, DC: U.S. Department of Agriculture, Soil Conservation Service, 1976); Donald Steila, *The Geography of Soils* (Englewood Cliffs, NJ: Prentice-Hall, 1976); J. S. Clayton et al., *Soils of Canada*, 2 vols. (Ottawa: Canada Department of Agriculture, 1977).

FIGURE 2-15   Desert vegetation in southern Arizona. Spindly shrubs are characteristic, but they often occur in considerable quantity and variety. (TLM photo).

either quite wet, quite dry, or quite rocky. They occur most widely in northern Quebec and Labrador but are also found in some of the High Arctic islands, in scattered localities in the West, and in southern Florida.

Histosols (H) represent the only order composed primarily of organic rather than mineral matter. They are often referred to by such terms as *bog*, *peat*, or *muck*. They can be found in any climate, provided that water is available. They occur most extensively in subarctic Canada, particularly south of Hudson Bay and in the Great Bear Lake area.

Inceptisols (I) also occur in widely differing environments. They are moist soils with generally clear-cut horizonation. Leaching is prominent in their formation. They lack illuvial horizons and are primarily eluvial in character. They are widespread in tundra areas of Canada and Alaska, where they are associated with permafrost, and are also notable in the Appalachians, the Lower Mississippi Valley, and the Pacific Northwest.

Mollisols (M) are mineral soils with a thick, dark surface layer that is rich in organic matter and bases. Their agricultural potential is generally high. They are chiefly found in subhumid or semiarid areas and are the principal soils of the western Corn Belt and the Great Plains, as well as in interior portions of the Pacific Northwest.

Spodosols (S) have a conspicuous subsurface horizon of humus accumulation, often with iron and aluminum. They are commonly moist or wet and heavily leached. They are often associated with the soil-forming process called podzolization and have limited agricultural potential. Their most extensive occurrence in North America is in southeastern Canada and New England, but they are also found in such divergent locations as northern Florida and the Great Slave Lake area.

Ultisols (U) are thoroughly weathered and extensively leached and so have experienced considerable mineral alteration. Typically they are reddish in color owing to the considerable amounts of iron and aluminum in the surface layers. Their principal locations are in the southeastern quarter of the United States.

Complex (X) soil associations are delineated on the map in many mountain areas and in much of California. Their variety is too complicated to allow generalization at this scale.

Areas with *little or no soil* (Z) are recognized in rugged mountain areas or where permanent icefields exist.

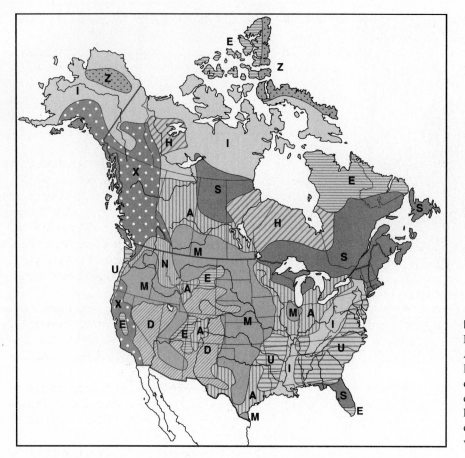

FIGURE 2-16 Major soils regions of North America: (A) Alfisols, (D) Aridisols, (E) Entisols, (H) Histosols, (I) Inceptisols, (M) Mollisols, (N) Andisols, (S) Spodosols, (U) Ultisols, (X) complex soil regions, and (Z) areas with little or no soils (after maps of Soil Geography Unit, Soil Conservation Service, U.S. Department of Agriculture).

This soil classification differs from most previous ones in that it is generic (based systematically on observable soil characteristics) rather than genetic (based on soil-forming conditions and processes). The resultant distribution pattern is less easy to comprehend because soils with similar characteristics sometimes are found in widely differing environments; for example, Inceptisols dominate in both southern Louisiana and the Northwest Territories. Nevertheless, genesis is not completely ignored, for soil properties are directly related to soil development. Thus the pattern of Figure 2-15 reflects some environmental relationships. The zonation of soils in the eastern and northern parts of North America is generally in east–west bands whereas that in the western part of the continent is banded north–south, emphasizing the roles of topography, climate, and vegetation in soil formation.

THE PATTERN OF WILDLIFE

Unlike the environmental elements previously discussed, the subcontinent's faunal complement is generally insignificant

in the total geographic scene. The spread of civilization mostly has been hostile to native wildlife, resulting in contracting habitats and decreasing numbers. A few species—such as opossum, coyote, armadillo, and raccoon—have withstood the human onslaught and actually expanded their ranges, but such examples are limited. The larger carnivores (e.g., grizzly bear, wolf, mountain lion) have particularly suffered and continue to dwindle despite significant mitigation endeavors in some areas. Interestingly enough, most ungulate species (such as deer, elk, moose, and pronghorn) have prospered under enlightened wildlife management efforts and the decline of native predators; thus, they now occur in much greater numbers than they did a century ago.

People also have influenced wildlife patterns in other ways. They have introduced new species, such as nutria, ring-necked pheasants, feral pigs, and feral horses, and artificially rearranged the distribution of native species—for example, expanding the range of mountain goats by deliberate translocation.

Wildlife in most parts of the subcontinent is an inconsequential element in the landscape and attracts little atten-

FIGURE 2-17 Wildlife usually is inconspicuous in the landscape, but not in the case of this Rocky Mountain bighorn sheep in Alberta's Jasper National Park (TLM photo).

tion from students of geography (Fig. 2-17). But in some areas, usually sparsely populated and with limited economic potential, wildlife assumes a more important role. Where such conditions pertain, the discussion takes place in the appropriate regional chapter.

ECOSYSTEMS

In summing up our discussion of environmental patterns, it is perhaps most useful to turn to the concept of ecosystems. As a term, *ecosystem* is a contraction of the phrase "ecological system." The ecosystem concept is functional, encompassing not only all the organisms (plants and animals) in a given area but also the totality of interactions among the organisms and between the organisms and the nonliving portion of the environment (soil, rocks, water, sunlight, atmosphere, etc.), which can be thought of as nutrients and energy. Thus an ecosystem is a biological community expressed in functional terms.

It is both a virtue and a complication that the ecosystem concept can be used at almost any scale. In other words, we can speak of a world ecosystem, or the ecosystem of a drop of water, or any level of generalization between these extremes. For our purposes, it is appropriate to consider broad-scale ecosystems that encompass large parts of the continent.

One of the most carefully constructed regionalizations of ecosystems was developed by representatives of the U.S. Forest Service and the U.S. Fish and Wildlife Service in the late 1970s and refined in the early 1990s. Called "Ecoregions of North America," it represents an effort to incorporate all major environmental aspects (topography, climate, soils, flora, and fauna) into a single hierarchy of ecosystem regions. Figure 2-18 depicts the two highest levels of this hierarchy.

The *Polar Domain* encompasses the high-latitude portions of the continent, where long, cold winters provide the dominant environmental factor. This domain includes four divisions: *Tundra, Tundra Mountains, Subarctic,* and *Subarctic Mountains.*

The *Humid Temperate Domain* includes the eastern half of the conterminous states, the southern fringe of Canada, and the entire Pacific Coast of the continent. It is subdivided into 11 divisions.

The *Dry Domain* consists of the interior West of the United States and a small adjacent part of Canada. It contains 7 divisions.

The *Humid Tropical Domain* includes only the southern tip of Florida (the *Tropical Savanna Division*) and the Hawaiian Islands (*Tropical Rainforest Mountains Division*).

Another recent effort by a geographer of the U.S. Environmental Protection Agency delineates ecoregions for the conterminous states on a somewhat different basis (including land use as well as environmental parameters) and for a different purpose (to assist managers of aquatic and terrestrial resources in recognizing more efficient management options). This system is not hierarchical, and it identifies 76 ecoregions in the 48 conterminous states (Fig. 2-19).

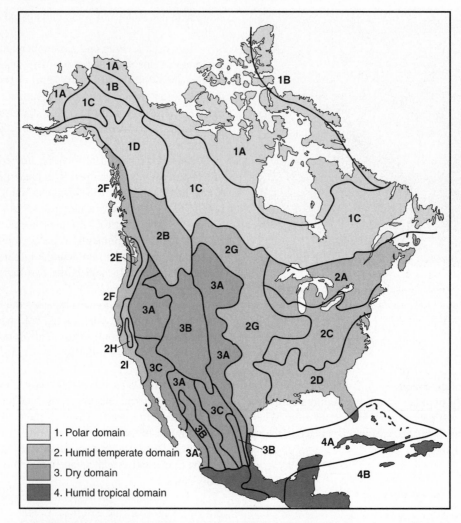

FIGURE 2-18 Major Ecoregions of North America.

1	Polar Domain	2E	Humid Maritime Division
1A	Tundra Division	2F	Humid Maritime Highlands Division
1B	Tundra Highlands Division	2G	Subhumid Prairie Division
1C	Subarctic Division	2H	Mediterranean Division
1D	Subarctic Highlands Division	2I	Mediterranean Highlands Division
2	Humid Temperate Domain	3	Dry Domain
2A	Humid Warm-Summer Continental Division	3A	Semiarid Steppe Division
2B	Humid Warm-Summer Continental Highlands Division	3B	Semiarid Steppe Highlands Division
2C	Humid Hot-Summer Continental Division	3C	Arid Desert Division
		4	Humid Tropical Domain
2D	Humid Subtropical Division	4A	Tropical Savanna Division
		4B	Tropical Rainforest Highlands Division

(After R. G. Bailey, *Ecoregions of the United States* (map), Ogden, Utah: U.S. Department of Agriculture, Forest Service, Intermountain Region, 1976; R. G. Bailey, *Description of the Ecoregions of the United States*, Washington, D.C.: U.S. Department of Agriculture, Forest Service. 2nd ed., 1995; R. G. Bailey and Charles T. Cushwa, *Ecoregions of North America* (map), Reston, VA: U.S. Fish and Wildlife Service, 1981.)

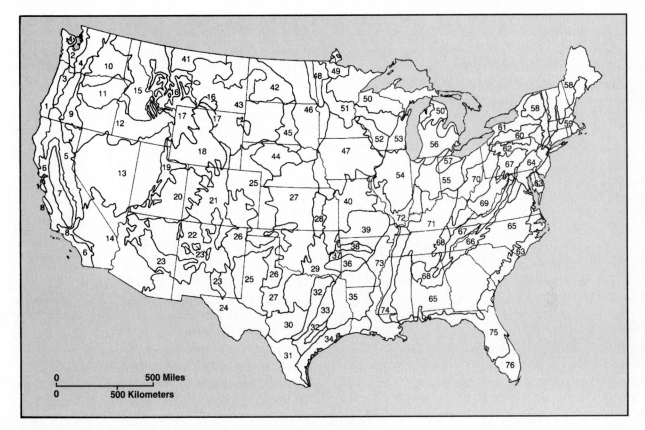

FIGURE 2-19 Ecoregions of the Conterminous United States

1	Coast Range	26	Southwestern Tablelands	52	Driftless Area
2	Puget Lowland	27	Central Great Plains	53	Southeastern Wisconsin Till Plains
3	Willamette Valley	28	Flint Hills	54	Central Corn Belt Plains
4	Cascades	29	Central Oklahoma/Texas Plains	55	Eastern Corn Belt Plains
5	Sierra Nevada	30	Central Texas Plateau	56	Southern Michigan/Northern Indiana Till
6	Southern and Central California Plains and	31	Southern Texas Plains		Plains
	Hills	32	Texas Blackland Prairies	57	Huron/Erie Lake Plain
7	Central California Valley	33	East Central Texas Plains	58	Northeastern Highlands
8	Southern California Mountains	34	Western Gulf Coastal Plain	59	Northeastern Coastal Zone
9	Eastern Cascades Slopes and Foothills	35	South Central Plains	60	Northern Appalachian Plateau and Uplands
10	Columbia Basin	36	Ouachita Mountains	61	Erie/Ontario Lake Plain
11	Blue Mountains	37	Arkansas Valley	62	North Central Appalachians
12	Snake River Basin/High Desert	38	Boston Mountains	63	Middle Atlantic Coastal Plain
13	Northern Basin and Range	39	Ozark Highlands	64	Northern Piedmont
14	Southern Basin and Range	40	Central Irregular Plains	65	Southeastern Plains
15	Northern Rockies	41	Northern Montana Glaciated Plains	66	Blue Ridge Mountains
16	Montana Valley and Foothill Prairies	42	Northwestern Glaciated Plains	67	Central Appalachian Ridges and Valleys
17	Middle Rockies	43	Northwestern Great Plains	68	Southwestern Appalachians
18	Wyoming Basin	44	Nebraska Sand Hills	69	Central Appalachians
19	Wasatch and Uinta Mountains	45	Northeastern Great Plains	70	Western Allegheny Plateau
20	Colorado Plateaus	46	Northern Glaciated Plains	71	Interior Plateau
21	Southern Rockies	47	Western Corn Belt Plains	72	Interior River Lowland
22	Arizona/New Mexico Plateau	48	Red River Valley	73	Mississippi Alluvial Plain
23	Arizona/New Mexico Mountains	49	Northern Minnesota Wetlands	74	Mississippi Valley Loess Plains
24	Southern Deserts	50	Northern Lakes and Forests	75	Southern Coastal Plain
25	Western High Plains	51	North Central Hardwood Forests	76	Southern Florida Coastal Plain

(After James M. Omernik, "Ecoregions of the Conterminous United States," *Annals*, Association of American Geographers, 77 (1987), map supplement. By permission of the Association of American Geographers.)

SELECTED BIBLIOGRAPHY

ATWOOD, WALLACE W., *The Physiographic Provinces of North America*. Boston: Ginn & Company, 1940.

BIRD, J. BRIAN, *The Natural Landscapes of Canada: A Study in Regional Earth Science*. Toronto: Wiley Publishers of Canada, Ltd., 1972.

————, "Recent Developments in Canadian Geomorphology," *Canadian Geographer*, 36 (Summer 1992), 172–181.

BRYSON, REID A., AND F. KENNETH HARE, EDS., *World Survey of Climatology:* Vol. 2, *Climates of North America*. New York: American Elsevier Publishing Company, 1974.

CLAYTON, J. S., ET AL., *Soils of Canada*. Ottawa: Research Branch, Canada Department of Agriculture, 1977.

FALCONER, A., ET AL., *Physical Geography: The Canadian Context*. Toronto: McGraw-Hill Ryerson, Ltd., 1974.

FENNEMAN, NEVIN M., *Physiography of Eastern United States*. New York: McGraw-Hill Book Co., 1938.

————, *Physiography of Western United States*. New York: McGraw-Hill Book Co., 1931.

FRENCH, HUGH M., AND OLAV SLAYMAKER, EDS., *Canada's Cold Environments*. Montreal: McGill-Queen's University Press, 1993.

GERSMEHL, PHILIP J., "Soil Taxonomy and Mapping," *Annals*, Association of American Geographers, 67 (1977), 419–428.

HARE, F. KENNETH, AND M. K. THOMAS, *Climate Canada*. (2d ed.). Toronto: Wiley Canada, Ltd., 1980.

HUNT, CHARLES B., *Natural Regions of the United States and Canada*. San Francisco: W. H. Freeman & Company, 1973.

IVES, J. D., "Glaciers," *Canadian Geographical Journal*, 74 (April 1967), 110–117.

KUCHLER, A. W., *Potential Natural Vegetation of the Conterminous United States*. New York: American Geographical Society, Special Publications 36, 1964.

LOVELAND, THOMAS R., ET AL., "Seasonal Land-Cover Regions of the United States," *Annals*, Association of American Geographers, 85 (June 1995), 339–355.

MARKHAM, CHARLES G., "Seasonality of Precipitation in the United States," *Annals*, Association of American Geographers, 60 (1970), 593–597.

NELSON, J. G., M. J. CHAMBERS, AND R. E. CHAMBERS, EDS., *Weather and Climate*. Toronto: Methuen Publications, 1970.

PALMER, TIM, *America by Rivers*. Covelo, CA: Island Press, 1996.

PIELOU, E. C., *After the Ice Age: The Return of Life to Glaciated North America*. Chicago: University of Chicago Press, 1991.

PIRKLE, E. C., AND W. H. YOHO, *Natural Regions of the United States* (4th ed.). Dubuque, IA: Kendall/Hunt Publishing Company, 1985.

RITCHIE, J. C., *Postglacial Vegetation of Canada*. New York: Cambridge University Press, 1988.

SCOTT, GEOFFREY A. J., *Canada's Vegetation: A World Perspective*. Montreal: McGill-Queen's University Press, 1995.

SEARLE, RICK, "Not Just a Pretty Face: Glaciers Are More than Scenic Backdrops; They're Barometers of Environmental Change," *Nature Canada*, 20 (Spring 1991), 34–39.

THORNBURY, WILLIAM D., *Regional Geomorphology of the United States*. New York: John Wiley & Sons, 1965.

TRENHAILE, ALAN S., *The Geomorphology of Canada: An Introduction*. Toronto: Oxford University Press, 1990.

TULLER, S. E., "What Are 'Standard' Seasons in Canada?" *Canadian Geographical Journal*, 90 (February 1975), 36–43.

VALE, THOMAS R., *Plants and People: Vegetation Change in North America*. Washington, DC: Association of American Geographers, 1982.

VISHER, S. S., *Climatic Atlas of the United States*. Cambridge: Harvard University Press, 1954.

WILLIAMS, MICHAEL, *Americans and Their Forests: A Historical Geography*. New York: Cambridge University Press, 1989.

ZWINGER, A. H., AND B. E. WILLARD, *Land Above the Trees: A Guide to American Alpine Tundra*. New York: Harper & Row, Publishers, 1972.

Population

3

The areally vast and physically varied subcontinent of North America was sparsely populated until relatively recent times. Its aboriginal peoples were numerically few, geographically scattered, and culturally diverse. The penetration and settlement of the continent by Europeans signaled an almost total change in its human geography; the Native Americans were decimated and displaced, and in a few short decades almost every vestige of their lifestyle was erased. The contemporary human geography of the United States and Canada, then, has been shaped almost completely by non-native people. Although the present population is mostly European in origin, it also contains other important elements. The saga of the blending of these elements is a chronicle of great complexity that is treated only briefly in this book (Fig. 3-1).

MELTING POT OR POTPOURRI?

At the time of establishment of the first European settlements it is probable that the total population of the area now occupied by the United States and Canada did not exceed 5 million.[1] The population today is about 60 times that figure.

[1] Estimates from various reputable scholars range from a low of 1 million to a high of 18 million. Hard evidence to support any of these estimates is difficult to come by.

It has often been stated that the United States and Canada are prime examples of a melting pot wherein people of diverse backgrounds are shaped in a common mold, thus becoming new citizens of new countries. There is some factual basis to this image, for tens of millions of immigrants settled in the two countries and, with the passage of time, many of their ethnic distinctions were blurred into obscurity. But the melting pot concept is only partially apt; in reality, the population of the two countries consists of an imperfect amalgam of diverse groups. In other words, the melting pot contains a lumpy stew.

By and large, the United States and Canada are countries strongly influenced by northwestern Europe. The historic ties to Great Britain are evident in cultural traits of language, religion, political system, technological achievement, and many other areas. Yet these ties were greatly modified by the importance of the French in Quebec and parts of the American South, the Spanish in southwestern United States, and the Africans in southeastern United States.

Each group of immigrants has been changed by the North American cultures, and, through time, has altered the North American landscape. New groups arrive and the exchange of cultural traits takes place. This sets into motion the acculturation process, and a slightly different North America

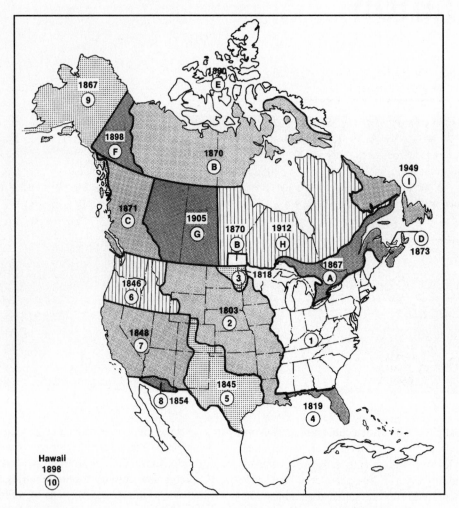

FIGURE 3-1   Territorial evolution of the United States and Canada (simplified).

United States
1   Territories and claims of the original 13 states
2   Louisiana Purchase
3   Red River cession
4   Purchase of Florida
5   Annexation of Texas
6   Oregon compromise
7   Mexican cession
8   Gadsden Purchase
9   Purchase of Alaska
10 Annexation of Hawaii

Canada
A   Formation of the Dominion of Canada
B   Acquisition of Northwest Territories and creation of Manitoba
C   Creation of British Columbia
D   Unification of Prince Edward Island
E   Addition of the Arctic islands
F   Formation of Yukon Territory
G   Creation of Alberta and Saskatchewan
H   Expansion of Manitoba, Ontario, and Quebec
I   Annexation of Newfoundland

emerges. The imprint of Latin American, Caribbean, and Asian immigrants is presently developing throughout the subcontinent.

However long, complete, or painful the process of assimilation may be, the melting pot concept is probably less apt today than ever before. The disparities among Korean merchants, Hmong tribesmen, Ukrainian Jews, Nicaraguan refugees, French socialites, Filipino peasants, and Haitian boat people probably are greater than any country has ever confronted. Yet the pluralistic society of the United States and Canada seems to be more tolerant of diversity than once was the case. No longer is there a concerted effort to homogenize the population; that precept has been overtaken by reality.

THE PEOPLING OF THE NORTH AMERICAN SUBCONTINENT

The current population mix of the United States and Canada is continually being modified by the influx of immigrants and the egress of emigrants as well as by the rate of natural increase. It is a blend of varied origins, and its diverse patterns through the years continue in dynamic flux today. There were five major original source regions for the peopling of these two countries: North America itself, Europe, Africa, Asia, and Latin America.

The Native Americans

The indigenous population of the continent at the time of European contact consisted of a great variety of tribes, thinly scattered (Fig. 3-2). The largest concentrations were in the area presently called California, in the Southwest, adjacent to the Gulf Coast, and along the Atlantic coastal plain.

The contrast with the relatively large population of *Indians* in Latin America is very marked and apparently has long been a feature of the Americas. There is nothing in

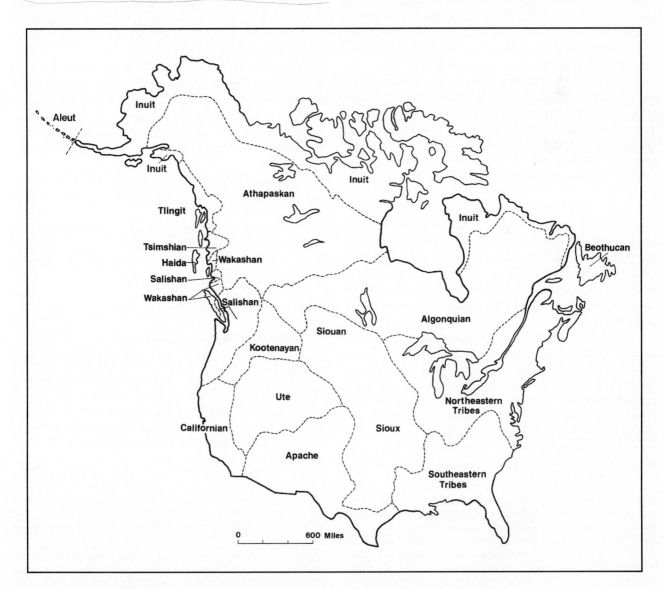

FIGURE 3-2   Generalized distribution of major tribal groupings in aboriginal North America.

North American archaeological evidence to suggest any developed Indian civilizations comparable with those of Mexico, parts of Central America, or the central Andes. The thin population density probably reflected a reliance on simple hunting and fishing or primitive agricultural economies.

It is generally accepted that the "first Americans" were relatively recent (within the last 35,000 years or so) arrivals from Asia, having entered the Western Hemisphere via Alaska and diffused widely throughout the New World. There were many physical and cultural variations among the Indians; their diversity was at least as great as that of the Europeans who later were to overwhelm them. Their only common physical attributes were black hair, brown eyes, and some shade of brown skin. There was much variety in both material and nonmaterial aspects of Indian culture, and hundreds of mutually unintelligible dialects (divided by scholars into six major linguistic groups) were spoken.

Inuit[2] and *Aleuts* are thought to be Asian immigrants of much more recent vintage, having crossed from Siberia as little as 1000 to 3000 years ago. The Inuit spread all across the Arctic from the Bering Sea to Greenland, hunting and fishing for a living. The Aleuts, apparently branching off from the mainstream of Inuit life, developed a distinctive culture of their own, based on a fishing economy in the Aleutian Islands.

Initial relations between Europeans and Native Americans were not always unpleasant. Many tribes developed profitable trading patterns with the Europeans. Indeed, such tribes as the Iroquois in the East, Crees south of Hudson Bay, Comanches in the Great Plains, and Apaches in the Southwest acquired metal weapons and horses and established short-lived empires at the expense of less fortunate neighboring tribes.

Within a few years or a few decades, however, the insatiable appetite of the land-hungry Europeans led to inevitable conflict. Although some tribes were fierce, brave, and warlike, white settlers subdued and relocated them in relatively short order. The Native Americans were cruelly decimated by warfare, but introduced diseases were even more potent destroyers. For example, a smallpox epidemic wiped out more than 90 percent of the Mandans in the mid-1800s at the same time that cholera was eliminating 50 percent of the

Kiowas and Comanches.[3] An "Indian Territory" was established between Texas and Kansas, and nearly 100,000 Indians from east of the Mississippi were crowded onto the overutilized hunting grounds of the Plains Indians.[4] Even this area was sporadically whittled down to size until it was finally thrown open to white settlement and became the state of Oklahoma.

Although there were occasional later outbursts by renegade Apaches in the Southwest, the last significant Indian conflict ended in the Wounded Knee massacre of 300 Sioux in 1890. By the turn of the century only several tens of thousands of Indians survived north of Mexico.

Presently, most Native Americans either live on reservations or have abandoned tribal life entirely. They have a rapid population growth rate but represent less than 1 percent of the total population of the two countries. In the United States some half a million Native Americans live on 314 reservations (including pueblos in New Mexico, colonies in Nevada, and rancherias in California); the largest concentrations are in Arizona, New Mexico, and South Dakota. Another 1.5 million or so Native Americans live outside reservations, the largest numbers by far being in Oklahoma and California.

In the early 1990s, for the first time in history, the Native American population of the United States became more than 50 percent urban.

In Canada in the mid-1990s there were about 800,000 Indians, of whom nearly two-thirds were status Indians.[5] About half of the status Indians live on nearly 2300 reserves (mostly small) that are scattered widely across the country.

In addition, some 90,000 Inuit and Aleuts live in Alaska, and approximately 35,000 Inuit reside in Canada.

The Pattern of European Immigration

North America has been by far the principal destination of European immigrants. It is estimated that between 1600 and 1995 nearly 90 million people left Europe to settle elsewhere, and more than four-fifths of them went to the United States and Canada. Although the social process of immigration has varied little during North American history,

---

[2] The term *Eskimo* was generally accepted until recently as the name for the indigenous people of the North American tundra. In the last few years many Eskimo people objected to this label as having been applied to them by outsiders. They have not reached total agreement on terminology, but the term "Inuit" is now the most widely approved appellation for the tundra dwellers of North America. It is universally used in Canada and is slowly coming into favor in Alaska. In this book "Inuit" will be used in preference to "Eskimo." The matter of native names will be explored more fully in Chapter 20.

[3] William T. Hagan, *American Indians* (Chicago: University of Chicago Press, 1969), p. 94.

[4] The continual forced rearrangement of tribes caused great misery and frustration, which is indelibly echoed in the pointed question of the Sioux chief Spotted Tail: "Why does not the Great Father put his red children on wheels, so he can move them as he will?"

[5] Status Indians are those registered with the federal government as Indians according to the terms of the Indian Act. Nonstatus Indians are native people self-identified as Indians but not registered for the purposes of the Indian Act.

the pattern of immigration can be clarified if chronological divisions are established. These divisions are strongly affected by the impact of major immigration legislation that, through time, favored immigrants from particular parts of Europe, and for long periods inhibited immigration from Asia and Latin America.

Before 1815

This was the period of primary European colonization of North America, when nearly all migrants—with the important exception of early French settlers in Acadia (Nova Scotia) and Lower Canada (Quebec)—were of Anglo-Saxon ethnic stock and followers of reformed churches subsequent to the Reformation. They varied from High Church Anglican aristocracy to Puritan and Lutheran peasant dissenters. The majority were English, but there were important groups of Germans, Dutch, and folk from Ulster, Scotland, and Wales.

In total numbers they probably did not greatly exceed 1 million, but their descendants in all parts of the United States and Canada now form an important segment of the population. More significant than their actual numbers is the effect they had in establishing the foundations of the social and economic patterns of the continent for generations. Different regional frameworks were established in the Maritimes, Lower Canada, Upper Canada, New England, the Middle Colonies, and the Southern Colonies, many aspects of which are still evident today.

After the American Revolution, immigration from England declined drastically, and the most numerous of the newcomers, in most years, were Scotch-Irish. During the 1790s European immigration to North America often exceeded 10,000 per year, but the flow slowed appreciably during the Napoleonic Wars and almost ceased during the War of 1812.

1815 to 1860s

This period represented the first of the great waves of immigration from Europe, a movement unprecedented in world history. Some 6 million immigrants were involved; the rate accelerated from about 200,000 in the 1820s to nearly 3 million in the decade of the 1850s. Compared with the resident population at the time, this influx was enormous.

These were people drawn by the economic opportunities available in a virgin land or driven from their homes by religious or political persecution or by revolution. Every country in Europe was represented, but the great majority were from the North Sea countries. More than half the migrants in this period came from the British Isles, especially Ireland. Germany was second only to Ireland as a source of immigrants; smaller numbers came from France, Switzerland, Norway, Sweden, and the Netherlands. Thus the immigrant mix of this period was less exclusively British, and there also was a larger proportion of Catholics (principally Irish and German) than previously.

1860s to 1890s

The second great wave of European immigration occupied the three decades following the American Civil War. In composition and character it was much like the first wave, but it was nearly twice as great in magnitude. Most migrants were still from northwestern Europe—Germany (250,000 German immigrants coming to the United States in 1882 constituted the largest number from any single country in any single year prior to the twentieth century) and the British Isles—and from Scandinavia, Switzerland, and Holland.

A significant feature of this period is that it was the only lengthy segment in Canadian history in which net out-migration occurred. Some 800,000 Canadian-born people moved south of the international border during this era.[6] A postwar wave of prosperity in the northern states coincided with an economic depression in Canada and drew a great many people, particularly French Canadians, to U.S. cities. In addition, farmlands in the Midwest were available for settlement, a further inducement to Canadian immigration. In the later years of this period the Canadian prairies were opened for settlement; but even the flood of Swedes, Ukrainians, Mennonites, Finns, Hungarians, and other Europeans to the Canadian west did not compensate for the southward flow.

1890s to World War I

Numerically this was the most significant period, for in these two decades an average of nearly 1 million immigrants per year came from Europe to North America. Perhaps of equal significance was the change in origin of the flow. From a predominantly northwest European source in the 1880s (87 percent in 1882, for example), the tide shifted to a predominantly southern and eastern European source after the turn of the century (81 percent in 1907).[7] Italy, Austria-Hungary (which included most of eastern Europe), and Russia were the homelands of the bulk of the immigrants,[8] but considerable numbers also

---

[6] T. R. Weir, "Population Changes in Canada, 1867–1967," *The Canadian Geographer*, 11 (1967), 201.

[7] Maldwyn Allen Jones, *American Immigration* (Chicago: University of Chicago Press, 1969), p. 179.

[8] Peak-year immigration to the United States was 286,000 from Italy in 1907; 340,000 from Austria-Hungary in 1907; and 291,000 from Russia in 1913.

came from such countries as Greece and Portugal. The change in Canadian immigration sources was almost as dramatic because the Canadian government had instituted a policy to induce and broaden the base of immigration. In 1896 and during the succeeding decade there were twice as many immigrants to Canada from continental Europe as from the British Isles.

This was the period of the melting pot in North America, but the assimilation of vast numbers of people of different cultures and languages was a slow and difficult process. By the time of World War I there was considerable agitation in the United States, and some in Canada, to slow the pace of immigration.

World War I to 1960s   Immigration to North America decreased markedly during World War I and then began a rapid climb in 1919. Concern mounted that "the greatest social experiment in human history" had become the "melting-pot mistake." Immigrants had contributed to material progress by opening new lands and providing a cheap labor pool for factories, mines, forestry, and construction, but it was becoming clear that the rate of influx was getting out of hand. There was further concern, particularly in the United States, about the "mix" of the immigrants, with increasing demand for restricting immigration from the Latin and Slavic portions of Europe and from Asia.

As a result, legislation was enacted in the United States in the early 1920s to restrict the immigrant flow. Canada did not promulgate significant restrictions for another decade. The basis of the restrictions was the ethnic composition of the U.S. population according to the census of 1900; thus the countries of northwestern Europe were given relatively large annual quotas (65,000 for the United Kingdom and 25,000 for Germany, for example), whereas other countries received much smaller allotments (6000 for Italy, 850 for Yugoslavia, 0 for Japan). Africa was ignored in the quota system, and Asians were prohibited entirely by separate legislation that originated in the late nineteenth century.

The total volume of immigration to the United States declined rapidly (down from 800,000 in 1921 to less than 150,000 by the end of the decade), whereas Canadian immigration maintained a relatively steady pace (between 100,000 and 150,000 per year during most of the 1920s). The Great Depression of the 1930s and war in the early 1940s reduced immigration to insignificant totals for a decade and a half.

Although the quota system was still operating in the United States, immediately after World War II there was a significant upturn in immigration to the United States and Canada. The heaviest influx to both countries was from Great Britain, Italy, and Germany. Bulking large in the postwar migrant flow was the admittance of displaced persons and refugees (without regard to quotas in the United States) in considerable numbers, which significantly increased the influx from eastern Europe until the Iron Curtain was closed.

Late 1960s to Present   Drastically altered immigration laws in both countries (since 1965 in the United States and 1967 in Canada, with further significant changes in both countries in the late 1970s and early 1990s) ushered in a new period in the immigration history of North America. In both countries immigration restrictions were greatly liberalized, "universalized," and made more complicated.[9]

Canadian immigration laws are broadly similar to those in the United States. Immigrant goals are based on annual evaluations of the nation's economic conditions and demographic "needs," with an emphasis given to the reunification of families.

These revised immigration policies resulted in major changes in the flow of migrants to both Canada and the United States. In the first place, total immigration increased significantly in the latter and moved upward erratically in the former. During the 1980s and 1990s legal immigration to the United States averaged 590,000 people annually, which is one-third higher than the average for the 1970s and is substantially more than at any other time since the early 1920s (Fig. 3-3). In addition, refugees and asylees[10] were admitted at a rate of more than 90,000 each year, and illegal immigrants were estimated to total at least 500,000 annually.

Another notable result of the changed immigration policies has been a marked decrease in the European component of total immigration (Fig. 3-4). Whereas 80 percent of the immigrants to Canada in the early 1960s were from Europe, they now constitute only about 20 percent of the total. The decrease has been even more striking in the United States—from 50 percent European in the early 1960s to 10 percent European in the late 1990s. Although Britain is still the single leading source of European immigrants to Canada, the great bulk of European immigrants to North America now emanate from Mediterranean countries, particularly Portugal, Italy, and Greece.

[9] U.S. immigration laws and regulations are considered to be second only to tax laws and regulations in their length and complexity.

[10] Foreigners seeking to escape from oppression in their homelands are admitted to the United States under special legislation, especially the Refugee Act of 1980, that is separate from normal immigration quotas. This procedure has particularly benefited people from Cuba, Southeast Asia, Haiti, Central America, and Hungary.

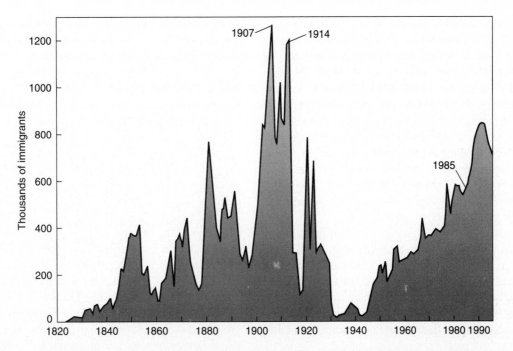

FIGURE 3-3   The historical sequence of legal immigration to the United States, 1820–1995 (based on data from the U.S. Department of Justice, Immigration and Naturalization Service).

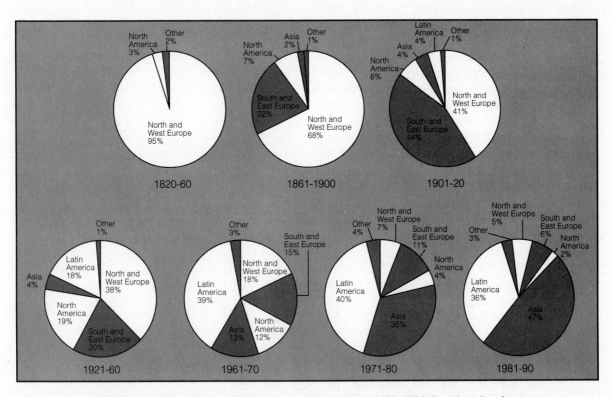

FIGURE 3-4   Sources of immigrants to the United States, 1820–1990 (based on data from the U.S. Department of Justice, Immigration and Naturalization Service).

Legislation enacted by the U.S. Congress in 1990 produced a major restructuring of national immigration policies, permitting at least 45 percent more foreigners to enter the country in each of the next 3 years and about 35 percent more in every year thereafter. The previous immigration law, which had been in effect for a quarter of a century, emphasized family reunification to such an extent that about 85 percent of visas had gone to Asians and Latin Americans. The new policy does not diminish the absolute number of "family" visas. However, by raising overall numbers substantially, it also reserves a general quota for people from "low-immigration" regions, which is expected to provide a significant boost to immigration from Europe and Africa.

During the early 1990s nearly 1 million legal immigrants entered the United States each year (Fig. 3-5). Europeans made up about 10 percent of the legal immigrant flow, whereas Asians constituted about one-third and Hispanics about one-half of the total. At the same time Canada was absorbing about 200,000 immigrants annually: more than half from Asia, one-fifth from Europe, and one-eighth from Latin America.

Involuntary Immigration: The African Source

Soon after the first white settlers occupied the land of coastal Virginia, the problem of labor for clearing the forests, cultivating the soil, and harvesting the crops arose. Because land was free, few settlers would consider working for others when they could have their own land. The local Native American population was too small and resisted subjugation easily since they were far more at home in their environment than the white colonizers. At first, indentured workers—Englishmen who temporarily sold their services for the price of ship passage to the New World—met the labor requirements.

But they did not prove satisfactory because they were not numerous enough and because it was difficult to keep them as workers once they reached the frontier.

To help solve the labor problem, Negro slaves were imported from West Africa, initially via already established slaving areas in the West Indies. The first African slaves in the English colonies landed at Jamestown in 1619. By the 1630s slaves were being brought in each year. They were not too popular at first, and by the end of the century there were fewer than 10,000 in the tobacco colonies. But about then the British government began to restrict the sending of convicts (another source of labor) to America, and cotton became an important crop as a result of Eli Whitney's invention of the cotton gin. Thus slaveholding began in earnest, and direct slave trade with the Guinea coast of West Africa developed.

On the eve of the American Revolution there were about half a million slaves in British North America, nine-tenths in the southern colonies. Despite their status, the slaves exerted a significant influence on Southern life, from language and social customs to agricultural techniques. By 1808, when further importation of slaves was prohibited, nearly 20,000 were being brought into the United States every year.

Slavery was abolished at the time of the Civil War in the 1860s, and few African immigrants have entered North America since then, until very recently. The descendants of those involuntary immigrants of the seventeenth to nineteenth centuries numbered nearly 9 million by 1900, however, and more than 30 million by 1990.

Throughout most of the seventeenth, eighteenth, and nineteenth centuries the African American population was concentrated in the southeastern part of the United States. In the twentieth century, however, a major movement northward and westward occurred. As recently as 1950, African

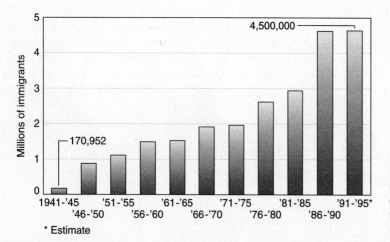

* Estimate

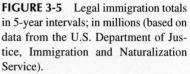

FIGURE 3-5   Legal immigration totals in 5-year intervals; in millions (based on data from the U.S. Department of Justice, Immigration and Naturalization Service).

Americans lived in approximately equal numbers in three situations: one-third in the rural South, one-third in the urban South, and one-third in cities of the North and West coasts. But the continued rural-to-urban movement has significantly decreased the proportion of nonurban African Americans. Most are now city dwellers, with the largest concentrations in New York, Chicago, Los Angeles, Philadelphia, Washington, and Detroit. Ever since the mid-1970s there has been a major and continuing migration of African Americans from the cities of the Northeast to the cities of the South.

Only a small number of blacks reside in Canada, largely in Toronto, Montreal, Halifax, and Windsor. Recent immigration from several black Caribbean nations—particularly Jamaica, but also Trinidad, Haiti, and Guyana—has made a large proportional increase, however.

The Irregular Sequence of Asian Immigration

Until the 1960s the immigration of people from Asia to North America was quite limited in number and extremely sporadic in occurrence. Occasionally, Asian workers were imported or came of their own volition in large numbers, but throughout most of history their entry was either severely curtailed or totally prohibited.

The earliest Asian immigration was the largest prior to the 1960s; some 300,000 Chinese came to California between 1850 and 1882, when the passage of the Chinese Exclusion Act halted the flow. During this same period several thousand Chinese entered British Columbia; originally they relocated from California and later directly from China. The great majority of these immigrants were males; after working in railway construction or gold mining for a while, many returned to China. Most remained on this continent, however, settling in "Chinatowns" in San Francisco, Los Angeles, Seattle, Vancouver, New York, and Chicago.

The first Japanese immigrants came to Hawaii as contract laborers to work on sugar-cane plantations in the 1880s. After Hawaii became a territory of the United States in 1898, the Japanese were free to come directly to the mainland, and several tens of thousands did so during the early years of this century. A few thousand Japanese also immigrated to British Columbia.

The quota system stopped Japanese immigration to the United States after World War I, and Canada effectively stopped Chinese immigration by legislation in 1923.[11]

---

[11] Some 4300 Chinese had immigrated to Canada in 1919, a larger total than from any other countries except the United Kingdom and the United States. The Chinese Immigration Act of 1923 resulted in such restriction that an average of only one Chinese immigrant per year entered Canada during the next two decades. See W. H. Agnew, "The Canadian Mosaic," in *Canada*

The new immigration policies of the 1960s once again opened the door to Asian settlers and resulted in a veritable flood of Asian immigrants. In the first few years under the new laws more Chinese (from Taiwan and Hong Kong) and Filipino immigrants entered the United States than citizens of any other Eastern Hemisphere countries; there was also a great upsurge of immigrants from Southeast Asia, Korea, and India (Fig. 3-6). Overall, Asians have constituted more than one-third of total immigrants to the United States in recent years, compared with only one-fifteenth in the mid-1960s. The Asian proportion in Canada is similar, with notable flows from Hong Kong, Vietnam, India, and the Philippines.

The great majority of all Asian immigrants have settled on the West Coast, particularly in California. Most are urban dwellers, joining the swelling Chinese, Japanese, Filipino, and Korean minorities of Los Angeles, San Francisco, Seattle, and Vancouver, with some spillover into Arizona and Alberta and conspicuous nodes in New York, Chicago, and Toronto.

It should be noted that Hawaii has long had a significant Asian population. The majority of the contemporary populace of that state is of Asian extraction, particularly Japanese and Chinese but with large numbers of Filipinos and Koreans.

Latin American Immigrants

The ancestors of Mexican Americans were the first Europeans to settle what is now the southwestern United States. A majority of their descendants still reside in the borderlands from Texas to California. Between 1598 and 1821, parts of the states of California, Arizona, New Mexico, Colorado, and Texas were settled by Spanish-speaking peoples from New Spain, which is present-day Mexico. After 1821, portions of this region became part of the Republic of Mexico, and each of the settled areas developed its own regional identity. The transformation of these areas from Spanish outposts to Mexican provinces to Mexican American subregions represents the planting, germination, and rooting of Mexican American cultures in the United States. The origins of many of the features associated with Mexican American culture, from place names to architecture and numerous social customs, can be traced to this early period of settlement in the United States borderlands.

Throughout the twentieth century there has been a fluctuating pattern of immigration to North America from Latin America. With the exception of the massive influx from Cuba, it has been primarily a move toward economic betterment, often on a short-term rather than permanent basis.

One Hundred: 1867–1967, ed. Dominion Bureau of Statistics (Ottawa: Queen's Printer, 1967), p. 89.

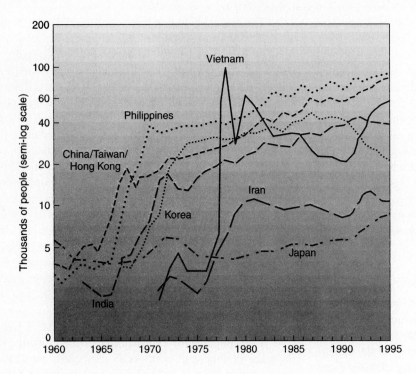

FIGURE 3-6   Principal components of recent Asian immigration to the United States.

Latin Americans were totally exempted (as were Canadians) from the quota provisions of U.S. immigration laws in the 1920s, which meant that they essentially enjoyed unrestricted immigration most of the time.

By far the largest, most continuous, and most conspicuous flow has been from Mexico. Legal immigration from Mexico began about the turn of the century, and several million legal immigrants have entered since that time.

In addition, a great number of undocumented Mexicans ("guesstimates" range from 2 million to 10 million) have come into the United States in recent years. Most Mexican immigrants have settled in the southwestern border states, Texas to California, plus Colorado, although there is a significant concentration in Chicago.

Immigration from the West Indies is also of long-standing duration. Puerto Rico has been the principal source because of its political affiliation with the United States. Since 1975 Jamaica has furnished more immigrants to the United States than any other country except the Philippines, Italy, and Greece. More than three-quarters of a million Cubans entered the United States in the 1980s and 1990s (over 150,000 in 1980 alone), mostly with refugee status. Several thousand Haitian "refugees" also came to the United States in the 1980s and 1990s. Despite strong official efforts at dispersal, some 75 percent of these Cuban and Haitian immigrants settled in southern Florida, chiefly in the Miami area.

THE CONTEMPORARY POPULATION

The 296 million people of the United States and Canada represent about 6 percent of total world population. They occupy a land area of some 8.3 million square miles, or 14.5 percent of the land area of the planet. This population is very unequally divided between the two countries, with about 10 people in the United States for every 1 person in Canada. The 10 to 1 ratio has been maintained throughout the past century and indicates that the rate of population growth in the two countries has been approximately equal for several generations.

Distribution

The principal population concentrations are in the northeastern quarter of the United States (Fig. 3-7) and adjacent parts of Ontario and Quebec (Fig. 3-8). The most notable clusters are in the Megalopolis Region of the Atlantic seaboard, around the shores of Lake Erie, and around the southern end of Lake Michigan. In the southeastern quarter of the United States and in parts of southern Ontario there is a moderate population density, fairly evenly distributed except for agglomeration around urban nodes. In the Atlantic Provinces an irregular pattern of moderate density alternates with patches of sparse population. In the central plains and prairies there is

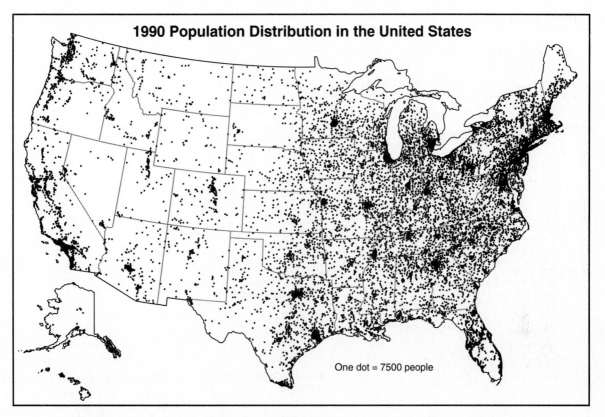

FIGURE 3-7   Distribution of population in the United States (courtesy U.S. Bureau of the Census).

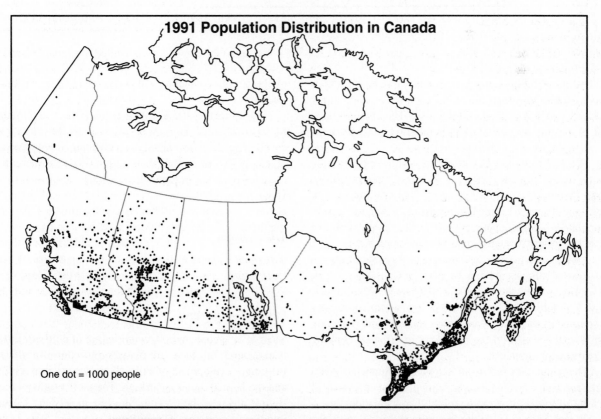

FIGURE 3-8   Distribution of population in Canada (courtesy Statistics Canada).

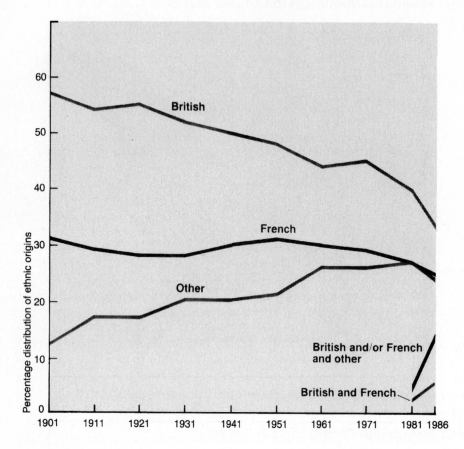

FIGURE 3-9 Changing Canadian ethnicity. Multiple responses were permitted beginning in 1981 (after Pamela Margaret White, *Ethnic Diversity in Canada*, 1986 Census of Canada; and recent estimates by Statistics Canada).

a generally decreasing density from east to west, with obvious concentrations along the major river valleys and at the eastern edge of the Rocky Mountains. The Intermontane West is sparsely populated, with population concentrations around cities.

The Pacific Coast has moderate-to-heavy population density in the valleys, with conspicuous agglomerations around the six principal urban areas. The vast expanses of central and northern Canada and most of Alaska are almost unpopulated; indeed, 72 percent of Canada's population live within 150 miles (240 km) of the international border.

As it has for some years, the greatest *absolute* population growth continues to be in the states of California, Texas, and Florida and in the provinces of Ontario, Alberta, and British Columbia. The fastest *rates* of increase are in Alaska, the southwestern states (especially Nevada and Arizona), Florida, and Texas, and the provinces of Alberta, Ontario, and British Columbia.

The population shift from Snowbelt to Sunbelt states is absolutely clear-cut. For several years every state from Virginia to California grew more rapidly than the national average, whereas no state in the north-central or northeastern part of the country (except New Hampshire) achieved that distinction.

During the 1980s many rural areas lost population. More than half of the rural counties across the nation experienced population decreases, in sharp contrast with the decade of the 1970s, when less than 20 percent of American rural counties declined in population. Depopulation was most severe in the agricultural states of the Midwest and Great Plains; in Iowa, for example, all but six counties lost population during the decade. Energy-producing states also experienced either decline or very slow growth, as did some northeastern states that endured a continuing dwindling of manufacturing. Four states (Iowa, North Dakota, West Virginia, and Wyoming) decreased in population during the decade.

Ethnic Components

Most of the people in both countries are of northern European background, but there are many significant minorities, and the rapid development of a pluralistic society in both Canada and the United States is striking. The most prominent minority in Canada is the French, who live mostly in Quebec and in adjacent portions of New Brunswick (Fig. 3-9). In recog-

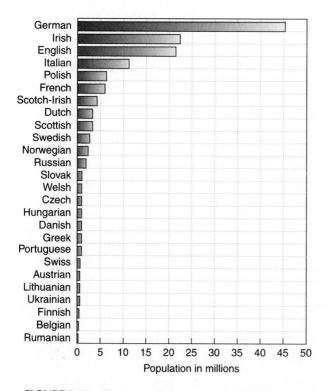

FIGURE 3-10   European ethnic components of U.S. population ancestry, 1990 (data from U.S. Census of Population, 1990).

for the most part they are relatively inconspicuous as minorities. As shown in Figure 3-10, people of German ancestry constitute by far the largest proportion of the U.S. population of European ancestry, with twice the numbers of the second most numerous ancestry group (Irish). People of European ancestry are sometimes prominently associated with particular areas, such as Scandinavians in Minnesota and Wisconsin, or Ukrainians in the Canadian prairies, but a detailed consideration of their distribution is beyond the scope of this general treatment. Some hints can be derived from Table 3-1, however, which lists the specific states with the largest populations of the principal ancestral groups. New York, the traditional "melting pot" state and port of entry for European immigrants, is seen to be especially prominent with regard to people of eastern European, southern European, and West Indian extraction. California, with its enormous total population, mirrors the ethnic diversity of the nation, with particular concentrations of people of Asian and Hispanic origin.

The group defined by the U.S. Census Bureau as "Hispanic origin" is a European ancestry minority that is sufficiently prominent in terms of population concentration and distinctive cultural attributes to merit particular attention. The number of people involved is unknown, for the Census Bureau readily admits that this group was underenumerated in official counts. In the mid-1990s, a conservative population estimate was about 25 million citizens and legal aliens of Hispanic origin, with several million others leading an undocumented existence as illegal aliens. Of this total, about two-thirds of the legal residents and well over half of the illegal ones are of Mexican origin. About 9 of every 10 Hispanics of Mexican origin live in the five southwestern states of California, Arizona, New Mexico, Colorado, and Texas.

The second largest component of the Hispanic minority consists of emigrants from Puerto Rico, of whom there are nearly 3 million in the United States. Puerto Ricans primarily inhabit the metropolises of the Northeast, with the greatest concentration by far in New York City (Fig. 3-11). There are slightly more than 1 million people of Cuban extraction in the United States, chiefly in southern Florida.

Although most Hispanic Americans are fluent in English, the majority consider Spanish to be their mother tongue and more than three-quarters of them speak Spanish in their homes. Most are at least nominally Roman Catholic, and they tend to be disadvantaged socially, economically, and politically, although they have developed considerable political leverage in south Texas, New Mexico, southern Arizona, and southern Florida.

Blacks are only a tiny fraction of the Canadian population, but they make up more than 12 percent of the popu-

nition of the long-standing historical significance of the Franco-Canadians, Canada is officially a bicultural nation. The relative position of French Canadians, however, is on the wane. For the first century of Canada's existence as a nation, French Canadians constituted a steady 30 percent of the national population total, but in the last few years this proportion has decreased noticeably. Indeed, both British and French ethnicity is on a declining trend. By the mid-1990s only one-third of all Canadians were of distinctly British origin, fewer than one-fourth were of distinctly French origin, and more than one-fourth were of *neither* British nor French origin (the remaining fraction of the populace was of "multiple" origin). Canadian ethnic diversity varies considerably by region: Newfoundland is the most dominantly British (89 percent), with Prince Edward Island and Nova Scotia not far behind; Quebec is most dominantly French (78 percent); the Northwest Territories, with its sizable population of native Canadians, has the highest proportion of people with non-British, non-French origins (64 percent).

Although people of German, Italian, Dutch, Scandinavian, Polish, Russian, and other European nationalities are significant components of the population in both countries,

TABLE 3-1
States with largest population of specific ancestry groups, as self-identified in 1980 census (in percentages)

California		New York		Pennsylvania	
English	10%	Afro-American	7%	Welsh	13%
German	9	Italian	23	Slovak	34
Irish	9	Polish	14	Ukrainian	20
French	10	Russian	24	Croatian	21
Scottish	16	Hungarian	14	Serbian	20
Mexican	44	Puerto Rican	49		
American Indian	11	Greek	17	**Michigan**	
Dutch	10	Austrian	21		
Swedish	13	Rumanian	23	Finnish	18%
Spanish/Hispanic	20	Indian	18	Belgian	17
Danish	16	Jamaican	54		
Portuguese	31	Dominican	78	**Massachusetts**	
Swiss	13	Colombian	34		
Chinese	38	Syrian	16	Canadian	17%
Filipino	44				
Korean	27	**Illinois**			
Lebanese	11			**Minnesota**	
Vietnamese	32	Czech	11%		
Armenian	38	Lithuanian	15	Norwegian	21%
Iranian	35				
Hawaii		**Ohio**		**Florida**	
Hawaiian	67%	Slovene	45%	Cuban	57%

Notes: All specific ancestry groups proclaimed by at least 100,000 persons are listed. Under each state the ranking is in order of number of people who proclaimed each ancestry group. Percentage figure refers to the proportion of total U.S. population of that ancestry group living in the listed state.

FIGURE 3-11 Ethnic neighborhoods are notable in many North American cities. This is the well-known sidewalk market along 125th Street in New York City's Spanish Harlem, where the bulk of the population is of Puerto Rican origin (TLM photo).

lace of the United States. As in the past, African Americans are more numerous in southern states than elsewhere; more than half the nation's African Americans live in the South. Moreover, the proportion is now increasing. After the decades of massive out-migration to the North and West, a reverse trend has developed since the mid-1970s. More African Americans are now moving to the South from the West, and particularly from the North, than are leaving the South.

The African American population has become highly urbanized. The long-continuing rural-to-urban flow has slowed, but only because there are few remaining blacks in rural areas. Washington, Detroit, Baltimore, New Orleans, Atlanta, Newark, Birmingham, and Richmond now have an African American majority population within their political limits, and in a dozen other large cities the African American proportion is above 40 percent. In absolute terms, New York City contains far more African Americans than any other city: 2.2 million, or 25 percent of its total population. In most instances the African American population is concentrated in central areas and is a very minor element in the suburbs. More than 25 percent of the total population of *all* central cities in the United States is African American, whereas less than 5 percent of the suburban population is African American.

Other ethnic groups are prominent in the contemporary population only in the major West Coast cities and a few western farming areas and in the largest eastern cities. The total population of Asian origin in North America is estimated at nearly 10 million. About one-fourth of this total is Chinese, and about one-fifth is Filipino, followed by Japanese, Indian, Korean, and Vietnamese. Hawaii had long been the major North American domicile for people of Asian origin, but the great flood of immigrants since the late 1960s has given California the principal concentrations of all significant Asian minorities except the Japanese, who are still slightly more numerous in Hawaii. Other states with significant Asian-origin populations are New York, Texas, and Illinois.

The changing ethnicity of the populace is highlighted by data from the 1980s (Fig. 3-12). During that decade in the United States the population of Asian extraction grew by about 99 percent, with about two-thirds of that increase due to immigration. During that same period the Hispanic population (not counting undocumented aliens) expanded by 53 percent, about half from immigration and half from natural increase. The Native American population grew by about 38 percent, the African American populace by 13 percent, and the non-Hispanic white population by 7.5 percent. Even so, about three-quarters of the nation's population consists of whites of non-Hispanic origin.

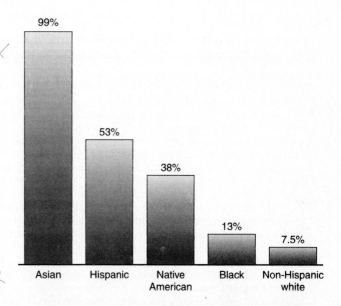

FIGURE 3-12   Proportional population increase by major ethnic groups, United States, 1980–1990 (based on data from the U.S. Department of Justice, Immigration and Naturalization Service).

Related Cultural Characteristics

The cultural geography of North America is diverse, complex, and imperfectly understood. No attempt will be made to explore it in systematic fashion here, although the accompanying map of culture areas (Fig. 3-13) is offered as a summary.

It is, however, appropriate to discuss a few details about certain nonmaterial culture elements that are closely associated with population: language, religion, and politics.

Language   In terms of language, the United States is one of the least complex areas in the world. Well over 90 percent of the population is fluent in English, and no sectional dialect is different enough to cause any problems in intelligibility. However, pluralism of the population has resulted in bilingualism with significantly large minority populations that have a high growth rate. These conditions are effecting changes in various urban areas. In areas near the Mexican border a large proportion of the people uses Spanish as either a primary or secondary language. In southern Louisiana, French is important, normally in a creolized form, but only in remote rural areas is English not dominant.

On some of the larger Native American reservations, particularly in Arizona, Indian languages are in everyday use. The larger metropolitan areas have various ethnic

A CLOSER LOOK   Hispanic American Capitals

The decision to make a film about the life of Tejano pop star Selena prompted the producers to open casting calls for the lead role in four cities chosen because of their large Hispanic populations: San Antonio, Los Angeles, Miami, and Chicago. With the addition of New York, these cities represent America's Hispanic capitals. A Hispanic American capital is not only a demographic stronghold but a center identified as a significant core area of Hispanic cultural and economic influence. These capitals are considered the top Hispanic media markets by opinion molders in journalism, by universities, and by Spanish language media owners.

Some 22.3 million Hispanics made up 9 percent of all Americans in 1990; just a decade earlier, Hispanics totaled only 15 million. Hispanic Americans constitute the fifth largest concentration of Spanish-speaking people in the world after Mexico, Spain, Argentina, and

Colombia. Yet, in the United States persons of Hispanic origin are quite diverse, representing many different nationalities. Mexican origin is the largest subgroup, counting nearly two-thirds of all Hispanics, followed in rank by Puerto Rican, Central American (especially Salvadorean and Guatemalan) and South American (chiefly Colombian), other Hispanic (mostly Spanish Americans and Spaniards), Cuban, and Dominican (Fig. 3-A).

Just as Hispanic Americans are a plural ethnic block, so too their geography is regionally varied. Almost every county in the United States contains some Hispanic population, and only a handful of counties in Montana, North and South Dakota, and Nebraska completely lack Hispanic Americans. The concentration of Hispanics, however, is geographically uneven, with the most pronounced regional agglomerations along the border with Mexico, especial-

ly in selected counties of Texas, California, Arizona, and New Mexico, as well as portions of Colorado, Florida, and Washington; smaller concentrations are evident in a few other states, mostly in the West.

Hispanic Americans are overwhelmingly concentrated in metropolitan areas; in 1990 they were more urban than the population of the United States as a whole. Metropolitan areas with large numbers of Hispanics are mostly located in Texas (9) and California (9), followed by smaller numbers of metro locations in several other states. Five cities are especially significant Hispanic American places: Los Angeles, San Antonio, Chicago, New York, and Miami (Fig. 3-B).

Los Angeles has the greatest number of Hispanics, but the vast majority of this group are of Mexican ancestry. The Mexican heritage of the City of Angels dates to the Spanish colonial era. Recently, Central American immigrants, particularly Salvadoreans and Guatemalans, have occupied distinctive quarters in parts of the central city and now represent the second largest concentration of Hispanics in the metropolitan area. San Antonio, although smaller than Los Angeles, is also a recognized center of Mexican population. In fact, the Alamo City has the highest percentage of Hispanic origin population among large metropolitan areas; some 48 percent of its residents are of Mexican ancestry. Unlike in Los Angeles, however, the Mexican population of San Antonio is not challenged by other Hispanic subgroups. Chicago has even more Hispanics than San Antonio, and like that city, it is a stronghold of Mexican origin population, who account for nearly 70 percent of all Hispanics in the city. Chicago became a Mexican immigrant destination during the early twentieth century. The Hispanic population of the Windy City has been supplemented by later migrations of Puerto Ricans and Cubans, especially, and recently, Central Americans, mostly Guatemalans.

The East Coast cities of New York and Miami are smaller nodes of Hispan-

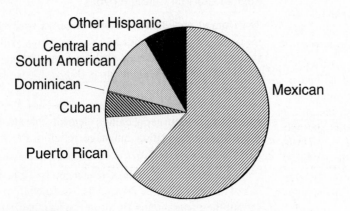

FIGURE 3-A
Persons of Hispanic origin in the United States, 1990

Subgroup	Total (millions)	Percentage of All Hispanics
Mexican	13.5	60.3
Puerto Rican	2.7	12.2
Central & South American	2.6	11.8
Other Hispanic	1.9	8.5
Cuban	1.0	4.6
Dominican	0.5	0.5

Source: 1990 Census of Population and Housing, Summary Tape Files 1C and 3C.

Hispanic American Capitals 1990

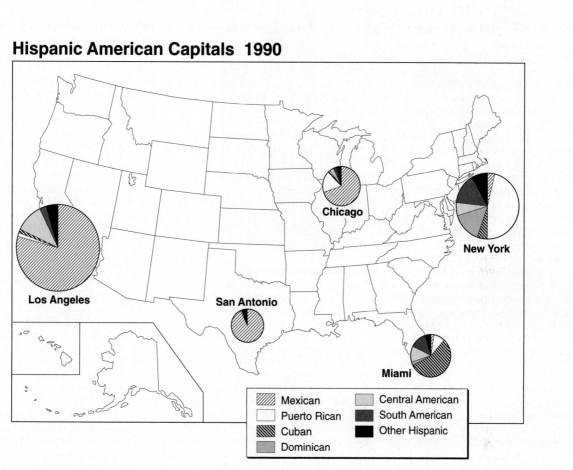

FIGURE 3-B
Hispanic American Capitals, 1990

Metro*	Hispanic Population	Percentage of Total Metro Population
Los Angeles	4,714,405	33
New York	2,704,960	15
Miami	1,055,368	33
Chicago	868,167	11
San Antonio	616,878	48

* Consolidated Metropolitan Statistical Area, except San Antonio, which is Metropolitan Statistical Area.
Source: 1990 Census of Population and Housing, Summary Tape File 3C.

ic residents compared with Los Angeles. However, these cities have much greater pluralities of Hispanic peoples than any of the western or central capitals. New York's Hispanic population is nearly half Puerto Rican, sometimes called "Nuyoricans," whereas Miami's Hispanic population is more than half Cuban. The remainder of the Hispanic American residents in these cities are also extremely diverse, with significant percentages of Dominicans and Colom-bians in New York, and Nicaraguans and Colombians in Miami. With some 800 overseas hemispheric flights a week out of Miami International Airport, it is

(continued)

FIGURE 3-C   The plurality of Hispanic nationalities in Miami means that one can purchase almost any major Latin American city newspaper daily on the streets of the Florida city. Miami is truly America's "Gateway to Latin America" (photo by D. D. Arreola).

Cuban concentrations in New York and Miami are also a consequence of historical and political bonds that stem especially from the 1950s and 1960s. Proximity has also been important; Puerto Rico is east of New York and well connected by frequent and inexpensive air travel, whereas Cuba is only some 90 miles from southern Florida. The recent concentrations of Central American Salvadoreans, Guatemalans, and Nicaraguans has been spurred both by political refugee movements from source countries and gravitation to the rich labor and business markets of cities like Los Angeles and Miami. Colombians and Dominicans, like Puerto Ricans, also have sought locations in the nearby cities of Miami and New York, respectively. New York's Dominican community is the second largest concentration of Dominicans in the world after Santo Domingo.

In a recent travel account, an Englishman, thrilled at his first visit to America, stepped off the airplane in New York and gazed around at the tall buildings. Stopping a passerby, he said, "I can scarcely believe I am in America. Is this really New York City?" The woman smiled and nodded; "Sí señor," she said.

Professor Daniel D. Arreola
Arizona State University, Tempe

little wonder that the Florida city is called "The Gateway to Latin America" (Fig. 3-C).

What explains the particular geographic concentrations found in the Hispanic American capitals? Historical political ties between Mexico and the southwestern United States as well as migrant and immigrant labor connections to a nearby source area partly explain the strong presence of Mexican origin populations in western and central capitals like Los Angeles, San Antonio, and Chicago. Puerto Rican and

enclaves where some non-English language dominates. This is most conspicuous in Spanish Harlem and other parts of New York City where Puerto Ricans and Dominicans have settled, in several big-city Chinatowns, in areas of Italian and Polish settlements, in the Cuban settlements of Florida, and in Southeast Asian communities in large western cities.

The linguistic pattern in Canada is much more heterogeneous; English and French are dominant, although many people prefer another language. Census statistics show that English is the mother tongue for 62 percent of the population but is the principal language for 66 percent. French is the mother tongue for 24 percent (but the principal language for only 23 percent), whereas the mother tongue of 14 percent of the population is neither English nor French (Italian, German, Chinese, and Ukrainian are the leaders). Canada is officially a bilingual country, and this cultural dualism has posed a major stumbling block in any attempt at establishing a national identity.

Another dimension of Canadian languages was shown in a study that identified 53 distinct indigenous (Indian and Inuit) languages that are still spoken.[12] The total number of speakers of indigenous languages, however, was estimated at only 154,000, which is about the same as the number of Canadians who claim Dutch as a mother tongue. Moreover, only three indigenous languages—Cree, Ojibwa, and Inuktitut (the principal Inuit tongue)—are considered to have enough speakers so that the languages are not in danger of dying out.

Religion   Religious affiliation is more varied (Fig. 3-14). In the United States approximately one-half of the population profess some branch of Protestantism, with Baptists and Methodists as the largest denominations. The

---

[12] *Indigenous Languages of Canada,* Commissioner of Official Languages, Ottawa, 1983.

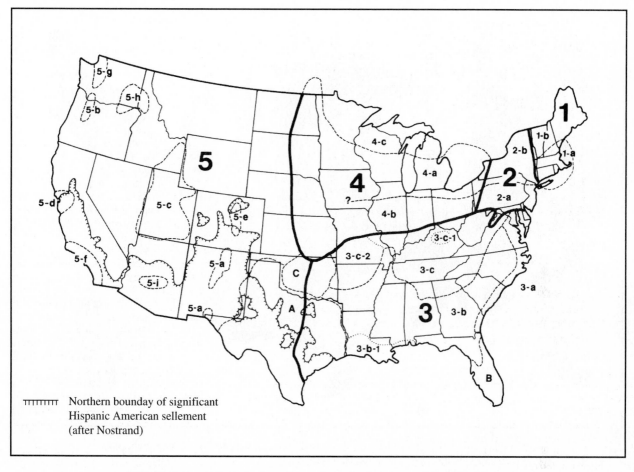

ᴛᴛᴛᴛᴛᴛ   Northern bounday of significant
Hispanic American sellement
(after Nostrand)

FIGURE 3-13   Zelinsky's view of the principal culture areas of the United States.

1   New England
   1-a   Nuclear New England
   1-b   Northern New England
2   The Midland
   2-a   Pennsylvanian Region
   2-b   New York Region, or New England Extended
3   The South
   3-a   Early British Colonial South
   3-b   Lowland, or Deep South
      3-b-1   French Louisiana
   3-c   Upland South
      3-c-1   The Bluegrass
      3-c-2   The Ozarks
4   The Middle West
   4-a   Upper Middle West
   4-b   Lower Middle West
   4-c   Cutover Area

5   The West
   5-a   Upper Rio Grande Valley
   5-b   Williamette Valley
   5-c   Mormon Region
   5-d   Central California
   5-e   Colorado Piedmont
   5-f   Southern California
   5-g   Puget Sound
   5-h   Inland Empire
   5-i   Central Arizona
Regions of uncertain status of affiliation are
  A   Texas
  B   Peninsular Florida
  C   Oklahoma
(After Wilbur Zelinsky, *The Cultural Geography of the United States*, © 1973, pp. 118–119. Reprinted by permission of Prentice-Hall, Inc.)

southeastern states, the Midwest, and much of the West are dominantly Protestant. Roman Catholics constitute about one-third of the total population, with particular concentrations in the Southwest, southern Louisiana, parts of New England, and many larger cities of the Northeast. Only about 3 percent of the population are Jewish and they are distinctly concentrated in the large cities; half the nation's Jews live in New York City, with other major con-

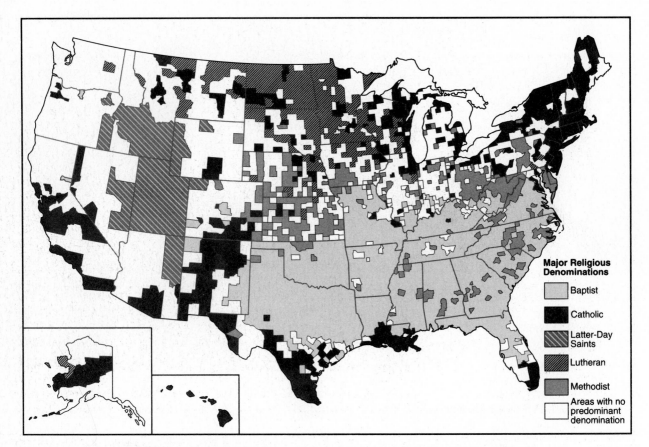

FIGURE 3-14 Religious affiliation in the United States, as indicated by church membership. In each case only the leading denomination is shown.

centrations in Los Angeles, Philadelphia, Chicago, and Boston.

Some 45 percent of the Canadian people profess Roman Catholicism; Quebec and New Brunswick are Catholic strongholds. About 40 percent of the remainder of the population are Protestant, with the two dominant denominations being United Church of Canada and Anglican.

In both Canada and the United States the number of Muslims and other non-Christians is rapidly growing, largely due to increasing immigration. Also, the number of people who are not adherents to any religious faith is increasing significantly. The latest Canadian census shows that one-eighth of all Canadians have no religious affiliation; informal data in the United States indicate that one-tenth of the people are non-religious.

Politics Political affiliations have changed markedly in the past three decades in both countries. Two parties, Democratic and Republican, dominate the political scene in the United States at federal, state, and local levels. Democratic strongholds have traditionally been in the southeastern states, in the big cities of the Northeast, and among certain minority groups, particularly African Americans and Jews. Republican strength has been concentrated in the Midwest, in the interior states of the West, and in rural New England. Significant regional trends in the last few years have been the resurgence of Republicans in the Southeast and of Democrats in New England and parts of the Midwest.

In contrast to the relatively rigid two-party system that has dominated the U.S. political scene, a number of political parties have been seen to rise and decline in Canada. In the mid-1990s there were five parties of significance at the federal level:

1. The Progressive Conservative Party (the Tories) is a mainstream conservative party that has governed Canada since 1984.

2. The Liberal Party has been the principal opposition party for many decades, except for periods when it was in power.

The Rapid Rise of Legalized Gambling

Gambling is an activity that is as old as civilization itself. In the United States legalized gambling in the form of lotteries was widespread in the eighteenth century, usually authorized by local governments to finance the construction of roads, colleges, or churches. For many decades, however, most forms of gambling have been illegal in this country. Gambling had the reputation of a "pariah" industry that was perceived as being associated with organized crime and as breeding corruption. Throughout the nation gambling is illegal except where it is regulated by the state. Certain forms of gambling, such as charity bingo games and parimutuel betting on horse and dog races, have been legal in some states for many years, but only very recently has big-time gambling become widely legalized.

The state of Nevada legalized casino "gaming" (the preferred term in the industry) and various other forms of gambling in 1933, and for nearly half a century had the U.S. market all to itself. In 1976 voters in New Jersey approved casino gambling specifically for the rundown resort town of Atlantic City, and by 1984 gambling revenues in Atlantic City had surpassed those of Las Vegas.

Meanwhile, the Seminoles in Florida opened the first high-stakes bingo hall in 1979. In 1988 the U.S. Congress enacted legislation that allowed Native American tribes to open casinos on reservation land, and by the mid-1990s more than half of all Native American tribes operated some form of gambling on their lands, with a gross revenue of well over a billion dollars. The largest and most flamboyant of these casinos is in Connecticut, where the tiny (200-member) Mashantucket Pequot tribe has been enormously successful in the operation of Foxwoods Casino (each of Foxwoods' 3000 slot machines takes in about $350 per day, in comparison with an average figure of about $100 for slot machines on the Las Vegas "strip").

In a flurry of activity in the late 1980s and early 1990s, several other states legalized casino gambling. In two states, casinos are permitted only in a very limited number of old mining towns (Deadwood in South Dakota and Blackhawk, Central City, and Cripple Creek in Colorado). In a number of other states, mostly in the Midwest, casinos are permitted only on riverboats. By the mid-1990s half of the 50 states permitted casino gambling, more than 500 casinos were in operation, and gambling companies provided employment for a million people. Nevada has the greatest number of casinos (about 300), Colorado and South Dakota have more than 50 each, and Mississippi has about three dozen. More than 120 of the casinos are operated on Native American lands, where it is estimated that $27 billion is wagered annually.

Another form of legalized gambling that has been mushrooming rapidly is the state-run lottery. New Hampshire organized the first modern-day lottery in 1964, and now some 40 states are in the business. Americans spend an estimated $25 billion annually on lottery tickets (more than four times the annual expenditure on movie tickets). States usually earmark their share of the lottery take for certain specific uses, with education being the principal beneficiary.

As the legitimization of gambling continues to spread, it seems clear that Americans will have ever-increasing opportunities to risk their money in legal fashion. It is equally clear that the intensification of competition will diminish the profits for most gambling enterprises.

TLM

---

3. The New Democratic Party evolved out of Depression-era socialist currents on the prairies and has long been Canada's third-largest party at the federal level.

4. The Reform Party was founded by disgruntled western Canadian conservatives in 1987; it stands for conservative economics and reduced social welfare programs.

5. The *Bloc Quebecois* came into being in 1990. It is a single-issue party, focusing on the sovereignty and interests of Quebec.

TRENDS AND QUESTIONS

The population of North America will undoubtedly continue to grow, but the predicted rate of growth is a matter for considerable debate. Projections based simply on the changing age structure of the population would indicate an upsurge in the rate of increase, for the number of young adults in the population is growing rapidly and these are the prime childbearing years. But fertility has been declining at a record rate, presumably as a result of changing attitudes by young adults, and this is a huge imponderable for prognostication. Some authorities have gone so far as to predict the possibility of achieving zero population growth within this century, which is considered by many as a most worthy goal. It is logical to expect a continued high rate of net immigration to both countries, but natural increase is the principal source of population growth and it is difficult to predict natural increase rates.

Until recently there was a continuing long-term decline in the rate of population increase in both countries. During the 1950s the U.S. population grew by 19 percent (30 percent in Canada); in the 1960s the U.S. growth rate was 13 percent (18 percent in Canada); during the 1970s the rate declined to 8 percent in the United States (13 percent in Canada). In the 1980s the U.S. rate increased slightly, to 10 percent, whereas Canada's rate dipped to 8 percent, the lowest for any decade since confederation. The U.S. population is currently increasing at the slow rate of about 1 percent annually, whereas Canada's growth rate has accelerated to about 1.5 percent annually, which is the highest in the industrialized world.

The regional pattern of population change is more clear-cut. The westward movement of people in both countries has been pronounced, if irregular, for years, and it gives every evidence of continuing, although perhaps differing in detail. Although their growth rates slowed in the 1980s, Ontario and British Columbia continue to be the Canadian leaders. California persists as the state with the largest absolute population gains on an annual basis, although its rate of growth is exceeded by all of the western states except Wyoming. Most other western states are growing at a rate that is more than twice the national average.

Growth in the so-called Sunbelt states (the southern tier from Virginia to California and Hawaii) has been notable and relatively continuous in recent years, reflecting the desire of an affluent and footloose population to settle in warmer areas. Related movements to scenic, high-amenity western states, such as Colorado, Alaska, Idaho, Utah, and Washington, are also conspicuous.

One of the most striking, and unexpected, demographic trends in the United States and to a lesser extent in Canada was a net in-migration to nonmetropolitan areas during the 1970s. For well over a century there had been a pronounced rural-to-urban movement of people in both countries. Especially since the farm population reached its peak during World War I, most rural counties have experienced a net emigration of people. In hundreds of counties this outmovement exceeded the natural increase (excess of births over deaths), resulting in actual population declines. During the 1970s, however, a trend referred to as "nonmetropolitan population turnaround" occurred, wherein small towns and rural areas experienced notably faster rates of population growth than did metropolitan areas.

Such "counterurbanization" had never before taken place and persisted for the entire decade; a nonmetropolitan growth rate of 14.4 percent easily surpassed a metropolitan growth rate of 10.5 percent. Explanation of the turnaround seems to involve a variety of factors, including more jobs in mining, decentralization of manufacturing and service employment to nonmetropolitan areas, slackening in the exodus from agriculture, the growing function of rural areas as recreational and/or retirement communities, increasing attractiveness of amenity-rich locations, desire for a simpler life away from the traumas of cities, and spillover of residential sprawl around urban areas. It seems clear that different factors pertain to different areas. Whatever the causes, the results were remarkable, if shortlived. Areas previously noted as sources of population flight, such as southern Appalachia, the Arkansas Ozarks, California's Sierra Nevada, and north-country Michigan, became destinations for settlement.

In the 1980s, however, the turnaround was reversed. The urban-to-rural movement dwindled. There was a continuing farm crisis, a recession, a restored reputation for cities, and a waning popularity among urbanites for an alternative lifestyle. During the 1980s the majority of nonmetropolitan counties actually lost population. Both the United States and Canada became increasingly urbanized.[13] Even so, there has not been a reversion to the pell-mell urban growth that typified the pre-1970 years. This underscores the need to be cautious about generalizations concerning demographic trends; even the most clear-cut geographical patterns may be subject to radical change.

Immigration continues to play an important role in the demography of both countries. In the mid-1990s more foreigners were arriving to settle in the United States than ever before—some 3000 per day. The consonant figure for Canada was about 600 per day. Indeed, a tidal wave of immigrants has been engulfing both countries since the mid-1980s. This trend seems likely to continue. Certainly the United States will remain the world's major destination for immigrants.

---

[13] One state—New Jersey—is now officially totally urban. Every county in the state is classed as a metropolitan area because each has a high-density population cluster of at least 50,000 people.

SELECTED BIBLIOGRAPHY

ALLEN, JAMES P., AND EUGENE J. TURNER, *We the People: An Atlas of America's Ethnic Diversity.* New York: Macmillan, 1987.

Anonymous, "Counting the Uncountable: Estimates of Undocumented Aliens in the United States," *Population and Development Review,* 3 (December 1977), 473–481.

BEALE, CALVIN L., *The Revival of Population Growth in Nonmetropolitan America.* Washington, D.C.: U.S. Department of Agriculture, Economic Research Service Report 605, 1975.

BEAN, FRANK D., AND MARTA TIENDA, *The Hispanic Population of the United States.* New York: Russell Sage Foundation, 1990.

BIGGAR, JEANNE C., "The Sunning of America: Migration to the Sunbelt," *Population Bulletin,* 34 (March 1979), 126–137.

BOSWELL, THOMAS O., AND TIMOTHY C. JONES, "A Regionalization of Mexican Americans in the United States," *Geographical Review,* 70 (January 1980), 88–98.

BRADLEY, MARTIN K., ET AL., *Churches and Church Membership in the United States, 1990.* Atlanta: Glenmary Research Center, 1992.

DAVIS, GEORGE A., AND O. FRED DONALDSON, *Blacks in the United States: A Geographical Perspective.* Boston: Houghton Mifflin Company, 1975.

DENEVAN, WILLIAM M., *The Native Population of the Americas in 1492.* Madison: University of Wisconsin Press, 1976.

DRIVER, HAROLD E., *Indians of North America.* Chicago: University of Chicago Press, 1961.

FREEMAN, GARY P., AND JAMES JUPP, EDS., *Nations of Immigrants: Australia, the United States and International Migration.* Melbourne: Oxford University Press Australia, 1992.

FUGUITT, GLENN V., AND CALVIN L. BEALE, *Changing Patterns of Nonmetropolitan Distribution.* Madison: Center for Demography and Ecology, University of Wisconsin, ca. 1986.

GOBER, PATRICIA, "Americans on the Move," *Population Bulletin,* 48 (November 1993), 40 pp.

GRAFF, THOMAS O., AND ROBERT F. WISEMAN, Changing Concentrations of Older Americans," *Geographical Review,* 68 (October 1978), 379–393.

HAGAN, WILLIAM T., *American Indians.* Chicago: University of Chicago Press, 1961.

HALVORSON, PETER L., AND WILLIAM M. NEWMAN, *Atlas of Religious Change in America, 1952–1990.* Atlanta: Glenmary Research Center, 1994.

HANSEN, MARCUS LEE, *The Immigrant in American History.* New York: Harper & Row, Publishers, 1964.

HELWEG, ARTHUR W., AND USHA M. HELWEG, *An Immigrant Success Story: East Indians in America.* Philadelphia: University of Pennsylvania Press, 1990.

JASSO, GUILLERMINA, AND MARK R. ROSENZWEIG, *The New Chosen People: Immigrants in the United States.* New York: Russel Sage Foundation, 1990.

JONES, MALDWYN ALLEN, *American Immigration.* Chicago: University of Chicago Press, 1960.

JONES, RICHARD C., "Undocumented Migration from Mexico: Some Geographical Questions," *Annals of the Association of American Geographers,* 72 (March 1982), 77–87.

JOSEPH, ALUN E., PHILIP D. KEDDIE, AND BARRY SMIT, "Unravelling the Population Turnaround in Rural Canada," *Canadian Geographer,* 32 (Spring 1988), 17–30.

KEDDIE, PHILIP D., AND ALUN E. JOSEPH, "The Turnaround of the Turnaround? Rural Population Change in Canada, 1976–1986," *Canadian Geographer,* 35 (Winter 1991), 367–379.

KOSINSKI, LESZEK A., "How Population Movement Reshapes the Nation," *Canadian Geographical Journal,* 92 (May–June 1976), 34–39.

LIEBERSON, STANLEY, AND MARY C. WATERS, *From Many Strands: Ethnic and Racial Groups in Contemporary America.* New York: The Russell Sage Foundation, 1988.

LONG, LARRY, *Migration and Residential Mobility in the United States.* New York: The Russell Sage Foundation, 1988.

LOUDER, DEAN R., AND ERIC WADDELL, EDS., *French America: Mobility, Identity, and Minority Experience across the Continent.* Baton Rouge: Louisiana State University Press, 1993.

LUCIUK, LUBOMYR Y., AND BOHDAN S. KORDAN, *Creating a Landscape: A Geography of Ukrainians in Canada.* Toronto: University of Toronto Press, 1989.

MARSDEN, L. R., "Is Canada Becoming Overpopulated?" *Canadian Geographical Journal,* 89 (November 1974), 40–47.

MARTIN, PHILIP, AND ELIZABETH MIDGLEY, "Immigration to the United States: Journey to an Uncertain Destination," *Population Bulletin,* 49 (September 1994), 47 pp.

MCHUGH, KEVIN H., "Hispanic Migration and Population Distribution in the United States," *Professional Geographer,* 41 (November 1989), 429–439.

MCKEE, JESSE O., ED., *Ethnicity in Contemporary America: A Geographical Appraisal.* Dubuque, IA: Kendall-Hunt Publishing Co., 1985.

NASH, GARY B., *Red, White, and Black: The Peoples of Early America,* (2d ed.). Englewood Cliffs, NJ: Prentice-Hall, 1982.

NOBLE, ALLEN G., ED., *To Build a New Land: Ethnic Landscapes in North America.* Baltimore: Johns Hopkins University Press, 1992.

O'HARE, WILLIAM P., KELVIN M. POLLARD, TAYNIA L. MANN, AND MARY M. KENT, "African Americans in the 1990s," *Population Bulletin,* 46 (July 1991), 40 pp.

PRICE, JOHN A., *Indians of Canada: Cultural Dynamics.* Englewood Cliffs, NJ: Prentice-Hall, 1979.

RICHARDSON, BOYCE, *People of Terra Nulliuis: Betrayal and Rebirth in Aboriginal Canada.* Seattle: University of Washington Press, 1994.

ROSS, THOMAS E., AND TYREL G. MOORE, EDS., *A Cultural Geography of North American Indians.* Boulder, CO: Westview Press, 1987.

ROY-SOLE, MONIQUE, "Keeping the Metis Faith Alive," *Canadian Geographic,* 115 (March–April 1995), 36–49.

SAUER, CARL O., "European Backgrounds," *Historical Geography Newsletter,* 6 (Spring 1976), 35–58.

SCHICK, FRANK L., AND RENEE SCHICK, *Statistical Handbook on U.S. Hispanics.* Phoenix: The Oryx Press, 1991.

SHUMWAY, J. MATTHEW, AND RICHARD H. JACKSON, "Native American Population Patterns," *Geographical Review,* 85 (April 1995), 185–201.

SNIPP, C. MATTHEW, *American Indians: The First of This Land.* New York: Russell Sage Foundation, 1989.

TAYLOR, J. GARTH, "Trying to Preserve Our Aboriginal Cultures," *Canadian Geographic,* 100 (December 1980–January 1981), 52–58.

TENNANT, PAUL, *Aboriginal People and Politics.* Vancouver: University of British Columbia Press, 1990.

TROVATO, FRANK, AND CARL F. GRINDSTAFF, EDS., *Perspectives on Canada's Population: An Introduction to Concepts and Issues.* Toronto: Oxford University Press, 1994.

The North American City

4

The keynote of North American geography for many decades involved the proliferation and spread of urbanism. An almost invariable growth of individual cities occurred, along with a continuing increase in the number of cities. During the 1970s a remarkable countertrend set in: many cities experienced a declining growth rate, some cities actually diminished in population, and there was an overall faster rate of population expansion in nonmetropolitan than metropolitan areas. As seen in Chapter 3, that aberrant trend has now disappeared, and urban growth has been resumed, if less exuberantly. An urban way of life is an incontrovertible fact for most people in contemporary North America. By the mid-1990s there were 60 metropolitan areas in the United States and Canada with populations exceeding 1 million and another 140 with populations in excess of 250,000.

Despite the dominance of cities in the lifestyle of North Americans (Fig. 4-1), this book does not emphasize urbanism in its regional treatment for several reasons. The most important is that most North American cities are quite similar. They have developed at approximately the same time and in roughly the same fashion in countries characterized by a relatively high standard of living and in which people, ideas, and money are shifted easily from one region to another (mobility of

population, pervasiveness of mass media, and fluidity of capital).

There are some exceptions to this generalization. Several cities have a character of their own; nobody would accuse New Orleans, Quebec City, or San Francisco of being undistinctive. Nevertheless, the vast majority of North American cities within the general size-category have a sameness in appearance, morphology, and function that is almost bewildering to a visitor from another continent where cities have grown up more individualistically over a much longer period of development.

Even though the two nations have been politically separate for more than two centuries, cities on both sides of the international border tend to be remarkably alike. In both countries it is the urban areas that absorb most of the continuing high rate of population and economic growth; in both countries the automobile has assumed a dominant role in shaping city form and patterning urban life; and in both countries the similarity in the standard of living and personal motivation and aspiration is reflected in urban function and morphology. Canadian cities are spread in a long east–west band that has a very narrow north–south breadth, and each tends to be more like American cities of similar longitude than like Canadian cities significantly farther east or west;

FIGURE 4-1   Metropolitan areas in the United States and Canada that have a population exceeding 1 million.

thus Calgary is more like Denver than like Toronto, and Winnipeg resembles Minneapolis more than it does Vancouver.

There are, however, some differences between cities on opposite sides of the border. These differences reflect variation in political and settlement history, ethnic mix, and urban institutional patterns of the two countries, among other things. For example, Canadian cities are denser and more compact, have less variation between their central-city and suburban populations, have less neighborhood homogeneity of income, make greater use of public transit, experience much less violent personal crime (although property crimes are about equal in both countries), have a notably different ethnic composition (especially in central cities), have had less central-city decline, and have much less local government fragmentation (which makes the urban area less cumbersome to govern and finance). These are not major structural differences, but taken together, they are significant in many cities.

Hence the concept of a North American city is an imperfect one; there are recognizable variations on both sides of the international border. The commonalities, however, are much more notable than the differences. This chapter consid-

ers North American cities in general, commenting on the major characteristics of their urban geography. Each regional chapter that follows has a section devoted to the leading cities of the region or to atypical aspects of urbanism in the region or to a particular urban theme that is pertinent to the region.

HISTORICAL DEVELOPMENT OF NORTH AMERICAN CITIES

Prior to 1800

In early colonial days small towns sprang up along the Atlantic seaboard, mostly in what are now New England and the Middle Atlantic states in the United States and in the Maritime Provinces and Lower St. Lawrence Valley in Canada (Fig. 4-2). This development was natural because these areas were nearest England and France, particularly the former, whence came immigrants, capital goods, and many consumer goods. Accordingly, merchandising establishments were more advantageously located in port cities from which goods could be more readily distributed to interior settlements.

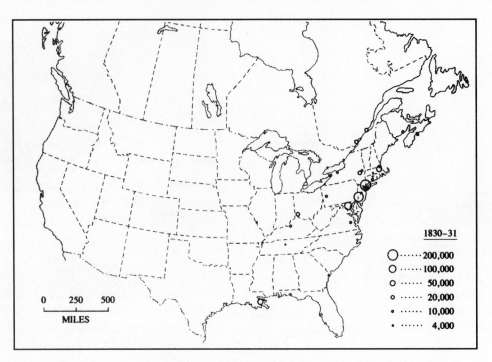

FIGURE 4-2 The largest cities in the United States and Canada in 1830–1831. The minimum-size city shown has a population of 4000 (after Figure 3.1 from *The North American City,* 4th ed., by Maurice Yeates. Copyright © 1990 by HarperCollins Publishers, Inc. Reprinted by permission of Addison–Wesley Educational Publishers, Inc.).

Here, too, were the favored locations for assembling raw materials for export and for performing what little processing was necessary for shipment abroad. A number of small ports existed, but Baltimore, Boston, Philadelphia, New York, and Montreal soon began to exert dominance (although for several decades Quebec City grew almost as rapidly as Montreal).

Urban growth was less impressive in the colonial South, where life centered around the plantation rather than the town. The local isolation and economic self-sufficiency of the plantation were inimical to the development of towns. Thus, nearly all southern settlements were located on navigable streams, and each planter owned a wharf accessible to the small shipping of that day. Both Charleston and Savannah were founded early and developed various urban functions, but after a short time neither rivaled the North Atlantic cities in urban development.

At the time of the first census of the United States in 1790, not a city in North America had as many as 50,000 inhabitants, and only Baltimore, Boston, Charleston, New York, Philadelphia, and Montreal exceeded 10,000. No city yet showed any indication of urban dominance; each Atlantic port served a small hinterland and was primarily oriented toward the sea and Europe.

1800–1870

Penetration of the interior and the use of inland waterways (Ohio River, lower Great Lakes, Erie Canal) began to produce a few inland urban centers: Richmond, Lancaster, Pittsburgh, Albany, and a bit later, St. Louis, Cincinnati, Louisville, Buffalo, and Rochester. But the Atlantic ports retained their regional primacy, and the Louisiana Purchase added another primary port, New Orleans.

Montreal had grown to dominate a large share of the trade of the interior of the continent and challenged New York, Philadelphia, and New Orleans (Fig. 4-3). The partitioning of British North America into Upper Canada (Ontario) and Lower Canada (Quebec) and the choice of York (later to be named Toronto) as capital of Upper Canada added a significant dimension to the urban scene in the north. For Toronto, the "law of initial advantage operated fully, and by 1830 all rivals to regional control had been subdued."[1]

Still, rural dominance was clear-cut until the 1840s, when railways began to develop, the factory system became

[1] Donald P. Kerr, "Metropolitan Dominance in Canada," in *Canada, A Geographical Interpretation,* ed. John Warkentin (Toronto: Methuen Publications, 1968), p. 540.

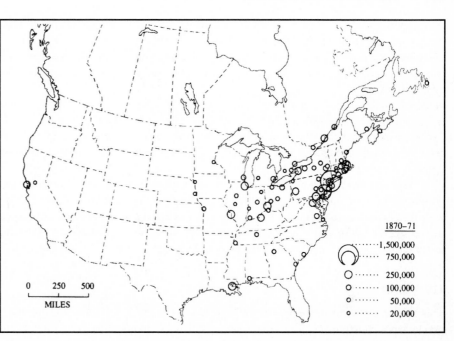

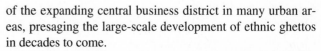

FIGURE 4-3 The largest cities in the United States and Canada in 1870–1871. The minimum-size city shown has a population of 20,000 (after Figure 3.2 from *The North American City,* 4th ed., by Maurice Yeates. Copyright © 1990 by HarperCollins Publishers, Inc. Reprinted by permission of Addison–Wesley Educational Publishers, Inc.).

established, and the industrial function of cities began to grow. The mechanization of spinning and weaving had set the pace in the previous two decades, but other types of manufacturing were oriented mostly toward households and workshop, often in rural locations, until the 1840s.

The large port cities grew especially in size with the building of canals, roads, and railways to the interior. A series of regional rail networks developed, with the larger networks converging at important inland waterway connections. Each important coastal port began to organize its own railway and push it inland. Of the major American ports, Boston alone was forced to content itself with connecting lines; its Boston and Albany Railway never got beyond the Hudson River. There was little railway development at this time in Maritime Canada. The Grand Trunk Railway, built after 1850 from Chicago through Toronto and Montreal, reached the Atlantic in the United States at Portland, Maine. Consequently, Canada's Atlantic port cities grew very slowly.

The midcentury period was a transition time for city development. Before then, industrialization was quite subordinate except in the five great Atlantic port cities that dominated trade relationships between the North American agricultural economy and Europe. The mercantile syndrome continued to pervade most other cities for some years, but industrial development and urban growth became much more closely linked. Labor supply was enhanced by the attraction of farm youths to cities and by accelerating immigration; after 1840 foreign immigrants began to concentrate on the edge

of the expanding central business district in many urban areas, presaging the large-scale development of ethnic ghettos in decades to come.

New York City rose to undisputed continental primacy at this time, partly because the opening of the Erie Canal cemented its western trade advantages but also because it was able to control much of the external trade of the South. "Indeed, it was largely because the merchants of New York and their itinerant factors controlled the cotton trade that urbanization in the South was extremely slow."[2] Thus Charleston and New Orleans were unable to wrest control of the cotton trade from New York. Growing from roughly equal in size with Philadelphia and Boston at the turn of the century, New York's population exceeded half a million by 1850; this population was more than twice that of any other city on the continent.

Inland regions of urbanization were limited mostly to the Ohio Valley (Cincinnati, Pittsburgh, Louisville) and upstate New York (Albany, Troy, Rochester, Buffalo), although Toronto almost kept pace with Montreal's *rate* of growth with nearly a 50 percent increase during the 1850s.

The growth rate of Canadian cities declined sharply during the 1860s, apparently owing to largescale emigration to the United States. Urban populations were booming in the Midwest. Water transport helped Cincinnati and St. Louis to

[2] David Ward, *Cities and Immigrants: A Geography of Change in Nineteenth Century America* (New York: Oxford University Press, 1971), p. 29.

grow rapidly, but by 1870 they were being challenged by the swiftly growing Great Lakes cities of Detroit, Milwaukee, Cleveland, and especially Chicago. There was also continued fast growth in New England and the Middle Atlantic states, especially near New York. Brooklyn was the third largest city in the nation, and Newark and Jersey City were both sizable. In the South only New Orleans had continued major growth and numbered nearly 200,000 people, whereas the only western city that had developed to more than 25,000 was San Francisco, with a population of 150,000.

Thus the development of a more-than-regional transportation system, combining regional rail networks with inland waterways, revolutionized urban development in this period. Industrial growth was an important stimulant in the larger cities, but major industrial development came later.

1870–1920

This was an era of maturing for North American cities, the previous period having been a formative one. National transportation systems were completed. National accessibility was extended to the South, the Southwest, and the Far West (Fig. 4-4). The remaining agricultural lands of the West, from Texas to British Columbia, were opened. A variety of major

mineral deposits was developed. But the principal stimulus to urban growth was industrial development. The economy of the two countries changed from a commercial-mercantilistic base to an industrial-capitalistic one.

The coastal cities were increasingly important, but much of the growth was concentrated in the industrial belt of northeastern United States and southern Ontario and Quebec. The geographical division of labor, a basis for present-day regionalism, was beginning to be apparent by the end of the Civil War period. There followed a quarter century of accelerated westward movement, rapid population increase, heavy immigration, and burgeoning urban growth. Cities found their functions multiplying, but the growth of manufacturing was usually at the core.

The big cities became bigger, but there were also developments among smaller centers. Toronto, only half the size of Montreal at the beginning of the period, began to capture trade territory to the north and west and by the turn of the century had approached parity in size; the growth rates of the two cities have been quite similar ever since. Winnipeg began to grow in the Prairies, and Los Angeles experienced the early stages of its spectacular population increase. There were boom times in Florida urban areas, in Appalachian coal towns, and in cities of the Carolina Piedmont.

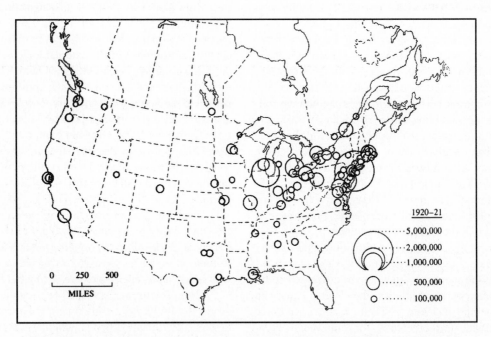

FIGURE 4-4 The largest cities in the United States and Canada in 1920–1921. The minimum-size city shown has a population of 100,000 (after Figure 3.3 from *The North American City,* 4th ed., by Maurice Yeates. Copyright © 1990 by HarperCollins Publishers, Inc. Reprinted by permission of Addison–Wesley Educational Publishers, Inc.).

This was a tumultuous period in North American urban development, but at its conclusion our basic urban system was firmly established.

1920–1970

By 1920 half the population of North America was urban, and the proportion continued to increase. There was substantial and continued growth in the older urban areas of the United States and Canada, but the most flamboyant developments were in new cities far removed from the traditional urban centers. The most spectacular growth occurred on the West Coast, from San Diego to Vancouver, and similar trends took form in Florida, Texas, the desert Southwest, and on the inland side of the Rockies (Colorado and Alberta).

By the beginning of the 1960s approximately 25 percent of the world's population lived in cities of 20,000 or more inhabitants; the comparable figure for North America was 55 percent, the highest proportion of any populous part of the globe.[3] Thus the concentration of urban population on this subcontinent not only is of recent vintage and sizable magnitude but also has occurred at a remarkably rapid rate (Fig. 4-5).

[3] Derived from data in Leroy O. Stone, *Urban Development in Canada* (Ottawa: Dominion Bureau of Statistics, 1967), p. 16.

This was a period in which the automobile shaped newer cities and reshaped older ones, and the remarkable development of air travel reinforced many interrelationships of the existing urban system besides adding some new dimensions. Typical characteristics included continued agglomeration, urban sprawl, freeway construction, central-city decay, urban renewal, massive air pollution, suburban high rises, concerted desegregation efforts, planned industrial districts, and extensive neighborhood and regional shopping centers.

1970 to Present

Significant changes in urban population trends appear to have ushered in a new period of development in recent years. It began as a time of counterurbanization; the 1970s experienced a reversal of the metropolitan growth syndrome, something that had never happened before. For the first time nonmetropolitan population grew at a faster rate than metropolitan population: 15 percent versus 10 percent between 1970 and 1980. For the first time there was a decline in metropolitan dominance: 75 percent of the U.S. population resided in metropolitan areas in 1980 compared with 76 percent in 1970. This counterurbanization trend was muted in Canada. During the 1970s the metropolitan population of Canada increased only from 55 to 56 percent.

After a decade, the metropolitan population turnaround turned around again. Beginning about 1980, the rate of met-

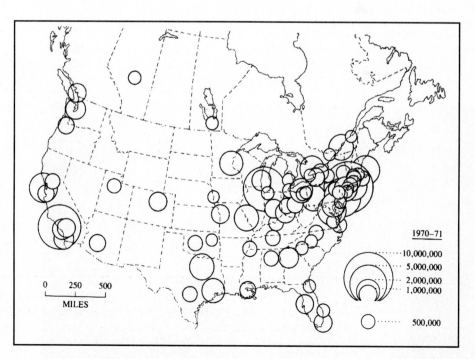

FIGURE 4-5 The largest metropolitan areas in the United States and Canada in 1970–1971. The minimum-size metropolitan area shown has a population of 500,000 (after Figure 3.4 from *The North American City,* 4th ed., by Maurice Yeates. Copyright © 1990 by HarperCollins Publishers, Inc. Reprinted by permission of Addison–Wesley Educational Publishers, Inc.).

ropolitan population growth once again exceeded that of non-metropolitan population, and metro growth has surpassed nonmetro growth at an accelerating rate ever since, although not as strikingly as in the pre-1970 period.

Sunbelt cities have experienced the fastest growth in recent years, but even the declining urban areas of the Northeast had resumed a sluggish growth pattern by the mid-1990s. Almost everywhere the central cities are decreasing in population; it is primarily the suburbs that are growing.

Characteristics of this most recent period are discussed at greater length later in this chapter.

URBAN MORPHOLOGY: CHANGING PATTERNS

Viewed from the air, a typical North American city appears as a sprawling mass of structures of varying size, shape, and construction, crisscrossed by a checkerboard street pattern that here and there assumes irregularities. The general impression is one of stereotyped monotony. The pattern of form and structure is so repetitive that one can anticipate a characteristic location of specialized districts and associations of activities within them. The stylized arrangement and predictable interrelations make it possible to formulate broad generalizations about North American urban anatomy that are particularly valid if confined to cities of similar size, function, and regional setting.

The Pattern of Land Use

When considered in detail, the pattern of land use varies with every city. There are, however, such basic similarities that general patterns can be described and, to some extent, explained. The resulting generalizations are broadly valid for most cities whether they are older and slower-growing cities with rigid zoning restrictions, such as Buffalo, or newer, burgeoning cities with only limited land-use zoning regulations, such as Tucson.

Commercial Land Use Most cities are primarily commercial centers. Their attraction as a place for people to live is largely predicated on the concentration of commercial or business activities. For North American cities as a group, more than half of all jobs are in commercial fields: wholesale trade, retail trade, finance, insurance, real estate, and various kinds of services. Although the proportion of a city's land area devoted to these activities is small—generally less than 4 percent of the total—the structures in which the activities take place are often conspicuous and involve the tallest and most obtrusive buildings in the urban area.

The *central business district* (CBD) is the commercial heart of the city. It normally occupies an area near but slightly removed from the original town site (Fig. 4-6). It usually has a geographically central position in relation to the urban area as a whole; in some cases, however, it may be situated well off center, particularly where prominent physical features, such as coastline, mountain front, or river, are involved. The CBD is normally characterized by the greatest intensity of urban activity: highest daytime population density, most crowded sidewalks, most-used surface streets, focus of mass transit routes, principal concentration of taxis, and greatest concentration of high-rise buildings. It also contains the most valuable land in the city and is the principal location of office space, restaurants, theaters, hotels, government offices, financial institutions, corporation headquarters, and auto parking facilities.

By the 1960s most North American CBDs were experiencing a decline in mass retailing and personal services due

FIGURE 4-6 A view across the central business district of Detroit (John Sotilden/Visuals Unlimited photo).

to the almost overwhelming economic challenge of sparkling new outlying shopping centers. In some CBDs this challenge was met by concerted efforts at downtown retail revival. Downtown merchants began to develop extensive, often climate-controlled shopping malls within the CBD. Significant initial efforts included San Francisco's Ghirardelli Square and Chicago's Brickyard. Frequently another dimension was added by locating many of the facilities underground—multiple underground levels beneath large buildings or street-level plazas, with direct connections both to major buildings on the surface and to subsurface rapid transit facilities, if any existed. They provided an almost fully self-contained environment for urbanites that was a long overdue and eminently logical adjustment to winter in northern cities. Montreal's pioneering example in Place Ville Marie (opened in 1962) stimulated similar developments in many other cities, even in such mild-winter locations as Los Angeles.

Montreal continues to set the pace for underground urban activities (Fig. 4-7). By the 1990s there existed a 7-mile (12-km) network of tunnels and shopping malls that connected 1000 stores and 100 restaurants and bars with theaters, hotels, office buildings, residential high rises, and rail, bus, and subway stations. A somewhat different style of subsurface development has been produced in Seattle and Atlanta, where antiquated sections of the central city had literally been buried by subsequent construction. Portions of these historic districts have been unearthed and refurbished as underground touring, entertainment, and shopping centers, primarily aimed at tourist business.

In the other direction, vertical expansion has been pioneered by development of "skyways" in Minnesota's Twin

FIGURE 4-8 Skyways (elevated pedestrian walkways) connect many of the buildings in downtown Minneapolis (courtesy Greater Minneapolis Convention & Visitors Association).

Cities (Fig. 4-8). These are enclosed walkways that connect downtown buildings at the second-floor level and constitute the largest network of elevated indoor sidewalks and concourses in the world. In the three decades since their inception, the skyway networks have expanded to encompass nearly 50 blocks of downtown Minneapolis and more than 30 blocks of downtown St. Paul (as of 1990). Their success, which is not unvarnished,[4] has led to similar programs on a more modest scale in several other cities (e.g., Edmonton, Calgary, Winnipeg, Cincinnati, Houston).

Despite the varied fortunes of the retail-service function, most CBDs persist prominently as the core of the city. Although some offices were lured to the suburbs, the 1970s were a boom period for downtown office-building construction; ever-larger and more complex skyscrapers sprouted in the downtown skyline. Some buildings experienced considerable difficulty in finding sufficient tenants, but the construction trend carried through the 1980s.

In the 1990s, however, things began to change. Corporate downsizing has reduced the downtown work force in cities all across the subcontinent. Banks and other familiar downtown institutions are merging or going out of business,

FIGURE 4-7 A Metro train in an underground station in Montreal (Thomas Kitchin/Tom Stack & Associates photo).

---

[4] The principal objections to skyways are: (1) They draw people and businesses from the ground level, leaving that level impoverished and stagnant; (2) they have a negative effect on urban architecture and reduce the stimulating vitality of a varied environment; and (3) the hours of operation are uneven in Minneapolis (not so in St. Paul, where the city owns the skyways, and maintains uniform hours), so that some segments may be closed while others are open.

and so are the big law and accounting firms. The gleaming office towers are beginning to show vacancies. It appears that the economic foundation of the CBD will probably be provided by government—local, state, and federal—offices.

Marginal to the CBD is the so-called transition zone. This is a discontinuous area of irregular shape and unpredictable size that has an almost continually changing land-use pattern. Its commercial prominence is often more oriented to wholesaling than to retailing, and industrial activities, in the broad sense, are usually notable. Still, much of the land in a typical transition zone is occupied by residences; this is a characteristic location for slum and ghetto development. In general, the transition zone is seedy and dilapidated, although some sections may be uncharacteristically bright and even prosperous owing to public or private urban renewal and slum clearance projects. Grandiose high-rise office buildings or apartment houses sometimes tower above the general obsolescence.

Another significant proportion of a city's commercial land use is found along *string streets,* which are usually major thoroughfares of considerable length, lined on both sides by varied businesses. The development may be patchy and discontinuous, but in larger cities the extent of string-street commercial zones may be measured in consecutive miles. Characteristically the businesses along a string street are small and diverse; however, there may be concentrations of specific types of enterprises, the best-known of which is "automobile row," where new and used car lots are clustered. A growing trend is the construction of high-rise office buildings along string streets, away from the CBD.

A modern expression of these commercial string developments is associated with outlying freeway locations. "The suburban freeway corridor now houses a complete mix of the business establishments regularly frequented by the geographically mobile middle- and upper-class residents of the modern metropolis."[5] These freeway ribbons permit relative ease of redevelopment due to their linear form and low density; thus incremental redevelopment is easier than in CBDs. They, of course, depend wholly on the automobile for customer access.

The most remarkable change in commercial land use in North American cities since World War II is the emergence of planned *suburban shopping centers.* Continually increasing amounts of a city's retail and service business are being carried out in these outlying centers, which are geared to the automobile era, with much more acreage devoted to parking spaces than to shopping areas. Apparently the first complete shopping center opened in Kansas City's Country Club dis-

trict in the early 1920s. In the early decades such centers emerged gradually; later, however, the planned shopping plaza blossomed full-grown at birth.

Planned shopping centers in the United States grew from about 100 in 1950 to a total of nearly 40,000 in the mid-1990s, although the pace of new construction has slowed considerably in recent years. Indeed, in the 1980s there was more expansion of existing shopping centers than the building of new ones.

It is estimated that only about 10 percent of all North American shopping centers are enclosed and climate controlled, but the great majority of the larger centers do have this characteristic. They are enclosed under a single massive roof that towers high enough to encompass three or four levels of walkway- and escalator-connected shops of varying sizes that are clustered around one or more spacious atria containing fountains, waterfalls, resting areas, and other attractions for the weary shopper.

At the time of this writing, the largest enclosed shopping center in the world is the West Edmonton Mall in Edmonton, Alberta, which is considered to be the world's first *mega-mall,* with 5.2 million square feet (470,000 m^2) literally under one roof (Fig. 4-9). It contains more than 600 shops, 19 theaters, an ice rink, a 7-acre (2.8-ha) water park, a 3000-bird aviary, an enormous amusement park, a church, a hotel, jobs for 18,000 people, and parking space for 12,000 vehicles. The mall contains nearly one-fourth of Edmonton's retail space and accounts for more than 1 percent of all retail sales in Canada. It hosts 15 million visitors annually, many of whom travel from distant parts of Alberta to spend a weekend at the mall.

The largest shopping mall in the United States is the Mall of America in Bloomington, Minnesota (a suburb of Minneapolis), which has 4.2 million square feet (380,000 m^2), although the Del Amo Fashion Plaza in Torrance, California (a suburb of Los Angeles), has 17 percent more actual retail space.

Three other notable trends in commercial (primarily retail) land use have developed in recent years—the mini-mall, the discount center, and the factory outlet complex.

Mini-malls are small, L-shaped corner buildings of 5 to 15 retail shops fronted by a small parking lot (Fig. 4-10). They first appeared in the late 1970s when the oil crisis persuaded many corner service station owners to sell their properties to the major oil companies, who in turn sold the parcels relatively cheaply and in bulk to developers. Thousands of mini-malls have sprung up across the country, with the greatest concentration in southern California (some 2000 in Los Angeles County alone). Criticized for creating parking and traffic problems and for egregious architecture, the mini-mall phenomenon may have run its course. By the mid-1990s, new

---

[5] Thomas J. Baerwald, "The Emergence of a New 'Downtown,'" *Geographical Review,* 68 (July 1978), 308.

FIGURE 4-9   One of the "boulevards" in the gigantic West Edmonton Mall (courtesy West Edmonton Mall).

FIGURE 4-10   Mini-malls have proliferated in most American cities, particularly in California. This scene is in west Los Angeles (TLM photo).

construction had virtually ended, vacancy rates were increasing, and rents were dropping.

The *discount warehouse* is another form of retail center that experienced phenomenal growth in the 1980s. These are enormous, high-roofed, barnlike structures that are surrounded by acres of parking spaces. Some (such as Wal-Mart, the leading retailer in North America) purvey general merchandise, whereas others (such as Home Depot) deal in specialized goods; many emphasize volume buying. Thousands of discount warehouses are now in business.

Factory outlet complexes consist of one to four dozen specialized stores, each of which sells merchandise from a single manufacturer. They are usually located adjacent to a major highway on the far outskirts of an urban place. At first (early 1980s) the complexes were usually sited in small towns, but the current trend is toward locations in the outer suburbs of larger cities (Fig. 4-11).

Residential Land Use   By far the most extensive use of land in North American cities is for residences, which occupy 30 to 40 percent of an average city's area.[6] Single-family dwellings (normally separate but sometimes attached to one another as row houses) are not always numerically in the majority, but they occupy more than three-fourths of the area devoted to housing.

[6] Jerome D. Fellmann, "Land Use and Density Patterns of the Metropolitan Area," *Journal of Geography,* 68 (May 1969), 265.

FIGURE 4-11   A factory outlet complex at Bradenton, Florida (Myrleen Ferguson/PhotoEdit photo).

Residences are scarce within the CBD, although some housing units are often found on the upper floors of buildings. The greatest residential density in any city normally occurs in and near the transition zone, where grand old homes of the past (frequently converted to rooming houses) are mixed with vast expanses of low-quality residences and occasional redevelopment pockets of high-rise apartments. Moreover, there has been a notable trend, particularly in the older cities of eastern United States and Canada, toward *gentrification.* This involves the purchase and restoring and/or refurbishing of old homes, usually by upwardly mobile young adults. The result is a shifting of population back toward, and a reinvigoration of life in, the central city.

The transition zone functions more traditionally as a tenement section that may include a large proportion of the city's slums and ghettos. The inhabitants of such areas are normally blue-collar workers with relatively low incomes, and a large proportion is likely to consist of ethnic minorities, particularly African Americans.

The *African American ghetto*[7] has almost within the space of a single generation become one of the two most rapidly expanding spatial configurations in large cities of the United States (the suburb is the other). Ghettos have been a prominent part of the North American metropolitan scene for many decades, but the rapid expansion, consolidation, and conspicuous social isolation of the African American ghetto in recent years has produced what amounts to a new urban

[7] Harold M. Rose, "The Origin and Pattern of Development of Urban Black Social Areas," *Journal of Geography,* 68 (September 1969), 328.

subculture. In most cities of the United States and in the few Canadian cities where African Americans reside in any numbers there is very strong de facto segregation between areas of white and African American households; this holds true in central cities as well as in suburbs, in the North or the South, and in large cities or small. But it is in central-city locations that African Americans find easiest access to housing and it is here that ghetto formation is pronounced and growing.

These ethnic enclaves usually occupy the least desirable parts of the city and strongly tend to be blighted zones except in newer cities where there is a prevalence of relatively new single-family residences located in the path of ghetto expansion (as in Denver, Phoenix, and some California cities). African American ghettos tend to be poverty areas, although there is often a concentration of African Americans with an income above poverty level. The ghetto is normally but not always a forced development; nevertheless, it performs the important function of providing a sense of community to its residents. Ghettos obviously offer more disadvantages than advantages, but with white abandonment of the central city as a place of residence, territorial dominance has clearly been relinquished to ghetto residents.

Other types of ghettos are also found in and around the transition zone of North-American cities, but rarely are they as large or conspicuous as the African American ghettos. Most notable are Hispanic ghettos, usually referred to as *barrios.* They are most prominent in many cities of the Southwest, in New York and other cities of the Northeast where Puerto Ricans and Dominicans are concentrated, and in Miami and other cities of southern Florida that contain large Cuban minorities.

With the general outward shift of population distribution in cities, there tends to be a similarly centrifugal displacement of residential zones based on economic factors. The vast expanse of middle-class housing, normally situated beyond the transition zone, has an inner boundary that moves outward with pressures from the central city. This usually creates a "gray" area that serves as buffer between middle- and lower-class housing, attracts an upwardly mobile segment of the central-city population, and is often a determinant of the direction of ghetto expansion.

There is also a general centrifugal gradation in population density throughout the residential areas from the transition zone to the outer suburbs. Multifamily housing is more common near the city center, and lowest population densities are in the outer areas, where relative remoteness reduces the price of land and subdivision ordinances require greater spacing between houses. There are, however, numerous variations from this pattern. The unremitting monotony of detached single-family dwellings has been significantly leavened by garden apartment complexes, cluster housing, and contemporary townhouse variations of the old row-house form. Such variety produces not only higher population densities but also diversity of residents. The old stereotypes of "suburban sameness" are less valid today than they were a couple of decades ago, when an entire school of social criticism was nurtured on attacking the aesthetics of suburbs.

In the suburban fringe there are initially independently planned street systems in separate communities or subdivisions, often with curvilinear pattern and considerable sprawl. These discrete areas are subsequently joined, and the open spaces among them are filled in with further urban (usually residential) development, producing irregular but not necessarily displeasing patterns of urban sprawl. The disparate densities between close-in and more remote residential areas tend to lessen with time as the settlement pattern ages and intensifies.

The *suburb* has a special place in the contemporary folk history of North America; it is The Place that connotes status, security, comfort, and convenience—the calm of country living with the amenities of the city within easy reach. The surge to the suburbs is nothing new; since at least the 1920s there has been a well-established tradition for the maturing generation, provided that it had the income and the means of internal transportation, to move to the edge of the city and establish yet another peripheral band of housing. The post–World War II expansion of the suburbs, however, is on

a heretofore undreamed-of scale. The more than 100 million North American suburbanites of the mid-1990s amounted to about 60 percent of the total urban population (Fig. 4-12).

Although historically considered bedroom zones for a population that commuted to the central cities to work, suburbs have increasingly become more complex in structure and function as their size and extent have burgeoned. Jobs followed people to the suburbs, and numerous nodes of high-density commercial and industrial development—high-rise office buildings, sprawling industrial parks, and immense shopping centers—scattered throughout suburbia are now typical. The usual movement of people, moreover, is from one suburb to another rather than commuting to the CBD, and a certain amount of reverse commuting has become established as central-city blue-collar workers increasingly must go to the suburbs to find work.

Suburbanites are generally but by no means universally well-off financially and are somewhat insulated from the decay and social trauma of the central city. To move to the suburbs is not to escape the problems of the city, because congestion, soaring taxes, crime, drugs, and pollution tend to follow, although generally to a lesser extent.

Indeed, many older suburbs are trapped in the same downward trajectory as are the central cities; they are evolving from garden city to crabgrass slum. The 1990 census showed that more than one-third of U.S. suburban cities had

FIGURE 4-12 Some suburban housing developments seem to sprawl endlessly, as here in Fort Lauderdale, Florida (TLM photo).

experienced significant declines in median household income since 1980. The relentless erosion of jobs and tax resources means that many older suburbs are starting to fall into the same abyss of disinvestment into which their center cities fell years previously. All of which means that there is a continuing trend of urbanites moving centripetally outward. The inner suburbs' losses are the outer suburbs' gains.

Industrial Land Use   Although land devoted to industrial usage occupies only an average of 6 to 7 percent of the city's area,[8] in most cities the significance of factories as employment centers is so great that industrial activity is critical to the local economy. Industrial areas may be widely scattered, but generalizations can be made about their location pattern. Most cities have one or more long-established and well-defined factory areas near the CBD, often containing several large firms as well as many smaller ones. These districts were usually established during the era of railway dominance and are characterized by the presence of rail lines and flat land.

If there is a functional waterfront (ocean, canal, or navigable river or lake) in the city, another old industrial area is likely to be located there. Heavy industry may be congregated in such an area, typically primary metals plants, oil refineries, and chemical plants. Often these areas were originally swampy or marshy and were then reclaimed by drainage, landfilling, or both.

Planned industrial districts are the product of more recent years. They are variously located, but usually the site was chosen with care so that ample space would be available and access to a major transport route would be ensured. Many planned industrial districts were sited along railway lines, but later the critical site factor became a prominent road or highway, for motor trucks are used more often than railways in transporting goods to and from factories. Perimeter highways or beltways are particularly attractive to the builders of planned industrial parks.

Factories may be found in many other kinds of locations in North American cities. The principal industrial areas, however, tend to fall into one of the preceding categories. As a useful generalization, it can be stated that most cities have experienced a decentralization of industrial land use; the suburban share continues to grow while the central-city share continues to decline.

Transportation Land Use   A surprisingly large amount of land in most cities is devoted to transportation of

one sort or another, including the storage of vehicles. This is the second greatest consumer of city space, exceeded only by residential land use. Indeed, in the "newer" North American cities it is not unusual for streets and parking areas to occupy half the total area of the CBD.

Parking continues to be a prominent need in our society. A cardinal rule of urban development is that an American is very reluctant to walk more than 600 feet (180 m) before getting into a car. Most modern buildings—particularly shopping centers and supermarkets—resemble islands surrounded by an ocean of asphalt. On average, local laws require four parking spaces for every 1000 square feet (90 m^2) of office space. Many projects have more space for cars than for people.

Relatively small proportions of the city area are required by other modes of transport, although airports can be very expansive, and new container ship terminals require extensive dockside storage space.

Other Types of Land Use   Many other types of activities are carried on in a city, but their locational patterns are less predictable. Parks and other types of green spaces are found in all cities and in some cases occupy a large share of the total city area. Institutions of various kinds tend to be widespread but scattered—for example, schools, cemeteries, museums. Government office complexes are relatively insignificant in most cities but may be particularly prominent in a national (Washington and Ottawa) or state (such as Albany or Sacramento) capital or in a city that is a significant regional headquarters for federal activity (such as Denver or San Francisco). Vacant land is another category that occupies varying amounts of space; even the most crowded cities have a certain amount of land that is not being used at present for any purpose.

The Pattern of Transportation

There are two different but overlapping facets to transportation in cities: internal movement within the city and external movement to or from the city. For either facet, the dominant fact of life is the pervasiveness of the automobile. Of all the money spent in the United States for freight transportation, nearly four-fifths is for motor trucking, and more than 90 percent of the total outlay for passenger movement goes to automobiles and buses. Although facilitating the movement of goods and people, the massive increase in rubber-tired transport threatens to overwhelm the system of streets and highways. The wastes of congestion become progressively worse despite every effort to facilitate traffic flow. In the New York

---

[8] Fellmann, "Land Use and Density Patterns of the Metropolitan Area," p. 265.

area, for example, 33 percent of the trucks do not move at all on a given day, and 20 percent of the remainder move empty.

Internal Transport   Movement within North American cities depends primarily on the traffic flow of streets and highways. A large and increasing share of total city area must be devoted to routeways and storage lots for cars and trucks. Essentially every North American city has a rectangular grid as the basic pattern for its street network—at least in the older portions of the urban area. The pattern has nearly always been modified by subsequent departures from the original layout; in many cases, there is a series of separate grids, adjusted to topography or to surveying changes, that are joined in variable fashions. The grid scheme, with its right-angled intersections, dominates the layout of North American CBDs. The streets were usually established in the preautomobile era and are normally too narrow to facilitate traffic flow; thus they engender massive downtown congestion and tax the ingenuity of traffic specialists to devise techniques to unclog the streets.

Away from the city center there is usually a greater diversity in the street pattern, particularly in newer subdivisions. Even so, it is rare for regularity to be maintained over a very large area because varied sequences of development and annexation lead to heterogeneity in planning.

Superimposed on the pattern of surface streets in all large North American cities as well as many small ones is a network (or the beginning of a network) of freeways or expressways (Fig. 4-13). These high-speed, multilaned, controlled-access traffic ways are laid in direct lines across the metropolis from one complicated interchange to another (Fig. 4-14), making functional connection with the surface street system at sporadic intervals by means of access ramps.

Modern freeways carry a large share of travel in most cities, permitting rapid movement (except at rush hours) at about one-third the accident risk of surface streets. Characteristically the freeway network of a city radiates outward from the CBD. Even though functional connection to a freeway is restricted, its route is often an axis along which urban development takes place at higher densities than in the intervening wedges, with the most intensive development likely to occur near access ramps and freeway interchanges.

Almost every large city has one or more beltways, often but not always a freeway, that roughly circle the city at a radius of several miles from the CBD. Probably the most famous example of this phenomenon is Boston's Circumferen-

FIGURE 4-13   Freeways usually cut directly across all other forms of land use without regard to the previous transportation route pattern. This is a typical freeway scar across north Dallas (TLM photo).

FIGURE 4-14 Freeway interchanges are complex and expansive. This is the interchange of the Santa Monica and Harbor Freeways in Los Angeles (TLM photo).

tial Highway (Route 128), but the pattern is now common, from Miami's Palmetto Expressway to Seattle's Renton Freeway. Such perimeter thoroughfares often serve as magnets that attract factories and other businesses to locate alongside them. This very attractiveness inhibits the proper functioning of the beltway concept. These routes are intended to move traffic around urban centers, but they frequently become so overloaded with local traffic that they generate a need for new circumferential routes still farther out to carry the through traffic.

One approach to dealing with urban traffic congestion is a revival of interest in constructing toll roads. In previous eras most toll roads were longhaul intercity highways (such as the Pennsylvania Turnpike and the Ohio Turnpike), but contemporary development focuses on urban outskirts to enable suburban dwellers to reach their jobs and other destinations. Such commuter routes often do not connect with the CBD, as that is not necessarily the most needed connection. Indeed, the 1980 Census discovered that 25 million Americans commuted from one suburb to another to reach their jobs, but only half that many commuted from suburbs to central cities, and the gap has continued to widen since then. New commuter-related toll roads have recently been built or extended in a dozen cities, particularly in Florida and Texas, and most notably in California's Orange County (just south of Los Angeles).

Despite freeways and toll roads, however, congestion continues to choke the cities of the continent. It is often suggested that the only hope for urban survival is *mass transit* with emphasis on *rapid transit*. Unfortunately, the panacea effect seems to be overestimated. Mass-transit patronage in the United States has absolutely declined more than one-third

over the last half century, in a period when the urban population has virtually tripled. Patronage of rapid transit, on the other hand, has declined only slightly over the same period, although urban governments either subsidize or own (and operate at a loss) all systems that are in operation.

About 300 U.S. cities provide urban transit service. The largest concentration, by far, is in New York City (Fig. 4-15). The next largest concentrations are strikingly different in their mix: Chicago's is approximately divided among commuter railroads, rapid rail transit, and buses, whereas that of Los Angeles is nearly all bus service. The next largest concentrations are in Boston, the San Francisco Bay area, northern New Jersey, Philadelphia, and Atlanta.

Public transit continues to lose riders in most large U.S. cities, especially in the Northeast. Larger Canadian cities, however, are experiencing increasing ridership. New light rail systems are operating in Edmonton, Calgary, and Vancouver, and the subway systems of Montreal and Toronto are being expanded (Fig. 4-16).

Enthusiasm for rapid transit in the United States is variable (for example, eagerly accepted by voters in Pittsburgh, thoroughly rejected by citizens in Houston), but the last quarter century has seen a flurry of construction. San Francisco's BART system, opened in 1972, was the first of the modern, high-tech rail systems. Since then more than a dozen other major cities have opened new rapid transit systems (most conspicuously in Washington, D.C., and Atlanta) or have lines under construction (even in Los Angeles!). Some are doing better than others, but all require large government subsidy.

North American urbanites are generally still wedded to their automobiles, and their willingness to shift to transit fa-

FIGURE 4-15   Passenger miles of rail and bus transit systems for U.S. metropolitan areas with more than 100 million passenger miles traveled in 1991. The size of each circle represents the total passenger miles of travel by rail and bus transit for the metropolitan area or consolidated metropolitan area. The gray slice is for commuter railroads, the red slice is for other rail transit, and the white slice is for bus transit (after Bureau of Transportation Statistics, U.S. Department of Transportation, *Transportation Statistics, Annual Report, 1994*, p. 29).

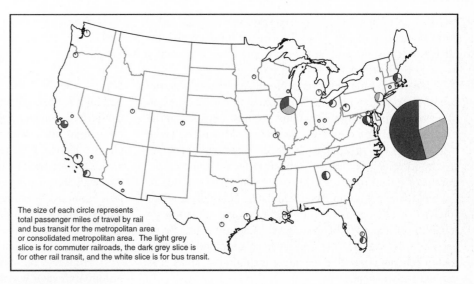

The size of each circle represents total passenger miles of travel by rail and bus transit for the metropolitan area or consolidated metropolitan area. The light grey slice is for commuter railroads, the dark grey slice is for other rail transit, and the white slice is for bus transit.

FIGURE 4-16   New rapid transit systems have been established in several North American cities in recent years. This light rail transit car is in Toronto (TLM photo).

cilities is limited. Moreover, the relatively low population densities in most cities means low volumes of traffic, which works against the feasibility of substantial investments in rapid transit facilities.

There is no way to avoid the fact that rapid transit systems are relatively idle most of the time and in demand only at rush hours. Furthermore, who is served by rapid transit systems? New lines in Chicago and Toronto were found to draw 90 percent of their passengers from bus lines and only 10 percent from automobiles. Most systems accommodate commuters rather than the poor, the aged, and the handi-

capped, who, some would argue, need public transit most of all.

If not mass transit, and not rapid transit, then what will prevent the North American city from grinding to a halt someday under the sheer bulk of its street and freeway traffic?

External Transport   The movement of people and goods into and out of cities is accomplished in a great variety of ways, although auto and truck transportation dominate. Cities are hubs in the cross-country highway networks of the United States and Canada, with routes converging to join the internal street system of the hubs. Despite the construction of bypasses and beltways, there is much mixing of a city's internal and external roadway traffic, with each contributing to congestion for the other. Even the building of the unprecedented ($70 billion in construction over a 21-year period) Interstate Highway System in the United States has done little to improve traffic within cities; it has immensely facilitated cross-country travel but failed to alleviate urban congestion, which was one of its principal objectives.

In general, railways were important factors in the founding and growth of North American cities, but in most cases they are relatively less significant today. Nevertheless, railroad facilities are still quite conspicuous in most cities. Rail lines converge on cities in the same fashion and sometimes in the same pattern as highways. For large cities, there are also railway belt lines to facilitate the shifting of rail cars from one line to another. Major passenger and freight terminals are usually located in the transition zone near the CBD, but in some cases the latter have been shifted to more distant sites. Most cities have a single passenger terminal ("Union

Station"). An important specialized feature of railway transport is the classification yard where freight trains are assembled and disassembled; yards of this kind are now usually located on the very outskirts of the urban area.

There are more airports in North America than in the rest of the world combined. The continuing rapid expansion of air travel means that most cities are now engaged in the construction of new or in the expansion of old airports. There is little in the way of a predictable location pattern except that an airport is usually located several miles (in some cases, dozens of miles) from the CBD, where an extensive area of flat and relatively cheap land is available. Normally a major freeway or other roadway thoroughfare (and sometimes a rapid transit line) is designed to give the airport a direct connection with the CBD. The long-predicted development of intracity helicopter travel in the larger metropolitan areas is as yet relatively insignificant.

Water transportation may be important to the economy of many ocean, river, and Great Lakes ports, but the amount of space used for port facilities is usually small compared with the total area of the city. Piers, docks, and warehouses are normally the most conspicuous permanent features along a port's waterfront, although the recent rapid change to containerization of cargo has led to enormous aggregations of container vans in open spaces adjacent to the docks.

Pipeline transportation is highly specialized and relatively inconspicuous in most cities. Internal networks distribute water and gas and collect sewage, but these are ubiquitous and largely underground features. External pipeline systems are normally associated with liquid or gaseous fuels, bringing petroleum or gas into the city for either refining or distribution to consumers. In any case, most pipelines are buried, and the prominent landscape features associated with this activity are huge storage tanks at the terminals.

Vertical Structure

The building of skyscrapers and other high-rise[9] buildings is not peculiar to North America, but the concept achieved its first real prominence in New York City, and the vertical dimension of the North American skyline has continued to be significant in any consideration of city form. In the past, the vertical structure of cities was predictable. Within the CBD would be an irregular concentration of tall buildings, with a rapid decrease in height centrifugally in all directions to the very low profile that characterized the vast majority of the urbanized area. The only significant exception to that generalized scheme was New York, with its prominent dual con-

centration of skyscrapers: the major one in midtown Manhattan and the secondary, but still very impressive, one in lower Manhattan.

Later the pattern of high-rise building became more diffuse. The CBD still has the conspicuous skyline, but tall buildings are being built ever more widely throughout the urban area. Secondary aggregations of high-rise structures are often associated with major suburban shopping centers, planned industrial districts, and even airports. Tall buildings are also being built increasingly along principal string-street thoroughfares, usually in a very sporadic pattern . Los Angeles's famous Wilshire Boulevard, for example, is now marked along its entire 15-mile (24-km) length from the CBD to the Pacific Ocean by an irregular string of high-rise office and residential buildings.

High-rise buildings became practical after the first primitive elevator was invented in 1852, but the skyscraper is a creature—and a symbol—of the twentieth century. On this continent the greatest concentration of tall buildings has always been in the older, more crowded cities of the Northeast; the newer, less intensively developed cities of the West tended to sprawl outward rather than upward. But since World War II, this pattern has changed; prominent skylines now sprout from such plainsland cities as Dallas, Denver, and Winnipeg, and the trend extends to almost all large cities of the subcontinent. Only a few, such as Washington, D.C., continue to restrict building height.

New York City still contains the greatest concentration of skyscrapers in the world; two-thirds of the world's tallest buildings and more than half of the nation's buildings over 500 feet (150 m) in height are located on Manhattan Island. The first of the world's "superskyscrapers" was erected in Lower Manhattan—the twin-tower, 110-story, 1350-foot (405-m) high World Trade Center; its 10 million square feet (0.9 million m^2) of office space are served by nearly 200 elevators. The nation's tallest office facility, however, is the 1450-foot (435-m) Sears Tower in Chicago.[10] Indeed, Chicago is the only other North American city with a skyline that includes a number of very tall buildings, but they total only about 20 percent of New York's. The construction of skyscrapers continues apace, and the present roster of cities with tallest buildings (Boston, Chicago, Dallas, Houston, New York, Pittsburgh, San Francisco, and Toronto) will undoubtedly change from year to year (Fig. 4-17).

The skyscraper is a visible symbol of high land values[11] and of congestion. Although the skyscraper permits

[9] Although legal definitions vary, any structure over 75 feet (22.5 m) generally is considered to be "high rise" for purposes of fire and other codes.

[10] The tallest human-built structure on the continent is the CN Tower in Toronto, the topmost point of which is 1815 feet (545 m) above its base.

[11] A prestigious new skyscraper can command a rent in excess of $1000 per square foot (0.09 m^2).

FIGURE 4-17   The tallest structure in a North American city is the CN Tower in Toronto (TLM photo).

use areas, such as airports or planned industrial districts, but, in general, their pervasiveness can be seen in cities throughout North America.

Building ordinances and zoning restrictions, which tend to be similar from city to city, are another reason for the visual similarity of cities in the United States and Canada. Many examples could be cited, but perhaps the most prominent is the requirement that residences be set back from the street; thus in most parts of most cities front yards are required even though their functional role is largely a thing of the past.

A more detailed look at cities shows their many differences in appearance. Every city has a certain visual uniqueness on the basis of street pattern, architecture, air pollution, degree of dirtiness, and a host of other elements. But often such distinctiveness is a function of site (slope land versus flat land or coastal versus inland, for example), regional location (as the widespread adoption of "Spanish" architecture in the Southwest), or relative age.

The federal program of *urban renewal* in the United States, along with its other consequences, had a marked effect on the appearance of many cities. There was no similar national program in Canada, although provincial and municipal authorities inaugurated various urban renewal efforts on a much smaller scale. The idea behind urban renewal is simple enough: communities acquire large parcels of slum property (using the power of eminent domain where necessary) and sell or lease them for massive public or private redevelopment, using mostly federal funds for capital requirements. Slum clearance was often involved, but some projects provided for the conservation and rehabilitation of areas that did not require demolition.

An overriding consideration in urban renewal philosophy was the acceptance of the need for federal aid to revitalize the economic base and taxable resources of cities. Several thousand urban renewal projects were carried out under the program, with erratic results. The principal objections are that more low-income housing is removed than replaced and that costs are much greater than benefits. In any event, the replacement of slums by modern high-rise buildings and green spaces has changed the face of many American cities, from Boston's West End to Los Angeles's Bunker Hill. The greatest emphasis on urban renewal has been in the Northeast, particularly in the cities of Pennsylvania.

Extensive urban renewal activity was phased out in both countries in the 1960s. Government-sponsored successor programs—called Community Development Block Grants in the United States and Neighborhood Improvement Programs in Canada—operate on a much smaller scale.

A problem that most waterside cities have faced with continuing despair is the decay of their traditional waterfront

many more people to live and work in a restricted area, it also adds to traffic confusion. The streets of most cities were designed for smaller populations, lower buildings, and more limited movement, so they cannot carry the present traffic load without friction and delays. Traffic slows to a snail's pace in the very places where speed and promptness are most desired. The diffusion of high-rise building away from the CBD is a partial response to this problem.

General Appearance

Most North American cities are visually similar, a generalization that is a logical outgrowth of the morphological similarity that is chronicled on preceding pages. Within the CBD, tall buildings dominate the scene. Elsewhere in the urban area, even in the transition zone, the most conspicuous visual element consists of trees, generally rising above low-level residential and commercial rooftops. The visual dominance of trees is interrupted wherever there are extensive special-

FIGURE 4-18   Old waterfront areas have been renovated and redeveloped in many cities. This is part of the Inner Harbor in Baltimore (TLM photo).

area as the water transport function diminished or changed. Waterfronts often became seedy, disreputable, inhospitable districts that were largely avoided by the citizenry and ignored by the authorities. Despite the shining example of riverfront development (Paseo del Rio) begun in San Antonio in the 1930s, there were few concerted efforts at waterfront rehabilitation until the 1980s. Stimulated by the success of Baltimore's Harborplace, opened in 1980 (Fig. 4-18), many other cities initiated waterfront development projects in the form of shopping-recreational-residential complexes, often focused on "festival markets" that mixed trendy shops and eateries with tourist attractions. Many North American waterfronts—such as Chicago's Navy Pier, Boston's Lewis Wharf, Halifax's Historic Properties, Jersey City's Liberty Park, and Toronto's Harbourfront—have been transformed into areas of considerable charm and attractiveness. All such developments have not been successful, however, and hard lessons have been learned in such places as New York's South Street Seaport, Flint's Water Street Pavilion, and Toledo's Portside.

URBAN FUNCTIONS: GROWING DIVERSITY

Besides being a morphological form that occupies space, a city is a functional entity that performs services for both its population and the population of its hinterland. In doing so, it may also provide services for people in more distant re-

gions or even in foreign lands. Every North American city is multifunctional in nature, being involved in several different kinds of economic activities.

The Commercial Function

Most cities came into being as trading centers, and the commercial aspect of the urban economy is almost invariably a major one regardless of city size or location. This function encompasses both retail and wholesale trade. In almost every major North American metropolitan area between 22 and 27 percent of the work force is employed in trade. There is a close correlation between a city's population and its component of retail trade, for sales from most stores are usually made to local residents. Wholesale trade, on the other hand, may vary more widely from a predictable norm. Many small cities have a minor wholesale trade element, whereas others, particularly larger metropolises, may be well situated to supply not only the retailers within the city but also those in smaller cities within their hinterland. Cities with gateway positions (either transportational or entrepreneurial gateways), such as Atlanta, Dallas, Memphis, Kansas City, Denver, Minneapolis–St. Paul, and Winnipeg, have an unusually large wholesale trade component in their economy.

The Industrial Function

A certain amount of manufacturing is carried on in any city worthy of the name, although it may largely consist of such

FIGURE 4-19 Many contemporary industrial areas are designed in attractive parklike settings. This scene is in Phoenix (TLM photo).

prosaic factories as printing shops, ice-cream plants, and bakeries. Larger cities generally have a much broader range of manufactures (Fig. 4-19). Some urban areas, of course, may specialize in one of several kinds of manufacturing; then this function becomes unusually important in the city's economy. Manufacturing employs from 15 to 25 percent of the work force in most major metropolitan areas, although there are significant variations above and below this range. Since about 1970, proportional employment in manufacturing (i.e., the percentage of the work force that draws its paychecks from factories) has been on a continual decline in virtually every major urban place in the United States and Canada, as both countries move toward a more service-oriented economy. No significant city has as much as 30 percent of its labor force employed in factories and in no million-population city does this figure exceed 25 percent. Many smaller cities have at least one-third of their total employment in manufacturing, especially in southern New England, the Carolina Piedmont, southern Michigan, and eastern Pennsylvania. Washington, D.C., and Honolulu are at the other end of the scale, with fewer than 5 percent of their employees working in factories.

The Service Function

The second fastest-growing component of the urban economy is the service sector, which includes a variety of professional, personal, and financial services, such as those provided by attorneys and real estate dealers. Most services do not produce tangible goods; thus the service output is usually incapable of being stored and is consumed at the time it is produced. Except for the rapidly growing, high-tech oriented, computer-based professional services, service activities tend to be the least mechanized and least automated part

of the economy; therefore a rapid increase in service output means a significant growth in the number of available jobs. Total employment in the service component of the economy now exceeds that of the trade component in both the United States and Canada.

The Administration Function

The rapid increase in government employment in both the United States and Canada has overshadowed all other growth sectors of the economy. Government employment expanded by approximately 50 percent in the 1950s, and roughly another 50 percent between 1960 and 1970, although the growth rate slowed somewhat in the 1970s and 1980s. The proportion of the labor force employed in government activities varies widely from city to city. It is expectably high in the national capitals of Washington, D.C., and Ottawa and in such small-city state and provincial capitals as Lansing, Austin, Charlottetown and Fredericton. But there are many categories of government employment, and a significant proportion of the jobs in this sector are in municipal and county governments; thus most North American cities have a prominent component of employment in public administration.

Other Urban Functions

Two other urban functions are common: transportation-communication and contract construction. Employment in both has been increasing through the years but at a slower rate than most other sectors of urban economy; consequently, the relative position of the two slowly continues to decline. In most cities each activity employs about 5 percent of the labor force. A few other specialized categories of employment are found in some cities, although only "mining" (which generally means office employees of mining companies) shows up enough to be prominent in individual cases.

URBAN POPULATION: VARIETY IN ABUNDANCE

Whatever else it is, a city is primarily a group of people who have chosen to live in an urban environment because of economic opportunities, amenities that are close at hand, or inertia. The pell-mell rate of urbanization that characterized the 1950s and 1960s slowed down subsequently, partly because of a trend toward nonmetropolitan living but partly because the vast majority of the populace already lived in cities.

We have already noted that the population of central cities in metropolitan areas, particularly the largest metropolitan areas, stagnated or declined in recent years, with all or most growth being confined to the expanding suburbs. Indeed, it was during the 1950s that many large cities had their last great growth, having actually lost population within their political limits since that time. One notable result is that contemporary urbanites have more living space than they did a decade or two ago: The population per square mile in North American urbanized areas decreased by more than 50 percent in the last half century.

The population structure of cities has many aspects. One of the most important is ethnicity.

North American cities have a remarkably varied array of ethnic mixes. The variety generally follows a regional pattern, but some cities show significant variations from the regional trend. It should be noted that it does not require a very

large number of people of a particular ethnic group to put a significant imprint on a city or on that city's image.

The situation in Canadian cities is quite different from that in U.S. cities. Every Canadian city of any size is at least 40 percent British or 40 percent French in terms of the lineage of the population. This bicultural bifurcation identifies the two main groups in any ethnic classification of Canadian urban areas. Figure 4-20 illustrates the magnitude of the bifurcation. The average Canadian urbanite is an Anglo-Saxon Protestant, normally of British descent but sometimes of German or Dutch extraction. In the province of Quebec, on the other hand, urban dwellers are overwhelmingly French; only in Montreal is the population less than 80 percent French.

There is a varied although largely European ethnic mix among the non-British, non-French minority in Canadian cities; however, there is no large, underprivileged, vocal ethnic group around which racial antagonisms cluster. Canadian urban minority groups tend to be upwardly mobile in both social and economic status. Ethnic ghettos occur in Canadian cities, but they are not—except for some French Canadian areas in predominantly Anglo-Canadian cities—particularly conspicuous in most cases. The most clustered and easily identified small ethnic minorities are probably the Chinese (particularly in Vancouver and Toronto), Hungarians, Greeks, Portuguese, and West Indians (Fig. 4-21).

In cities south of the international border African Americans are often a prominent minority that is all the more conspicuous because of the racially segregated housing pattern that usually prevails and the generally low economic level of the group as a whole. We have noted that African

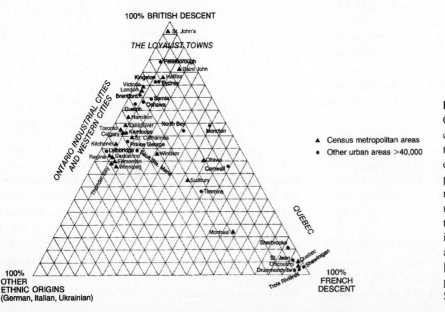

FIGURE 4-20   The ethnic pattern of Canadian cities (1971). This chart indicates the proportion of the city's population composed of each of three different ethnic origins. The closer a city is to a point of the triangle, the more homogeneous is its origin. Note the extreme homogeneity of most Quebec cities, and the spectrum of British and other origins in the remainder of the country. (from J. and R. Simmons, *Urban Canada*, 2d ed. [Toronto: Copp Clark Publishing Company, 1974], p. 40. By permission of Statistics Canada).

FIGURE 4-21   A sidewalk market scene in North America's largest Chinatown, in New York's Lower East Side (TLM photo).

Americans live predominantly in ghettos throughout the nation. The central-city ghettos continue a rapid outward expansion, but only very recently has the African American populace begun to deconcentrate from the central cities, due to an increased rate of African American suburbanization and a declining central-city growth rate. African Americans are particularly concentrated in large cities; one-third of all black Americans reside in 10 cities and more than one-fourth are concentrated in only six metropolitan areas.

Hispanic Americans are another prominent urban minority, although their numbers are fewer and their distribution is less widespread than that of African Americans. In cities of the Southwest, from California's Central Valley to the Piedmont of north-central Colorado to the east Texas metropolises of Dallas and Houston, Hispanics are frequently found in large numbers. There are other concentrations of this minority in Chicago (largely of Mexican origin), the major cities of Megalopolis (primarily Puerto Ricans), and various urban areas in Florida (mostly Cubans). Although less ghettoized than urban blacks, Hispanics are often clustered in distribution, and their generally low economic status is likely to be reflected in slum housing.

URBAN ILLS: MASSIVE MALADJUSTMENTS

The big city everywhere is the object of criticism. Critics insist that all cities are ailing and are not good places in which to live. They point to smog, crowding, strain on family life, snarled traffic, segregation of minorities, crime, drugs, youth gangs, violence, impersonality, and a host of other urban evils. The long-continued growth of our cities begets massive growing pains.

And urban sprawl continues apace. For example, a two-decade (1966–1986) study of the 70 largest cities in Canada showed that 744,500 acres (301,400 ha) of rural land—an area three times the size of urbanized Toronto—were converted to urban uses. Similar conditions undoubtedly prevail in the United States.

As urban areas expand, provision of such necessities as water, sewerage, paved streets, utilities, refuse collection, police and fire protection, schools, and parks is a continuing headache, particularly when more than one municipal governing body is involved. Also, as the flight to the suburbs continues, it is generally accompanied by a degeneration of much of the core of the city; the results are intensified slums, loss of merchandising revenue, and a decline in the tax base.

As the metropolitan area expands, local transportation becomes more complicated. As many more cars drive many more miles on only a few more streets, relatively speaking, traffic congestion becomes intense, the journey to work lengthens, and parking facilities become inadequate. The big city must maintain constant vigilance to keep from choking.

The maintenance of enough good domestic water also challenges the exploding metropolis. In subhumid regions cities must sometimes reach out dozens or hundreds of miles to pipe in adequate water; even such humid-land cities as New York and Boston must extend lines farther and farther to tap satisfactory watersheds.

Where humans congregate, the delicate problem of pollution is accelerated. Rare indeed is the stream in any urban area that is not heavily infiltrated by inadequately treated liquid waste from home and factory. The shocking condition of American waterways has caused some civic groups to wage stringent cleanup campaigns, with emphasis on adequate sewage treatment. The result has been heartening improvements in such infamous rivers as the Ohio and Philadelphia's Schuylkill—improvements showing that this problem can be solved in other areas. Today the menace of atmospheric pollution is recognized as a major problem (Fig. 4-22). The highly (and justifiably) publicized smog of Los Angeles is the most striking instance, but "smust" in Phoenix, "smaze" in Denver, and smoke in Montreal are further examples of an undiminishing phenomenon in most large cities. Industrial vapors and burning refuse contribute to pollution, but automobile exhaust fumes are generally believed to be the major cause. The air pollution problem will undoubtedly get worse before it gets better.

A CLOSER LOOK   Cities and Towns with the Greatest Ethnic Diversity

Most Americans are interested in how the local areas they know are like or unlike other areas, and the rating of cities as to their quality of life has been popular topic for conversation and news reporting. One characteristic of places is their ethnic composition. Because different ethnic groups have long been concentrated in certain regions and cities and because migrations have elaborated older settlement patterns, ethnic variety has become especially intensified in certain cities and towns. Let's ask the question: What places in the United States have the most ethnic diversity? My colleague, Professor Eugene Turner, and I have studied this matter, and our statistical analyses permit some fairly precise answers.[a]

In modern America there appear to be five basic ethnic groupings: white, African American, American Indian, Asian, and Hispanic. Although elsewhere we have examined patterns of diversity when ethnicity is defined in terms of more detailed groupings like Asian Indian and Mexican, here we focus on only the five categories. Diversity can be measured by a statistic called the *entropy index,* which calculates the relative evenness of ethnic group sizes and gives the highest possible value when all groups are present in equal numbers. (However, the index tells us nothing about the social status of the groups or how much the different groups mix socially or occupy the same neighborhoods.) Computer tapes from the U.S. Census of 1980 contain detailed information of the ethnic identities of local populations. Because cities and towns are the most basic local units with which people easily identify, this research analyzes the diversity in all the 2903 urban places with over 10,000 inhabitants. The accompanying map shows those 60 places that ranked highest (Fig. 4-A).

[a] For a more detailed presentation of this research see James P. Allen and Eugene Turner, "The Most Ethnically Diverse Urban Places in the United States," *Urban Geography,* 10 (1989), 523–539.

Perhaps most Americans would say that ethnic diversity is greatest in the largest cities, and indeed, some of the most diverse urban places are those largest cities (shown in bold type on the map). Los Angeles and San Francisco rank 8th and 10th in diversity among all places, and New York City ranks 27th, although two cities of over a million people each—Philadelphia and Detroit—are much less diverse than one might expect from their size.

If we next look at the locations of the highly diverse places, the map shows that over half are in California, with many of the rest in Texas, Florida, and New Jersey. These states (together with New York City) have been the leading destinations for immigrants in recent years, with California alone receiving about a quarter of all immigrants to this country. Because Asians and Latin Americans make up such a large portion of our recent immigrants, places that become homes to immigrants obviously become more diverse.

Our research demonstrated that highly diverse urban places are often not large and typically represent several different types of places: older industrial cities, cities in metropolitan areas (including newer suburban cities), farm labor towns, and military installations.

The older industrial cities include Hartford, Chicago and East Chicago; all five New Jersey cities; and Lynwood, Richmond, San Pablo, and Pittsburg in California. These places were once prominent manufacturing centers, and those cities still provide homes for many workers and former workers, both whites and African Americans, as well as Puerto Ricans and Mexicans depending on the region of the country. For example, Chicago and East Chicago first received large numbers of black workers from the South and Mexican workers from Texas when jobs in steel mills, oil refineries, etc., opened as the United States geared up for World War I and the supply of European immigrant laborers was cut off by German submarine attacks in the Atlantic Ocean. Hartford's ethnic diversity is particularly due to the many Puerto Ricans and Jamaicans who

began to settle there in the 1940s; and more recently Cubans, Filipinos, and Asian Indians have moved into New Jersey cities.

Most of the highly diverse California places are independent cities in either the San Francisco or Los Angeles areas, just as the most diverse places in Florida are part of the greater Miami area. In California these places have received many Asians in recent decades, as well as whites, African Americans and people of Mexican origin, while in Florida there have been fewer Asians and Mexicans but more African Americans, Cubans, and Puerto Ricans. For example, Daly City, California, was one of the first suburban destinations for Chinese from San Francisco's Chinatown, and the Japanese in Gardena originally worked on farms and operated nurseries in what was formerly an agricultural area but is now home to many workers with jobs in nearby industries. Cerritos is a new independent suburb in the Los Angeles area, unusual in having by far the highest median income of all the 60 highly diverse places. In Florida the populations of Carol City and Pinewood have approximately equal numbers of blacks and whites in addition to many Puerto Ricans and Cubans.

Two places—Immokalee in Florida and Delano in California—are diverse primarily because so many inhabitants counted in the census were agricultural laborers, and in both Oxnard and Stockton former or current farm workers accentuate the ethnic diversity. Farm workers represent a racially varied group, with people of Mexican origin predominating in California and often in Florida but with blacks typically numerous among East Coast migratory workers and Filipinos often retired from or occasionally still part of California crews.

The last group of highly diverse places are military installations and areas associated with them. Army posts are often especially diverse because blacks usually represent over 30 percent of the population, soldiers of Mexican or Puerto Rican origin make up another minority, and some soldiers have married Asian women while overseas. Navy

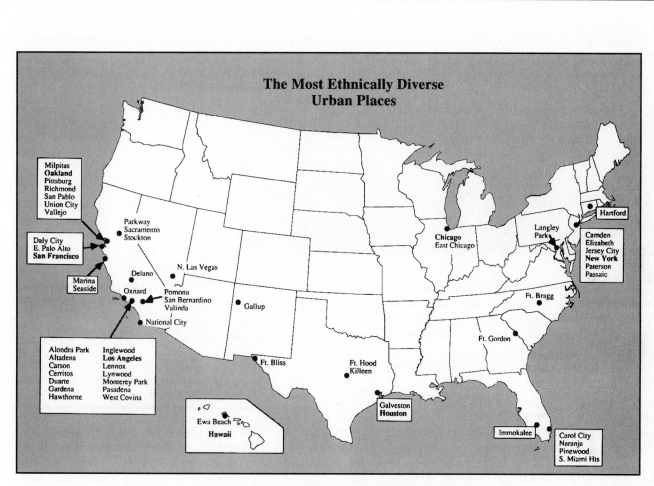

The Most Ethnically Diverse Urban Places

Milpitas
Oakland
Pittsburg
Richmond
San Pablo
Union City
Vallejo

Daly City
E. Palo Alto
San Francisco

Marina
Seaside

Parkway
Sacramento
Stockton

Delano

N. Las Vegas

Oxnard

Pomona
San Bernardino
Valinda

National City

Alondra Park   Inglewood
Altadena   **Los Angeles**
Carson   Lennox
Cerritos   Lynwood
Duarte   Monterey Park
Gardena   Pasadena
Hawthorne   West Covina

Gallup

Ewa Beach
Hawaii

Ft. Bliss

Ft. Hood
Killeen

Galveston
Houston

Chicago
East Chicago

Langley
Park

Ft. Bragg

Ft. Gordon

Immokalee

Hartford

Camden
Elizabeth
Jersey City
New York
Paterson
Passaic

Carol City
Naranja
Pinewood
S. Miami Hts

FIGURE 4-A   The most ethnically diverse urban places.

bases are also sometimes highly diverse, typically with attached Filipino communities because the U.S. Navy has long provided a special avenue of advancement and immigration for young men who first sign up in the Philippines as stewards.

Army posts are evident on the map as forts, such as Fort Gordon and Fort Bragg, but other places are not so readily identified as being connected with the military. In California, many people stationed at Fort Ord live in adjacent Marina and Seaside. Also, National City is integrally tied to the large San Diego navy base, Oxnard has two navy installations, and Vallejo is associated with a major naval shipyard.

Altogether, this research has shown that several different types of places, including many small cities and towns, often exhibit a high degree of ethnic diversity.[b] However, there are no data to indicate whether interethnic relations in any place are relatively harmonious or whether there are substantial tensions

[b] At the other end of the scale, the least diverse places in the United States tend to be small in size, nearly all white, and often non-metropolitan. The 10 least diverse urban places are Berlin, N.H.; Saco, Me.; Swampscott, Mass., Nanticoke, Shamokin, Warren, and Whitehall, Penn.; Morton, Ill.; Monroe, Wis.; and Sun City, Ariz.

between groups because of cultural or social class differences or economic competition between groups for jobs and homes. Perhaps you could suggest some hypotheses concerning what local situations might tend to produce interethnic harmony as opposed to tensions and suspicions in a town or city. Do you think that a list of most ethnically diverse places will have changed much during the 1990s, or do immigrants and minorities tend to settle in places that are already diverse?

Professor James P. Allen
California State University,
Northridge

FIGURE 4-22   Where did Los Angeles go? It was here yesterday. A contemporary smog scene (Frank Hanna/Visuals Unlimited photo).

Wherever people live close together, social friction escalates and the city is the seat of continually burgeoning social problems. Crime statistics increase, with the annual rate climbing above one crime for every 30 urbanites and one violent crime for every 250 urbanites. Similar, or worse, trends can be found for gang membership, substance abuse, and alcoholism.

Coping with the ills of the city and planning to avoid unending escalation of these problems are immensely complicated by the fragmentation of administrative responsibility that is so widespread in the United States, although less so in Canada. Broad planning is inhibited and effective implementation of general solutions is prevented by the multiplicity of municipalities, townships, counties, school districts, and other special districts (everything from cemetery districts to mosquito abatement units). There are more than 85,000 units of government in the United States, most of them empowered to levy taxes. The average 1-million-population metropolitan area in the United States encompasses nearly 300 separate governmental jurisdictions. A few attempts at formation of a metropolitan government for large cities and their suburbs have been mooted, but as yet nothing very effective has resulted.

The situation is less chaotic in Canada, primarily because provincial governments have considerable power to regulate municipal institutions. Canadian cities are not surrounded by as many small independent municipalities as are their counterparts in the United States, and urban problems can be approached on a broader front. The most far-reaching attempt at metropolitan government was established in Ontario in 1953, when Metro Toronto was created. It has been quite successful in providing integrated services to the metropolitan area for which high capital expenditures were necessary; other aspects of its operation have been less satisfactory. Metro Toronto is at least an innovative guidepost for the future; subsequent experiments along the same line, as in Winnipeg Unicity, will be watched with interest by urban administrators throughout North America.

URBAN DELIGHTS: THE PROOF OF THE PUDDING

If cities are such bad places, why do so many people live there? North Americans still flock to urban areas, and most metropolitan areas continue to increase in size (Fig. 4-23). Clearly the number of urban critics is significantly smaller than the number of people who choose to live in urban areas. Disadvantages may be legion, but the attractions are also numerous. Most notable, probably, is the ready availability of material and nonmaterial satisfactions. A vast quantity of goods and services can be purchased, a variety of entertainment may be sampled, and a plethora of mass media provides almost unlimited information and mental stimulation. For many urbanites, however, it appears that the opportunity for social interaction may be even more important than economic advantage or amenity attraction: In cities one can expect to find like-minded people with whom to interact (in a process that Vance called "congregation"). For most people, this combination of positive attributes apparently far outweighs the detrimental aspects of urban life.

URBAN DICHOTOMY: CENTRAL CITIES VERSUS SUBURBS

For several decades North American urban areas were dichotomized into the cores and the suburbs. It has been only

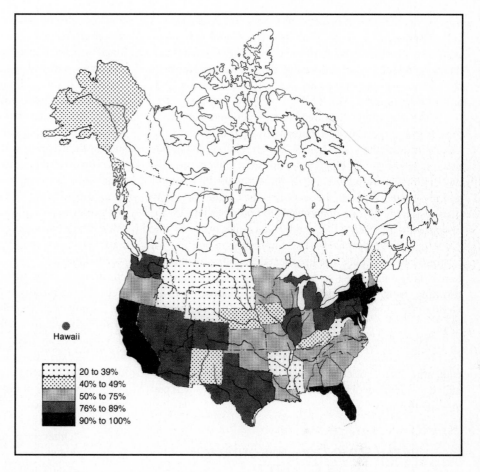

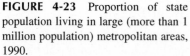

Hawaii

20 to 39%
40% to 49%
50% to 75%
76% to 89%
90% to 100%

FIGURE 4-23 Proportion of state population living in large (more than 1 million population) metropolitan areas, 1990.

in the last few years, however, that the suburban peripheries have experienced such spectacular growth, not only in residential population but also in many other functions that were traditionally restricted to the central areas. As a result, North American metropolitan areas "have undergone a remarkably swift spatial reorganization from tightly focused single-cores to decentralized multinodal" systems.[12] At first there was merely a cautious outward drift of store clusters in the wake of new residential subdivisions, but the almost instantaneous success of these retailing pioneers encouraged bolder innovations, which were epitomized by the development of immense regional shopping complexes that attracted other kinds of employers to their vicinity. The transportational convenience of suburban freeway and beltway locations encouraged successive waves of retailing, wholesaling, manufacturing, and service-oriented activities to abandon the CBD and shift to the suburbs.

Thus suburbs no longer function merely as bedroom communities but as "integrated complexes of tall office struc-

tures, industrial parks, regional shopping malls, and residential subdivisions."[13] The process involves a reclustering of office, manufacturing, retail, and service activities into more or less distinct suburban nucleations.

Central cities thus provide a continually decreasing variety and magnitude of functions. "The leading residual activity is office functions, a set of interdependent activities requiring face-to-face contact and the external economies of the CBD's specialized services."[14] Increasingly, however, office activities are also moving to the suburbs—not that the CBDs are dying. On the contrary, a resurgence of downtown construction activity in recent years has increased the availability of white-collar jobs in central cities. In a sense this represents the "suburbanization" of the CBD—a "diffusion toward the downtown of low-perceived-density, autoconvenient landscape elements and forms initially tested and found successful and tasteful in suburbia."[15] A significant propor-

[12] Peter O. Muller, "Toward a Geography of the Suburbs," *Proceedings*, Association of American Geographers, 6 (1974), 36.

[13] Joseph S. Wood, "'Suburbanization of Center City," *Geographical Review,* 78 (July 1988), 325.

[14] Muller, "Toward a Geography of the Suburbs," 38.

[15] Wood, "Suburbanization of Center City," 325.

tion of this downtown investment is seen as "defensive," a civic responsibility to the wider urban community to maintain the economic viability of the core. "These new investments represent what could well be the last major defense of the central city against the persistent and ever-growing challenge of the suburbs."[16]

The crux of the dichotomy revolves around money and jobs. A very large share of suburban growth is beyond the political limits of the central cities, which severely undercuts their tax base and payrolls. Thus there are relatively prosperous suburban communities and destitute central cities. Suburbanites largely ignore the central cities, and inner-city blue-collar workers are often unable to find replacement jobs downtown and must face either unemployment or the prospect of long-distance reverse commuting to outlying industrial concentrations. The heightened unemployment problem is only one aspect of the increasing social unheavals resulting from inner city economic stagnation, or what George Sternlieb called the "defunctioning" of the central city.

Three-fifths of our metropolitan population now reside in suburbs. Urban analyst Peter Muller concluded that the economic decline of central cities "has become irreversible" and that "suburban dominance within the metropolis will persist and intensify."[17]

URBAN TOMORROW: THE OUTREACH OF CITY LIFE

What will tomorrow's city be like? Many radical designs and grandiose predictions have been made, but only one thing is clear: The urban place of the future will be a combination of elaborate planning and unstructured eventuality.

The surest bet for the short run is the continuance of urban sprawl. The areal expansion of individual cities will result in the increasing coalescence of adjacent metropolitan areas and the creation of more supercities around the western end of Lake Ontario, around the margin of Lake Erie, around the southern end of Lake Michigan, and along the coast of southern California. The metropolis expands and so do the suburbs and ghettos.

There is, however, a growing but erratic enthusiasm for "slow growth," whereby legal impediments to unbridled expansion are enacted by municipal governments or by voter approval. The first major court decision approving growth

controls dates from the early 1970s, when a New York appellate court upheld municipal regulations in Ramapo, N.Y. Since then a variety of laws and regulations have been enacted or approved widely across the country. The thrust of these initiatives is varied: limiting residential construction, restricting new office space, reducing commercial or industrial zoning, creating "green belts" where no construction is allowed. Apparently nothing can stop urban expansion, but actions such as these can slow it down.

Innovations in the use of urban space will surely proliferate, although surrealistic cities of the future may still be a few decades away. Many cities have experimented with pedestrian malls and walkways, a feature that is likely to continue. The advantages of separating pedestrian and vehicular traffic are clear, and in intensive-use areas such separation is worth almost any cost. Increasingly, buildings will also be constructed above transportation routes. The concept of air rights above railway tracks, highways, and transport terminals is now well accepted and will be resorted to more and more.[18] Skyscrapers will continue to grow; buildings of 250 to 300 stories are freely predicted. Subterranean expansion is also to be expected—shopping areas and underground delivery systems for people, goods, and garbage.

New forms of city structure continue to develop. Here are some examples.

1. "Outer city" development, also called "satellite downtowns." This represents a centrifugal force in urban development in which a sprawling urban zone develops peripheral to an established metropolis. Prime examples are Orange County near Los Angeles, "Silicon Valley" south of San Francisco, Nassau–Suffolk on outer Long Island, the area between Dallas and Forth Worth, and the "gold coast" north of Miami. These are not commuter locations; they represent a clear displacement of economic and political power away from the adjacent metropolis and become job magnets in their own right.

2. "Community associations." These are clustered congregations of homeowners that function as small democratic units that assess fees and levy fines, contract with maintenance and security firms, and perform many duties normally associated with small-town governments. It is estimated that some 20 million Americans already belong to such asso-

[16] Gerald Manners, "The Office in Metropolis: An Opportunity for Shaping Metropolitan America," *Economic Geography,* 50 (April 1974), 95.

[17] Peter Muller, "Suburbia, Geography, and the Prospect of a Nation without Important Cities," *Geographical Survey,* 7 (January 1978), 16.

[18] Prominent examples of the use of air rights include one of New York's largest office buildings, the Pan-American Building, over a railway station; Chicago's Merchandise Mart, the world's most spacious office building, over railway lines; and part of the grounds of the United Nations complex over a major Manhattan parkway.

FIGURE 4-24   One of the most famous of the entirely planned retirement communities is Del Webb's Sun City near Phoenix (courtesy Del E. Webb Development Company).

ciations, ranging from huge planned communities (e.g., Sun City, Arizona) to small clusters of condominiums (Fig. 4-24). There are now more than 60,000 such associations in the country (almost all spawned since 1970). The development of planned communities is guided chiefly by economic and security factors; they are usually access controlled. The greatest proliferation thus far has been in Florida, where perhaps a fourth of the population lives in condominiums or planned communities.

3. "Movable towns." Much of our population is footloose and shifts from place to place, often following the seasons. This rootless population frequently resides in mobile homes or campers and may live in several different "towns" in a single year. These movable towns are particularly notable along the Colorado River in western Arizona, in the Lower Rio Grande Valley of Texas, and in parts of Florida. Residents are often senior citizens.

Regardless of futuristic trends, all indications are that the relative importance of the major metropolis is going to decline as single-city foci of the past become metamorphosed into polycentric metropolitan forms in which multifarious outlying activity concentrations rival, or outstrip, the traditional CBD. The terms "galactic city" or "suburban downtown" are increasingly used to refer to such a loose, separated network of activity clusters (to which journalist Joel Garreau has applied the name "edge cities").[19] The familiar residential suburbs are replaced by virtually autonomous, self-contained agglomerations of commercial, residential, social, and

[19] Joel Garreau, *Edge City: Life on the New Frontier*. New York: Doubleday & Company, 1991.

FIGURE 4-25   The complexity of urban North America is indicated in this view of Manhattan Island, looking northward from the World Trade Towers of lower Manhattan (TLM photo).

recreational facilities that sprout on the peripheries of major cities. These centers typically emerge at the intersections of beltways and radial highways, which give such nodes the same accessibility advantages formerly associated with the central-city downtown. Thus the lost vitality of the central city is reborn in the edge cities, which has the effect of turning the metropolis inside out (Fig. 4-25).

It also seems clear that the city of the future will be a place where, increasingly, communication is substituted for the physical movement of people. This portends further deconcentration because the activity centers presumably can be located wherever there is access to global computer and satellite networks. Thus is ushered in the "wired metropolis" or "transactional city."

SELECTED BIBLIOGRAPHY

ABLER, RONALD, JOHN S. ADAMS, AND KI-SUK LEE, *A Comparative Atlas of America's Great Cities: Twenty Metropolitan Regions.* Minneapolis: University of Minnesota Press, 1976.

ADAMS, JOHN S., ED., *Urban Policy-Making and Metropolitan Dynamics: A Comparative Geographical Analysis.* Cambridge, MA: Ballinger Publishing Company, 1976.

ANDERSON, WILLIAM P., AND YORGOS Y. PAPAGEORGIOU, "Metropolitan and Non-Metropolitan Population Trends in Canada, 1966–1982," *Canadian Geographer,* 36 (Summer 1992), 124–144.

BORCHERT, JOHN R., "America's Changing Metropolitan Regions," *Annals,* Association of American Geographers, 62 (1972), 352–373.

BOURNE, LARRY S., AND DAVID F. LEY, EDS., *The Changing Social Geography of Canadian Cities.* Montreal: McGill-Queen's University Press, 1993.

BROADWAY, MICHAEL, "Differences in Inner-City Deprivation: An Analysis of Seven Canadian Cities," *Canadian Geographer,* 36 (Summer 1992), 189–196.

BRUNN, STANLEY D., AND JAMES O. WHEELER, EDS., *The American Metropolitan System: Present and Future.* Washington, D.C.: V. H. Winston & Sons, 1980.

BUNTING, TRUDI E., AND PIERRE FILION, *The Changing Canadian Inner City.* Waterloo, Ont.: University of Waterloo, 1988.

————, EDS., *Canadian Cities in Transition.* Toronto: Oxford University Press, 1991.

CERVERO, ROBERT, *America's Suburban Centers: The Land Use–Transportation Link.* Winchester, MA: Unwin Hyman, 1989.

CONZEN, MICHAEL P., "American Cities in Profound Transition: The New City Geography of the 1980s," *Journal of Geography,* 82 (May–June, 1983), 94–101.

————, "The Maturing Urban System in the United States, 1840–1910," *Annals,* Association of American Geographers, 67 (1977), 88–108.

DUNN, EDGAR S., JR., *The Development of the United States Urban System.* Baltimore: Johns Hopkins University Press, 2 vols., 1980 and 1983.

EWING, GORDON O., "The Bases of Differences Between American and Canadian Cities," *Canadian Geographer,* 36 (Fall 1992), 266–279.

FORD, LARRY R., *Cities and Buildings: Skyscrapers, Skid Rows, and Suburbs.* Baltimore: Johns Hopkins University Press, 1994.

————, "Reading the Skylines of American Cities," *Geographical Review,* 82 (April 1992), 180–200.

GARREAU, JOEL, *Edge City: Life on the New Frontier.* New York: Doubleday & Co., 1991.

GERECKS, KENT, ED., *The Canadian City.* Montreal: Black Rose Books, 1991.

GORRIE, PETER, "Farewell to Chinatown," *Canadian Geographic,* 111 (August/September 1991), 16–29.

GOSS, JON, "The 'Magic of the Mall': An Analysis of Form, Function and Meaning in the Contemporary Retail Built Economy," *Annals,* Association of American Geographers, 83 (March 1993), 18–47.

GRIFFIN, DONALD W., AND RICHARD E. PRESTON, "A Restatement of the 'Transition Zone' Concept," *Annals,* Association of American Geographers, 56 (1966), 339–350.

HARRIS, CHAUNCY D., "A Functional Classification of Cities in the United States," *Geographical Review,* 33 (1943), 89–99.

HART, JOHN FRASER, "The Bypass Strip as an Ideal Landscape," *Geographical Review,* 72 (April 1982), 218–223.

HART, JOHN FRASER, ED., *Our Changing Cities.* Baltimore: Johns Hopkins University Press, 1991.

JONES, KENNETH G., AND JAMES W. SIMMONS, *The Retail Environment.* New York: Rutledge, 1990.

KNOX, PAUL L., *The Restless Urban Landscape.* Englewood Cliffs, NJ: Prentice Hall, 1993.

LAI, DAVID CHUENYAN, *Chinatowns: Towns Within Cities in Canada.* Vancouver: University of British Columbia Press, 1988.

LANG, MICHAEL H., *Gentrification Amid Urban Decline: Strategies for America's Older Cities.* Cambridge, MA: Ballinger Publishing Company, 1982.

LAW, CHRISTOPHER M., WITH E. K. GRIMES, C. J. GRUNDY, M. L. SE-NIOR, AND J. R. TUPPER, *The Uncertain Future of the Urban Core.* New York: Routledge, Chapman and Hall, 1988.

MAYER, HAROLD M., "Cities: Transportation and Internal Circulation," *Journal of Geography,* 68 (1969), 390–408.

MERCER, JOHN, "On Continentalism, Distinctiveness, and Comparative Urban Geography: Canadian and American Cities," *Canadian Geographer* 23 (Summer 1979), 119–139.

MITCHELSON, RONALD L., AND JAMES O. WHEELER, "The Flow of Information in a Global Economy: The Role of the American Urban System in 1990," *Annals,* Association of American Geographers, 84 (March 1994, 87–107)

MULLER, PETER O., *Contemporary Suburban America.* Englewood Cliffs, NJ: Prentice-Hall, 1981.

————, "Transportation and Urban Growth: The Shaping of the American Metropolis," *Focus,* 36 (Summer 1986), 8–17.

MURPHY, RAYMOND E., *The American City: An Urban Geography* (2d ed.). New York: McGraw-Hill Book Co., 1974.

PALM, RISA, *The Geography of American Cities.* New York: Oxford University Press, 1981.

RIGBY, DAVID L., "Technical Change and Profits in Canadian Manufacturing: A Regional Analysis," *Canadian Geographer,* 35 (Winter 1991), 353–366.

ROSE, HAROLD M., "The Development of an Urban Subsystem: The Case of the Negro Ghetto," *Annals,* Association of American Geographers, 60 (1970), 1–17.

SEMPLE, KEITH R., AND W. RANDY SMITH, Metropolitan Dominance and Foreign Ownership in the Canadian Urban System," *Canadian Geographer,* 25 (Spring 1981), 4–26.

SIMMONS, JAMES, AND ROBERT SIMMONS, *Urban Canada* (2d ed.). Toronto: Copp Clark Publishing Company, 1976.

SMITH, NEIL, AND PETER WILLIAMS, EDS., *Gentrification of the City.* Boston: Allen & Unwin, 1986.

SUI, DANIEL J., AND JAMES O. WHEELER, "The Location of Office Space in the Metropolitan Service Economy of the United States, 1985–1990," *Professional Geographer,* 45 (February 1993), 33–43.

WARD, DAVID, *Cities and Immigrants: A Geography of Change in Nineteenth Century America.* New York: Oxford University Press, 1971.

————, "The Emergence of Central Immigrant Ghettoes in American Cities: 1840–1920," *Annals,* Association of American Geographers, 58 (1968), 343–359.

WOOD, JOSEPH S., "Suburbanization of Center City," *Geographical Review,* 78 (July 1988), 325–329.

YEATES, MAURICE, *North American Urban Patterns.* New York: John Wiley & Sons, 1980.

YEATES, MAURICE, AND BARRY GARNER, *The North American City* (3d ed.). New York: Harper & Row, Publishers, 1980.

Regions of the United States and Canada

It is convenient and, in many ways, useful to refer to Canada and the United States as discrete units that are sufficiently similar to support generalizations that apply to both nations. But such unitary consideration may, in fact, be misleading or inaccurate. Generalizations about either nation usually force a vast number of unlike areas into the same category. It is more meaningful to think of these countries as composed of a number of parts that are more or less dissimilar. Geographers call these parts *regions*. Although the concept of regions leaves something to be desired, it continues "to be one of the most logical and satisfactory ways of organizing geographical information."[1]

It is well known that people vary—in speech, customs, habits, mores, and other ways—from region to region; however, there is no simple explanation for such variations (Fig. 5-1). In some cases, it may be attributable to aspects of the environment, whereby an element or complex of the physical realm exerts a pervasive influence on the population. Such effects appear to be most significant in areas where human activity is hindered by environmental extremes, such as deserts, mountains, or swamplands. In other cases, the regional varia-

tion of people is more closely related to culture than to nature. Whereas land is relatively changeless, humankind is the active agent; movements of people, changing stages of occupance, or different assessments of the resource base may occasion significant variations in the geography of regions (Fig. 5-2).

The United States and Canada are nations that consist of many political units, but they are also large portions of Earth's surface and include many varying regions. Each region differs from the others in one or more significant aspects of environment, culture, or economy; indeed, the magnitude of regional differences is often vast and complex. This book treats the continental expanse by regions, with broad sweeps of the brush. It is hoped that such treatment will provide not only a sounder comprehension of the geography of the two countries but also a fuller appreciation of the notion of regionalism, a vital concept to those who seek a better understanding of the world.

Even in such affluent and economically integrated nations as Canada and the United States regional considerations have long been recognized in business and government. Federal Reserve banking districts were formalized as early as 1913, and there are now more than 50 U.S. government bureaus in the executive branch alone that are organized with regional divisions. Businesses have similarly created region-

[1] Peter Haggett, Andrew D. Cliff, and Allan Frey, *Locational Analysis in Human Geography,* 2d ed. (New York: John Wiley & Sons, 1977), p. 451.

FIGURE 5-1   A pristine environment is rare in North America. This scene, from Wind Cave National Park in South Dakota, is a reminder of the way things were before the blessings of civilization were introduced (TLM photo).

FIGURE 5-2   Where people congregate, the landscape is totally changed. This urban sprawl is in Los Angeles (TLM photo).

al offices, mail-order houses, and magazines to cope with the varied problems and opportunities faced in operations covering such extensive areas as the United States and Canada. By working with large units, regional geography can make a more meaningful contribution to understanding national life. The often unsatisfactory artificiality of political boundaries becomes particularly apparent when attention is focused on regions.

Geography, which links the data of the social sciences with those of the natural sciences, is the logical discipline for dealing with regions. It sees in the region not only the physical, biological, social, political, and economic factors, but it also synthesizes them. In short, it considers the region in its totality—not merely the elements that are there but also the processes and relationships that have operated, are operating, and presumably will operate in the future.

THE GEOGRAPHICAL REGION

Geographers generally recognize two kinds of regions: *uniform* regions and *nodal* regions. The former possess significant aspects of homogeneity throughout their extent, whereas the latter are very diverse internally and are homogeneous only with respect to their internal structure or organization. A nodal region always includes a focus, or foci, and a surrounding area tied to the focus by lines of circulation. A city and its hinterland illustrate, in crude fashion, the concept of a nodal region.

The prevailing view of this book is that at the macroscale the geography of the continent can best be understood by dealing with uniform regions; consequently, the ensuing discussion and the regional divisions are based solely on the concept of uniform regions. A recognizable uniform region should have some characteristics that provide a measure of distinctiveness.

A uniform region normally contains a core area of individuality in which regional characteristics are most clearly exemplified. The core possesses two distinct qualities that may be blurred in the periphery:

1. It differs noticeably from neighboring core areas.
2. It exists as a recognizable and coherent segment of space defined by the criteria whereby it is selected.

Beyond the core lies a marginal area. Regional boundaries are usually not lines but rather transitional zones that assume the character of adjoining regions or cores. The width may vary from a few feet to many miles; thus the field geographer making a reconnaissance survey seldom knows at what point one region is left and another entered. At some point, of course, the geographer passes from one to the other, but the human eye usually cannot perceive it at the moment of change. The distinguishing features of one region melt gradually into those of the neighboring region except perhaps along a mountain front or the shore of a large body of water.

THE PROBLEM OF REGIONAL BOUNDARIES

Consider, for example, the problem of delimiting the boundaries of one of the most universally recognized regions of the subcontinent, the central flatland known as the Great Plains in the United States and the Prairies in Canada. The eastern boundary of this region is the least exact; in this book it is considered to be an irregular north–south zone extending from Texas to Manitoba. Several criteria are used in positioning the boundary line shown in Figure 5-3, but most significant is the change from a predominantly Corn Belt–type agriculture on the east to a predominantly Wheat Belt–type agriculture on the west. To locate this boundary at any particular point is an exercise in frustration. Southeasternmost Nebraska is clearly Corn Belt and southwesternmost Nebraska is clearly Wheat Belt, but where is the boundary?

In an east–west transect of this area one passes slowly from a corn-and-soybeans-dominated crop pattern to a wheat-and-grain sorghum-dominated crop pattern before reaching the midpoint of the transect. But then corn appears significantly again and there is an erratic alternation of the two crop combinations for many tens of miles until the wheat-and-grain sorghum pattern clearly prevails in southwesternmost Nebraska; thus the eastern boundary of the Great Plains—as exemplified by this single transect—that is portrayed as a black line in Figure 5-3 is actually a broad transition zone that approaches 300 miles (480 km) in width in some places.

The northern boundary of the region is somewhat easier to delimit but is still far from precise. Once again, several criteria are used, but the most prominent criterion is the change from grass and grain in the south to forest in the north. In theory it is reasonable to expect an abrupt and conspicuous demarcation between such different vegetation associations, but actually the change is again transitional. A south–north transect through central Saskatchewan would find a significant interfingering of grass and grain with forest, the whole pattern being complicated by enclaves of forest well to the south and patches of grassland scattered deep in the forest. Again, the map shows a precise line as boundary and, again, the boundary (as generalized from this transect) is erratically transitional, embracing a zone that is in some places almost 100 miles (160 km) in breadth.

The western boundary of the region is much more clear-cut, but even it is imprecise. The principal criterion for pinpointing the boundary is the change from flat land on the east to sloping land on the west; this change is particularly pronounced where the Great Plains meet the front ranges of the Rocky Mountains. An east–west transect in northern Col-

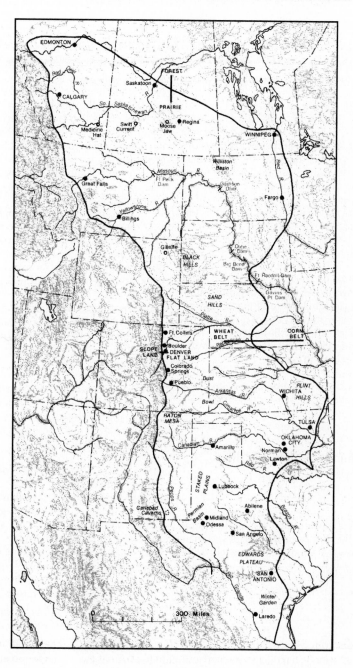

FIGURE 5-3 The Great Plains and Prairies Region, showing the boundary transects discussed in this chapter.

orado finds essentially flat land around Denver becoming virtually all slopeland west of Golden, only 25 miles (40 km) away. But the intervening area fits well in neither region, for the pervasive flatness of the plains is interrupted by the gentle slopes of the Colorado Piedmont, the somewhat steeper slopes of scattered foothills, and the varied terrain of mesas and hogbacks before it is replaced by the more precipitous slopes of the mountain ranges. Thus even one of the most

abrupt regional boundaries that can be found has a transition zone several miles in width.

THE PROBLEM OF REGIONAL STATISTICS

Although many geographers agree that regional geography is a fundamental part of their subject—that it is even the heart

A CLOSER LOOK   Regions Are Devised, Not Discovered

It is important to keep in mind that regions are not naturally occurring phenomena awaiting discovery. The landscape does not consist of a God-given mosaic of regions, awaiting recognition by an adroit researcher. Rather, regions are concepts devised in the human mind for some purpose, and they are useful only insofar as they serve that purpose.

The regional system used in this text is designed for pedagogic use. Its purpose is to facilitate learning by the artful generalization of regional characteristics. It is but one of many systems that might be invented.

Anyone can play the regionalization game; imagination is all that is needed. However, the success of any system of regionalization will depend in large measure upon the care with which it is conceived and crafted.

Presented here is a sample of another kind of regional system, designed by a newspaper columnist to entertain readers. It is indicative of the principle that the regional concept has many uses.

A Plan for California in the Nineties: Break It Up
Robert A. Jones

Welcome to the '90s. If you have been reading the predictions for California over the next decade, perhaps you share my fear that we are fast approaching the end of civilization as we know it. Vicious water wars are scheduled to break out between the north and south, the last redwoods will be chopped down, and 19,000 more people will be arriving each week to enjoy it all. Very gloomy.

So I have a modest proposal. Let's face the fact that California has grown far too big, that it makes about as much sense for California to be a single state as it does for the Soviet Union to be a single country. Let's deal with the reality that the cotton farmers of Visalia don't give a fig for the TV execs in Burbank, and vice versa. Let's break up California.

I'm not just talking about the old strategy of drawing a line between the north and the south. Things have gotten much trickier than that. We need a Plan for the Nineties, and here it is.

As you can see from the accompanying map [(Fig. 5-A)], the Plan provides for three separate states in the new Californias, plus "Oregon." We will get to "Oregon" later. Right now let's take the states one at a time:

• In the south, we must recognize that Los Angeles has become a separate world in California, a city-state only dimly aware of the nether regions to the north. A recent survey showed that Los Angeles makes approximately 500 times more telephone calls to

New York than it does to Fresno. The fact that this survey surprised no one is evidence of L.A.'s estrangement from its geographic neighbors.

Creating a state of Los Angeles would liberate the region from the nattering influence of the environmentalists in Northern California. We in the south could get on with making Los Angeles the richest and ugliest city on Earth. We will require lots of desert to convert into subdivisions, and that has been provided. Ditto with coastline. We get the unspoiled stretch from Santa Barbara to San Simeon so we can make it look like Redondo Beach.

If anyone gets nostalgic about open space, the Plan offers a rental program from the state of Lettuce.

Since Los Angeles itself would be too crowded to accommodate the state capitol, we might want to declare San Bernardino a tear-down site and build a new one from scratch, Brasilia-style. As for the name, we should probably recognize the new realities of our time and call it "Sony." They might even chip in on construction costs.

• In the Central Valley of California, the state of Lettuce would provide a sense of place to our heartland. The valley has always shared more with Nebraska than coastal California. This way the farmers could listen to

of the discipline—students of regional geography usually face an enormous handicap for precision study because of the unavailability of quantitative data. Statistics produced by the various data-collecting agencies (government and otherwise) normally are applicable only to political units (states, provinces, counties, townships, and other civil divisions) and so are difficult to apply to the more functional and less precise geographical regions. As a partial response to this problem, more refined statistical units, such as enumeration districts, urbanized areas, metropolitan statistical areas, consolidated metropolitan statistical areas, state economic areas,

industrial areas, and labor market areas, have been designated; statistics applicable to such areas are often more meaningful for the regional geographer to use. The difficulty of fitting statistics to regions still persists, but improvements continue to be made.

DETERMINATION OF REGIONS

Dividing a continent into regions is a matter of "scientific generalizing." It serves the dual purpose of facilitating the as-

Tammy Wynette and eat chicken-fried steak in peace. Lettuce would be all theirs.

To provide some needed revenge for all the cultural slights suffered over the years, Lettuce would also get the Sierra Nevada. When the coastal folks got sick of their cities, the people of Lettuce could rent them chunks of the Sierra for breath-taking user fees.

• Around the Bay Area we would create what you might call a boutique state. Ecofornia would be small because not much would happen there. San Francisco could convert entirely to tourism and stop worrying about its declining position in California. Within this tiny empire, San Francisco would be forever the center of things.

For territory, Ecofornia would acquire Big Sur to the south and Napa/Sonoma to the north. This would provide some degree of employment diversity. Anyone who got tired of mixing Irish coffees for Pennsylvania optometrists in San Francisco could go to the country and mix Irish coffees at a bed-and-breakfast.

As for the far north, the harsh truth is this region does not belong to California and never has. Does anyone know what goes on in Alturas? That's what I thought, so let's give the far north to Oregon. In return all we ask is that they leave a few redwoods standing, just for old time's sake.

A brilliant plan, I hear you saying, just brilliant, but reality-wise a little unlikely. I understand these doubts. Just keep in mind that they were saying the same thing six months ago in Prague.

(Reprinted from the *Los Angeles Times,* January 3, 1990. By permission.)

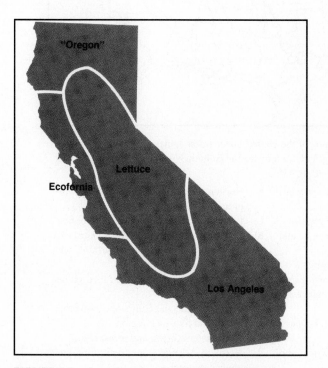

FIGURE 5-A   A suggested regionalization of California.

simulation of large masses of geographic information by students and organizing data into potentially more meaningful patterns for researchers. One recognizes a region by noting the intimate association existing between the people of an area and their occupance and livelihood pattern within that area. Similarity of interest and activity sometimes indicates a similarity of environment but may be totally unrelated to environmental implications.

Geographers deal with both natural and cultural landscapes. Everyone who has traveled, even slightly, has noted that the natural landscape changes from one part of the country to another. When two unlike areas are adjacent, the geographer may separate them on a map by a line and thereby recognize them as separate natural regions. Similarly, the cultural (human-created or human-modified) landscape may be studied and even delineated into separate cultural regions.

Whenever people come into any area, they promptly modify its natural landscape, not in a haphazard way but according to the culture system that they bring with them. They cut down the forest, plow under the native grass, raise do-

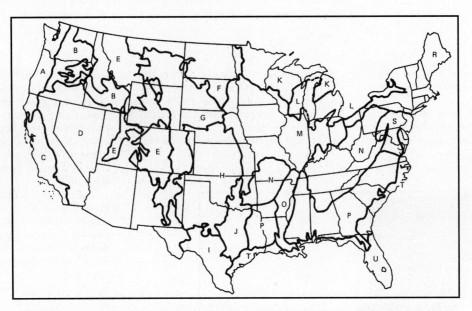

FIGURE 5-4 Land resource regions of the United States (after James A. Lewis, *Landown-ership in the United States 1978*, Agriculture Information Bulletin No. 435. Washington, D.C.: U.S. Department of Agriculture, 1980, p. 43).

A Northwestern forest, forage, and special-ty crop region

B Northwestern wheat and range region

C California subtropical fruit, truck, and specialty crop region

D Western range and irrigated region

E Rocky Mountain range and forest region

F Northern Great Plains spring wheat re-gion

G Western Great Plains range and irrigated region

H Central Great Plains winter wheat and range region

I Southwestern plateaus and plains range and cotton region

J Southwestern prairies cotton and forage region

K Northern lake states forest and forage re-gion

L Lake states fruit, truck, and dairy region

M Central feed grains and livestock region

N East and Central general farming and forest region

O Mississippi delta cotton and feed grains region

P South Atlantic and Gulf slope cash crop, forest, and livestock region

R Northeastern forage and forest region

S Northern Atlantic slope truck, fruit, and poultry region

T Atlantic and Gulf coast lowland forest and truck crop region

U Florida subtropical fruit, truck, crop, and range region

mesticated animals, erect houses and buildings, build fences, construct roads and railroads, put up telephone and telegraph lines, dig canals, build bridges, and tunnel under mountains. All this constitutes the cultural landscape of a region.

Regions can be defined, recognized, and delineated on the basis of single or multiple characteristics, which can be either simple or complex in concept and determination. In preceding chapters, especially Chapter 2, we noted a variety of regional systems. Four other systems are displayed here as samples of the extraordinary diversity that is possible when dealing with the regional concept.

1. Figure 5-4 shows land resource regions as recog-nized by the U.S. Soil Conservation Service. The re-gions are based on agricultural-pastoral-forestry use patterns.

2. Figure 5-5 portrays water resource regions of the United States as recognized by the U.S. Department of Agriculture. These are essentially watershed re-gions.

3. Figure 5-6 shows "popular regions" as determined by "names of metropolitan enterprises," based on research by cultural geographer Wilbur Zelinsky.

FIGURE 5-5   Water resource regions of the United States (after James A. Lewis, *Landown-ership in the United States, 1978,* Agriculture Information Bulletin No. 435. Washington, D.C.: U.S. Department of Agriculture, 1980, p. 43).

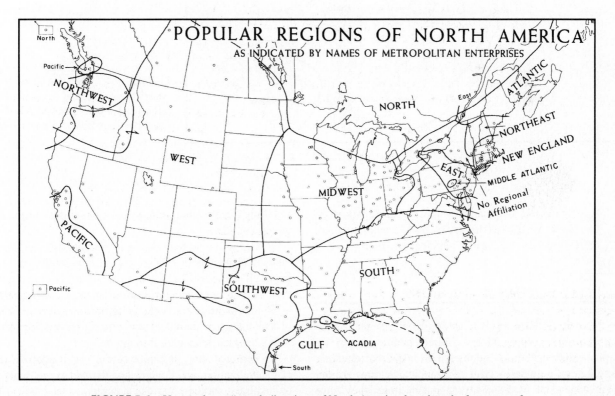

FIGURE 5-6   Vernacular or "popular" regions of North America, based on the frequency of regional and locational terms in the names of enterprises listed in telephone directories of the metropolitan areas of the United States and Canada (after Wilbur Zelinsky, "North America's Vernacular Regions," *Annals*, Association of American Geographers, vol. 70, March 1980, p. 14; by permission of the Association of American Geographers).

FIGURE 5-7 The regions of North America as depicted by a popular writer (after Joel Garreau, *The Nine Nations of North America.* Boston: Houghton Mifflin, 1981).

4. Figure 5-7 demonstrates the regional pattern of North America as devised by popular journalist Joel Garreau, on the basis of personal research.

CHANGING REGIONS

In several disciplines closely related to geography the regional systems are fixed by nature; climatic, pedologic, physiographic, and vegetation regions are all based on relatively static natural boundaries. But geographical regions are not fixed; instead of hard-and-fast boundary lines, they have ever-changing ones. When people push wheat culture farther north in Canada or farther west in Kansas, they are responsible for changes in geographical regions because raising wheat may be of such significance in the regional totality that

a shifting of cultivation limits requires a similar shifting of regional boundaries.

Geographical regions are in continuous evolution. An inherent characteristic of a geographical region is that it is dynamic. It changes with time as people learn to assess and use their natural environment in different ways, with changing economic, political, or social conditions or with the advance of technology. It may contract, expand, fragment, or drastically change its character through the years.

Regional analysis can be carried out at different levels of generalization, and conclusions derived at one scale may be invalid at another. Various geographers have pointed out that a change in scale often requires a restatement of the problem and there is no basis for presuming that associations existing at one scale will also exist at another. One implication of this situation is that a hierarchy of regions can be designed. The creation of such a hierarchy is a useful exercise of geographical scholarship. It is beyond the intent of this book to do so; nevertheless, understanding some regions appears to be enhanced by the designation of subregions, and subregions are delimited in some chapters.

REGIONS OF THE UNITED STATES AND CANADA

In this book the subcontinent is divided into 15 regions. The criteria for making the regional divisions are both multiple and varied. In some cases, physical considerations were dominant, and in others cultural factors played a more important role. But in all instances, the broad regions as delineated reflect as accurately as possible the basic features of homogeneity inherent in various parts of North America at present. In general, the principal criteria used in determining regional boundaries are the socioeconomic conditions currently characteristic, which have often been decisively influenced by the physical environment and historical development. The regions, as numbered in Figure 5-8, are as follows:

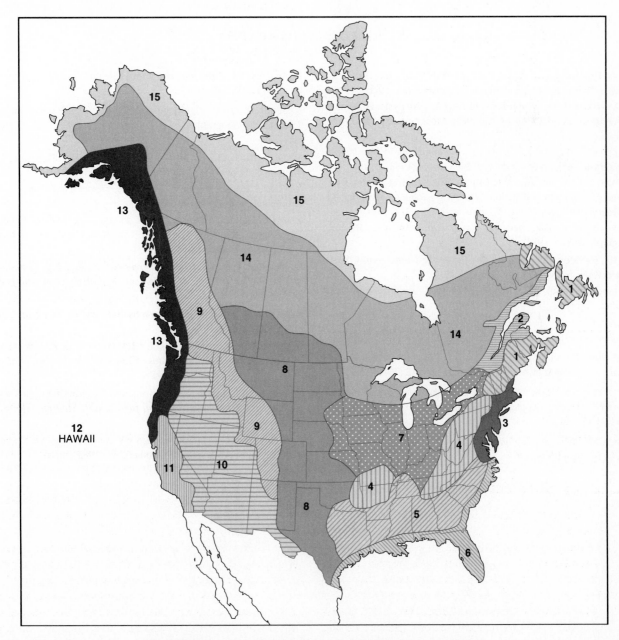

FIGURE 5-8 The geographic regions of the United States and Canada.

1. the Atlantic Northeast
2. French Canada
3. Megalopolis
4. the Appalachians and the Ozarks
5. the Inland South
6. the Southeastern Coast
7. the Heartland
8. the Great Plains and Prairies

9. the Rocky Mountains
10. the Intermontane Region
11. the California Region
12. the Hawaiian Islands
13. the North Pacific Coast
14. the Boreal Forest
15. the Arctic

SELECTED BIBLIOGRAPHY

AYERS, EDWARD L., ET AL., *All Over the Map: Rethinking American Regions*. Baltimore: Johns Hopkins University Press, 1996.

BERRY, BRIAN J. L., "Approaches to Regional Analysis: A Synthesis," *Annals,* Association of American Geographers, 54 (1964), 2–11.

BRADSHAW, MICHAEL, *Regions and Regionalism in the United States*. Jackson: University Press of Mississippi, 1988.

BROWNING, CLYDE E., AND WIL GESLER, The Sun Belt—Snow Belt: A Case of Sloppy Regionalizing," *The Professional Geographer,* 31 (February 1979), 66–74.

GARREAU, JOEL, *The Nine Nations of North America*. Boston: Houghton Mifflin, 1981.

GASTIL, RAYMOND D., *Cultural Regions of the United States*. Seattle: University of Washington Press, 1975.

GEULKE, LEONARD, "Regional Geography," *Professional Geographer,* 29 (1977), 1–7.

HARRIS, R. COLE, "The Historical Geography of North American Regions," *American Behavioral Scientist,* 22 (1978), 115–130.

HART, JOHN FRASER, "The Highest Form of the Geographer's Art," *Annals,* Association of American Geographers, 72 (1982), 1–29.

JAMES, PRESTON, "Toward a Further Understanding of the Regional Concept," *Annals,* Association of American Geographers, 42 (1952), 195–222.

JENSEN, MERRILL, ED., *Regionalism in America*. Madison: University of Wisconsin Press, 1965.

KOHN, CLYDE F., "Regions and Regionalizing," *Journal of Geography,* 69 (1970), 134–140.

KRUEGER, RALPH R., "Unity out of Diversity: The Ruminations of a Traditional Geographer," *Canadian Geographer,* 24 (1980), 335–348.

McDONALD, JAMES R., *A Geography of Regions*. Dubuque, IA: William C. Brown Co., Publishers, 1972.

MINSHULL, ROGER, *Regional Geography, Theory and Practice*. London: Hutchinson Libraries, 1967.

RAY, D. MICHAEL, *Dimensions of Canadian Regionalism*. Ottawa: Department of Energy, Mines and Resources, Policy Research and Coordination Branch, 1971.

SCHWARTZ, M. A., *Politics and Territory: The Sociology of Regional Persistence in Canada*. Montreal: McGill-Queens University Press, 1974.

STEEL, R. W., "Regional Geography in Practice," *Geography,* 67 (1982), 2–8.

WALTER, BOB J., AND FRANK E. BERNARD, "Ash Pile or Rising Phoenix? A Review of the Status of Regional Geography," *Journal of Geography,* 77 (1978), 192–197.

WONDERS, WILLIAM C., "Regions and Regionalisms in Canada," *Zeitschrift der Gesellschaft für Kanada-Studien,* 1 (1982), 7–43.

ZELINSKY, WILBUR, "North America's Vernacular Regions," *Annals,* Association of American Geographers, 70 (1980), 1–16.

The Atlantic Northeast

6

A sharp contrast exists between the predominant rural character of the Atlantic Northeast and the highly urbanized region to its south. The Atlantic Northeast is a region in which the forest and the sea have been pervasive influences on the lifestyle of its inhabitants. There is a sense of history and tradition, reflecting the relatively limited resources of the land and the relative richness of the adjacent ocean.

Included within the region are the less populated, rougher parts of northern New England and New York State, all the Maritime Provinces except that portion of New Brunswick that is predominantly French Canadian in population and culture, and the island of Newfoundland with a portion of the adjacent Labrador coast that is oriented toward "Newfoundland-type" commercial fishing (Fig. 6-1). Interior Labrador is not included because of its sparse population and similarity to the Boreal Forest Region. Northern New Brunswick and adjacent Quebec are excluded because of the significant cultural differences in French Canada.

The most difficult decision to be made in delimiting the Atlantic Northeast as a region is the separation of northern and southern New England. Throughout most of New England's history its people showed a greater degree of regional consciousness, perhaps, than did people in any other part of North America. Their traditions, institutions, and ways of living and thinking exhibited considerable uniformity; however, in recent years there has been an increasing divergence between the way of life of the urbanites in southern New England and the people of the small towns and rural areas that characterize northern New England. Furthermore, the economic and psychological orientation of much of southern New England is increasingly dominated by New York. It seems clear that urbanized southern New England is more logically a portion of the megalopolitan region to the south than it is of the rurally oriented region to the north. Accordingly, the southern boundary of the Atlantic Northeast is considered to lie just north of the urbanized areas of Portland in Maine, the Merrimack Valley in New Hampshire, the small cities of western Massachusetts, and the Mohawk Valley of New York State.

A REGION OF SCENIC CHARM AND ECONOMIC DISADVANTAGE

The Atlantic Northeast is a region that abounds in scenic delights: the verdant slopes of the Adirondacks, the rounded skyline of the Green Mountains, quaint covered bridges and village greens, sparkling blue lakes, Maine's incredibly rockbound coast, the tide-carved pedestal rocks of the Bay of

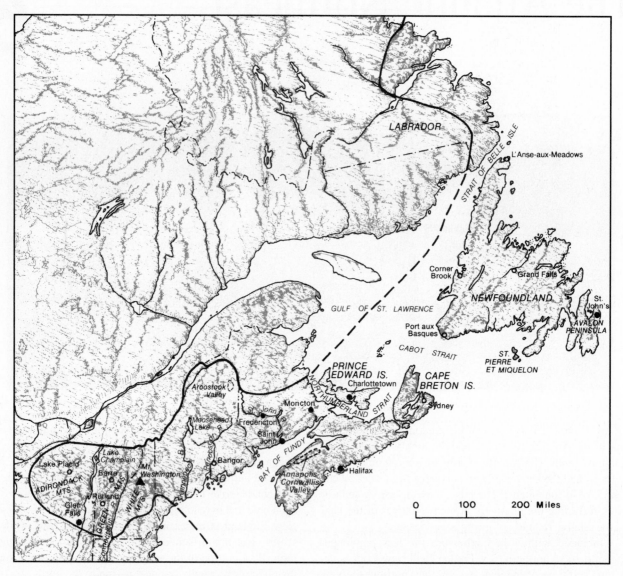

FIGURE 6-1   The Atlantic Northeast Region (base map copyright A. K. Lobeck; reprinted by permission of Hammond, Inc.).

Fundy, the classical symmetry of Prince Edward Island's farm landscape, the ordered orchards of the Annapolis–Cornwallis Valley, the splendid harbors of Halifax and St. John's, the Old World charm of Newfoundland outports. Yet this region of delightful views has suffered longer and more continually from economic handicaps than any other on the continent (Fig. 6-2).

Much of the problem is environmental. Soils are generally poor, the growing season is short and cool, the winter is long and bleak, mineral resources are scarce, and second-growth timber grows slowly. Timber, fish, and running water were the chief physical advantages, but all three have handicaps. The readily accessible forests have long since been cut, and second-growth replacement takes many decades. Moreover, various parasites have taken an inordinate toll of the timber resources in recent years. Fishing was the mainstay of the economy for a great many years and has continued in declining importance, but the sea's contribution to the regional economy has been erratic and unstable. The rushing streams of the region were early sources of hydraulic and later hydroelectric power; but their productive capacity is limited, and today the Atlantic Northeast is a power-deficient region.

FIGURE 6-2   Much of the Atlantic Northeast is still covered with forest, as in this scene near Truro in Nova Scotia (TLM photo).

This region experienced slow economic growth even in periods of national affluence. Its resources are too small and too scattered to tempt external investors. The region has been described as one of "effort" rather than of "increment."[1] Even government infusion of capital tends to be easily dispersed without causing much growth.

The population of the region is scattered, and there are only a few nodes of urban-industrial concentration. Essentially it is a region of emigration. The rates of population increase for the portions of four states and four provinces included in the region are consistently among the lowest recorded on the continent. The trend has been for Maritimers to move to Ontario, and northern New Englanders to move south. And no large influx of immigrants has occurred to compensate. Northern New England has absorbed a steady stream of French Canadians into its forest industries, but generally the region has not shared in the continent's high rate of net immigration or net population growth.

Northern New England is economically depressed, and the four Atlantic Provinces have a long record as Canada's economic problem region. The extremes are all there, including lower income per person, lower goods output per capita, lower average investment of new capital, and higher unemployment. The Atlantic Provinces as a whole average 20 percent below the per capita income for all Canada, a disparity that has persisted for decades.

Many study committees have been established in both countries to prescribe remedies for the ills of the region. Various assaults on economic deprivation have been mounted and many palliatives tried. Some local benefits have resulted from these efforts, but the region as a whole continues to suffer.

THE PHYSICAL SETTING

The Atlantic Northeast is a region of slopeland, forest, and coast; of bare rock, cold waters, and leaden skies; and of thin soils, swift streams, and implacable tides. It is a land where beavers still build dams, moose feed on lake bottoms, and Atlantic salmon come upstream to spawn.

Surface Features

The coastal area consists primarily of low rounded hills and valleys. Most of the region is traversed by fast-flowing streams, and much of it is dotted with small lakes. The coastal area has been slightly submerged; consequently, ocean waters have invaded the lower valleys, creating bays or estuaries. Often, branch bays extend up the side valleys. The

[1] David Erskine, "The Atlantic Region," in *Canada: A Geographical Interpretation,* ed. John Warkentin (Toronto: Methuen Publications, 1968), p. 233.

coast is characterized by innumerable good, if small, harbors. Superficially the coasts of Maine and Nova Scotia appear to be fiorded, but they are probably drowned normal river valleys that were slightly modified by ice action. The restricted and indented nature of much of the coastline gave rise to some of the greatest tidal fluctuations to be found anywhere

(Fig. 6-3). Many rugged and rocky headlands extend to the water's edge, especially in Maine and Newfoundland. Beaches are relatively small and scarce.

All mountains of the upland area are geologically old and worn down by erosion (Fig. 6-4). The Green Mountains of Vermont have rounded summits, the result of the great ice

FIGURE 6-3 The phenomenal tidal fluctuations of the Bay of Fundy serve as significant erosive agents on softer rocks. The 20-foot-(6-m-) high evergreen trees growing on this tide-carved pedestal rock along the New Brunswick coast give an indication of scale (TLM photo).

FIGURE 6-4 The physical landscape of this region is dominated by bare rock, green forest, and blue water. This view is of Mount Penobscot, near the coast of central Maine (TLM photo).

sheets that covered ridges and valleys alike. The highest peaks are less than 4500 feet (1350 m) above sea level. The White Mountains are higher and bolder and were not completely overridden by the ice, as evidenced by several cirques (locally called "ravines"); their highest points are in the Presidential Range, where Mount Washington reaches an elevation of nearly 6300 feet (1890 m). Northeastward the mountains become less conspicuous, although their summits remain at an approximate elevation of 5000 feet (1500 m).

The Adirondacks, although an older upland mass, underwent changes at the time of the Appalachian mountain-building movement that caused a doming of the upper surface; furthermore, they were eroded profoundly during glacial times. Although not as high as the White Mountains, the Adirondacks cover more area.

The mountains of eastern Canada are lower and more rounded, having been subdued through long periods of erosion. Their general elevation is slightly more than 2000 feet (600 m) above sea level.

The extensive upland is composed largely of igneous and metamorphic rocks—granites, schists, gneisses, marbles, and slates—so valuable that this has become the leading source of building stones on the continent.

The stream courses of the upland area were altered by glaciation. Many water bodies, such as Lake Placid, Lake Winnipesaukee, and Moosehead Lake, characterize the region. They have been of inestimable value in the development of the tourist industry.

The Aroostook Valley, occupying the upper part of the St. John River drainage, is the result of stream erosion in softer rocks. The Lake Champlain Lowland and the Connecticut Valley were severely eroded by tongues of ice that moved southward between the Green Mountains and the Adirondacks and between the Green Mountains and the White Mountains. Many valleys and lowlands are underlain by sedimentary rocks.

Newfoundland consists of a combination of moorland and forest, with an abundance of rocks, ponds, and shrubby barrens. It has been described as "a queer dishevelled region where the Almighty appears to have assembled all the materials essential to a large-scale act of creation and to have quit with the job barely begun. Ponds are dropped indiscriminately in valleys and on hilltops, rocks strewn everywhere with purposeless prodigality."[2]

Much of the island is a rolling plateau 500 to 1000 feet (150 to 300 m) above sea level, with elevations above 3000 feet (900 m) in the Long Ranges. The coastline is severely in-

[2] Edward McCourt, *The Road across Canada* (Toronto: Macmillan of Canada, 1965), p. 16.

dented; the juxtaposition of bay and peninsula is the most conspicuous feature of the littoral landscape. The coast of southern Labrador is the rugged, elevated, fiorded edge of the Canadian Shield and is strewn with small offshore islands.

Climate

The upland area lies within the humid continental climatic regime, with the Atlantic Ocean exerting little influence because of the prevailing west-to-east movement of air masses and fronts in these latitudes. The immediate coastal areas, however, experience sufficient maritime influence to have a somewhat milder and more equable climate than might be expected. Nova Scotia's mean January temperature is about the same as that of central New York despite being 2 to 5 degrees of latitude farther north. Winters, although long and cold, are not severe for the latitude. Temperatures may fall below zero, however, and snow covers the ground throughout most of the winter. Spring surrenders reluctantly to summer because of the presence of ice in the Gulf of St. Lawrence and because of the cold Labrador Current. Summers are cool, temperatures of 90°F (32°C) being extremely rare. The growing season varies from 100 to 160 days in Nova Scotia.

The precipitation of 40 to 55 inches (102 to 140 cm) is well distributed throughout the year. Southeast winds from the warm Gulf Stream blowing across the cold waters between the Gulf of Maine and Newfoundland create the summer fogs that characterize the coasts of New England, New Brunswick, and Nova Scotia. Newfoundland itself is notorious for its foggy coasts (Fig. 6-5).

The growing season is short, averaging less than 120 days. Summers are cool and winters extremely cold, temperatures dropping at times to 30° below zero Fahrenheit (−34°C). The abundant precipitation is evenly distributed throughout the year.

In Newfoundland and south coastal Labrador the climate is largely the result of a clash between continental and oceanic influences, with the former dominating. The winters are much colder than those in British Columbia or Britain in the same latitude. Altitude is a factor, as shown by the replacement of forest by tundra at elevations exceeding about 1000 feet (300 m).

No point in Newfoundland is more than 90 miles (144 km) from the sea, but the ocean's relative coldness and the prevailing westerly winds, which bring continental influences, do not permit much amelioration of the temperatures. In the Gulf of St. Lawrence all harbors freeze over in the winter, the Strait of Belle Isle sometimes being completely blocked by ice. The bays on the Labrador coast and large areas of the adjacent sea freeze solid by October or November.

A CLOSER LOOK   The Stormiest Mountain in North America

Mount Washington, in the Presidential Range of New Hampshire's White Mountains, is the highest peak in northeastern United States, with a summit elevation of 6288 feet (1917 m) above sea level. Although there are many peaks higher in both altitude and latitude on the continent, Mount Washington is a virtually unanimous choice as the stormiest mountain in North America.

On the peak of Mount Washington average snowfall exceeds 20 feet (6 m), and in some years more than twice that amount is received. The annual average temperature is below freezing, with an absolute minimum of 47° below zero Fahrenheit (−44°C). Only once in history has the temperature on Mount Washington exceeded 70°F (16°C).

It is the wind, however, that gives Mount Washington a special place in weather lore. Year-round the wind blows at an average gale-strength of 35 miles per hour (56 km/h). It exceeds hurricane strength [74 mph (119 km/h)] nearly one-third of the time. And for a few minutes on April 12, 1934, anemometers at the Mount Washington Observatory recorded steady winds of 231 mph (372 km/h)—the strongest winds ever measured anywhere.

Despite this daunting combination of wind, cold, and storminess, Mount Washington is readily accessible for much of the year via an automobile toll road and a steam cog railway. It has been a premier cold-weather testing laboratory for science and technology for more than a century. It has become the world's premier study site for deicing technology, and it is used as a natural laboratory for cloud physics research and for development and testing of instruments, aircraft components, and structures required to withstand high winds and icing conditions.

TLM

FIGURE 6-5   Fog is commonplace in the region, particularly in Newfoundland. This is Logy Bay on the Avalon Peninsula (TLM photo).

Summers everywhere are cool because the Labrador Current, laden with ice floes and icebergs, moves southward along the east and south coasts.

Natural Vegetation

Most of the land included within this region was once covered with trees. The principal treeless localities were small areas of dunes, marshes, meadows, bogs, and exposed mountain summits except in Newfoundland, where only one-third of the island is forested because temperature, wind, and moisture conditions are unfavorable for tree growth.

The New England section was originally covered with a relatively dense forest of mixed deciduous and coniferous species. Even today more than 80 percent of the land is forested by a mixture of northern hardwoods, with white pine, spruce, and fir. Originally white pine was the outstanding tree; attaining a height of 240 feet (73 m) and a diameter of 6 feet (2 m) at the butt, it dwarfed even the tall spruce. It was sometimes called the "masting pine" because the larger trees were marked with the Royal Arrow and reserved for masts for the Royal Navy. Maine is still referred to as the "Pine Tree State." Most of the existing forest is second-growth; the original timber was cut for lumber or fuel or cleared for agriculture many decades ago. Only in the more remote parts of the uplands, primarily in Maine, are there still virgin stands of trees.

In the Maritime Provinces a mixed forest cover is also still widespread. Although forest has been thoroughly removed from Prince Edward Island, most of the land is still forested in Nova Scotia; forest clearing was even less common in New Brunswick. Some large areas of relatively pure hardwood or softwood species may be found, but mixed growth is much more common. Spruce, hemlock, fir, pine, maple, and birch are the typical trees.

The forests of Newfoundland are much more predominantly coniferous, with balsam fir and black spruce dominating. Birch, poplar, and aspen are the principal hardwood species. Nowhere are the trees large; as a result, little lumbering is carried on, although pulpwood cutting is important. The coast of Labrador is almost completely lacking in forest; but the sheltered stream basins support some tree growth.

Soils

Owing to differences in parent rock, slope, drainage, and previous extent of glaciation, the soils of the Atlantic Northeast are varied; nevertheless, the dominant soils throughout the region are Spodosols. The only other soil order represented importantly in the region is Alfisol. These are soils that developed under cool, moist conditions and are thus leached and acidic in nature. Usually there is a layer of organic accumulation near the surface and, more characteristically, considerable accumulation of compounds of iron and aluminum. A layer of clay accumulation is often found in the Alfisols of the valleys.

Agricultural productivity of these soils is undistinguished. Those that are derived from shales, especially in New Brunswick and northern Nova Scotia, are heavy and poorly drained. The sedimentary floored lowlands give rise to soils that are more fertile for farming; the shales of Vermont and the sandstones of the Annapolis Valley and Prince Edward Island yield more productive agricultural soils than are found elsewhere. Rockiness, poor drainage, and peat formation are major and widespread handicaps to crop and pasture development. Alluvial soils, occurring on narrow-valley floodplains, are extremely important to agriculture even though their total acreage is small.

Fauna

There is nothing out of the ordinary about the wildlife in most of the region; it has a "north woods" environment that contains a predictable faunal complement. One anomalous situation does prevail in Newfoundland. This large island was singularly lacking in a number of the common mainland terrestrial species. There were, for example, almost no rodents, especially the smaller varieties. Three very characteristic denizens of the north woods, porcupine, mink, and moose, were also missing. To remedy these deficiencies, exotic animals were introduced at various times during the past century. The most spectacular success resulted from the introduction of six moose about the turn of the century; the moose is now more numerous and widespread than the native caribou and has yielded more than 100,000 legal kills to hunters. Chipmunks and mink were also successfully introduced, and shrews were brought to Newfoundland to control a larch-destroying sawfly.

In recent years the moose has expanded its range naturally in New England, moving south and west. Maine and New Hampshire now have large moose populations, and even Massachusetts harbors enough moose so that the opening of a hunting season is being considered.

SETTLEMENT AND EARLY DEVELOPMENT

The original inhabitants of the Atlantic Northeast Region were mostly Indians of the Algonquian language family. The Micmac tribe was dominant throughout the Maritimes, where some of their campsites dated to 5000 years ago, and on the south coast of Newfoundland, where they apparently settled in the 1700s. Beothuk Indians lived elsewhere in Newfoundland; they were wiped out in the early 1800s by starvation, disease, and fighting with European fishermen. A variety of other Algonquian tribes lived in northern New England; most were village dwellers, but some were seminomadic.

The New England Segment

The first important settlement in what is now New England was founded at Plymouth in 1620. All the early colonies, including those before and immediately after the landing of the Pilgrims, were planted on the seaboard. The coast was, then, the first American frontier. Its settlements were bounded by untamed hills on the west and by the stormy Atlantic on the east. Beckoned by the soil and the sea, its shore-dwelling pioneers obeyed both, and their adjustments to the two envi-

A CLOSER LOOK   Acid Rain

One of the most vexing and perplexing environmental problems is the rapidly increasing intensity, magnitude, and extent of *acid rain*. This term refers to a phenomenon that involves the deposition of either wet or dry acidic materials from the atmosphere on the earth's surface. Although most conspicuously associated with rainfall, the pollutants may fall to earth with snow, sleet, hail, or fog, or in the dry form of gases or particulate matter.

Sulfuric and nitric acids are the principal culprits recognized thus far. Although there is no universal agreement on the exact origin and processes involved, evidence indicates that the principal human-induced sources consist of sulfur dioxide emissions from smokestacks, and nitrogen oxide exhaust from motor vehicles. These and other emissions of sulfur and nitrogen compounds are expelled into the air, where they may be wafted hundreds or even thousands of miles by winds. During this time they may mix with atmospheric moisture to form sulfuric and nitric acids that sooner or later are precipitated (Fig. 6-A).

Acidity is measured on a pH scale based on the relative concentration of active hydrogen ions. The scale ranges from 0 to 14; the lower end representing extreme acidity (battery acid has a value of 1) and the upper end extreme alkalinity (lye has a value of 14). It is a logarithmic scale, which means that a difference of one whole number on the scale reflects a 10-fold increase or decrease in absolute values. Rainfall in clean, dust-free air has a pH of about 5.6; thus it is slightly acidic because of the reaction of water with carbon dioxide to form a weak carbonic acid. Increasingly, however, precipitation with pH less than 4.5 (the level below which most fish perish) is being recorded, and an acid fog with a record low of 1.7 (8000 times more acidic than normal

rainfall) was experienced in southern California in 1982.

Many parts of Earth's surface have naturally alkaline conditions in soil or bedrock that buffer or neutralize acid precipitation. In most of the Atlantic Northeast, however, the buffering capability is limited because of shallow soils and crystalline bedrock. Consequently, the deleterious effects of acid rain were first detected in this region. Later the problem spread much more widely.

The most conspicuous damage is done to aquatic ecosystems. Several hundred lakes in eastern Canada and the United States have become biological deserts in the last three decades, primarily due to deposition from acid rain (Fig. 6-B).

One of the great complexities of the situation is that much of the pollution is deposited at great distances from its source. Downwind locations receive unwanted acid deposition from upwind

FIGURE 6-A   Acid rain is not readily recognized by the average person (courtesy *Alberta Environment Views*, March/April 1982, p. 3).

ronments laid the foundations for the land and sea life of the nation.

As population became more dense in maritime New England and Canada, the more venturesome settlers trekked

farther into the wilderness. As long as the French controlled the St. Lawrence Lowlands, the Indians of the upland remained entrenched in this so-called neutral ground, thus restricting white settlements to the seaboard. In the

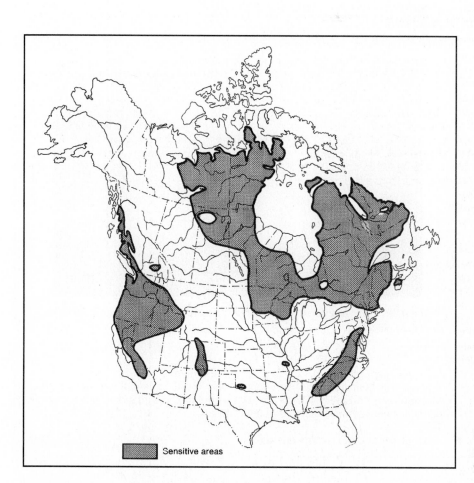

FIGURE 6-B Areas particularly sensitive to acid rain due to a scarcity of natural buffers.

sources. It is known, for example, that the Midwest is a major source of toxic emissions that cause damage in the Northeast. But it is understandably difficult to persuade people in Illinois to finance expensive cleanup costs that will benefit forests and lakes in Maine.

One of the thorniest issues in North American international relations is the Canadian dissatisfaction with the U.S. government's approach toward mitigation of acid rain. In general, acid rain is viewed by Canadians as their gravest environmental problem, and it is believed that about half the acid rain that

falls on Canada comes from U.S. sources, particularly older coal-burning power plants in the Ohio and Tennessee River valleys.

In Canada, the federal and provincial governments have bitten the bullet by legislating a requirement for 50 percent reduction in acid rain emissions. In 1990 the United States finally promulgated comparable legislation with amendments to the Clean Air Act. This paved the way for a bilateral agreement between the United States and Canada, signed in 1991, that codified the principle that countries are responsible for the

effects of their air pollution on one another. Specifics of the agreement include a permanent cap placed on sulfur dioxide emissions in both countries, a schedule for reducing emissions of nitrogen oxides, and stricter emission standards for new motor vehicles.

It is estimated that this program will require an annual expenditure of about $4.5 billion in the United States and about $500 million in Canada. Although these are awesome numbers, it is clear that the costs of not developing such a program would be exponentially greater.

TLM

Adirondacks hostile Iroquois kept the English confined to the Hudson and Mohawk Valleys until the close of the Revolutionary War. Feeling that at last the power of the Indian had been broken after the conquest of Canada by the British, pioneers from the older parts of New England penetrated the upland. By the 1760s most lower valleys in New Hampshire and Vermont were occupied. The clearing of the forest for farms led to an early development of logging and lumbering,

A CLOSER LOOK   L'Anse-aux-Meadows—North America's Earliest European Settlement

Long before Columbus made his historic voyage to the New World in 1492, Norse Vikings had "discovered" North America by voyages from their colonies in Iceland and Greenland. Although there are many hints and inferences of a Viking presence as far south as Virginia and as far west as Minnesota, the only authenticated site of a Norse settlement in North America is near the most northeasterly point of the island of Newfoundland, at a place called L'Anse-aux-Meadows ("the cove of the jellyfish") on Epaves Bay.

Discovered in 1960, the locale is set aside as a Canadian National Historic Park and has been declared a World Heritage Site. The site contained the foundations of seven sod homes, a smithy, four small boat houses, and associated work sheds and cooking pits (Fig. 6-C). The houses and artifacts are clearly of Norse origin, and carbon dating established that they were from about the year A.D. 1000, the time of Viking exploratory voyages.

The settlement apparently was not of long duration, probably serving as an over-wintering station and used as a

FIGURE 6-C   A restored Viking hut at L'Anse-aux-Meadows, Newfoundland (TLM photo).

base for exploring and seeking resources. It is believed that L'Anse-aux-Meadows was occupied on from 2 to 10 occasions between the years 1000 and 1011.

Until some other settlement site is discovered, L'Anse-aux-Meadows is the only viable candidate for identification as the fabled Vinland of Leif Ericsson's Sagas.

TLM

which could be carried on in winter, when farm work was not feasible. The logs, dragged on the snow to frozen streams, were floated to mills in the spring when the ice melted; thus a supplemental source of income was provided for the pioneers, and it continued to be important until the latter part of the nineteenth century.

Maritime Canada

The first permanent settlement in North America north of Florida was at Port Royal on the Bay of Fundy in 1605. Here the French found salt marshes that needed no clearing. This environment was attractive to people from the mouth of the Loire, whose forebears for generations had reclaimed and diked somewhat similar land. These French called their new home *Acadie* (Acadia). The Acadians converted the river marshes into productive farmland that characterized the cultivated area almost exclusively for a century.

The French population grew rapidly after the Treaty of Utrecht in 1713. Louisburg was fortified to guard the mouth of the St. Lawrence and the fishing fleet. The Acadians remained until the outbreak of hostilities between the British and French preceding the Seven Years' War in the 1750s. Then more than 6000 were rounded up and banished; they were scattered from Massachusetts to South Carolina. Some fled into the forests of what are now New Brunswick, Prince Edward Island, and Quebec; some made their way to Quebec City, the Ohio Valley, and even to Louisiana; and others joined the French in St. Pierre, Miquelon, and the West Indies. Many starved. The reason most commonly given for their expulsion was fear on the part of England that such a heavy concentration of French in this part of the continent was a menace to English safety.

The first significant arrival of British settlers in the Atlantic Provinces was in 1610 in Newfoundland. Despite a hard and lonely life, oriented almost exclusively toward export of salted codfish, increasing numbers of English and Irish settlers occupied the numerous bays of southeastern Newfoundland through the seventeenth century. By 1750 British settlers and New Englanders were beginning to come in large numbers to Nova Scotia and New Brunswick. Shortly afterward 2000 Germans founded Lunenburg. Large migrations of Scottish people, principally from the Highlands, came after 1800, dominating the population of Nova Scotia and, to a lesser extent, Prince Edward Island.

THE PRESENT INHABITANTS

There is a remarkable homogeneity to the contemporary population of the region. The vast majority of the people are of white Anglo-Saxon ancestry. English is the only language of importance except along the northern margin of the region and in scattered pockets of French settlement; the religious affiliation over large areas is dominantly Protestant except in locales of French or Irish Catholic settlement; and most of the people who live in the region were born in the region, often the offspring of several generations in the region.

The taciturn, traditionally conservative Yankee stereotype is still dominant in northern New England. Yankees descended from antecedents who during a "period of poverty and struggle . . . beat down the forest, won the fields, sailed the seas, and went forth to populate Western commonwealths."[3] Similar British stock, often referred to as Loyalists, originally peopled much of the Atlantic Provinces of Canada and remains dominant today.

The population of the Atlantic Provinces is somewhat less uniform than that of northern New England because large blocks of immigrants settled in groups and generally tended to occupy the same areas for generations. The origins are usually English, Scottish, or Irish. For example, in eastern Newfoundland in general and the Avalon Peninsula in particular the population is about 95 percent Irish Catholic except for the city of St. John's. Similarly, most of Cape Breton Island is a stronghold of Highland Scots, and people of Scottish ancestry predominate throughout northeastern Nova Scotia, eastern New Brunswick, and Prince Edward Island.

The French component of the population, so important in the early days of Acadian settlement, is today largely marginal. A large proportion of the New Brunswick population is of French origin, but their dominance is in the northern part of the province, which is considered to lie outside this region.

With increasing distance from Quebec, the French element in the New Brunswick population rapidly decreases. Along the northern margin of New England, from Maine's Aroostook Valley to the northeastern New York counties of Clinton and Franklin, there are a significant number of French Canadians who migrated across the international border to work in forests and mills. Wherever the French live in the region, the traditional manifestations of their culture, typified by the French language and the Roman Catholic religion, are prevalent.

As a general rule, cities of the Atlantic Northeast have a more varied populace than do small towns and rural areas. The larger the urban place, the more cosmopolitan its population is likely to be; thus all cities of the region have various minority ethnic groups, although their numbers are few.

The people of this region are mostly longtime residents. Emigration takes place to Megalopolis, Toronto, and urban Alberta, but there is little immigration to balance it. This fact can be illustrated by comparing the birthplaces of the people of Atlantic cities with those of other Canadian cities. Of the larger Canadian metropolitan areas, only one (Quebec City) has a higher proportion of its population born in the same province than do the three Atlantic cities of St. John's, Saint John, and Halifax.

Much of a region's character is asserted by the attitudes, traditions, and values of its people and by the imprint of these intangibles on the regional landscape. The Atlantic Northeast is singularly rich in this regard. Although the elements are difficult to define, and the cause-and-effect relationships are obscure, some of its manifestations are conspicuous. As Edward McCourt has observed about "Newfies," the people of this region "have evolved their own mores, created their own culture, made and sung their own songs, and added to the language according to their need."[4]

One aspect of the cultural character of the region relates to the form and charm of the small units of settlement, the hamlet and village. The name of the village is likely to sound as pleasant as it appears. Thus travelers in central New Hampshire can stop at the hamlet of Sandwich; if unsatisfied there, they can drive on to the next village, Center Sandwich; or the next, North Sandwich. Vivid place names persist throughout the region. Consider a selection from Newfoundland: Harbour Grace, Heart's Delight, Witless Bay, Maiden Arm, Uncle Dickies Burr, Come By Chance, Mount Misery, Hit or Miss Point, Right-in-the-Road Island, Holy Water Pond, Damnable Bay, Lushes Bight, Cuckhold's Head, Bleak Joke Cove, and Hug My Dug Island.

[3] J. Russell Smith and M. Ogden Phillips, *North America* (New York: Harcourt, Brace & Co., 1940), p. 113.

[4] McCourt, *The Road Across Canada*, p. 30.

The village form has a certain uniformity: the fishing hamlet clustered on a tiny beach about the head of a bay and the farming town rectangularly arrayed around the village green. Actually, village greens are present in almost every town in northern New England but very uncommon in Canada's Atlantic Provinces. White, tall-spired churches with their adjacent cemeteries are, however, the dominant edifices in villages throughout the region.

The rural landscape also has characteristic regional forms. The covered bridge, for example, is a famous feature in northern New England (Fig. 6-6) and is often carefully cultivated as a tourist attraction. Covered bridges are common throughout the Maritimes and are more prevalent in New Brunswick than in any part of New England. House-and-barn arrangements are also interesting. The attached house and barn is notable in New Hampshire and Maine. In Vermont, on the other hand, the house and barn are often located adjacent to the road but on opposite sides, presumably to take advantage of the snow-clearing efforts of the road crews.

AGRICULTURE

Although Newfoundlanders were fishermen from the start, farming usually was the first occupation of settlers elsewhere in the region. In most places it was small-scale and difficult but was long a dominant activity, reaching its peak in about the 1820s and diminishing greatly in relative importance after then (Fig. 6-7).

New England

Largely because of the huge nearby urban markets in Megalopolis and the relatively rugged physical geography of northern New England, dairy farming has been the principal farm activity for more than a century, and will likely continue to be so for many years to come. With excellent transportation facilities, the emphasis is on whole-milk production. In recent years there has been considerable increase in dairying in northern Vermont, western New Hampshire, and central Maine.

In addition to hay, corn is widely grown as a feed crop for the dairy herds. It is mostly cut green for silage. The better soils of Vermont are reflected in corn production statistics; Maine and New Hamphire combined yield only two-thirds as much silage corn as Vermont.

Poultry farming, long a notable farm activity in the subregion, has declined precipitously during the last two decades. Total output has diminished by about half during this period, although broiler raising is still prominent in central Maine, and egg production is widespread (Fig. 6-8).

FIGURE 6-6   A covered bridge over a tidal stream in southern New Brunswick (TLM photo).

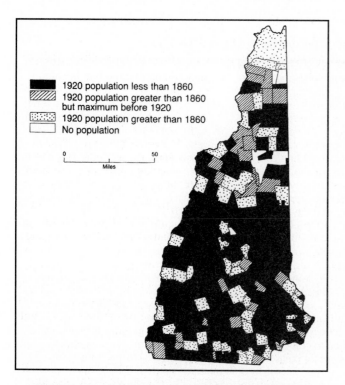

FIGURE 6-7 The decline of agriculture in the nineteenth century in New England was accompanied by population decrease. This map, showing population change in New Hampshire from 1860 to 1920, is representative of all of northern New England (after William H. Wallace, "A Hard Land for a Tough People: A Historical Geography of New Hampshire," *New Hampshire Profiles*, April 1975, p. 28).

Aside from corn, field crops are reltively minor on the farm scene except in a few specialty areas. Berries and apples are moderately widespread, and there has been a continuing modest increase in nursery and greenhouse products. On the

negative side, potato growing, traditionally the largest single component of crop farming in New England, has experienced an enormous decline. Potato acreage has dropped by more than 50 percent in the last two decades, with the largest part of the decrease in northern Maine, an area long noted for potato monoculture.

The Atlantic Provinces

Cultivated land in Nova Scotia lies mostly in the western lowland, the granitic interior and the Atlantic Coast tending to discourage agriculture. The only outstanding area of commercial farming is the Annapolis–Cornwallis Valley in the southwest, which produces a large proportion of Canada's leading fruit crop, the apple (Fig. 6-9) The output is mostly sold as processed apple products rather than being marketed fresh. Forage crops, vegetables, and potatoes are secondary crops that are widely grown. Nova Scotia is also notable for its output of wild blueberries, most of which are found in Cumberland County (the county adjacent to the New Brunswick border). Half of Canada's total blueberry production is in Nova Scotia, providing jobs for 6000 people.

In New Brunswick the area in farms is about equal to that in Nova Scotia, but the area in field crops is nearly twice as great. That the province is not outstanding agriculturally seems proved by the fact that forest still covers nine-tenths of the land. An exception is the St. John River valley, which is farmed as intensively as any area in the Atlantic Northeast Region. The upper part of the valley is a major potato-growing area, whereas the lower part concentrates on forage and feed crops for the numerous local dairy herds.

About two-thirds of the total area of Prince Edward Island is in farms, a much higher ratio than in any other political unit in North America. The countryside is a delight to the eye, with alternation of field and woodlot, unbelievably green

FIGURE 6-8 Enormous broiler "hotels" are not unusual in northern New England. This is near Waterville, Maine (TLM photo).

A CLOSER LOOK   The Decline and Rebirth of a New Hampshire Hill Town

The historical geography of Sandwich, New Hampshire, typifies the changing character of New England's hill towns over the past 250 years. Sandwich is located on the southern flank of the White Mountains, almost exactly in the center of the state. The southern half of the town is composed of rolling hills ranging from 500 to 1000 feet (150–300 m) above sea level.[a] Mountains with relief ranging from 1000 to 3000 feet (300–900 m) rise to the north and south.

The township of Sandwich was granted by the provincial governor in 1763, but settlement did not begin until 1767. Like most towns in central New Hampshire, Sandwich reached its maximum population in 1830, when it had 2744 residents. But the peak of agriculture did not come for another half-century. The amount of land improved for farming reached its maximum in 1880, when it amounted to 51 percent of the area of the town. Rapid decline began in the decade that followed and continued for the next half-century. By 1930 only 9 percent of the land was being used for agriculture. This was the era of aging farmers, decaying houses, and fields reverting to woods. More than 300 farmhouses were abandoned, many

[a] In New England, the term "town" refers to what in other regions of the United States is called a township. Land in New England was initially granted in rectilinear parcels, most of which approximated 36 square miles, which were designated as "townships." As time passed, "township" was shortened to "town." The town was also the basic governmental unit in colonial New England and remains so today. Thus, a New England town is a political unit that usually contains forests, farms, and villages, although some towns such as Boston and Hartford have become cities.

roads were discontinued, and entire neighborhoods were abandoned as people moved away or died.

A recurrent theme in the sequent occupance of any place is that each phase of its development sets the stage for the next. In Sandwich the decline of agriculture set the stage for the recreation and summer resident activity that were to dominate the twentieth century. Land values fell so greatly that city folk could buy land and houses at very low prices. This was the circumstance that permitted middle-class businesspersons and professionals to acquire old farms and houses for summer use, something that would not have been possible had farming remained viable.

Valuing Sandwich for its aesthetics rather than its productivity had begun by 1850. Travelers began to come to its hotels not for business but for pleasure and to escape the heat, noise, and smells of the rapidly growing cities. As the hotel trade grew, the opportunity to profit from summer visitors by providing food and lodging was perceived, and Sandwich entered the boarding house era. At the turn of the century, Sandwich had at least 30 boarding houses catering to a largely white-collar, middle-class clientele. Those who came to Sandwich to board, liked it, and could afford it, acquired or built houses for summer residence. So began a practice that became an increasingly important feature of life in Sandwich throughout the following century. The rescue and restoration of old houses began in 1890, and by 1910 some 20 houses were owned by summer residents. Contemporaneous with the restoration of old houses was the erection of new dwellings built specifically for summer use. The earliest of these were built in 1891, and by 1910 some

20 new summer homes had been built. At the time of World War II, Sandwich had about 150 summer homes (Fig. 6-D).

The population of Sandwich reached its nadir with 615 people in 1950, little more than one-fifth the 1830 maximum. Numbers grew slowly in the 1950s and 1960s, but the next two decades brought the largest 20-year increase in 160 years, and in 1990 population reached 1019.

Agriculture has continued its century-long decline. Dairy and poultry farming, long the mainstays of Sandwich agriculture, disappeared during the 1970s and 1980s. Today, only horticulture and horse farms remain viable. Farm land has declined to less than 2 percent of the area of the town, and open land remains only in small patches along the roads and around the villages, like beads on a string. Most of the remaining fields are kept open, not for farming, but to preserve a view. Indeed, a significant number of fields have been newly cleared for this purpose in the past two decades.

The second half of the twentieth century may appropriately be called the era of recreation. In Sandwich, summer residents, an important element since the beginning of the century, have been joined by an increasing number of weekenders and short-term visitors. But lacking significant public accommodation and eschewing large-scale resort development, Sandwich has continued in a pattern established long ago: summer homes. The number of seasonal houses increased fourfold between 1940 and 1970 and amounted to 58 percent of all housing. Growth in seasonal houses became less important in the 1980s as two other groups began to have a major impact on the town.

cultivated land, neat farmsteads, and a frequent view of water (Fig. 6-10). The only significant commercial crop is potatoes; Prince Edward Island is one of the two leading potato-growing areas in Canada. Dairying is the only other farm activity of note.

Agriculture in Newfoundland is quite limited; it provides only 1 percent of the province's gross domestic product (GDP). There are about 750 farms in Newfoundland, half of

them small, and most located near St. John's. Dairying is by far the leading farm activity.

FOREST INDUSTRIES

The Atlantic Northeast, the subcontinent's pioneer logging region, possessed an almost incomparable forest of tall

FIGURE 6-D   The Harrison Atwood house (c. 1800) at the foot of Mt. Israel (2,630 feet or 877 m) in Sandwich, New Hampshire. This place was farmed for 150 years but had become a summer residence by the 1930s. Today it is the year-round residence of former Bostonians who operate a market garden and do maintenance work for retirees (Photo by Bill Finney; courtesy of the Sandwich Historical Society).

Retirement became a major force shaping Sandwich in the years that followed World War II. Many who first came as summer visitors decided to spend their last years there, and by 1990 retirees made up nearly a quarter of the town population, demonstrating that it is not only the Sunbelt that appeals to the elderly. Retirees have brought with them a large infusion of capital and income that has greatly stimulated the local economy, increasing the demand for all kinds of goods and services, and contributed importantly to a building boom that has employed many local artisans.

Not all those who have moved to Sandwich seeking the good life are retired. The other major change of the past quarter century has been an influx of young people, people in their productive years. Job opportunities have improved dramatically since the 1950s, allowing those who grew up in Sandwich to stay and a large number of newcomers to find jobs. Although the economy of Sandwich has expanded, the growth of the regional economy has been more important. Today more than half of the labor force travels out of town to work, and like commuter towns everywhere, Sandwich has a morning and evening rush hour.

A recurrent theme of human geography is that the same area means different things to different people. Sandwich, a small town in the hills of New Hampshire, illustrates the differing value placed on the land by Yankee farmers, summer folk, retirees, and working people seeking the good life. We should also be reminded that each phase of the sequent occupance of Sandwich has set the stage for the next.

Professor William Wallace
University of New Hampshire
Durham

straight conifers and valuable hardwoods. Perhaps nine-tenths of the region was covered in forests. For 200 years or more after the landing of the Pilgrims in 1620, the settlers uninterruptedly continued the removal of trees.

The heyday of forest industries in New England is long since past. There is still considerable woodland, but much of the good timber is inaccessible and much of the accessible timber is of poor quality. Even so, a moderate quantity of sawtimber is cut each year, although no state in the region ranks among the 20 leaders in annual lumber production. Pulpwood production is more significant; in most years only a half dozen states yield more than Maine.

Most forest land in Maine is owned by large corporations. In the northern two-thirds of the state some 75 percent of the land is in only 20 ownerships. The operations of the logging companies enhanced forest recreation by permitting

FIGURE 6-9 An apple harvest scene in the Annapolis Valley of Nova Scotia (photo by Winston Fraser).

FIGURE 6-11 Clearcutting is now a widely used logging technique in northeastern North America. This area is near Fredericton, New Brunswick (TLM photo).

FIGURE 6-10 The rural landscape of Prince Edward Island consists mostly of green meadows and pastures, interspersed with small clumps of trees. This scene is near Kensington (TLM photo).

increased accessibility over logging roads, improved wildlife habitats by opening up the dense forest stands, and provided artificial lakes. But the damming of Maine's few remaining wild rivers is a source of great controversy among conservationists, and debates over the value of wildland development policies will probably continue.

After a long period of decline, lumber production is now increasing in northern Maine, although pulpwood is still the principal product. Much of the logging operation shifted to virtual clearcutting in the 1980s (Fig. 6-11), largely as a re-

sponse to the increasing depredations of the spruce budworm, which reduces a mature fir (its preferred host) to a dead stick in 5 or 6 years.[5] Throughout the Atlantic Northeast budworm infestations are at abnormally high and growing levels.

Logging is also significant in the Canadian portion of the region, especially in New Brunswick, which produces nearly half of all the lumber and pulpwood in the Atlantic Provinces. Moreover, its sawmills and pulp mills are generally larger than those in the other provinces. Nova Scotia's forest land, unlike that of New Brunswick and Newfoundland, is largely in private ownership; there is a roughly even balance between output of lumber and pulpwood, with processing primarily in small mills. In Newfoundland most of the productive forest land is owned or leased by three large pulp and paper companies; commercial output is heavily oriented toward pulpwood (Fig. 6-12), although there is also a great deal of subsistence logging for firewood. The limited forest land of Prince Edward Island is found chiefly in small, individually owned woodlots. For the Atlantic Provinces as a whole, the cutting of pulpwood is nearly twice as great as timber harvesting for all other purposes combined.

FISHING

Fish was the first export from the New World. From Newfoundland to Cape Cod lie offshore banks that were one of the richest fishing grounds in the world. These banks were frequented by Scandinavian, Portuguese, Dutch, English, and French fishermen before the period of colonization in America. As early as 1504 Breton and Norman fishermen were catching cod in the western North Atlantic; and by 1577

[5] Bret Wallach, "Logging in Maine's Empty Quarter," *Annals,* Association of American Geographers, 70 (December 1980), 551.

FIGURE 6-12 A wood-chipping machine in action near Corner Brook, Newfoundland (TLM photo).

France had 150 vessels, Spain 100, Portugal 50, and England 15 fishing for cod on the banks.[6]

Early fisheries concentrated largely on cod, which was salted, pickled, and especially dried for export to European,

[6] R. H. Fiedler, "Fisheries of North America," *Geographical Review,* 30 (April 1940), 201.

largely Mediterranean, and tropical markets. Maximum production of dried salted cod was reached in the 1880s, but changing fishing conditions and market requirements caused a gradual decline after that. During the present century freezing replaced drying, salting, and pickling as the major technique of preserving fish enroute to market.

The Fisheries

Although the regional catch includes a great variety of species, lobster is by far the most important. Lobster is the most valuable fishery product in every political unit in the Atlantic Northeast. Lobsters traditionally have been caught in baited traps (called pots) as they crawl around in shallow water looking for food (Fig. 6-13). Traps are still much in use, but an increasing share of the total catch is taken by deep trawling. Open season on these crustaceans is very short, typically only 2 or 3 months per year.

Groundfish (demersal or bottom-feeding species)—such as cod, haddock, pollock, and turbot—mostly are taken by trawling, although a considerable quantity are caught in shallow coastal waters by hook and line and by the use of traps. Pelagic (surface-feeding) species, such as herring and mackerel, are caught with a variety of techniques, but particularly with seines.

Crabs usually are taken, like lobsters, in baited pots. Most other shellfish (shrimp, scallops, clams, oysters, etc.) are trawled.

FIGURE 6-13 Piles of lobster pots awaiting the beginning of the season on a coastal pier in Maine (TLM photo).

Since the mid-1980s, aquaculture has become an increasingly important aspect of the commercial fishing scene. Atlantic salmon have been the big success story in regional aquaculture. In about a decade they had become a $100 million a year industry in the sheltered waters around the southern tip of New Brunswick, where more than 150 producers raise salmon in shallow-water holding pens. In the mid-1990s the first cod farms began operations, and their success (or failure) will be watched with great interest. Lobster aquaculture is also developing. By the early 1990s there were several dozen locations where lobsters in "pound" were held captive in the sea. The latest development is the "dryland pound," wherein lobsters are retained indoors in drawers, in near-freezing seawater, for up to a year; this allows fresh lobsters to be marketed year-round.

The Special Case of Newfoundland

Newfoundland, from its earliest days, was a staple product colony, the staple being codfish. For three centuries the bulk of the populace consisted of fishermen. They sold their catch to local merchants, who passed it on to exporters in St. John's, who depended on distant and unstable markets. The result was a precarious economy with continuing poverty for most fishermen and an uncertain opulence for the St. John's exporters. Catching cod in inshore waters remained the unchallenged base for Newfoundland's limited economy until the end of the nineteenth century.

In this century Newfoundland's seaward outlook weakened, and the contribution of fishing to the provincial income declined steadily. However, until the early 1990s there were still thousands of inshore fishermen in Newfoundland, living primarily in small villages (called *outports*) at the heads of innumerable bays scattered around the island (Fig. 6-14). Most of their fishing is done within the bays or slightly beyond them; they set cod traps in the bay mouths, use a variety of nets, and used hook and line farther out. Before being marketed in Europe, most of the cod was cleaned and then dried on racks (called *flakes*) in the uncertain Newfoundland sun. In more recent years, however the great bulk of the catch has been filleted, frozen, and sold in the United States.

The commercial fisheries of the Atlantic Northeast have a long history of erratic instability, which shows no sign of changing. After three decades of continuing decline, a revival was triggered in the late 1970s, particularly in Canada, when the governments of the United States and Canada declared a 200-mile offshore jurisdiction, which significantly reduced foreign competition in this rich fishing area, especially over most of the Grand Banks, which are now under Canadian control.[7]

By the beginning of the 1990s, however, a serious decline in numbers, particularly of the once abundant cod, caused another major retrenchment in the industry. Apparently, Canada's Atlantic fishing industry had become too efficient. Sophisticated fishing vessels were catching many more fish than the fishery could handle. Most of the problem was generated by offshore draggers that tow a huge bag of net along the bottom, scooping up everything in its path. A

[7] In 1984 a long and acrimonious controversy between the United States and Canada over the "ownership" of Georges Bank (second most valuable of the fishing banks) in the Gulf of Maine was resolved by the International Court of Justice. The compromise solution gave the United States control of about 75 percent of the prolific waters, with the other 25 percent going to Canada.

FIGURE 6-14 A representative Newfoundland outport, The Battery (TLM photo).

dragger collects much more than the target fish (the "by-catch" mostly is killed in the operation and is simply discarded); it also damages the sea bottom environment in the process.

In 1992 it was announced that the Atlantic cod (*Gadus morhua*) stock, Newfoundland's lifeblood for more than four centuries, had collapsed. The government declared a 2-year moratorium on commercial cod fishing, which has since been extended indefinitely. Canada has established a Fisheries Resource Conservation Council to review and implement all necessary conservation measures for East Coast Canadian fisheries. The Council has mandated a variety of other closures, especially for groundfish. By 1994 similar closures for cod, haddock, and flounder fisheries were imposed for New England fishermen, and much of Georges Bank and waters off Cape Cod were totally closed to commercial fishermen.

This economic catastrophe has put some 50,000 Canadians and several thousand Americans out of work. Newfoundland, which already had the lowest per capita income among the provinces, has been hardest hit. About one-fourth of Newfoundland's work force is unemployed, and in many of the outports the unemployment rate exceeds 75 percent. The federal and provincial governments have initiated several short-term income replacement and retraining programs, but the long-term outlook is grim.

MINING

There is a long history of mining in the region. The mineral industries have been cursed with irregular productivity, however, and prosperity has been limited. Still, mining is an important contributor to regional income, some notable expansion of output has occurred in recent years, and the immediate prospects of offshore petroleum production are impressive.

Coal has been mined in Nova Scotia for more than a century, chiefly on the north shore of Cape Breton Island. The last four decades, however, have been a period of erratic decline.

Another traditional mining activity of the region is the production of *iron ore* on the flanks of the Adirondacks, where initial output began about 1800. Production has been erratic for the last century, but considerable reserves of magnetite are still present.

Perhaps a dozen mining enterprises are scattered over Newfoundland, Nova Scotia, and New Brunswick. Their fortunes ebb and flow with time.

The *quarrying* of building stone has for many decades been the only significant "mining" activity in northern New England. The hard-rock complex underlying most of this sub-region has long been an important supplier of granite, marble, and slate, particularly from several locations in Vermont. Demand for these high-quality stones has decreased in recent years, however, and the prominence of quarrying has declined.

Major reserves of *petroleum* and *natural gas* have been found in the treacherous waters far offshore from the Atlantic Provinces. The only production thus far is from the Hibernia oil field, which is about 170 miles (270 km) southeast of St. John's. Staggering engineering problems—deep producing horizons, shifting currents, frequent storms, much fogginess, the danger of collision with floating icebergs—had to be overcome before commercial production began in the late 1980s.

The offshore excitement also extends to Nova Scotia. Deep-water exploration revealed evidence of oil over a wide area south and east of this province, and in 1983 promising natural gas fields were found near Sable Island, about 200 miles (320 km) east of Halifax (Fig. 6-15).

RECREATION

Tourist-oriented recreation has been a mainstay of northern New England's economy since World War II. The relatively cool summers are attractive to most tourists and the snowy winters are popular with skiers. Moreover, the area is immediately adjacent to the huge population of Megalopolis and there are well-developed highways to enable urbanites to reach secluded rural retreats and wildlands in a relatively short time.

FIGURE 6-15 An oil-drilling platform in the harbor at Halifax awaiting movement to an offshore location (Link/Visuals Unlimited photo).

The last three decades have seen a great upsurge in interest in owning second homes in northern New England. Rural land values, which had remained remarkably depressed considering their nearness to large population clusters, have now skyrocketed. Condominium clusters and other kinds of resort living have mushroomed on mountainsides and lakeshores.

Tourism is less significant and more seasonal in the Atlantic Provinces. Most visitors during the relatively short summer tourist season come from nearby localities in the Maritimes or from neighboring Quebec. There are six national parks in the Canadian section of this region, but the visitors seem to be particularly attracted to the beach resorts of Prince Edward Island. The two ferry systems connecting the island to the continent are drastically overcrowded during the summer.

URBANISM AND URBAN ACTIVITIES

We have seen that the Atlantic Northeast is largely a region of rural charm, rural activities, and economic impoverishment. The people of the region, as with almost all regions on the continent, are now mostly urbanites, however, and the endeavors that provide most of the jobs and contribute most to the economy of the region are urban-oriented. The primary activities of farming, mining, and fishing are quite important and are conspicuous in the landscape; but the secondary and tertiary activities of manufacturing, trade, services, construction, and governmental functions actually dominate the economy, and their locus is mostly in urban areas.

There are no large cities in the region. There are, however, many long-established urban places, and it is here that economic growth is concentrated and population growth is taking place. As of the mid-1990s, two-thirds of the inhabitants of the Atlantic Northeast were urban dwellers.

Much of this pattern of urban growth and rural stagnation is also common to other North American regions. Both economic opportunities and the material amenities of life tend to be prevalent in cities; consequently, immigrants from other regions or other continents usually settle in the cities. And as rural opportunities diminish, particularly in agriculture and fishing, the rural-to-urban population drift, especially among young adults, continues.

The industrial functions normally associated with cities are limited and specialized in the region. Among the cities of this region, Halifax alone is large enough to encompass a relatively full range of manufacturing activities, but it is by no means a major industrial center. Notable specialized industrial centers include Sydney–North Sydney in Nova Scotia, St. John's in Newfoundland, and Bangor in Maine.

Many of the more prominent manufacturing facilities in the Atlantic Northeast are located in small towns or even in rural localities several miles from any town. This is particularly true of woodprocessing plants, both those producing sawn lumber and those whose output is pulp and/or paper. Some major mills are located in medium-sized "cities," such as Corner Brook, Newfoundland; but more often the mill site is associated with much smaller urban places, such as Berlin, New Hampshire, or Millinocket, Maine. Fish-processing facilities, too, are often found in small coastal settlements rather than in larger cities.

The major urban centers of the region are ports, and most of the larger ones have experienced a considerable increase in traffic in recent years. *Halifax* is clearly the most important port in the region, besides being the largest city by far. It normally accommodates more overseas shipping than any other eastern Canadian city except Montreal. It has developed a major container terminal, as well as housing both Canada's largest naval base and largest coast guard base. Winter is its most active season, for Canada's St. Lawrence and Great Lake ports usually are closed by ice for several months. Halifax is not really a primate city for the region, but its magnificent harbor and reasonably diversified economic base ensure its relative prosperity (Fig. 6-16).

St. John's also has a splendid protected harbor, although its mouth (the Narrows) is occasionally choked with ice (Fig. 6-17). Its extreme foreland location provides little opportunity for service to anything more than a provincial hinterland, and even this port-of-entry and distribution function is challenged by smaller Newfoundland ports. But the harbor facilities have been modernized and St. John's is magnificently situated to serve as a base for offshore fishery fleets, offshore oil activities, and as the only effective storm port in a region of stormy seas.[8]

Saint John is somewhat smaller than St. John's, but its situation is quite different. It serves as one Atlantic terminus (Halifax and Portland are the others) of Canada's transcontinental railway system and thus has a busy season of general cargo activity in winter; however, its somewhat more remote location than Halifax from the Atlantic side and its less suitable harbor give it a secondary rank among Atlantic ports.

Moncton has long been the railway center of the Maritimes, and has recently become the telephone call-center of

---

[8] When storms, particularly hurricanes, come to the banks area or other parts of the western North Atlantic, less durable craft, particularly fishing vessels, often make for a protected port. St. John's is conspicuously the most suitable anchorage in the area and may become crowded with ships seeking refuge.

FIGURE 6-16   The waterfront in downtown Halifax (Link/Visuals Unlimited photo).

FIGURE 6-17   The central business district of St. John's, Newfoundland, is immediately adjacent to the harborfront (courtesy Tourism Newfoundland and Labrador).

A CLOSER LOOK   A Bridge to Prince Edward Island

Although situated virtually in the middle of the Atlantic Provinces, Prince Edward Island has a remote location because it is separated from the mainland by the broad waters of Northumberland Strait. Its surface transportation links with the rest of the continent consist primarily of a 9-mile (15-km) government-run ferry service to New Brunswick and a 16-mile (25-km) privately owned ferry service to Nova Scotia. These ferries transport some 2.5 million passengers and 1.1 million vehicles annually, but stoppages are not unusual (particularly due to winter ice), and lengthy delays are commonplace (most notably during the summer tourist season).

Ever since the time Prince Edward Island became a Canadian province in the 1870s there have been dreams and plans to provide a "fixed link" with the mainland. Tunnels, causeways, and bridges all have been considered, and construction of a combined causeway-bridge actually was begun in the 1960s, but the federal government soon cancelled the project because of high costs.

Debates about both the desirability and feasibility of a fixed link have raged for years. Proponents focus on convenience and lower transportation costs;

opponents worry about environmental degradation and the loss of an unhurried pastoral way of life. In 1988 a referendum was held on the island, resulting in a 60–40 vote endorsing construction of a fixed link.

Federal feasibility studies strongly favored the building of a toll bridge roughly parallel to the shorter ferry route, and construction started in June 1994. When completed (estimated in 1997), it will be the world's longest bridge over winter ice–covered waters (Fig. 6-E).

TLM

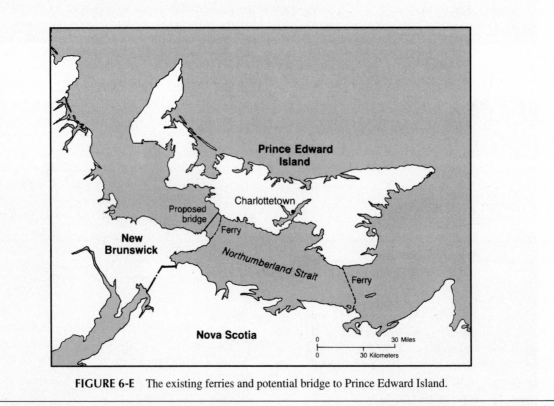

FIGURE 6-E   The existing ferries and potential bridge to Prince Edward Island.

Canada, as such major couriers as Federal Express, United Parcel Service, and Purolator have moved their customer call-center operations there.

These and other cities of moderate size in the Atlantic Northeast are listed in Table 6-1. Each regional chapter in this text includes a table that gives the population for the metropolitan area and the political city, as estimated for

1995 by the U.S. Bureau of the Census and by Statistics Canada. The cities in each table are arranged in alphabetical order; only the more important ones, based on population and lying within the region under consideration, are included.

On the regional maps at the beginning of each chapter, the following urban places are shown:

TABLE 6-1

Largest Urban Places of the Atlantic Northeast, 1995

Name	Population of Principal City	Population of Metropolitan Area
Bangor, ME	33,750	98,000
Burlington, VT	42,000	171,500
Charlottetown, PE	61,500	
Corner Brook, NF	34,000	
Dartmouth, NS	71,000	
Fredericton, NB	76,500	
Glen Falls, NY	52,000	129,500
Halifax, NS	125,000	340,300
Moncton, NB	110,000	
Saint John, NB	78,500	130,000
St. John's, NF	79,000	176,000
Sydney, NS	20,000	
Truro, NS	47,000	

1. Those with a 1995 population exceeding 250,000 are indicated by a large dot and their names are shown in all capital letters (as BOSTON).

2. Those with a population between 50,000 and 250,000 are shown by a large dot and their names are in capital and lowercase letters (as Halifax).

3. Those with a population of less than 50,000 that are mentioned prominently in the text are indicated by an open circle and their names are in capital and lowercase letters (as Bangor). To ensure legibility, cities and towns that are distinctly suburbs of larger urban places are not shown.

THE OUTLOOK

In spite of a long history and a distinguished heritage, the Atlantic Northeast has not been a favored region from an economic standpoint. Its natural resources are limited and its position in the northeast and southeast corners of the two countries involved cut it off from the mainstream of activity just enough to inhibit its commercial vitality. In few other significant parts of the continent are there so many part-time workers—part-time farmers, part-time fishermen, part-time loggers. The rest of their time is spent in another wage-earning capacity, in doing odd jobs, or in subsisting on welfare payments (the latter is often a significant portion of total income). There is no reason to expect a change in this broad pattern.

The region has a disproportionately large number of people engaged in marginal activities in the fields of farming, fishing, and logging, even though recent decades have shown a steady decline in employment in all three. A dearth of arable land, a restricted market, and a limit on crop options all circumscribe agricultural growth. There is a general trend toward larger-scale enterprises in farming, forestry, and fishing, which usually means more efficient production but fewer jobs.

The recent drastic decline in the fishing industry overshadows all other economic developments in the region. Near-future prospects are grim, but eventually the moratoriums should pay off with increased fish stocks. It is clear that careful management of the fisheries is essential, but it is worrisome that it was believed that careful management was already in place during the years leading up to the abrupt collapse of the fisheries.

Mining prospers erratically. Great hopes were focused on offshore oil and gas activity, but the reality has so far been much below the predictions. Still, the Grand Banks and Sable Island discoveries are being developed, and other favorable geologic structures east of Newfoundland, Nova Scotia, and Maine await exploration.

The Atlantic Provinces continue to be the poorest part of Canada, with a per capita income 40 percent below that of Ontario. An inordinate share of the budgets of these four provinces is still provided by federal assistance ("equalization payments"). To anticipate a significant change in the long-standing reality of regional (relative) poverty would be illogical.

Northern New England has undergone repeated economic readjustments. Pulp and paper production should continue to be important, and hydroelectric developments will probably continue; however, both activities are coming under increasingly severe scrutiny by advocates of wildland preservation. Agricultural specialization should continue but not on a large scale. The recent decline of both poultry raising and potato growing has been a shock to the rural economy. Only dairying persists as an element of stability in the agrarian scene.

Urban activities throughout the region can be expected to grow steadily—if slowly. The rate of expansion will remain well below the average growth rate for both nations, and most industrial growth will be in production of items for the local or regional market. But despite the prospect of increased urbanization, the Atlantic Northeast is never likely to be famous for its urban-industrial advantages. Rather, rural amenities are emphasized as the regional attraction. It has been said that "Vermont is the only place within a day's drive of New York that is fit to live in."[9] Or, stated differently, David Erskine has noted that the region's "compensation must be in a life that is freer, or lived in more attractive surroundings, or with closer family ties."[10]

[9] Joe McCarthy and the Editors of Time-Life Books, *New England* (New York: Time Incorporated, 1967), p. 163.

[10] Erskine, "The Atlantic Region," p. 280.

The lure of the Atlantic Northeast is being increasingly appreciated by nearby urbanites. The tourist, the second-home owner, the developer, and the speculator have landed in force. Short- and long-term visitors arrive in ever greater numbers during the summer, and winter recreation also continues to expand. The wider dissemination of such activities over the region augurs well for a more balanced economic gain.

SELECTED BIBLIOGRAPHY

BACKUS, RICHARD H., AND DONALD W. BOURNE, EDS., *Georges Bank.* Cambridge: MIT Press, 1987.

BAIRD, DAVID M., "Down the Saint John," *Canadian Geographic,* 109 (October–November 1989), 68–75.

BELLIVEAU, J. E., "The Acadian French and Their Language," *Canadian Geographical Journal,* 95 (October–November 1977), 46–55.

CALHOUN, SUE, "Snow Crab Bonanza Faltering in Acadia," *Canadian Geographic,* 108 (December 1988–January 1989), 50–59.

CAMERON, SILVER DONALD, "Almighty Cod! It Reigns Supreme Over Our Atlantic Fishery," *Canadian Geographic,* 108 (June–July 1988), 32–41.

———, "Lure and Lore of the Lobster," *Canadian Geographic,* 109 (June–July 1989), 66–72.

———, Net Losses: The Sorry State of Our Atlantic Fishery," *Canadian Geographic,* 110 (April–May 1990), 28–37.

———, "Port of Halifax Living Up to Its 'Greatest Harbour' Name," *Canadian Geographic,* 108 (December 1988–January 1989), 12–23.

CLARK, ANDREW H., *Acadia: The Geography of Early Nova Scotia.* Madison: University of Wisconsin Press, 1968.

———, *Three Centuries and an Island.* Toronto: University of Toronto Press, 1954.

CRABB, PETER, "Is Newfoundland One of Canada's Maritime Provinces?" *The Professional Geographer,* 33 (November 1981), 489–490.

DAY, DOUGLAS, ED., *Geographical Perspectives on the Maritime Provinces.* Halifax: St. Mary's University, 1988.

FORBES, E. R., AND D. A. MUIS, EDS., *The Atlantic Provinces in Confederation.* Toronto: University of Toronto Press, 1992.

HALLIDAY, H. A., "The Lonely Magdalen Islands," *Canadian Geographical Journal,* 86 (January 1973), 2–13.

HORNSBY, STEPHEN J., *Nineteenth Century Cape Breton: A Historical Geography.* Montreal: McGill-Queen's University Press, 1992.

IRLAND, LLOYD C., *Wildlands and Woodlots: The Story of New England's Forests.* Hanover, NH: University Press of New England, 1982.

LEBLANC, ROBERT G., "The Acadian Migrations," *Canadian Geographical Journal,* 81 (July 1970), 10–19.

LYNCH, ALLAN, "Blue Bonanza," *Canadian Geographic,* 115 (January–February 1995), 38–47.

MACPHERSON, ALAN G., ED., *The Atlantic Provinces.* Toronto: University of Toronto Press, 1972.

———, AND JOYCE MACPHERSON, *The Natural Environment of Newfoundland, Past and Present.* St. John's: Memorial University of Newfoundland, Department of Geography, 1981.

MCCALLA, ROBERT, *The Maritime Provinces Atlas.* Halifax: Maritext Limited, 1991.

MCCALLA, ROBERT J., "Separation and Specialization of Land Uses in Cityport Waterfronts: The Cases of Saint John and Halifax," *Canadian Geographer,* 27 (Spring 1983), 48–61.

MCGHEE, ROBERT, "The Vikings Got Here First, but Why Did They Leave?," *Canadian Geographic,* 108 (August–September 1988), 12–21.

MCLAREN, I. A., "Sable Island: Our Heritage and Responsibility," *Canadian Geographical Journal,* 85 (September 1972), 108–114.

MCMANIS, DOUGLAS R., *Colonial New England: A Historical Geography.* New York: Oxford University Press, 1975.

MCMANUS, GARY E., AND CLIFFORD H. WOOD, *Atlas of Newfoundland and Labrador.* Toronto: Breakwater, 1991.

REID, JOHN G., "The Beginnings of the Maritimes: A Reappraisal," *The American Review of Canadian Studies,* 9 (Spring 1979), 38–51.

ROBINSON, J. LEWIS, "Changing Settlement Patterns in Newfoundland," *Geographical Review,* 65 (1975), 267–268.

SCHMANDT, JURGEN, AND HILLIARD RODERICK, EDS., *Acid Rain and Friendly Neighbors: The Policy Dispute between Canada and the United States.* Durham, NC: Duke University Press, 1985.

WALLACE, WILLIAM H., "A Hard Land for a Tough People: A Historical Geography of New Hampshire," *New Hampshire Profiles* (April 1975), 21–32.

WALLACH, BRET, "Logging in Maine's Empty Quarter," *Annals, Association of American Geographers,* 70 (December 1980), 542–552.

WYNN, GRAEME, *Timber Colony: A Historical Geography of Early Nineteenth Century New Brunswick.* Toronto: University of Toronto Press, 1981.

French Canada

7

The most culturally distinctive region of the continent is French Canada. For more than three and a half centuries it has been occupied by people whose primary cultural attributes are different from those of the settlers of the other regions of the continent. It has been a Franco-culture island in an Anglo-culture sea, and this unique cultural expression has been maintained without significant external reinforcement. In the last two centuries there has been very little immigration, or even much tangible support, from France; yet the settled southern portions of Quebec continue to be dominated by a solidly French culture, and the areal extent of this influence has continued a slow expansion into adjacent parts of New Brunswick and Ontario, even into New England.

The predominant expression of French Canadian culture is, of course, the use of the French language. More than 80 percent of the people of the province of Quebec consider French their mother tongue; and if the cosmopolitan city of Montreal is excluded from the statistics, more than 90 percent of the people are Francophones (linguistically French), with many of the remainder (the English-speaking Anglophones and the allophones who speak a language other than French or English) being bilingual. Indeed, in most of the St. Lawrence Valley and estuary downstream from Montreal the proportion of French speakers exceeds 98 percent of the total population. Hand in hand with the French language in the region goes the Roman Catholic religion. Catholicism in Canada is not restricted to Francophones, but more than 85 percent of the inhabitants of Quebec profess to be followers of the Roman Catholic faith. It is in the French Canadian region, then, that the Catholic Church has its firm Canadian base, largely accounting for the fact that some 45 percent of the Canadian population professes to be Roman Catholic.

These two important social attributes—language and religion—are not the only nonmaterial elements of the distinctive French Canadian culture, but they are clearly the most conspicuous. Other intangible culture traits, such as cohesiveness of family life or dietary preferences, are less easily identified and quantified but may be equally significant in some areas.

Although recognizing the importance of nonmaterial culture elements, a geographer is continually seeking to identify and interpret expressions of culture in the landscape. In French Canada this search is amply rewarded. Physical manifestations of the dominant culture can be recognized in farmscapes and townscapes, fences and signs, field patterns and architectural styles, general appearance, place names, and many other facets of the landscape (Fig. 7-1).

FIGURE 7-1 One of the many manifestations of French Canadian culture is in architectural styles. The wraparound veranda with its decorative wooden trappings and the conspicuous dormers are typical of many homes in the region. This scene is near Levis, Quebec (TLM photo).

Consider, for example, the matter of place names in French Canada. The great majority of all places—towns, rivers, streets, mountains—are named for saints. Any map of the region shows this dominance: St. Laurent, St. Maurice, Ste. Anne de Bellevue, Ste. Foy, St. Hyacinthe, St. Jérôme, St. Félix-de-Valois. The principal areas where such names are not dominant are those in which a later spread of French settlement was superimposed on an already established Anglo framework, as in the Eastern Townships of Quebec or parts of New Brunswick.

The prominence of the church in the French Canadian way of life is manifested in the landscape, especially the rural landscape, by many features other than place names. Roman Catholic religious institutions are numerous and conspicuous. Churches, seminaries, monasteries, convents, shrines, retreat houses, cemeteries, and other edifices are prominent throughout the settled parts of the region. They are usually larger in size than other structures in the area and often solid and massive in style. Their prominence is somewhat subdued in cities, where secular buildings may also be large and massive. Indeed, the role of the church has diminished in the large urban centers of Quebec, whereas it is still a prominent social institution in the village and rural areas. The church continues as a distinct rural and small-town landscape and societal feature but is no longer a dominant urban social and political factor. In smaller towns and villages church-related structures commonly dominate the scene, with the built-up area clustering around a large stone Roman Catholic church.

The rural landscape, too, has its characteristic features. Most notable is the pattern of land ownership and field alignment. Most farms are long, narrow rectangles, and fields within the farm repeat the pattern on a smaller scale. The background to this unique pattern is discussed subsequently.

Centrally located within the fields are often long heaps of stones that were gathered by the farmer after years of winter frost-heave, and accumulated in piles. Often a cedar pole fence surrounds the field, although wire fences have been increasingly adopted in recent years, particularly in the upstream portions of the region. The farmstead, too, often has predictable characteristics. The buildings are likely to be constructed of unpainted wood, somewhat unkempt, gray, and bleak in appearance. Certain architectural styles are common: the farmhouse often has a lengthy veranda or porch, a certain amount of "gingerbread" on the exterior, and several high dormers; the barn is likely to be of the inclined-ramp variety with livestock housed on the lower floor and a wooden ramp leading to the second story, where machinery, tools, and feed are kept.

A CULTURALLY ORIENTED REGION

The designation of French Canada as a major continental region is primarily in recognition of its cultural uniqueness and the significance of this cultural imprint on its geography. Other factors, environmental and economic, contribute to regional unity, but it is the manifestations of French Canadian culture that are the distinctive shapers of the total geography of this region.

No other region, as delineated in this book, is recognized principally on the basis of its cultural components. Some other sections of the continent have important elements of non-Anglo culture, but in every case the designation of a culturally oriented region is felt to be unwarranted either because the culture does not sufficiently permeate the geography of the region or because the area and population involved are too small to justify separation as a region.

A case might be put, for example, that the southwestern borderlands of the United States—those parts of Texas, New Mexico, Arizona, and California that are close to Mexico—contain millions of people of Mexican extraction who have a Hispanic culture that is analogous in many ways to the French Canadian culture region in eastern Canada.[1] Although recognizing the validity of this assertion, the author feels that the form and function of life in the borderlands are clearly dominated by an Anglo pattern that is insufficiently different to justify regional recognition. Similar reasoning holds true for the Hawaiian Islands, with their significant Oriental and Polynesian culture complexes, although Hawaii is delineated as a separate region for other reasons.

True cultural distinction is also shown in certain parts of North America where the bulk of the population is aboriginal in origin. The Navaho–Hopi–Ute Indian complex of the Four Corners country, for example, has recognizable cultural uniqueness. Also, in large stretches of subarctic and arctic Canada and Alaska varying degrees of distinctive Indian, Inuit, and Aleut culture predominate. But in each instance regional recognition does not seem warranted on a largely cultural basis because of the small size of the population involved or its highly fragmented settlement pattern over broad areas where environmental factors seem more important as regional delineators.

FRENCH CANADA AS A REGION AND A CONCEPT

The region of French Canada, as delineated here, does not include all the province of Quebec and is not limited by the borders of Quebec (Fig. 7-2). It is considered to encompass

that portion of eastern Canada that is dominated by French Canadian culture except where significant areas of non–French Canadian culture or nonsettlement intervene. Thus the region includes most of the southern settled parts of Quebec: from the lower Ottawa River valley in the west; down the St. Lawrence Valley to include the Gaspé Peninsula, Anticosti Island, and the north shore of the Gulf of St. Lawrence in the east; as well as the relatively densely settled portion of the Shield that encompasses the Lake St. John lowland and the Saguenay River Valley.

It is also considered to include the French-dominated portions of the province of New Brunswick, largely the area north of 47° north latitude. In the seven counties of northern New Brunswick about three-fifths of the population speak French as a mother tongue, and in three of the counties more than 80 percent of the people are Francophone.

With a few minor exceptions in northeasternmost New York State, northernmost Vermont, and the Aroostook Valley of Maine, the area of French Canadian dominance ends abruptly at the international boundary; so the U.S. border can be considered the southern margin of the region.

As thus delimited, the region of French Canada includes more than 90 percent of the population of Quebec and about 35 percent of the population of New Brunswick. This amounts to about one-fourth of the population of Canada or about 7 million people as of the mid-1990s.

It is important to note that a few hundred thousand French Canadians live outside the region of French Canada just described. They are found throughout the settled parts of Canada, although usually in small numbers. The principal concentrations of French Canadians who live outside the French Canada Region are in the Abitibi–Timiskaming area of west-central Quebec,[2] in the iron mining towns of the Labrador–Quebec border country, in scattered mining communities of east-central Ontario, in southeastern Manitoba (especially the Winnipeg suburb of St. Boniface), and in various parts of the Maritime Provinces mentioned in the previous chapter.

The French Canadians living in these other parts of Canada maintain most of the same cultural attributes as those who live within the French Canada Region except that the

[1] See, for example, Richard L. Nostrand, "The Hispanic-American Borderland: Delimitation of an American Culture Region," *Annals,* Association of American Geographers, 60 (1970), 638–661.

[2] It is tempting to expand the boundaries of the French Canada Region so as to encompass the Abitibi–Timiskaming area because of its lengthy historic relationship with the French Canadians of the St. Lawrence lowland and because more than 150,000 French Canadians live there today. However, there are nearly 200 miles (320 km) of sparsely settled forest land between these two areas at their closest points; this seems to be too great a gap for logical expansion to include a relatively small outlying area.

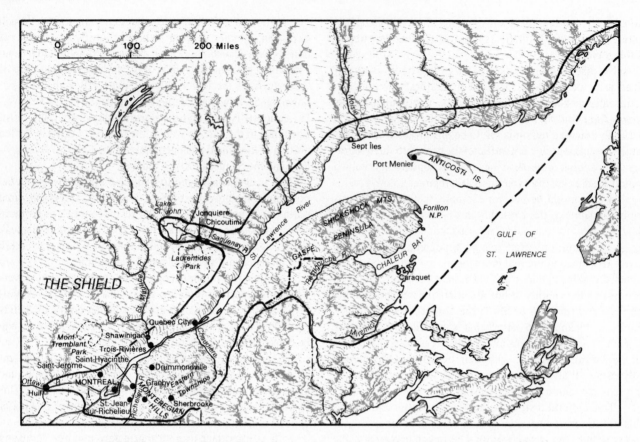

FIGURE 7-2   The French Canada Region (base map copyright A. K. Lobeck; reprinted by permission of Hammond, Inc.).

French language becomes decreasingly important with increasing distance from Quebec.[3]

THE REGION AND ITS PARTS

The location of the region is a great paradox; it is isolated from surrounding regions by natural barriers and yet has served throughout history as the principal thoroughfare connecting Canada with the Old World. To the north of French Canada is the rocky forested fastness of the Canadian Shield, largely unpopulated. To the south the Appalachian highlands serve as a very effective barrier, with the single important exception of the Lake Champlain lowland, to connection with the populous parts of the United States. To the east is the vastness of the Atlantic Ocean, ice-locked for part of the

year. Only to the west is there a relatively easy route to connect "Lower Canada" (as Quebec was formerly called) to "Upper Canada" and even here the zone of settlement is a narrow one, pinched between the southward extension of the Shield and the Adirondack Mountains.

Despite the difficulties imposed by white-water rapids on the rivers, an estuary and gulf with seasonal ice problems, and rocky impediments to land transport, the St. Lawrence corridor has been the major routeway providing the Canadian heartland with access to Europe and the rest of the world. The development of the St. Lawrence Seaway system since the 1950s has reinforced the importance of the corridor.

Even a region as small as French Canada has various parts, and the parts demonstrate differing characteristics. The St. Lawrence Valley is the central part and core of the region; it is a broad lowland in the southwest that narrows progressively downstream until the estuary of the river occupies almost all the flat land below Quebec City. Southeast of the

[3] Ludger Beauregard, "Le Canada Francais par la Carte," *Revue de Géographie de Montréal,* 22 (1968), 35.

valley and extending to the international border lie the gently rolling and hilly lands of the Eastern Townships, where early English settlers have been largely replaced by more recent French Canadian arrivals. South of the Gulf of St. Lawrence is the rocky, tree-covered peninsula of Gaspé, occupied by marginal farmers and hardy fishermen. The portion of the region in New Brunswick encompasses a peripheral circle of fishing, farming, and forestry around an interior that is almost totally unpopulated. The pattern of life in each section has much in common, but there are important differences, the most striking being between the upstream and downstream portions of the region.

THE ENVIRONMENT

In many ways, the environment of French Canada is similar to that of the Atlantic Northeast: rocky uplands, extensive forests, bleak winters, rushing streams, and limited mineral resources. There are also important differences: the most significant is the amount of flat land and relatively productive soil that provided the region's economy with a widespread agricultural base.

Topography

The St. Lawrence corridor consists of a long stretch of flat, valley-bottom land that varies greatly in width. It is broadest in the Montreal plain and on the right bank of the river between Montreal and Quebec City. Former beach terraces and strand lines, indicating relatively recent emergence of the lowland from beneath the sea, are commonplace. The surface materials of the lowland are mostly sands and other recent deposits of marine, fluvial, or glacial origin, which cover the bedrock foundation quite deeply in places. The major irregularities in the plain are the Monteregian Hills, scattered remnants of old volcanic stocks that rise several hundred feet above their surroundings; most famous is the westernmost, Mount Royal, in the heart of the city of Montreal.

On the left-bank side of the St. Lawrence River the flat land soon gives way to the sloping edge of the Canadian Shield, which rises rockily and in many places abruptly in steep hills or complex escarpments. The only significant extensions of the lowland on this northwest side of the St. Lawrence are where major tributaries, such as the Ottawa, the St. Maurice, and the Saguenay, have breached the Shield escarpment.

Southeast of the St. Lawrence there is much flat land that extends well into the Eastern Townships. But slopelands are more characteristic, with stretches of rolling hills becoming higher and more complicated mountain ranges near the

U.S. border. The mountains become higher toward the northeast, forming a sort of rolling plateau that in places is deeply dissected by rivers. Elevations of more than 4000 feet (1200 m) are reached in the rocky fastness of the Shickshock Mountains of the Gaspé Peninsula. Northern New Brunswick is a mixture of rolling lowlands and rocky hills, with major valleys carved by the Restigouche, Nipisguit, and Miramichi Rivers.

Climate

Winter is the memorable season in French Canada, both because of its coldness and its length. Qualitative descriptions may vary, but there is no dispute that there is a long period of low temperature in the region. Even the areas of mildest climate have a winter period of $5\frac{1}{2}$ or 6 months, with early frosts beginning in October and streams not running freely in the new year until April. January temperatures average below 20°F (−7°C) throughout the region, with Montreal recording a mean of 14°F (−10°C) and Quebec City 10°F (−12°C). Snowfall is heavy [100 inches (2540 mm) annually is not unusual even at sea-level localities], and weather changes are frequent, although warm periods are brief (Fig. 7-3). Storm tracks converge in the region in winter, and migratory cyclones and anticyclones pass with frequency.

But winter's icy grip is not an unmixed blessing. The deep snow cover allows both humans and logs to move more easily in the forest and facilitates the accumulation of logs on frozen waterways for floating downstream in the spring thaw. The "dead" navigation season on the St. Lawrence, occasioned by the winter freeze, has sometimes lasted from December to April. Today the energetic use of icebreakers permits the port of Montreal to function for all but a few days of the year; Quebec City's port is closed only occasionally, for high tides keep the ice broken and shifting in the estuary.

Summer weather is quite varied in the region. The upstream portion of the St. Lawrence Valley experiences much warmth and humidity and has a growing season of approximately 5 months. Higher elevations and the most easterly coastal areas (Gaspé, both sides of the estuary, and Anticosti Island) have mild-to-cool summers, with a frost-free period of less than 100 days. Summer is a time of generally abundant rainfall, which is a distinct agricultural advantage to the warmer upstream areas. Annual precipitation totals about 40 inches (1016 mm) at both Montreal and Quebec City.

Soils

Soils are quite variable, but most are heavily leached, acidic, deficient in nutrients, and poorly drained. Those that developed on glacial deposits are very stony, and sands and clays are prominent where marine deposits are the parent material.

FIGURE 7-3   Winter is a prominent fact of life in French Canada. Here the frozen St. Charles River in Quebec City serves as a temporary walkway and skating rink (courtesy Government of Quebec, Tourist Branch).

Spodosols dominate in upland areas and in downstream parts of the region. They are poor soils for agriculture—totally leached in the upper horizons, quite acidic, and ash gray in color. Alfisols are the chief soils of the upstream areas, especially on the terraces, and although somewhat leached and acidic in nature, they respond well to fertilizer and are important agricultural soils. Small areas of alluvial soils in the valley bottoms are the most productive for farming in the region.

Natural Vegetation

Originally almost the entire region was forested except for poorly drained, marshy areas in the lowlands and some higher, rocky upland slopes. Softwoods dominated, with magnificent stands of white pine and fir, particularly in downstream areas and northern New Brunswick. The upstream sections and much of the Eastern Townships were covered with a mixed forest in which maple, elm, beech, and birch were prominent. Most uplands are still forested, although logging has removed all the accessible timber at least once. The lowlands have long since been cleared for farming; yet extensive woodlots have been maintained in the agricultural areas throughout the region.

SETTLING THE REGION

At the time of European contact most of the St. Lawrence Valley and associated lowlands were occupied by Iroquoian tribes. Gaspé and northern New Brunswick were inhabited by the unobtrusive Micmacs and Malecites, who fished along the coast in summer and hunted in the forest in winter.

The Huron tribe occupied a territory in southern Ontario and was a very active trading tribe. Furs were the basis for their trading relationship with the French, with whom the Hurons acquired a trade monopoly from the earliest days of settlement. In 1649–1650 a series of Iroquois attacks, combined with appalling winter starvation, virtually eliminated the Huron people and wiped out their tribal identity.

During the subsequent decade the Iroquois annihilated or adopted most other small tribes in this region except for the Ottawas. The Iroquois signed a treaty with the French at Montreal in 1688 and for some decades continued to serve as a buffer between the settlements in Quebec and those in the American colonies to the southeast.

European settlement in the region dates from 1608 with the founding of Quebec City by Champlain.[4] "New France" was to be an agricultural colony, but throughout the early decades it was fur trading that attracted the most interest and enthusiasm and pulled French explorers ever deeper into the interior of the continent. Agricultural settlement moved slowly up the St. Lawrence and some of its major tributaries; Trois-Rivières was founded as a trading center in 1634, and the first settlers disembarked on Montreal Island in 1642.

[4] The city's proper name is simply Quebec. But in order to spare confusion, in this book Quebec refers to the province and Quebec City refers to the city.

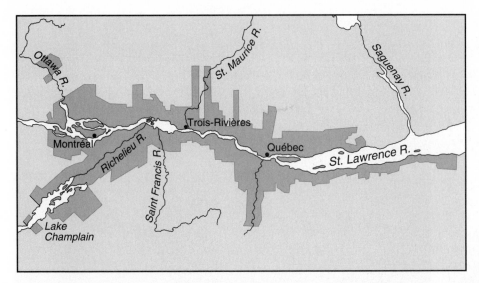

FIGURE 7-4 Extent of the seigneurial system in the St. Lawrence Valley at the end of the French regime in 1763.

The initial settlement and land ownership pattern in rural Quebec was quite different from that in most of the rest of the continent (Fig. 7-4). Today its imprint on the landscape is both unique and notable (see vignette).

In 1763 France gave up its claim to New France, and the 65,000 *Canadiens* came under British rule. A land survey was soon carried out and rectangular townships were laid out between the seigneuries of the St. Lawrence and the international border (originally called the Eastern Townships but now more frequently referred to as Estrie), with the area being thrown open to Anglo-Saxon colonists. Other townships were surveyed sporadically around the margin of the seigneurial territory until it was completely surrounded by lands earmarked for British settlement. Many of the early settlers were "Loyalists" immigrating after the revolution from the new American republic to the south; English, Scottish, and Irish immigrants also came. Farms were typically square or rectangular and the pattern was totally unlike the long-lot system, although farmsteads sometimes appeared to be aligned because they were built along access roads.

For the better part of a century after France's retirement from Canada, the French populace remained within the seigneurial domain; as the population expanded rapidly, it filled in the as yet unsettled parts of the seigneuries. But by the middle of the nineteenth century French Canadians were more than half a million strong and overflowed into Anglo-Saxon townships.

In the two decades prior to Canadian confederation (1867) the French population of the Eastern Townships increased by 120 percent, whereas the British population of the same area increased by 6 percent. The total numbers of the two groups were approximately equal by the date of confederation, but the continuing rapid increase of the French population soon greatly exceeded that of the British. On the other side of the St. Lawrence, on the edge of the Laurentians, the French Canadian settlement also rapidly expanded, filling in the interstices among the predominantly Irish settlers in those townships and in townships in northeastern Ontario. Soon there was a French Canadian majority on the edge of the Shield, and a "spillover" of farmers and loggers occupied the Saguenay Valley and Lake St. John lowland in considerable numbers (Fig. 7-5).

At the time of confederation the population of the French Canada Region was overwhelmingly rural. Montreal and Quebec City were the only urban centers of note, with populations of about 100,000 and 60,000, respectively. By the time of confederation the occupance of French Canada was virtually complete.

Few new areas of rural settlement were established after then except in the area northwest of Montreal and north of Ottawa—particularly in Terrebonne and Labelle counties—which reached its maximum extent in the 1930s and 1940s. In some other sections the rural population density increased in this century, but in most it actually declined because of farm abandonment. The major change in population distribution has been the growth of cities. French Canada today, like the rest of the country, is overwhelmingly urban.

THE BILINGUAL ROAD TO SEPARATISM

Virtually every city in the region, with the notable exception of Montreal, is overwhelmingly French. The cultural diversity of Montreal provides a dynamic focus for what is certainly the most significant social and political problem facing Canada today: the accommodation of a large and vibrant

FIGURE 7-5   Linearity rampant! The farmsteads are lined up along the roads, and the narrow fields extend perpendicularly in either direction. This typical long-lot pattern is southeast of Montreal (TLM photo).

French Canadian minority within a predominantly Anglo-Canadian nation. Probably the most important political fact in Canada is that a quarter of its citizens do not speak the majority language as a mother tongue, if at all. The special position of the French Canadian minority has been legally recognized since the founding of the Canadian nation; the British North America Act of 1867, which established the federation, included various irreducible obligations to the province of Quebec. Quebec was guaranteed its civil law, its religious liberty, jurisdiction over its educational system, and the equality of its language in both Parliament and federal courts of the nation as well as in the Legislature and courts of the province. It is unlikely, however, that the framers of this remarkably tolerant legislation anticipated the cultural tenacity of the French Canadians.

The cultural dichotomy persists, with foreboding overtones. Much lip service has been given to the principle of Canada as a bilingual and bicultural nation. In the real world of government and business, however, Canada functions as an English-oriented country with Quebec as a French-oriented enclave. French Canadian objections to the status quo have become increasingly strident since World War II, the summary complaints being that English is the language of business, so that Francophones must use English in order to advance, and that the Franco community is being assimilated by the Anglo community.

Quebec's *revolution tranquille* (quiet revolution) of the 1960s brought French- and English-speaking Canadians into direct large-scale competition for jobs and power in modern business and government. This effort was slow in attracting national attention, but the abrupt actions of a fanatic extremist group (the FLQ or *Front de Libération du Québec*) shocked the nation with bombing, kidnapping, and murder. Strong government and private efforts were then instituted to assuage the situation. The "special relationship" of Quebec to the confederation was heartily affirmed, bilingualization of the federal civil service was accelerated, and the prime minister (a French Canadian) actively supported not only bilingualism but a government policy of multiculturalism under a bilingual framework as well.

In the 1970s, however, political polarization based on linguistic polarization became increasingly pronounced (Fig. 7-6). A strict provincial language law that curtailed the use of English in government, business, and education was enacted, making Quebec, in effect, a unilingual province.[5] The *Parti Québécois* was elected to power, with a prominent plank in its platform being the eventual political secession of Quebec from Canada. In 1980 a referendum was held in Quebec on whether the provincial government should "negotiate a proposed agreement" of "sovereignty-association" between Quebec and Canada. The federal government urged a negative vote, whereas the provincial government obviously favored a positive vote. The electorate of Quebec rejected the proposal by a 60–40 majority.

For about a decade the matter of separatism for Quebec was less conspicuous, although public opinion became in-

[5] As a result of this legislation, New Brunswick is now the only province that is legally bilingual on the basis of provincial law, although federal law declares that the nation is legally bilingual.

creasingly polarized, and there was considerable "Anglo flight" of people and businesses out of the province. Population growth in Quebec slowed dramatically, and the provincial economy stagnated. The political issue was again brought to the fore in the early 1990s, and in 1995 a plebiscite on sovereignty was again held in the province, after an exhaustive and passionate campaign. The proposal failed by less than 1 percent of the total vote. The Quebec government leadership has made it clear that they will submit the issue to the electorate once again within a couple of years. The questions associated with possible secession are complex and murky, but the political unraveling of a united Canada is obviously occurring.

THE PRIMARY ECONOMY

Although most of the population is urban and the economy is now basically urban related, the image of French Canada has always been a rural one. In many parts of the region rural activities are prominent, and several primary industries contribute significantly to both employment and income.

Agriculture

There are still some 400,000 farmers in French Canada; most carry on a mixed farming enterprise with emphasis on dairy production, although the average number of cows per farm is small. Throughout the region more than half the cultivated acreage is in hay, with oats the other principal crop.

The average farm is 150 acres (60 ha) in size and is shaped in the traditional long-lot pattern everywhere except in parts of the Eastern Townships and northern New Brunswick. The wooden farmhouse, with its kitchen garden, faces the road. Cordwood is generally stacked beside the house, and in the rear are other farm buildings—garage, chicken house, woodshed, piggery, two-story inclined-ramp barn, a silo or two. The narrow rectangular fields usually contain a long linear "centerpiece" of stones that were gathered and stacked by the farmer over many years. The land is divided into several parts, with hay and pasture crops alternating in rotation with row crops (corn for silage, potatoes, other root crops such as rutabagas) or grain (mostly oats). The significant farm animals are dairy cattle, principally Holsteins or Ayrshires; in addition, there are usually poultry and hogs (Fig. 7-7). At the far end of the farm is often the woodlot and the sugar bush (sugar maple trees).

The most prosperous farm area in the region is in the upstream part of the St. Lawrence Valley and the southwestern corner of the Eastern Townships. Farms in this section are 25 percent smaller than the provincial average, but a much higher proportion of land is under cultivation and a much

FIGURE 7-6   This bilingual stop sign on the outskirts of Quebec City has been defaced by painting out the English word; it is a mild, visible expression of the deep passions that underlie French Canada's bilingual/bicultural problem

more intensive farm operation exists. Machinery is more numerous, fertilizer more frequently applied, and modern techniques more abundant. Most commercial orchards and all the sugar beets of the region are grown here. Other specialty products of significance include vegetables, poultry, tobacco, and honey. Still, most farms are dairy farms and the output of milk, butter, and cheese is enormous.

Forest Industries

Quebec has the largest volume of timber resources in Canada; however, most of it is located on the Shield, outside the French Canada Region. There has always been, and continues to be, production from the relatively small woodlots of the St. Lawrence Valley for lumber, pulpwood, posts, and firewood. This production is mostly small scale and only locally important. In the Eastern Townships, Appalachian uplands, Gaspé Peninsula, and northern New Brunswick there has been important commercial output of lumber and pulpwood since the middle of the last century. But large-scale logging enterprises are chiefly located on the Shield, and the great bulk of Quebec's production comes from there.

Forest exploitation in the past was restricted almost exclusively to winter. Trees were cut and hauled by various means to the rivers, where they were dumped on the ice. At the time of the spring thaw vast flotillas of rough logs came churning down to the mill sites, where they were caught and

A CLOSER LOOK   A Long-Lot Landscape

The most striking feature of the landscape in rural French Canada is the almost limitless array of long, narrow rectangles that subdivide the agricultural land. With the notable exception of the Eastern Townships, property lines and field boundaries replicate the pattern with faithful precision disdainful of topographic variation, throughout most of the region. Moreover, farmsteads are almost invariably positioned at the same ends of adjacent rectangles so that neighboring farmhouses exist in close proximity to one another with remarkable linearity of location. This distinctive contemporary landscape morphology is a heritage of the earliest days of French settlement in Lower Canada and survived with tenacious persistence and little change for three centuries.

The first element in the pattern was the establishment of a seigneurial system. In the seventeenth century the kings of France awarded land grants (called *seigneuries*) with feudal privileges to individual entrepreneurs (*seigneurs*). The seigneurs, in turn, were expected to subgrant parcels of land to peasant farmers (*habitants*). The seigneuries varied greatly in size, but each fronted on a river and extended inland for a mile or two in some cases and up to almost 100 miles (160 km) in others. A total of about 240 seigneuries were created, involving about 8 million acres (3.2 million ha) of land, mostly along the St. Lawrence River.

The typical land grant (called a *roture*) within a seigneury was a long, narrow rectangle, fronting for 150 to 200 yards (135 to 180 m) along the river and extending inland for a mile or more.[a]

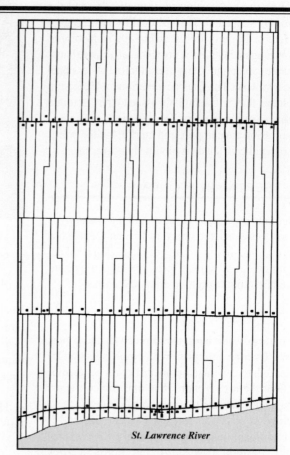

FIGURE 7-A   A hypothetical model of rang settlement patterns in Quebec (from *Canada Before Confederation: A Study in Historical Geography* by R. Cole Harris and John Warkentin, p. 74. © 1974 by Oxford University Press, Inc; reprinted by permission).

[a] R. Cole Harris, "Some Remarks on the Seigneurial Geography of Early Canada," in *Canada's Changing Geography,* ed. R. Louis Gentilcore (Scarborough, Ont.: Prentice-Hall of Canada, 1967), p. 31.

This gave each farm direct access to the river, which was the only transportation route in the early years. When all riverside rotures had been granted, a road would be built along their inland margin, paralleling the river, and a second rank (or *rang*) of rotures would be developed. In some cases, up to a dozen rangs were successively arrayed back from the river, separated from one another by parallel concession roads that ran without break for dozens or, occasionally, even hundreds of miles (Fig. 7-A).

The habitants invariably built their farmsteads at the end of their rotures adjacent to the river or road; thus an almost continuous string of individual

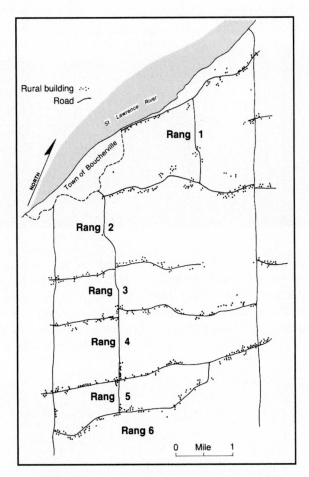

FIGURE 7-B An example of the rang–long-lot pattern of rural settlement in a portion of the St. Lawrence Valley. The land ownership pattern of such an area can be visualized as long, narrow properties running at right angles to the roads (map data from Army Survey Establishment 1:50,000 series, Beloeil 31H/11W sheet, 1965).

of the concession road, thus serving rotures that extended in opposite directions from the road. By the early 1800s many seigneuries contained six or seven rangs double along roads roughly parallel to the river and linked by occasional crossroads with no settlement along them.[c] The custom of equal inheritance rights resulted in increasing fractionization of the rotures along the lines of the original subdivision, with each succeeding fraction becoming narrower so that each farmer still had access to river or road (Fig. 7-B). The original "long-lot" farms were designed to be about 10 times as long as they were wide, but repeated linear subdivision sometimes created units that were virtually too narrow to farm economically.[d]

The French population grew slowly at the outset, reaching about 3000 by 1660, but the seigneurial domain continued to expand along the St. Lawrence, reaching downstream to the Gaspé Peninsula and upstream to the border of Upper Canada; it also extended up a number of tributaries, particularly the Chaudière, the Richelieu, and the Ottawa.[e]

TLM

ceedings, 17th International Congress of the International Geographical Union (1952), p. 723.

[c] R. Cole Harris and John Warkentin, *Canada Before Confederation* (New York: Oxford University Press, 1974), p. 73.

[d] Peter Brooke Clibbon, "Evolution and Present Pattern of the Ecumene of Southern Quebec," in *Quebec*, ed. Fernand Grenier (Toronto: University of Toronto Press, 1972), p. 17.

[e] This long-lot arrangement of properties and fields was developed in other areas of French settlement on the continent, notably in sections of the Maritimes, in Manitoba, in Louisiana, and along the Detroit River, but nowhere did it reach anything like the magnitude or the permanence of its extent in Quebec.

settlements grew up along transportation routes.[b] The most common settlement pattern was the *rang double,* in which houses were built on either side

[b] Deffontaines noted that at the end of the French rule a traveler could have seen almost every house in Canada by making a canoe trip along the St. Lawrence and Richelieu rivers. P. Deffontaines, "Le Rang: Type de Peuplement Rural du Canada Francais," *Pro-*

FIGURE 7-7 The dominant farm activity in French Canada is dairying. This Holstein herd is in the valley of the Ottawa River between Montreal and Ottawa (TLM photo).

processed. This seasonality made part-time logging an important source of income for many residents of the region, especially small farmers whose winter labors were few. But in the last two decades forest exploitation has become virtually a year-round operation, and opportunities for part-time employment are quite limited.

The rivers still serve as log thoroughfares, with bumpers and booms to guide the logs to the proper catchment areas in quiet water (Fig. 7-8). A large proportion of the cut nowadays is, however, transported by truck or train; thus the gathering of logs can go on every month. Still, the important milling sites are along the streams, especially in the St. Mau-

FIGURE 7-8 Many of the region's rivers are still heavily used for transportation and storage of logs. This is the St. Maurice River upstream from Trois-Rivières (TLM photo).

A CLOSER LOOK   Anticosti—the Largest Island Nobody Knows

Anticosti Island, located in the principal oceanic shipping lane between Canada and Europe, and only 325 miles (520 km) from Quebec City, is the fourth largest non-Arctic island in North America. Yet its population is only about 350, and very few people ever visit it. With an area of 3045 square miles (7890 km²), it is nearly half again as large as Prince Edward Island, which is 150 miles (240 km) due south of Anticosti (Fig. 7-C).

The island is a rocky, uplifted horst block, rolling to rugged in topography, with a coastline dominated by steep limestone cliffs and broad reefs. A large bog occupies the eastern end, but most of the island is covered with a fairly dense forest of spruce and balsam.

Anticosti was discovered by Jacques Cartier in 1534, and its first owner was Louis Jolliet, the noted Mississippi River explorer, who received it as a royal gift in 1680. It became a *seigneury*, and for more than two centuries was owned by absentee landlords, with only a handful of permanent residents. It was purchased in 1895 by a wealthy Frenchman to serve as a hunting reserve, and for three decades was the largest private game preserve in North America.

A pulp and paper company acquired Anticosti in 1926, and soon there was a pulpwood forestry boom that supported 3000 residents. Within a few years the logging operations declined, and the Quebec government purchased the island in 1974. It is now a provincial park, its principal attractions being wilderness, a herd of about 100,000 white-tailed deer, and numerous fishable streams.

Visitors must obtain a permit to set foot on the island, and another permit if they intend to travel around. There is one small village, called Port Menier, on Anticosti; it is linked to the mainland by daily air service and weekly boat service.

TLM

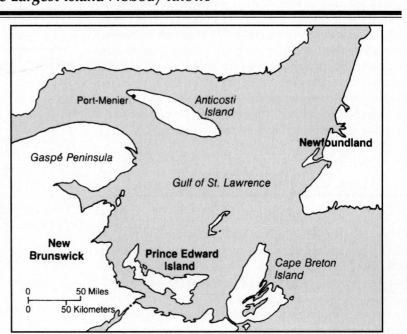

FIGURE 7-C   Anticosti Island.

rice and Saguenay Valleys and where major tributaries, such as the Ottawa and the St. Maurice, join the St. Lawrence; indeed, every river junction on the left bank of the St. Lawrence has a sawmill, pulp mill, or both nearby. Other types of wood-processing plants, such as those making doors, plywood, and pressed wood products, tend to be located in the major industrial centers.

Fishing

On the whole, fishing makes a trivial contribution to the regional economy. The principal species caught by French Canadian fishermen are cod, redfish, and herring. Less than 10 percent of the total value of Canadian Atlantic fisheries is landed at ports in French Canada, and less than 1 percent of the regional work force is employed in catching and processing fish. Nevertheless, around the Gaspé Peninsula and in northern New Brunswick fishing is a significant enterprise, and in many places fishing villages and fishermen's houses literally line the bay shores. The village of Caraquet, New Brunswick, for example, is said to be the longest town in Canada. It consists of an almost-continuous line of homes along the south shore of Chaleur Bay, some 20 miles (32 km) long and less than one block wide.

Mining

Mining is another primary activity with significance in only a limited part of the region. The world's largest commercial deposits of asbestos, with an annual output amounting to more than one-fourth of the world total, are located in the "Serpentine Belt," which extends in an arc northeastward from the Vermont border through the Eastern Townships. Economic hardship has come to the Serpentine Belt in the last few years, as concern for health hazards in the use of asbestos has significantly diminished demand for this mineral.

URBAN-INDUSTRIAL FRENCH CANADA

As in most of the continent, the true dynamism of this region's geography is found in its urban-industrial development. The rapid and significant growth of urbanism, based on solid industrial development, made the French Canadian as much an urbanite as the New Englander or the southern Californian.

Industrial output in the region, concentrated primarily in metropolitan Montreal, amounts to 25 percent of the Canadian total. More than half the Canadian production of tobacco products, cotton textiles, leather footwear, aircraft, and ships comes from French Canada. Moreover, most of Canada's aluminum manufacturing capacity is in the region, with the largest single plant at Arvida in the Saguenay Valley.

The largest industry in the region is pulp and paper. More than 60 large plants have riverside locations, mostly along the St. Lawrence, in the French Canada Region, which provides 40 percent of Canada's total output. There is a close functional relationship between the resources of the Canadian Shield and the pulp and paper factories of the Quebec lowland. The vast forests of the Shield are the source of the raw materials, and the numerous rivers of the Shield provide both transportation for the logs and hydroelectric power for the mills.

The urban system of French Canada includes many small cities, a few medium-sized ones, and two dominant metropolises (see Table 7-1 for a listing of the region's largest urban places). The medium-sized centers include the Trois Rivières–Cap-de-la-Madeleine conurbation of more than 100,000 people at the confluence of the St. Maurice and St. Lawrence Rivers; the notable industrial (especially aluminum refining and forest products) and commercial complex of Chicoutimi–Arvida–Jonquiere in the upper Saguenay Valley; Sherbrooke, the subregional center of the Eastern Townships; the prominent industrial city of Shawinigan in the St. Mau-

TABLE 7-1

Largest urban places of French Canada, 1995

Name	Population of Principal City	Population of Metropolitan Area
Chicoutimi, QC	62,000	168,500
Drummondville, QC	63,500	
Granby, QC	64,500	
Hull, QC	72,000	255,000
Jonquiere, QC	59,000	
Montreal, QC	1,050,000	3,324,500
Quebec, QC	169,500	690,000
Rimouski, QC	49,000	
Saint-Hyacinthe, QC	52,000	
Saint-Jean-sur-Richelieu, QC	74,500	
Saint-Jérôme, QC	59,500	
Shawinigan, QC	61,500	
Sherbrooke, QC	77,000	148,500
Trois-Rivières, QC	51,000	142,750

rice Valley; and Hull, essentially an industrial suburb of Ottawa but located on the Quebec side of the Ottawa River.

But it is the bipolar axis of Montreal and Quebec City that dominates the region. Quebec City is the hearth of French Canadian culture and serves as the political, religious, and symbolic center of French Canada. Montreal is French Canada's contribution to the world, a vibrant and exciting commercial, industrial, and financial node that shares with Toronto the primacy of all Canada.

Montreal

The city of Montreal spreads over more than one-third of the island of the same name that is located adjacent to the first major rapids on the St. Lawrence River. The city also overflows onto the nearby Île Jésus and eastward across the river into Chambly County. Its site is dominated by the hill of Mount Royal, which rises directly behind the central business district (Fig. 7-9).

Jacques Cartier discovered an Indian fort and settlement on Montreal Island. Soon a French town was sited there, because the strategic and trading potentials were superb. Montreal was a fur and lumber trading center for the French, but its principal business and industrial growth occurred under the British. It was long the largest city in Canada, besides being by far the nation's leading port (Fig. 7-10).

With its sprawling suburbs, numerous industrial districts, massive skyscrapers, and heavy traffic, Montreal has much the look of any other North American metropolis. But

FIGURE 7-9   Looking southeastward across the central business district of Montreal, from Mount Royal toward the St. Lawrence River (TLM photo).

FIGURE 7-10   During most winters the harbor of Montreal is closed by ice for a few weeks. Tugs and icebreakers help to extend the navigation season (TLM photo).

it is about two-thirds Francophone and is called the second largest French-speaking city in the world. The remainder of its population mix is quite varied, and the life of the city is cosmopolitan. Old World charm shows in the streets and squares of the older sections; the excitement of modern architecture dominates Place Ville Marie and its downtown surroundings; the attractive and efficient Metro subway system gives a fresh dimension to internal transport; and Mirabel airport is one of the largest and most modern in the world.

Montreal does, of course, have problems; they are profuse. Because of various economic and political factors, the most conspicuous being the imposition of French culture on local business by the provincial government, the population and economic growth of Montreal stagnated significantly in the mid-1970s and 1980s. Toronto has now far surpassed Montreal as both the leading population and financial center of the nation, and today the Montreal Stock Exchange ranks third behind those of Toronto and Vancouver. Some 300 significant business firms have shifted their operations from Montreal to another province, usually Ontario. Endemic unemployment, precarious employment, and an aging population have made Montreal "a city of renters." [6]

Both the city of Montreal and the province of Quebec are struggling with massive budget deficits, a significant part of which is a holdover from the huge debts incurred in constructing facilities for the 1976 Olympics.

Yet, imaginative, and usually expensive, efforts continue to be made for the purposes of improving the livability and ambience of the city and shoring up its economy. One of the most successful has been a downtown rejuvenation that may represent the continent's most successful approach to urban renewal. A covered-city (largely underground) development now links some 80 acres (32 ha) in a multilevel complex that connects 1000 stores and 100 restaurants and bars with theaters, hotels, office buildings, residential high rises, and rail, bus, and Metro stations.

The port of Montreal has more than maintained its position as Canada's busiest, primarily by aggressive development of container facilities. Montreal handles about half of all Canada's container traffic, and it is the third-ranking (after New York and Baltimore) east coast container port of the subcontinent.

Quebec City

Quebec City is much smaller than Montreal, and therefore much less important to Canada, although its significance to French Canada can hardly be overstated. Its imposing site crowns 300-foot (90-m) cliffs that rise abruptly from river

[6] Dean Louder (ed.), *The Heart of French Canada: From Ottawa to Quebec City.* New Brunswick, NJ: Rutgers University Press, 1992, p. 56.

level just upstream from the Île d'Orléans, where the St. Lawrence River opens out into its estuary. The historic fort of the Citadel still stands above the cliff ramparts, connected by Canada's most famous boardwalk with a castlelike hotel, the Château Frontenac, whose imposing turrets command a breathtaking view downriver (Fig. 7-11).

The old walled portion of the city (called Upper Town) with its narrow, cobbled streets adjoins the Citadel. Below the cliffs lies Lower Town, also with a large older section of narrow streets and European charm. More modern residential and commercial areas sprawl to the north and west. Industrial districts are chiefly close to the harbor, which is increasing its shipping through expanded container facilities and efforts to keep the port open all winter. Although the importance of manufacturing to the economy is growing, Quebec City's major functions are administrative, commercial, and ecclesiastical. Tourism is another important activity in this city, which is probably the most picturesque of the continent.

TOURISM

The unique cultural attributes of French Canada make it a most attractive goal for tourists from non-French parts of the continent, and the large population centers of southern Ontario and northeastern United States are near enough to make accessibility no problem. The landscape, architecture, institutions, and cuisine are the principal attractions; tourist interest centers on things to see rather than things to do except perhaps in Montreal.

There are a number of areas to attract the nature lover and outdoor enthusiast in and near the region. The provincial parks of Mont Tremblant and Laurentides, on the nearby

FIGURE 7-11 The spired turrets of Chateau Frontenac (on the right) and the sprawling stone walls of the Citadel (on the left) dominate the riverside bluffs of Quebec City. Modern office towers rise in the distance, beyond the walled premises of Upper Town (courtesy Government of Quebec, Tourist Branch).

Shield margin, draw many visitors, especially from Montreal and Quebec City; many come for winter sports. New national parks have been established at Forillon in the Gaspé and La Maurice north of Trois-Rivières.

Beaches are few and inadequate. But along the St. Lawrence estuary there is a surprising amount of tourist development, often consisting of small cottages situated above mud flats, and summer visitors from nearby inland areas are accommodated in large numbers. Still, the principal attractions for tourists in the region are cultural. Essentially all visitors go to Montreal and most also visit Quebec City; if there is time left over, they may visit some areas of rural charm.

THE OUTLOOK

French Canada has long been an economically disadvantaged region. Although not as serious as in the Atlantic Provinces, a high rate of unemployment and an even greater amount of seasonal unemployment has persisted. These factors are most notable in the eastern part of the region—the shores of the estuary, the Gaspé Peninsula, and northern New Brunswick. In the last few years the regional economic disparity has decreased somewhat, although much of the improvement has focused in the Montreal area. This trend is likely to continue, although slowly and irregularly.

Quebec's manufacturing expansion after World War II was dramatic. The region, overall, changed from agrarian to industrial in three decades, and the general advantages of market, labor, power, and transportation are still there. Most of the industrial prosperity and growth will continue to occur in the already established centers, especially Montreal, but also Quebec City, Trois-Rivières, Hull, and Shawinigan.

The tertiary sector of the economy—trade, finance, services—has grown rapidly and accounts for much of the improvement in average per capita income in the region. Such growth should continue, although it will be focused principally in Montreal; indeed, Montreal will continue as the dominant growth center in almost all phases of the regional economy despite the problems previously discussed.

The primacy of Montreal is such that the provincial economy has been summed up as "Montreal and the Quebec desert." Certainly the upstream part of French Canada participates much more fully in the advantages of urbanization and industrialization than does the rest of the region. This economic disequilibrium will undoubtedly persist and perhaps become even more unbalanced as time passes.

The French Canada Region today is in such a state of political uncertainty that economic and social conditions are difficult to forecast. If Quebec secedes from Canada and becomes a separate nation, the impact on both countries will be enormous.

SELECTED BIBLIOGRAPHY

BALDWIN, BARBARA, "Forillon—The Anatomy of a National Park," *Canadian Geographical Journal,* 82 (1971), 148–157.

BOAL, FREDERICK W. "One Foot on Each Bank of the Ottawa," *Canadian Geographer,* 37 (1993), 320–331.

CAYO, DON, "Islands in the Gulf," *Canadian Geographic,* 111 (1991), 30–43.

CERMAKIAN, JEAN, "The Geographic Basis for the Viability of an Independent State of Quebec," *Canadian Geographer,* 18 (1974), 288–294.

D'AMOUR, PIERRE, "The Richelieu," *Canadian Geographic,* 112 (1992), 20–35.

FRASER, WINSTON, "Anticosti: Big, Unknown, and Now a Park," *Canadian Geographic,* 103 (April–May 1983), 30–39.

GALT, GEORGE, "Westmount Holds On," *Canadian Geographic,* 103 (December 1983–January 1984), 8–19.

GILBERT, ANNE, AND JOAN MARSHALL, "Local Changes in Linguistic Balance in the Bilingual Zone: Francophones de l'Ontario et Anglophones du Quebec," *Canadian Geographer,* 39 (1995), 194–218.

GILL, ROBERT M., "Bilingualism in New Brunswick and the Future of L'Acadia," *The American Review of Canadian Studies,* 5 (Autumn 1980), 56.

GRENIER, FERNAND, ED., *Quebec.* Toronto: University of Toronto Press, 1972.

HAMILTON, JANICE, "Montreal's Underground City," *Canadian Geographic,* 107 (December 1987–January 1988), 50–57.

HARRIS, RICHARD C., *The Seigneurial System in Early Canada.* Madison: University of Wisconsin Press, 1966.

KAPLAN, DAVID H., "Population and Politics in a Plural Society: The Changing Geography of Canada's Linguistic Groups," *Annals,* Association of American Geographers, 84 (1994), 46–67.

LANG, K. M., "The Gaspé: A Naturalist's Wonderland," *Canadian Geographic,* 98 (February–March 1979), 36–39.

LE BOURDAIS, CELINE, AND MICHEL BEAUDRY, "The Changing Residential Structure of Montreal, 1971–81," *Canadian Geographer,* 32 (Summer 1988), 98–113.

LOUDER, DEAN, ED., *The Heart of French Canada: From Ottawa to Quebec City.* New Brunswick, NJ: Rutgers University Press, 1992.

SOBOL, JOHN, "Montreal's Mountain," *Canadian Geographic,* 111 (1991), 52–63.

SQUIRE, W. A., "New Brunswick's Hills and Mountains," *Canadian Geographical Journal,* 77 (1968), 52–57.

Megalopolis

8

The essence of the Megalopolis Region is the concentration of a relatively large population in a relatively small area. This is one of the smallest major regions in North America, encompassing only about 50,000 square miles (130,000 km²), or less than 1 percent of the continent. Its population of some 47 million people, however, is the second largest population of any region, amounting to 16 percent of the North American total.

It is a coastal region, and although not now primarily oriented toward the sea, it serves as the major western terminus of the world's busiest oceanic route that extends across the North Atlantic to Western Europe. Its role as eastern terminus of transcontinental land transportation routes and as two-way terminal (international to the east and transcontinental to the west) for airline routes is equally significant.

The region was a major early destination of European settlers. Thus most cities have a long history compared to others on the continent, and the region is rich in historical tradition. Yet it is pulsing with change.

It is the premier region of economic and social superlatives to be found in North America. It represents the greatest accumulation of wealth and the greatest concentration of poverty; it has the greatest variety of urban amenities and the greatest number of urban problems; it has the highest popula-

tion densities and the most varied population mix; and it is clearly the leading business and governmental center of the nation. Its economic and social maladjustments are legion and yet its attempts at alleviating them are imaginative and far-reaching.

Most of all, however, it is an urban region. Its geography is that of cities and supercities. It is one of the most highly urbanized parts of the world and its lifestyle is geared to the dynamic bustle of urban processes, problems, and opportunities.

EXTENT OF THE REGION

Megalopolis, the world's greatest conurbation, developed along a northeast-southwest axis approximately paralleling the Middle Atlantic and southern New England coast of the United States (Fig. 8-1). Its core is the almost completely urbanized area extending from metropolitan New York across New Jersey to metropolitan Philadelphia. It extends northeastward from New York City through a number of smaller cities to metropolitan Boston and southwestward from Philadelphia to Baltimore and Washington.

This metropolitan complex from Boston to Washington has been recognized as a major urban region for a number of

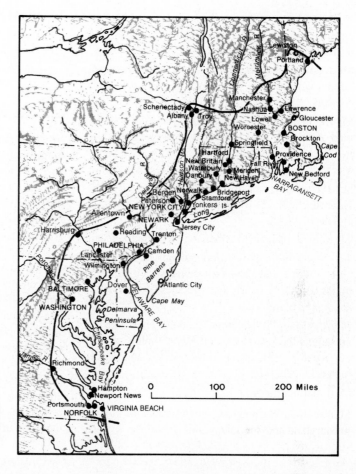

FIGURE 8-1 The Megalopolis Region (base map copyright A.K. Lobeck; reprinted by permission of Hammond, Inc.).

years. The concept of a unified Atlantic seaboard metropolitan region was notably publicized by Jean Gottmann's epic study of 1961, in which he adopted the ancient Greek name *Megalopolis* to apply to the region.[1] Since the time of Gottmann's work, continued urban expansion at either end has made it logical to extend the area under consideration northward in New York and New England, westward in Pennsylvania, and southward in Virginia.

The Atlantic coastline marks the eastern margin of the region, which means that fairly extensive rural areas, particularly in southeastern Massachusetts, southern New Jersey, the Delmarva Peninsula, and eastern Virginia, are included within the confines of Megalopolis. These lands, however, are used primarily for urban-serving agriculture or recreation and can thus logically be accepted as part of the region.

The western boundary of Megalopolis is considered to be where urban population densities and land-use patterns

fade and are replaced by rural densities and patterns to the west of the principal urban nodes of the region. This zone is fairly easy to demarcate in the south, where the Hampton Roads cities, Richmond, Washington, and Baltimore, are involved. West and northwest of Philadelphia the distinction is less clear, as is the case north of New York City and north of Boston. In the first-mentioned district the megalopolitan boundary is drawn to include most of southeastern Pennsylvania as far west as the middle Susquehanna Valley around Harrisburg. North of New York City the boundary is drawn to include that portion of the Hudson Valley as far as the Albany–Schenectady–Troy metropolitan area. North of Boston the extent of Megalopolis is considered to include the urbanized lower portion of the Merrimack Valley as far as Concord and the urbanized coastal zone as far as Portland and Lewiston.

As thus delimited, the Megalopolis Region encompasses a number of significant urban nodes and complexes, most of which have important interconnections while remaining relatively discrete urban units.

---

[1] Jean Gottmann, *Megalopolis, The Urbanized Northeastern Seaboard of the United States* (New York: Twentieth Century Fund, 1961).

1. In southern New Hampshire and northern Massachusetts are the old but still highly industrialized mill towns of the Merrimack Valley.

2. Metropolitan Boston, with its many suburbs and related towns, spreads over much of eastern Massachusetts.

3. Metropolitan Providence includes most of Rhode Island and small cities of related economic structure in adjacent Massachusetts.

4. The Lower Connecticut Valley area includes a number of closely spaced, medium-sized cities, especially Springfield, Hartford–New Britain, Waterbury, and New Haven.

5. The urbanized node of Albany–Schenectady–Troy.

6. The New York City metropolitan area sprawls over parts of three states and encompasses such adjacent major cities as Newark, Jersey City, Paterson, Elizabeth, and Stamford.

7. Metropolitan Philadelphia includes an extensive area along the lower Delaware River Valley, including Trenton and Camden in New Jersey and Wilmington in Delaware.

8. Metropolitan Baltimore.

9. Metropolitan Washington, which spreads widely into Maryland and Virginia.

10. Metropolitan Richmond.

11. The Hampton Roads urban complex includes the urbanized areas on either side of the mouth of the James River estuary.

CHARACTER OF THE REGION

Nowhere else on the continent is there such a concentration of the physical works of humans to dominate the land and the horizon for square mile after square mile (Fig. 8-2); yet there is a surprisingly rural aspect to much of the region. Despite the prominence of skyscrapers, controlled-access highways, massive bridges, extensive airports, and noisy factories, the green quietness of the countryside is also widespread.

There are many places in Rhode Island, a tiny state but one with the second greatest population density, where a person can stand on a viewpoint and look in all directions and see no evidence of people or their works; there is nothing but trees and clouds and sky (Fig. 8-3). In an extensive area in southern New Jersey, that most urbanized of states, birds nest, streams run pure, and forests are virgin. The deer population of the region is now greater than it was a century ago. Many areas actually have more woodland than they did 10 or 20 years ago because of farm abandonment. For Megalopolis as a whole it is estimated that only about 20 percent of the land is in urban use.

FIGURE 8-2 Megalopolis is a hive of urban activity, as indicated by this northeasterly view from the top of the Empire State Building (TLM photo).

FIGURE 8-3 Even in Megalopolis there are spacious areas of natural landscape, with little evidence of human imprint. This forested scene is in southeastern Massachusetts (TLM photo).

From the viewpoint of an orbiting satellite it would probably be the interdigitation of urban and rural land uses that would be the most prominent aspect of the geographical scene. But in no way is the region entirely urbanized; neither does it form any sort of a single urban unit. Instead, there are major nodes of urbanization that are growing toward one another, usually along the radial spokes of principal highways, with much rural land between the spokes. In some cases, particularly around New York, Philadelphia, and Boston, nodes coalesce; but in most parts of the region the nodes remain discrete, and a great deal of nonurban land is interlocked.

Another important facet of the regional character is its dependence on interchange with other regions and countries. Although the full range of urban functions occurs in multiplicity in Megalopolis, the region generates a monstrous demand for primary goods (foodstuffs and industrial raw materials), most of which it cannot produce. There is an almost equally significant need for the movement of regional products to extraregional markets; thus major transportation terminals, which are numerous, are loci of much activity and of extensive storage and processing facilities.

Most significant of all, in assessing the geographic character of Megalopolis, is probably the intensity of living that prevails there. This factor is shown in its crudest form by the high population density, both as a general average for the whole region and more specifically for the urbanized nodes. The average population density for the region is nearly 900 persons per square mile. New Jersey, in the heart of the region, is the first state to be classified as "totally urban" by the Bureau of the Census.[2] Its average density is nearly 1000 per square mile. Population density reaches remarkable levels in some cities, the highest being on Manhattan Island with a nighttime density of more than 80,000 and a daytime density of almost 200,000 per square mile.

Many other elements are involved in "intensity of living." Where people live close together, we also find a clustering of structures, activities, and movements. Such agglomeration is advantageous in providing concentrated opportunities for variable want satisfactions within a limited space. Yet it also engenders the handicaps associated with crowding: waste of space and time, frustration and psychological trauma, pollution and health problems, stifled transportation, and so on.

THE URBAN SCENE

The urban complex of Megalopolis is almost overwhelming in its magnitude and diversity. Within the region are over

---

[2] This means that every county in the state has reached the Census Bureau's definition of "urban"; that is, it includes a central city and a surrounding closely settled urban fringe that together have a population of 50,000 or more.

1000 places classed as urban by the Census Bureau, more than 4000 separate governmental units, and a population of about 47 million, of which nearly 90 percent is urban (see Table 8-1 for a listing of the region's largest urban places). For convenience, the region is subdivided into 11 urban groupings, to be considered successively from north to south (Fig. 8-4).

The Merrimack Valley

With a total population of about 800,000, the Merrimack Valley of New Hampshire and Massachusetts includes five small cities: Manchester, Nashua, Lowell, Lawrence, and Haverhill. These former mill towns, highly industrialized and specialized toward textiles and shoes, mirrored the economic pattern and problems of southern New England for many decades (Fig. 8-5). As New England's advantages for making these products declined, areas that could not find satisfactory replacements were subjected to increasing unemployment and economic distress.

After the prosperous years of the 1920s, the textile industry, which employed well over half of all industrial workers in the valley, declined to a fraction of its former magnitude. The manufacture of leather goods, primarily shoes, decreased much less precipitously and is still the leading type of manufacturing in the valley. Electrical machinery production was the principal growth industry and there has been increasing diversification. The valley is a low-wage area with generally depressed incomes, but during the last several years rapid economic and population growth have occurred, particularly in the New Hampshire portion.

Metropolitan Boston

The dominant city of New England is located at the western end of Massachusetts Bay, a broad-mouthed indentation that is generally well protected by the extended arms of Cape Ann and Cape Cod. The site of the original city was a narrow-necked peninsula of low glacial hills and poorly drained swamps. The Charles and Mystic Rivers flow into the bay, and much of the older part of the city now occupies land that had to be drained before construction could take place.

Shortly after the founding of Boston in 1630, other towns, such as Cambridge and Quincy, were established on better-drained land nearby. Boston with its numerous protected wharves was, however, the dominant settlement from the outset. By the end of the colonial period it was exceeded in size nationally only by Philadelphia and New York, and it maintained this third-ranking position for more than a century.

Boston's central business district was extended after further drainage projects were completed but remains concentrated chiefly near the original nucleus. Urban sprawl

TABLE 8-1

Largest urban places of the Megalopolis Region, 1995

Name	Population of Principal City	Population of Metropolitan Area
Albany, NY	96,000	876,000
Alexandria, VA	123,000	
Allentown, PA	110,000	615,000
Atlantic City, NJ	36,000	338,000
Baltimore, MD	730,000	2,500,000
Barnstable, MA	48,000	138,000
Bergen, NJ	78,000	1,302,000
Boston, MA	551,000	3,400,000
Bridgeport, CT	135,000	442,000
Brockton, MA	94,000	240,000
Chesapeake, VA	182,000	
Danbury, CT	65,000	201,000
Dover, DE	63,000	121,000
Elizabeth, NJ	110,000	
Fall River, MA	79,000	
Fitchburg, MA	40,000	140,000
Hagerstown, MD	36,000	128,000
Hampton, VA	141,000	
Harrisburg, PA	53,000	612,000
Hartford, CT	133,000	1,163,000
Jersey City, NJ	230,000	554,000
Lancaster, PA	61,000	438,000
Lawrence, MA	58,000	354,000
Lowell, MA	95,000	287,000
Manchester, NH	98,000	183,000
Nashua, NH	78,000	170,000
Nassau–Suffolk, NY		2,650,000
New Bedford, MA	91,000	175,000
New Haven, CT	125,000	531,000
New London, CT	29,000	293,000
New York, NY	7,315,000	8,560,000
Newark, NJ	265,000	1,930,000
Newport News, VA	187,000	
Norfolk, VA	256,000	1,550,000
Paterson, NJ	140,000	
Philadelphia, PA	1,545,000	4,955,000
Portland, ME	64,000	216,000
Portsmouth, NH	25,000	222,000
Portsmouth, VA	105,000	
Providence, RI	155,000	1,135,000
Reading, PA	79,000	350,000
Richmond, VA	203,000	930,000
Springfield, MA	156,000	586,000
Stamford, CT	108,000	336,000
Trenton, NJ	91,000	330,000
Virginia Beach, VA	440,000	
Washington, DC	570,000	4,451,000
Waterbury, CT	110,000	230,000
Wilmington, DE	71,000	535,000
Worcester, MA	168,000	490,000
Yonkers, NY	189,000	
York, PA	42,000	370,000

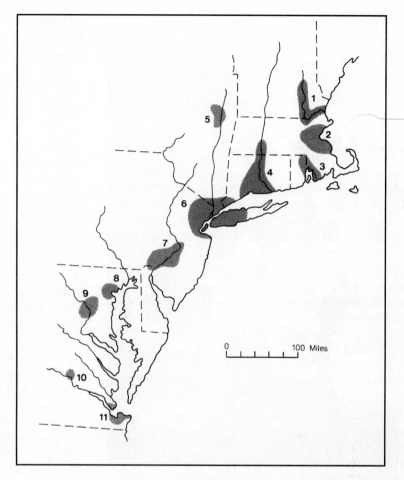

FIGURE 8-4 The urban complexes of Megalopolis: (1) the Merrimack Valley, (2) Metropolitan Boston, (3) the Narragansett Basin, (4) the Lower Connecticut Valley, (5) Albany–Schenectady–Troy, (6) Metropolitan New York City, (7) Metropolitan Philadelphia, (8) Metropolitan Baltimore, (9) Metropolitan Washington, (10) Metropolitan Richmond, and (11) Hampton Roads.

widely expanded the city to the north, west, and south of the original area and many ancillary settlements were absorbed into the metropolis.

Although early significant as an industrial center, the economy of Boston became increasingly oriented toward tertiary and quaternary activities—trade, services, and government employment—which provide about three-fourths of the jobs in the metropolitan area. The port of Boston is a busy one, but the lack of an extensive hinterland has resulted in a relative decline in its importance. It is a major importing port, but in total trade it is outranked by four other ports in the Megalopolis Region alone.

The industrial component of Boston's economy is a prominent one and, as with most cities, tells much about the distinctiveness of the city's function. Although textiles and leather goods were the traditional mainstays of manufacturing in New England, the former has not been important in Boston for several decades, and the latter has been on an erratic but declining trend for several years, even in the historic shoemaking suburbs of Brockton and Beverley.

Boston has long been a major industrial center, and, despite problems, continues to rank among the national leaders

in this regard. Virtually all cities of the northeastern United States have experienced significant relative declines in their industrial importance in the last quarter century, but Boston less than most. As traditional industries declined, eastern Massachusetts experienced a significant industrial resurgence in "high-technology" manufacturing, which involves electronics, pharmaceuticals, aerospace, instrumentation, weaponry, and especially computers. By the mid-1980s more than one-third of all manufacturing in Massachusetts consisted of the high-technology category, a higher proportion than that of any other state. Many new firms represent spinoffs from older companies that are attracted to the Boston area by the availability of skilled labor, managerial competence, inventive genius, venture capital, research universities, and the advantage of technological interrelationships (agglomeration economies) with other local firms. High-technology industries are now as firmly established in Boston as in California, a development that contributed (along with rapid growth in the service sector) to the lowest unemployment rate in the nation (2 percent in 1987). Since then, however, Boston's economy has lost much of its vigor. Many manufacturing jobs have disappeared, the unemployment rate is climbing, and

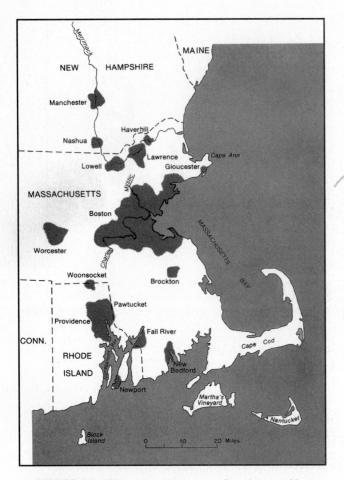

FIGURE 8-5   The metropolitan areas of southeastern New England.

"electronics parkway"). The pattern is now being repeated on Interstate 495, an outer beltway some 25 miles (40 km) from the central business district, with many planned industrial parks at intersections of the freeway and express arterials from downtown.

Boston has long been known as an elitist city and cultural center with many attractions for living, yet numerous problems exist. Lack of money to meet governmental expenses is at the root of many of them. Most of the city budget comes from property taxes and yet almost half of all municipal property, such as government buildings, churches, and educational institutions,[3] is untaxed. The city population in the past was dominated by Europeans, especially Irish and Italians. Recent years have seen a rapid influx of other ethnic groups, particularly African Americans and Puerto Ricans, and intergroup hostility has become a major social problem. African American and Puerto Rican residential areas are more highly segregated than in perhaps any other major North American city and the areas of lowest per capita income coincide precisely with these districts.

Even so, metropolitan Boston continues to grow, and the average family income is one of the highest of any large city. Imaginative attempts are underway to assuage two urban problems for which Boston has long been famous: reorganization and extension of mass transit facilities to prevent traffic strangulation of an auto-oriented city that is plagued by an infamous maze of narrow, twisting streets and a major start toward metropolitan government. Boston remains the primate city of northern and eastern New England and continues as the northern bastion of Megalopolis.

Narragansett Basin

New England's second largest city and second-ranking industrial center is Providence, situated where the Blackstone River flows into the head of Narragansett Bay. The Providence area, with its industrial satellites of Pawtucket and Woonsocket to the north and Fall River and New Bedford to the southeast, represents in microcosm the industrial history of southeastern New England, demonstrating graphically the economic displacement that occurs in a heavily industrialized area when its leading industry falters.

From colonial days southern New England was the center of the American cotton textile industry. Initial advantages included

1.   Excellent water power facilities.
2.   A seaboard location for importing cotton.

the state is mired in deficit financing. Still, there is a solid base of high-tech industry and considerable optimism for another economic turnaround.

As one of the oldest major cities on the continent and one that was quite prominent in colonial times, Boston maintains an air of historic charm in many of its older sections. Narrow twisting streets, colonial architecture, and tiny parks are widespread. But modernity is increasingly evident even in the central city, as evidenced by skyscrapers, freeways, an airport on reclaimed land in the bay, and even a multilevel parking garage under the famed Boston Common (Fig. 8-6). Boston also has one of the first beltways completed around a major American city; the Circumferential Highway (Route 128), at a radius of about 12 miles (19 km) from the city center, has become a major attraction for light industry and suburban office buildings (it is sometimes referred to as the

---

[3] Metropolitan Boston has more students per capita than any other American conurbation.

FIGURE 8-6 A waterfront view of downtown Boston (courtesy Greater Boston Convention and Visitors Bureau).

3. Damp air, which was essential to prevent twisting and snarling during spinning and to reduce fiber breakage in weaving.
4. Skilled labor.
5. Clean, soft water.
6. Location in a major market area.

After the 1920s, most of the cotton textile industry relocated in other areas, particularly in the southern Appalachian Piedmont, where labor, taxes, power, and raw materials were all less expensive. The greatest concentrations of cotton textile factories in New England were in the Narragansett Basin and the Merrimack Valley; their attrition has been a damaging blow to the local economy. The last cotton-weaving mill in Rhode Island ceased operation in 1968, and both cotton spinning and weaving are almost gone from adjacent parts of Massachusetts.

The woolen textile industry of New England prospered longer. In the early days there was local wool in addition to water power, skilled labor, and an appreciable nearby market. Moreover, Boston has always been the major wool-importing port of the nation. Rapid decline engulfed New England's woolen textile industry in the late 1940s and the downward trend has continued at a slower pace. The Narragansett Basin has been particularly hard hit. The textile industry once employed more than half of all manufactural workers in Providence; it now provides jobs for barely one-tenth of the total.

It should be noted that even with a declining rate of industrial employment, the Narragansett Basin cities are still heavily dependent on manufacturing. Indeed, Rhode Island is exceeded only by North Carolina and South Carolina in the proportion of labor force employed in factories. Providence itself, with a preeminent rank as producer of jewelry and silverware, is still among the two dozen leading industrial employment centers in the nation.

Connecticut Valley

The broad lowland drained by the lower Connecticut River contains a number of prosperous, medium-sized cities, with a total population of about 2.7 million (Fig. 8-7). The Springfield–Chicopee–Holyoke complex in Massachusetts and Hartford in Connecticut are located on the river; the other cities of New Britain–Bristol, Waterbury–Naugatuck, Meriden, New Haven, and Bridgeport are a bit farther west.

This is a long-standing industrial district of importance. Factories specialize in diversified light products requiring considerable mechanical skill: machinery, tools, hardware, firearms, brass, plastics, electrical goods, electronics equipment, precision instruments, watches, and clocks. These are mostly products of high value and small bulk, which require little raw material and can easily stand transport charges to distant markets.

This southwestern part of New England has adjusted fairly well to changes in the economy. The flight of textiles has not resulted in nearly so many abandoned factories; instead, other types of manufacturing have been attracted by the cadre of trained workers and the available buildings. The growth in electrical machinery production has been steady, partly because of the research and product development facilities of the district. Aircraft engine manufacture is significant in Connecticut even though airframe assembly is accomplished in other parts of the country. Various kinds of machinery and hardware and other light metal goods are produced in quantity in several cities.

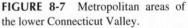

FIGURE 8-7   Metropolitan areas of the lower Connecticut Valley.

Although there is heavy dependence on manufacturing (Connecticut ranks fourth among the states in percentage of its labor force employed in factories), the economy of the subregion is broadly diversified and includes a notable concentration of the nation's insurance industry in Hartford. The commercial orientation of the valley is largely westward toward New York, which apparently is an important catalyst.

This is a district with no major cores. Nineteenth-century industrial cities have persisted but have not dominated. Instead, many nodes of specialized activities have emerged. The result is a noncentric population pattern, with much daily movement among a wide range of foci.

Albany–Schenectady–Troy Metropolitan Area

Situated at the commercially strategic confluence of the Mohawk and Hudson Rivers, the old Dutch settlement of Albany has become a metropolitan area of 850,000 people clustered about multiple nuclei in parts of four counties (Fig. 8-8). As the upstream end of the Hudson River axis and the eastern end of the low-level Mohawk corridor to the Great Lakes, the Albany area had major crossroads significance from its earliest days. By late in the 1600s Albany was the leading fur-

trading center for the English colonies and a century later it was declared the state capital. The completion of the Erie Canal in 1825 was a major stimulus to economic prosperity and population growth. Despite many changes in transportation orientation, the urban node still functions as a transshipment point between river or ocean traffic on one hand and canal or overland traffic on the other.

The area has been an important industrial center for many decades. Its initial raw material advantages (sand, limestone, and nearby iron ore) provided an early start for manufacturing, and the industrial component of the economy remained strong through the years. Troy has been noted as a center for making men's shirts; Albany was a wood-manufacturing center; and Schenectady was most famous for General Electric and the American Locomotive Company.

The area's industrial importance later declined, with a commensurate expansion in trade, services, and, especially, government employment.

Metropolitan New York

The metropolitan area of New York City, which occupies two dozen counties in parts of three states, is the core of

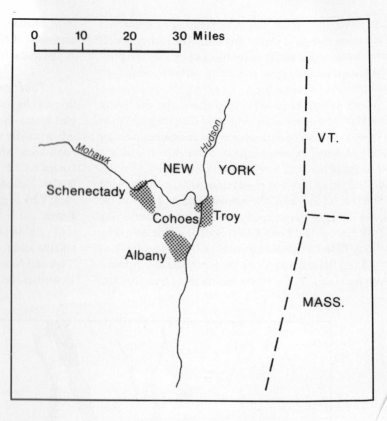

FIGURE 8-8   The Albany–Schenectady–Troy metropolitan area.

Megalopolis. It is the principal city of the United States and, in many ways, the most important if not dominant city in the world. The massing of people and activities around the mouth of the Hudson estuary represents Megalopolis at its most intensive.

This metropolis, census-defined as the New York Consolidated Metropolitan Statistical Area (CMSA), contains nearly 40 percent of the population of the Megalopolis Region, a total of about 18 million people. The CMSA's population, however, has not increased since the early 1970s. Despite this apparent stagnation, there have been massive changes. The white population has declined by 2.5 million, the African American population has increased by 2 million, and there has been commensurate growth in numbers of Hispanics (especially Puerto Ricans) and Asians (notably Chinese). Since 1970 the central city population has diminished by three-quarters of a million, but the CMSA population has remained about the same, so the suburbs clearly have continued to grow.

The economic primacy of New York cannot be satisfactorily explained in any simple fashion. It is the result of a continuum of complex interactions that are worldwide in scope and spread over several centuries in time. In elemen-

tary terms, it can be viewed as the result of an ₑ struggle among several competing ports to doᵢ juncture of two major trade routes: the North Aₗ ping lanes to Europe and the continental connₑ North American interior. New York's locatioₙ ly appear to be more advantageous than thatₒ petitors, notably Baltimore, Boston, Haliₓ especially Philadelphia. The others, howₑ various handicaps. Baltimore, Philadelₚ had to be reached by tortuous navigatiₒ and Montreal was iced in for sevₑₗ Both Boston and Halifax had lesₛ New York, and Boston, in particuₗ water in its harbor.

New York's major advₐ or. The availability of an eaₛ west via the Hudson River ₐ York a preeminence overₑ that was never relinquisₕ

New York has ₐ It is located at the mₒ well-protected harₖ which keeps the ₑ

lantic is direct and broad. The tidal range is so small that ships can come and go at almost any time. The harbor is well protected from storms and is never blocked by ice. The principal handicaps involve rapid and tricky currents and the occasional presence of dense fog.

A look at any large-scale map shows the exceeding fragmentation of the city's site (Fig. 8-9). This fragmentation was a boon to waterborne commerce but constitutes a major problem to internal communication. Manhattan Island is a narrow finger of land [13.5 miles (21.6 km) long by 2.3 miles (3.7 km) wide at its broadest point] situated between the wide Hudson River on the west, the narrow Harlem River on the north, and the East River (really a tidal estuary connecting New York Harbor and Long Island Sound) on the east (Fig. 8-10). Long Island extends eastward for 100 miles (160 km) between Long Island Sound on the north and the Atlantic Ocean on the south. Staten Island, nearly three times the size

of Ma...
southwe...
es add to the...

Function

tion and its location a...y shore to the
port routes, New York C...s, and marsh-
urban activity in the United S...re.
retail sales, wholesale sales, fore...s popula-
commercial activity, New York is the...air trans-
leader. Indeed, the gross regional pro...fields of
New York is greater than that of such count...d—of
Brazil.

Although with diminished dominance, New...itan
still the leading U.S. port. The bi-state Port Authority o...
York and New Jersey operates this largest port in the nati...
as well as the largest air terminal system in the world, the...

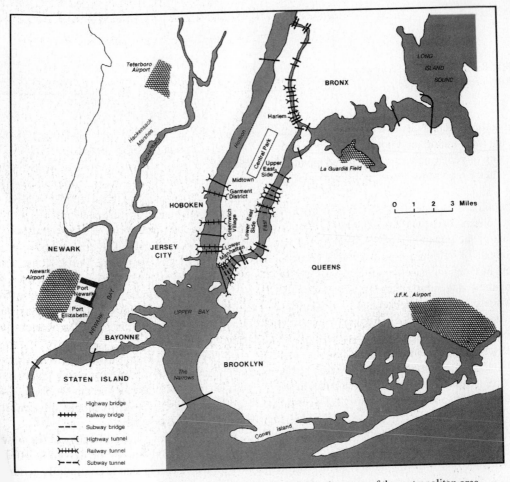

FIGURE 8-9   Manhattan Island and its connections with other parts of the metropolitan area.

FIGURE 8-10   The island of Manhattan extends southward from the distant Bronx mainland, sandwiched between the Hudson River on the left and the East River on the right. The twin towers of the World Trade Center dominate the Lower Manhattan scene at the southern tip of the island (courtesy Port Authority of New York and New Jersey).

most heavily used tunnels and bridges in the world, and the world's largest office complex (the World Trade Center in Lower Manhattan).

Although commerce contributed more to the city's primacy than industry, New York was long the leading manu-facturing center in the nation. Although no longer the leader, New York ranks in the top five cities in industrial employ-ment. However, the industrial structure is not highly diversi-fied. One type of manufacturing, the making of garments, predominates. More clothing is made in New York than in the next four largest garment-manufacturing centers combined. Most clothing factories are concentrated in an area of 200 acres (80 ha) in central Manhattan, where they occupy the upper floors of moderate-height buildings and are so inconspicuous as to pass unnoticed by the casual visitor.

New York City continues as the leading financial center of the world. The great money market is focused in the Wall Street district of Lower Manhattan, where a proliferation of financial specialists can provide external economies to businesses at all scales of operation. As a result of the great concentration of financial institutions, the velocity of demand deposits (relative frequency of use of a deposited dollar) is greater than elsewhere in the nation. Furthermore, the three securities exchanges (stock markets) in this district handle nearly nine-tenths of the organized stock and bond transactions of the country.

As a headquarters and managerial center, New York is also unsurpassed. It houses far more corporate headquarters than any other city despite recent decentralization tendencies that have seen literally thousands of offices, and tens of thousands of white-collar jobs, shifted to the suburbs and beyond. Still, the wide variety of business services and skills available in New York provides a spectrum of expertise that is not remotely approached by any other city. Indeed, the conglomeration of office space is easily the world's largest. Within 50 miles (80 km) of Midtown Manhattan is found 40 percent of the total U.S. office space; Manhattan alone has four times that of Chicago.

New York's entertainment, cultural, and tourist functions are outstanding. The city is national leader in theaters, museums, libraries, art galleries, mass media headquarters, higher education institutions, hotel rooms, and most other significant urban amenities.

Morphology   The political city of New York is composed of five boroughs: Manhattan, Bronx, Queens, Kings (Brooklyn), and Richmond (Staten Island). The metropolitan area, however, sprawls widely on the mainland and on Long Island (Fig. 8-11).

Manhattan Island is the nerve center of the city. The island consists of an ancient, stable rock mass that is rigid enough to support the tall buildings for which the city is famous. In the middle of Manhattan is Central Park, one of the largest and best-known green spaces in the world. Just to the south is the Midtown district, which contains the world's greatest collection of skyscrapers (Fig. 8-12). Slightly southwest of this area and across Times Square is the garment district, an area of moderate-sized buildings, clogged streets, and jampacked sidewalks. Farther south is the old "Bohemian" district of Greenwich Village, now fashionable again, and the more recent, art-oriented neighborhood of Soho. At the southern end of the island is Lower Manhattan, dominated by the Wall Street financial district and another cluster of skyscrapers. The southeastern section is called the Lower East Side, an area with a lengthy history of crowded tenements, low incomes, and ghettoized ethnic groups (Chinatown, Little Italy, the Bowery, the East Village, and so on). To the east of Central Park is the Upper East Side, an area known for its affluence. Northeast of Central Park is the expanding Puerto Rican ghetto of Spanish Harlem (Fig. 8-13),

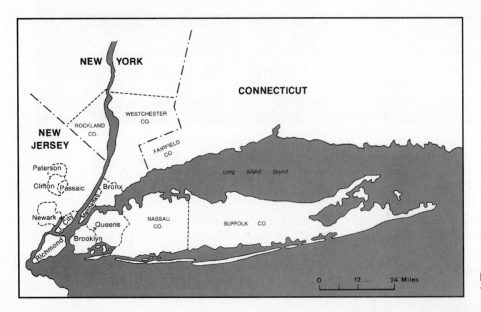

FIGURE 8-11 Metropolitan New York City.

FIGURE 8-12 The typical bustle of a Midtown Manhattan street corner (TLM photo).

and north of the park is the long-established African American Harlem district (Fig. 8-14).

Manhattan is a veritable hodgepodge of functional areas. It includes the greatest concentration of manufacturing facilities (the garment district), commercial buildings (Midtown), and financial institutions (Lower Manhattan) on the subcontinent, yet its residential function is a major one. Detached, single-family dwellings are unknown, and the 1.5 million inhabitants of the island live in multistoried apartments. Commuters swell this population total considerably during daylight hours; it is estimated that there are more than 2 million jobs between Central Park and the Battery (southern tip of the island) and that 1.5 million of those are filled by commuters.

FIGURE 8-13 A sidewalk market on the Lower East Side of Manhattan (Robert Brenner/PhotoEdit photo).

The *Bronx is* the only borough of the city that is on the continental mainland. A moderately hilly district, it is primarily residential in function. Close settlement in apartments characterizes the southern portion, but farther north there is a sparser density and some single-family homes. The population of the Bronx is approximately equal to that of Manhattan.

The borough of *Queens* occupies more than 120 square miles (310 km^2) at the western end of Long Island, extending from Long Island Sound across the island to Jamaica Bay. It is primarily a residential borough but also has large commercial and industrial districts. Individual homes are more commonplace than in the other boroughs, but multiple residences are in the majority. Kennedy International and La Guardia, two of the four principal metropolitan airports, are located in Queens. The borough's population is about 2.1 million.

There are more people in *Kings* (*Brooklyn*)—about 2.3 million—than in any other borough. It occupies the southwestern tip of Long Island and is closely connected to lower Manhattan by bridges (Fig. 8-15) and tunnels. The business center of Brooklyn is near the eastern end of the Brooklyn Bridge and would be an imposing central business district if it were not under the shadow of Manhattan skyscrapers. Brooklyn contains considerable industrial land, but it is primarily a residential area largely developed for multifamily apartments.

The borough of *Richmond* occupies Staten Island, which until the early 1960s had no direct connection with the

FIGURE 8-14 A street scene in one of the tenement districts of Harlem (TLM photo).

FIGURE 8-15 Many bridges span the waterways of New York City. This view from Lower Manhattan toward Brooklyn shows the Brooklyn (center) and Manhattan (left) Bridges (TLM photo).

rest of the city except by ferry. As a result, much of Staten Island is still suburban or semirural. The completion of the Verrazano Narrows Bridge (the world's longest suspension bridge) to Brooklyn changed all that, and the population of Richmond has now passed 400,000 and is growing steadily.

The urban spillover into *outer Long Island* has reached beyond Queens and spread throughout most of Nassau County (population 1.4 million) and deep into Suffolk County (1.3 million). The eastern end of Long Island is still rural, but urban sprawl reaches ever farther eastward along the main transport routes.

North and east of the Bronx, the metropolitan area extends into *mainland New York.* Westchester County, with its numerous elite residential areas, has a population of nearly

A CLOSER LOOK   New York City's Changing Landscapes

Since the founding of the United States, New York City has served as a mirror of America, a microcosm of the diversity of the nation's peoples and experiences, a gateway for immigrants, a center for theater and the arts, and a tocsin hinting at future trends. The Statue of Liberty, Ellis Island, Times Square, Greenwich Village, Soho, Rockefeller Center, Harlem, the Empire State building, the Upper West and Upper East Sides of Manhattan, and Broadway are well-known icons on the American cultural landscape. The montage that is New York is the product of decades, if not centuries, of continuous change, often reflecting the city's changing role in the national and world economies. In the 1970s, for example, New York suffered a painful and traumatic deindustrialization, losing its enormous garment industry to the lower-wage regions of the South and overseas. Unemployment soared, more than a million New Yorkers left town for greener pastures in the suburbs and elsewhere in the nation, and in 1975 the city government teetered on the brink of bankruptcy.

In the 1980s, New York rode a tidal wave of growth in finance and business services. New York stands, alone with London and Tokyo, as one of the world's premier "global cities," pivotal centers of the world economy, vast conglomerations of headquarters, banking firms, insurance, advertising, legal services, and accounting companies. Wall Street has long symbolized Manhattan's role in this regard, much as Detroit once personified the automobile industry. New York has long been home to "money center" banks such as Bank of America, Chemical, Citicorp, Chase Manhattan, and Morgan Guaranty. In 1994, finance employed roughly 740,000 people in the metropolitan region. The deregulation of the world's financial markets, the globalization of finance through telecommunications and electronic funds transfer systems, which linked national markets together in a virtually seamless web, mounting foreign (especially Japanese) investment, and surging international trade all

joined to lift New York out of the crisis of the 1970s. Investment banking flourished with the growth of stocks, bonds, equities, exotic new instruments, and a wave of mergers, acquisitions, and takeovers. The volume of stocks traded on the Dow Jones climbed to ever new heights, and with it, employment and salaries in the critical securities industry. Fueled by rising incomes and waves of "yuppies," housing prices climbed out of reach for many people. Even parts of Brooklyn and Queens, which were traditionally manufacturing centers, benefited from the boom.

In the 1990s, like London and Tokyo, New York suffered another downturn. The decade saw a worldwide recession in the city's "bread and butter," banking and finance, including Third World debt and the savings and loan crisis. Reduced military expenditures forced Pentagon-oriented producers on Long Island, such as Grumman, to reduce costs. Bone-crushing international competition led to widespread layoffs as many firms "downsized" in a quest to become "lean and mean." For the unskilled, with few other opportunities, this turn of events was particularly depressing. Overbuilding in the commercial real estate market led to a glut of office space and depressed rents, dampening the construction industry. Many large services firms began to uncouple their highly skilled headquarters and low-skilled clerical (back office) functions, moving the latter out of the city to cheaper locations on the metropolitan periphery. Given the increasing locational flexibility afforded by satellites and fiber optics systems, back offices have begun to relocate on a much broader, continental scale: many New York companies have moved clerical jobs to the Midwest and even Ireland and the Caribbean, decreasing the need for low-skilled labor at home. Problems associated with the urban underclass, predominantly minorities, which was long excluded from the boom of the 1980s, have re-emerged, including high school drop-outs, teenage pregnancies, crime, and drugs. One-quarter of New

Yorkers (almost 2 million people) live below the official poverty line and receive public assistance. The homeless population, estimated at roughly 100,000 people, continues to swell. In parts of the south Bronx, AIDS threatens one in six adults. New York's infrastructure is woefully out of date, including JFK airport, one of the nation's largest, as well as its vast labyrinth of bridges and tunnels; New York is a "first world city with a fourth world infrastructure." Politicians elected on antitax platforms at the national and state levels have deprived the city of badly needed revenues. The municipal government has, accordingly, slashed services, including the poorly funded and desperately overcrowded school system.

Despite these travails, New York continues to attract immigrants from abroad. In addition to its long-established African American population, new arrivals from the Third World have made New York a city populated by a majority of non-whites. The city's Latino/Hispanic population—traditionally Puerto Ricans—has been joined by growing pools of recent Latin American arrivals from the Dominican Republic, Cuba, Mexico, and Colombia. Other new New Yorkers arrive from the West Indies (Jamaica, Trinidad, Barbados, Haiti), Russia, India and Pakistan, Korea, Hong Kong and China, and Vietnam. In many communities, immigrants have dramatically reshaped the housing market and retail landscapes, contributing to a new round of prosperity and diversity.

Given the rapid and numerous changes New York has been through, the only sure guess is that the future will bring more. In "the city that never sleeps," New Yorkers are used to constant, frenzied change. Accordingly, the New York of the 21st century will undoubtedly generate more than its share of unexpected surprises.

Professor Barney Warf
Florida State University
Tallahassee

1 million. Rockland County (300,000) and Fairfield County (in Connecticut) are on the fringes of growth.

On the *Jersey side* of the Hudson, the urban agglomeration extends north, south, and west so that essentially all of northeastern New Jersey is encompassed. Urbanization is not complete in the area, but it is both expanding and intensifying. The Jersey City peninsula has a population of some 600,000; the Newark area, which occupies three counties, has more than 2 million; and the Paterson–Clifton–Passaic area claims over 1.5 million.

New Jersey is a long-maligned state that is now experiencing a prospering economy based on high-tech financial services, defense contracts, booming office construction, and mushrooming suburbs. The state's unemployment rate is well below the national average. Much of the prosperity is within the New York metropolitan area, where even the long-blighted Jersey shore of the Hudson River is now referred to as the "Gold Coast," with rotting piers and aging warehouses giving way to luxury condominiums and high-tech offices.

Of great significance to the entire metropolitan area are the reclaimed marshlands along the western shore of the Hudson estuary and New York Harbor, which are used for heavy industry and nuisance industries (such as oil refineries and chemical plants), railway marshaling yards, new sports facilities, and the predominant share of New York's present port facilities. The containerized cargo–handling facilities of *Port Elizabeth* and *Port Newark* are among the world's finest and most extensive, and most of New York's general cargo traffic funnels through these facilities. In addition, there are a scrap steel terminal, a pelletized lumber port, an automated (pipeline) wine terminal, and an expansive area for imported automobiles.

Metropolitan Philadelphia

Metropolitan Philadelphia, the sixth largest urban complex in North America, is not as densely populated or sprawling as New York. Its principal axis lies parallel to the lower course of the Delaware River, extending from Wilmington, Delaware, in the southwest to Trenton, New Jersey, in the northeast, a distance of some 50 miles (80 km) (Fig. 8-16). Its southeast-northwest dimension is less than half as great, reaching from Camden, New Jersey, up the Schuylkill Valley to Norristown, Pennsylvania.

William Penn selected and planned the initial settlement in 1682 on well-drained land a short distance north of the marshy confluence of the Schuylkill and Delaware Rivers. Penn's rectangular street pattern with wide lanes and numerous parks was well designed for a relatively small city. As Philadelphia grew, however, it absorbed other settlements, so that the street layout is heterogeneous and irregular

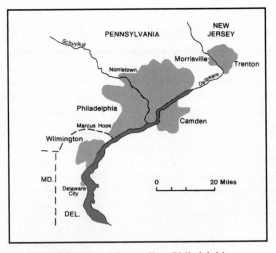

FIGURE 8-16   Metropolitan Philadelphia.

away from the present central business district. Hills, marshes, and valleys further complicated the pattern in outlying areas. Converging highways and railways produced a radial design for the major thoroughfares.

Much has been written about Philadelphia's misfortune in not having an easy access route to the interior of the continent. It is true that the ridge-and-valley section of the Appalachians, lying athwart the direction of westward penetration, inhibited Philadelphia's competitive position vis-à-vis New York. Nevertheless, Philadelphia was the largest city in North America for most of the eighteenth century, and the Pennsylvania Railroad, completed in the mid-1800s, gave Philadelphia a much more direct (if more costly) route to the Midwest than New York's Hudson–Mohawk corridor provided. For many years the "Pennsy" was one of America's busiest and most prosperous railways.

The harbor activities of the lower Delaware are administered by a tristate authority that coordinates the lengthy port facilities and maintains a dredged channel as far upstream as Morrisville. Both banks of the river are lined with piers and wharves downstream from Philadelphia for several miles, continuing on the right bank below Wilmington to New Castle and Delaware City. Oil refineries, chemical plants, shipyards, and other heavy industrial facilities are also numerous in riverside sites. Most of the port business consists of imported raw materials, particularly petroleum (from both overseas and coastwise traffic) and iron ore (largely from Venezuela and Canada). In comparison with New York, the Delaware River port complex is minor as an exporter.

Metropolitan Philadelphia is a major manufacturing center. Its industrial structure is extremely well balanced compared to other urban centers of Megalopolis; indeed, it is considered the most diversified in the United States. Philadelphia has large apparel and publishing industries, but unlike New York City, they do not dominate the manufactural scene. Also important are the electrical and nonelectrical machinery industries, which together provide jobs for almost one-fourth of the area's factory workers. Total manufacturing employment has declined for several years, but a steady increase in nonmanufacturing jobs has somewhat compensated for the decline.

The urban problems and opportunities of North America in general and Megalopolis in particular are clearly demonstrated in Philadelphia. Here is a classic example of an older industrialized city with a dwindling manufacturing job base, a city torn by racial antipathy, a city of run-down neighborhoods left to decay in the wake of white flight to the suburbs, a city of substandard housing for many minority residents, and a city with a decaying central business district and a ballooning municipal budget deficit.

Yet a combination of public works and private-sector investment has started to revitalize the city. The downtown area experienced some of the most efficient urban renewal on the continent. Modern bridges replaced ferries across the Delaware; acres of decaying buildings, just north of Independence Hall, were removed in favor of a graceful mall; and the modern complex of Penn Center buildings now stands where the "Chinese Wall" muddle of elevated railway tracks once congested downtown traffic flow. Several inner-city residential neighborhoods broke the grip of decay through private gentrification. Portions of the waterfront are being refurbished. There is even a perceptible increase in high-technology employment. Philadelphia seems to be on the way back (Fig. 8-17).

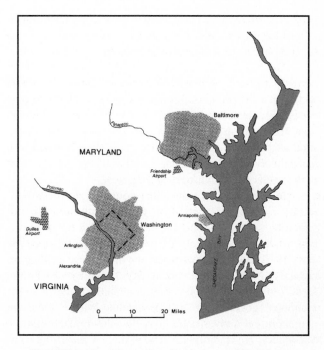

FIGURE 8-18   Metropolitan Baltimore and Washington.

Metropolitan Baltimore

Baltimore, a seaport on Chesapeake Bay, is located in what is almost the remotest corner of the bay (nearly 200 miles [320 km] from its mouth) where the estuary of the Patapsco River provides a useful deep-water harbor (Fig. 8-18). It was founded relatively late (1729) at a Fall Line power site, and its exports of agricultural produce (largely flour and tobacco) soon ensured its preeminence over Annapolis, an administrative center already eight decades old. By the end of the eighteenth century it was the fourth-ranking port in the nation; ever since it has maintained a high position among American ports, based in part on a productive hinterland, in part on excellent railway connections to the interior, and in part on prosperous local manufacturing. Even its distance from the open sea has been reduced by the construction of the Chesapeake and Delaware Canal across the narrow neck of the Delmarva Peninsula.

Although its exports (particularly grain and coal) are not inconsiderable, Baltimore is primarily an importer of bulk materials from both foreign and Gulf Coast sources; many of these materials are processed locally. Copper, sugar, and petroleum refining, as well as fertilizer manufacture, are significant port-related industries, but most prominent in this regard is the Sparrows Point steel complex. Baltimore is the nation's leader in the import of iron ore. In an effort to ex-

FIGURE 8-17   A downtown bike path in Philadelphia (Terry Wild Studios).

pand general cargo traffic, Baltimore's container-handling facilities have been greatly expanded, making Baltimore the second-ranking East Coast container port.

The city, which has been called "the most southerly northern city and the most northerly southern city," sprawls widely in all directions but is not restricted by adjacent urban centers and has considerable room for expansion. Although relatively near the larger Washington conurbation to the southwest, Baltimore's quite different economic orientation has minimized the disadvantages of the shadow effect except in the matter of air-transport development.

In common with many other northeastern cities, Baltimore experienced a period of demoralizing decay, epitomized by the fact that no new hotel or office building was constructed for a quarter of a century. However, since the early 1970s Baltimore has made one of the most striking recoveries of any North American city, fueled by a $1 billion facelift initiated by the business community rather than the municipal government, and proceeding with a focus on neighborhood revival rather than downtown rehabilitation. Showpieces are Harborplace (a $450 million waterfront complex of shops and restaurants), Charles Center (a 13-block downtown development), a new subway system, the new National Aquarium, an enormous convention center, and Oriole Park at Camden Yards, but the most innovative approach is "homesteading" (buy an old house from the city for $1, provided that you renovate it and live in it) and "shopsteading" (buy a defunct store for $100, provided that you restore it and live in it).

The dramatic revitalization of Baltimore has been very uneven, both in terms of area and people. The highly visible Inner Harbor development contrasts starkly with many decaying inner-city neighborhoods. Baltimore remains a highly segregated city; two-thirds of the census tracts are either more than 90 percent white or more than 90 percent nonwhite.

Metropolitan Washington

It is only 35 miles (56 km) from Baltimore to Washington; yet the rural area in between is still far from being completely suburbanized, and the two cities are remarkably unlike. Originally laid out in the 1790s, Washington was designed as a governmental center that was relatively centrally located to the population distribution of that time. The original District of Columbia was a square of land ceded approximately equally from Maryland and Virginia. The latter state was allowed to reannex its portion in 1846, so that the present 69 square miles (180 km^2) of the District is all on the Maryland side of the Potomac River.

Much of the original site was low lying and swampy, but extensive drainage and an elaborate municipal plan produced a city of orderly pattern and impressive appearance. The original street pattern combined the regular form of a rectangular grid with a diagonal network of avenues and traffic circles. Massive complexes of government buildings are mixed with attractive parks between the Potomac River and the central business district, with the landscaped mall that connects Capitol–White House–Washington Monument as the hub of activity. "This two-mile-long open space and its surrounding buildings are the core of monumental Washington—architectural symbol of the nation, principal mecca for millions of tourists each year, and breathing room for thousands of Washingtonians. It's also the location of 95,000 federal jobs—one-fourth of the metropolitan-area total."[4]

Both government offices and commercial districts are scattered over the metropolis, although square mile after square mile of residences, ranging from stately old mansions to modern high-rise apartments to monotonous row houses, dominate the surroundings (Fig. 8-19). In the last four decades runaway growth overwhelmed orderliness, and the metropolitan area sprawls well beyond the District's boundaries into the two adjoining states. Suburban growth outside the District has been more free-wheeling in Virginia than in Maryland because of stronger planning and growth controls in Maryland.

Overcrowding and congestion are a trademark of the Washington scene. The Capitol Beltway around the city is helpful, but routes in and out become traffic quagmires during rush hours, and inadequate bridging of the Potomac aggravates the problem. Washington is largely a white-collar city and most employment opportunities are downtown, not in the suburbs. Consequently, the daily commutation flow, on a per capita basis, is higher than anywhere else in Megalopolis. Successful attempts were made to alleviate this situation by locating some government complexes in the suburbs and by developing an elaborate new rapid transit system, the Washington Metro. The Metro is attractive, comfortable, and efficient.

An increasing proportion of the population reside in suburbs outside the District. In all suburban counties the population is growing rapidly, and probably nowhere in the eastern United States is the emergence of "suburban downtowns" more striking than in suburban Washington, especially in the Virginia suburbs. The metropolitan area now houses more than 4 million people. African Americans constitute nearly three-fourths of the population within the District proper,

[4] John R. Borchert, *Megalopolis: Washington, D.C., to Boston.* New Brunswick, NJ: Rutgers University Press, 1992, p. 48.

FIGURE 8-19 Moderate-height office buildings and multiunit residences are commonplace in and around Washington's central business district (TLM photo).

90 percent of the public school enrollment, and more than one-quarter of the total population of the metropolitan area. Unlike in most large cities, there is an increasing flow of African Americans to the suburbs, particularly southeastward into Prince Georges County.

The economic function of Washington is unlike that of any other major North American metropolis. The employment structure is unusual in four important respects:

1. Most obvious is the remarkably high percentage of federal government workers, amounting to about 20 percent of the District's work force. It is much higher than for any other major city on the country.

2. The proportion of females in the work force is abnormally high, reflecting the large number of clerical jobs in government offices.

3. The proportion of African Americans in the work force is also higher than in other large cities; the lack of hiring discrimination in federal employment accounts for much of the black population influx.

4. Only 4 percent of the jobs in metropolitan Washington are in manufacturing, which is by far the lowest proportion in the nation, and half of that 4 percent is employed in printing and publishing, indicating the great flood of government publications.

Two other facets of the city's economy are noteworthy. Its financial community has been expanding, largely as a result of increasing federal controls and regulations; in no other part of Megalopolis outside New York has the *relative* growth of financial services been as marked. Also, Washington continues to be a major tourist attraction; its many government offices, monuments, and museums are high on the list of sights to see for American and foreign visitors alike.

Metropolitan Richmond

The rapidly growing capital of Virginia is a Fall Line city on the James River, some 75 miles (120 km) south of Washington (Fig. 8-20). Its inclusion in the Megalopolis Region can-

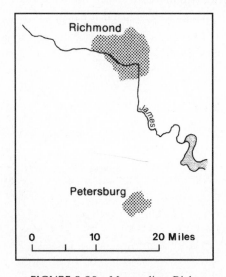

FIGURE 8-20 Metropolitan Richmond.

not be justified by contiguous urban land use, although the strip between Richmond and Washington is rapidly being suburbanized from both ends. Rather, it is the city's business orientation to the north and east and its efficient transport connections with Washington and Norfolk that suggest a logical extension of Megalopolis as far southwest as Richmond.

The city was founded in early colonial days and became a bastion of southern culture and Confederate politics. It experienced rapid growth as a commercial and, to a lesser extent, financial center in the mid-1900s, but more recently it has experienced an economic downturn, as is evidenced by an abundance of empty storefronts in the central business district. Its moderate level of industrial development is highly specialized toward production of cigarettes and chemicals, which together provide nearly half of total manufactural jobs.

Hampton Roads

Just inland from the mouth of Chesapeake Bay is the commodious harbor of Hampton Roads, situated where the north-flowing Elizabeth River enters the mouth of the James River estuary and slightly upstream of the latter's confluence with Chesapeake Bay (Fig. 8-21). The resultant natural harbor is one of the finest in the Western Hemisphere and around it developed one of the largest maritime, military, and ship-building complexes in the world. Clustered about the mouth of the Elizabeth River are the cities of Norfolk (Virginia's largest), Chesapeake, and Portsmouth, and on the tip of the peninsula across the mouth of the James River are Newport News and Hampton.

Military and port-related activities dominate the economy of this group of cities, sometimes nicknamed "Tidewa-ter." More than a dozen military bases are in or adjacent to the urbanized areas, and their payrolls constitute a large share of the total metropolitan income. The strategic location in the middle of the nation's Atlantic Coast and the splendid harbor make Hampton Roads a logical center for protective naval and air facilities.

The export of coal is the most prominent shipping activity of the Hampton Roads ports. The coal-handling facilities at Norfork and Newport News are among the largest and most automated in the world, and Hampton Roads handles about half of all U.S. coal exports.

Manufacturing, with one exception, is not of major importance in the Hampton Roads cities. The exception is ship-building, which employs more than half the industrial workers in the area and is dominated by the massive shipyards at Newport News.

Nearby recreational attractions on three sides of Hampton Roads generate much tourist as well as local business. A few miles northwest of Newport News are three outstanding historic restorations: Williamsburg, Yorktown, and Jamestown. East of Norfolk are the tidewater beach areas, focusing on Virginia Beach. To the northeast are the waters of Chesapeake Bay, with their fishing and boating interest, and the bridge-tunnel route that crosses the bay mouth to the relatively unspoiled southern reaches of the Delmarva Peninsula.

THE RURAL SCENE

Despite the predominance of an urban lifestyle, population, and economy, most of the land in Megalopolis is actually rur-

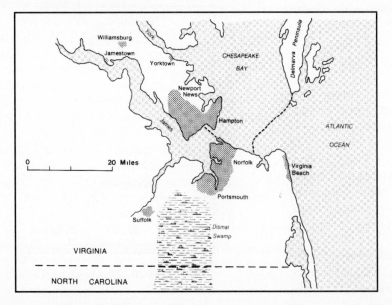

FIGURE 8-21 Hampton Roads.

al. It is estimated that some 80 percent of the total land area of the region supports nonurban land uses. This rural land is important for many things. It supplies a great volume of food-stuffs to the cities of the region, although most of the food for Megalopolis is actually produced elsewhere; its rivers, streams, springs, and lakes provide much of the region's water supply, although many urban watersheds extend far beyond the confines of Megalopolis. Perhaps most important of all, the rural lands provide breathing space and recreational areas for the 47 million Megalopolitans. The significance of green spaces becomes greater with the construction of each new high-rise apartment block, with each passing day of expanding urban sprawl.

The Coast, the Coastal Plain, and Beyond

The most pervasive aspect of the regional environment is the sea and its interface with the land. The long axis of Megalopolis parallels the Atlantic, and there are many hundreds of miles of coastline along the irregular shore. People turn to the ocean and its edge for much of their commerce and recreation and for some of their food. No part of Megalopolis is more than 100 miles (160 km) from the coast.

The most striking characteristic of the coast is its irregularity. Few parts of the North American coastline, and certainly no section with such a sizable population, have such an uneven, embayed, island-studded outline. Different parts of the present shoreline have varied origins, and their diversity of form is striking.

Each the principal embayments has a different shape and a different pattern of river flow into it. From north to south the embayments include Cape Cod Bay, protected from the stormy North Atlantic by the hooked peninsula of Cape Cod; Narragansett Bay, Rhode Island's island-dotted waterway; Long Island Sound, sandwiched between the Connecticut coast and the North Shore of Long Island; Lower New York Bay, the broad entryway to the Hudson lowland, between New Jersey and Long Island; Delaware Bay, the extensive estuary of the Delaware River; and Chesapeake Bay, the continent's most complex and second largest estuary (Fig. 8-22).

There are three prominent peninsulas along this stretch of coast. Cape Cod is a long, low-lying sandy hook that is world famous as a summer recreational area. Cape May is a peninsula at the southern tip of New Jersey that shelters Delaware Bay from the open ocean. The Delmarva Peninsula, largest on the East Coast north of Florida, encompasses parts of the three states for which it is named.

The numerous offshore islands, with one outstanding exception, are crowded summer vacationlands that are largely depopulated in winter. Most famous are the islands off the southeastern coast of Massachusetts: Nantucket, Martha's

FIGURE 8-22   A Chesapeake Bay scene (courtesy Maryland Department of Business and Economic Development).

Vineyard, the Elizabeth Islands, and Block Island. Many of the sandbar islands off the Long Island, New Jersey, and Delmarva coasts are also popular holiday spots. Long Island itself, of course, is much larger and more complex in its function.

The coastal plain is underlain by relatively unconsolidated sediments, most of which are of geologically recent vintage (Tertiary and Quaternary), although the inner margin of the plain has some older (Cretaceous) deposits. Beneath the sediments is a base complex of ancient igneous and metamorphic rocks, which reaches the surface in a very complicated pattern in southern New England, the New York City area, and northern New Jersey.

After a long period of gentle uplift of the continental shelf in Tertiary time, in which a series of broad open valleys developed, a significant drowning of the coastal plain occurred, presumably as a combined result of the weight of the Pleistocene ice load and the postglacial rise in sea level attributable to glacial meltwater.[5] Glaciation modified the northern part of the coastal plain, primarily by the laying down of extensive glacial and glaciofluvial deposits; indeed, the size of Long Island, Cape Cod, Martha's Vineyard, and Nantucket Island was significantly increased by the deposit of terminal moraines. The most recently developed terrain features of the region are ephemeral beaches, coastal dunes, and sandbars, with their associated shallow lagoons.

The resulting topographic pattern throughout the Megalopolis Region is one of exceedingly flat land sloping gently toward the sea. Along the inner (western) margin of

[5] William D. Thornbury, *Regional Geomorphology of the United States* (New York: John Wiley & Sons, 1965), p. 36.

the region there is a rise toward rolling land or occasional steep-sided hills, but in the plain itself only minor prominences appear above the uniform level; they are mostly in the north and are chiefly related to accumulations of glacial deposition. Hard-rock bluffs (the Palisades) lining the lower course of the Hudson River are exceptions to this generalization.

The megalopolitan coast is a well-watered region: its coastal plain is crossed by a large number of important rivers flowing southward or southeastward to the Atlantic. These rivers originate in the interior uplands of the Appalachian system and move swiftly off the crystalline rocks of the Piedmont onto the softer sedimentaries of the coastal plain. The lithologic change from hard rock to softer rock is usually marked by a steeper gradient, producing a rapid or small waterfall, and a line drawn on a map to connect these sites is often referred to as the *Fall Line.*[6]

To the west of the Fall Line the terrain is more irregular. The Appalachian Piedmont is mostly a gently undulating surface that reaches to only a few hundred feet above sea level in its highest parts. The underlying metamorphic and plutonic rocks are crystalline and ancient.

Along the inner (western) margin of the Piedmont is a lengthy lowland tract that developed on sedimentary rocks of mostly Triassic age. This so-called Triassic Lowland encompasses some of the finest agricultural land on the continent, in part because of the soils that developed there.

West of the Piedmont the Megalopolis Region spills over into the ridge-and-valley section of the Appalachians. Its eastern margin is a north–south trending hilly stretch that extends from the Watchtung Mountains of northern New Jersey to northern Virginia's Blue Ridge. Immediately westward is the Great Valley of the Appalachians, whose broad sweep is occupied by inland Megalopolitan cities from Poughkeepsie, New York, to Harrisburg and York in Pennsylvania.

Climate

The climate is classified as humid midlatitude and semimarine. The coast has a leeward location on the continent, relative to the general west-to-east movement of weather

systems, which mutes the maritime influence. Still, the effect of the adjacent ocean is to ameliorate both summer and winter temperatures, to lengthen the growing season, to ensure that the region's harbors remain ice free (with rare exceptions), and to increase the moisture content of the atmosphere.

Precipitation is the most consistently prominent feature of the climate. Most of the region receives between 40 and 45 inches (102 and 127 cm) of moisture annually. It is generally well distributed throughout the year, with a slight maximum in summer.

Storms of various kinds are not uncommon, and most bring considerable precipitation. Thundershowers are prevalent during the warmer months, and late summer occasionally brings a "northeaster," a rainstorm with high wind that may last for several days. Every few years in late summer or fall the region may be visited by an errant tropical hurricane that has worked its way north from the Caribbean; heavy rain and roaring winds may whip up the sea sufficiently to cause considerable coastal damage and occasional loss of life. Winter snowstorms in the northern part of the region are often abrupt, and the deep snow is occasionally paralyzing.

"Wild" Vegetation

Most of what is now the Megalopolis Region was originally covered with a mixture of woodland and forest in which deciduous species predominated but coniferous trees sometimes occurred in significant concentrations, especially on areas of sandy soil. Much of the forest and woodland was cleared for farming and other purposes by the middle of the nineteenth century. Since then, there has been a resurgence of trees, largely because of farm abandonment. The proportion of land in trees and brush has increased steadily for several decades—Connecticut, for example, was one-third wooded in 1850 but is two-thirds wooded today—and the trend seems likely to continue for some time.

The present vegetation associations of woods and brush are difficult to classify as natural vegetation; there has been too much human interference, both deliberate and accidental, in most areas. Many exotic species, especially shade trees and weeds, were added to the floristic inventory. Still, there are vast sectors where woods and forest predominate, and in a few localities the natural vegetation has not been changed or altered by humans.

Of particular interest is the so-called pine barrens area of New Jersey. Located in the most densely populated state and just southeast of the most densely traveled traffic corridor in the world, these 650,000 acres (260,000 ha) of pitch pine and oak have a population density of less than 25 people per square mile. The area has remained virtually intact for

---

[6] Several cities of varying sizes have riverside sites along the Fall Line between New Jersey and Georgia. Some scholars believe that a causative relationship is involved, for the rapids often served as both an early source of power generation and as head of navigation on the coastal plain rivers, thus providing economic incentive for town sites. This contention has been challenged, particularly by Roy Merrens, who has marshaled an impressive array of evidence to refute the notion that any functional relationships between river rapids and town sites along the Fall Line do in fact, exist. See H. Roy Merrens, "Historical Geography and Early American History," *William and Mary Quarterly,* 3rd series, 22 (October 1965), 529–548.

several hundred years, apparently because of lack of agricultural potential. It also has an extensive aquifer of pure, soft water. Many real estate schemes have threatened the barrens, but so far only the edges have been nibbled away.

Soils

The soils of Megalopolis vary widely in characteristics. Still, three principal kinds can be noted: sand or sandy loams in the better-drained parts of the coastal plain, hydromorphic soils in the numerous areas of marsh and swamp, and varied residual soils in scattered localities.

Spodosols, with their subsurface accumulations of iron, aluminum, and organic matter, dominate in southern New England. Thin, light-colored Inceptisols are typical of the New York–northern New Jersey area. Clayey or sandy Ultisols predominate in the southern part of the region.

Agriculture is well entrenched but not primarily because of the soils; indeed, few soils are inherently very fertile for crop growing, and high agricultural productivity is usually associated with heavy use of fertilizer. In many farming areas the natural soil is simply a medium through which to feed the crops. Even the relatively productive soils of the Connecticut Valley and parts of New Jersey are heavily fertilized wherever cash crops are grown.

Specialized Agriculture

The past few decades have seen a continuing attrition of farmland in the Megalopolis Region. Urban sprawl pushed many farms out of production, and woodland replaced cultivated land where marginal farms were abandoned for agriculture. The long-run trend has been for general farming to decline and specialty farming to increase and for total farm acreage to decrease while total value of output continues to rise. In other words, with the pressures from expanding urban land use on one side and increasing costs of farming on the other, the Megalopolitan farmer has tended to abandon the poorer land and the less valuable products and concentrate on high-value output. The remaining farms tend to be efficient and specialized, emphasizing the production of perishables either for local consumption or immediate canning or freezing.

Although the region has a generally favorable climate for agriculture and soils that are only moderately fertile but responsive to fertilizer, it is not the physical advantages that underlie farming in Megalopolis; indeed, there is practically nothing in the region that could not be grown as well or better elsewhere in the country. The great advantage is market, and most agricultural production is market oriented.

The typical farm of contemporary Megalopolis is engaged either in raising vegetables and fruits in a market-gardening type of operation or in dairying. Important specialty production involves meat-type chickens (broilers, fryers) on the Delmarva Peninsula, the historically significant but steadily declining tobacco farms of tidewater Maryland, the very productive general farms of the Pennsylvania Dutch country (see vignette), and horticulture (nursery and greenhouse plants) in a multiplicity of near-urban locations.

Commercial Fishing

Although the commercial fishing industry of Megalopolis has a long history of successful operation, its significance, both absolutely and relatively, continues to decline through the years. Generally, fishing fleets operate relatively close to home, in near North Atlantic waters, although some boats—particularly those from New England ports—regularly visit the "banks" area. Shellfish (primarily oysters, clams, and crabs), menhaden, and flounder are the principal catch (Fig. 8-23).

Menhaden, a fish little known to the general public, is taken in great volume. In most years it amounts to more than 20 percent by weight of the total U.S. catch, which ranks it second only to Alaskan pollock; but it is a low-value fish, rarely providing more than 4 percent of the total value of the U.S. catch. About one-third of the menhaden caught in the United States comes from the waters off the Atlantic coastal plain. Partly because of its extreme oiliness, menhaden is not generally considered a food fish. It is used mainly as a source of oil and meal in making stock and pet feed and commercial fertilizers.

The mid-Atlantic coast has a long history as a leading fishery for oysters and crabs, particularly from Chesapeake Bay. However, in the last few years the oyster catch has declined precipitously, owing to pollution, overfishing, and a mysterious disease. Chesapeake Bay still produces more crab meat than the rest of the nation combined, but this, too, is on a downward trend.

The total commercial fishery of the Megalopolis Region yields about one-fourth of the U.S. catch, whether measured by weight or value. Virginia is the third-ranking state in volume of catch, largely because of menhaden landings. Massachusetts is the sixth-ranking state in value of catch. The leading fishing ports of Megalopolis are New Bedford, Massachusetts (third-ranked nationally in value of landings); Portland, Maine (tenth-ranked nationally in value of landings); and Cape May, New Jersey (in the top twenty nationally in both value and quantity of landings).

RECREATION AND TOURISM

Tourist and recreational attractions of Megalopolis are numerous and varied, as might be expected in such a populous

A CLOSER LOOK   The Pennsylvania Dutch—A World Apart

In southeastern Pennsylvania lies an area that is remarkably different from all other parts of the Megalopolis Region. The area is generally referred to as Pennsylvania Dutch Country. In centers on Lancaster County but also includes most of Berks County to the east and York County to the west (Fig. 8-A). The original settlers were mostly *Deutsch*, or German, in nationality, and the term was corrupted to *Dutch*. Other settlers hailed from Switzerland, Austria, France, and Holland, but whatever their origin the majority shared a common bond of piety and came to America largely to escape religious persecution. Their descendants have clung remarkably to tradition as well as to land. Unlike farmers in many areas, generations pass through the same homesteads; sons follow fathers on the same soil.

This industrious, devout, primarily rural populace is referred to as "plain people." They are mostly pious Old Order Amish folk but include some Mennonites, Quakers, and Brethren (Dunkards). They are dedicated to a traditional way of life that preserves the peaceful, family-centered, home-oriented style of living for which the Pennsylvania Dutch are justly famous. For the most part, both adults and children dress uniformly in plain, solid-colored clothing that is without ornaments, buttons (they use hooks-and-eyes), collars, or embroidery. Women always wear aprons and plain bonnets and often capes. Married men wear full beards but not mustaches (because of their historical association with the military), whereas the women never cut their hair and wear it in a braided bun at the nape of the neck. They are opposed to such modern conveniences as electricity and telephones in the home, farm machinery, television, indoor plumbing, bicycles, and automobiles. They use horse and buggies for transport (Fig. 8-B) and are ingenious in their use of windmills and waterwheels for power. They seek a self-sufficient isolation from the temptations of the modern world and are totally opposed to such civilized delights as Social Security, insurance, lawyers, courts, and war. Their insistence on a

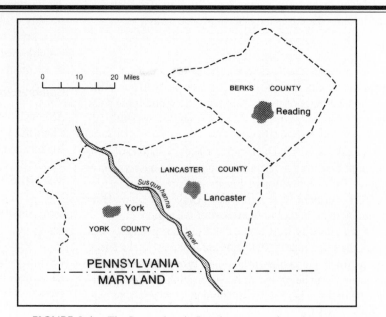

FIGURE 8-A   The Pennsylvania Dutch country of southeastern Pennsylvania.

FIGURE 8-B   A concentration of buggies at an Old Order Amish farm means either a church service or a wedding is underway (courtesy the Pennsylvania Dutch Convention and Visitors Bureau).

FIGURE 8-C   A representative Amish farm near Lancaster, Pennsylvania. Tobacco and corn are prominent crops (TLM photo).

close-knit education has been upheld, and their children generally attend a nearby, parochial one-room school through the eighth grade (an average of five new schools was built for Old Order children every year during the 1980s), after which education is handled in a family setting. Most plain people are trilingual, speaking Pennsylvania Dutch, English, and High German. Despite having large families, the proportion of plain people continues to diminish because of rapid growth of other components of the population. In the mid-1990s less than 10 percent of the three-county population was Old Order.

Few farming areas in North America have been as richly endowed by nature and as well handled by people as the Pennsylvania Dutch country. The gently rolling terrain, abundant summer rainfall, relatively long growing season, and fertile, limestone-derived soils provide a splendid environment for agriculture. The Pennsylvania Dutch farmers are legendary for their skill and industriousness, although at least one authority has challenged the widespread belief that the eighteenth-century *Deutsch* were better farmers than those from the British Isles.[a] In any event, the farm landscape provides ample evidence that this is an area of rich and productive agriculture. The fields are well tended, the abundant livestock is well fed, scientific techniques are widely used (with some ingenious devices to compensate for the lack of machinery on plain people's farms), and most farmsteads are virtual showplaces. Lancaster County is not only Pennsylvania's leading farm county but is also the nation's leading nonirrigated farm county.

Livestock is the principal agricultural output of the Pennsylvania Dutch country with dairy products, cattle, broilers, and eggs providing (in order) the principal farm income. For many years tobacco was the principal cash crop of the area, but it has been in a severe decline recently. The principal crops today are corn, vegetables, and potatoes.

The Pennsylvania Dutch country is an enclave of remarkable agricultural productivity in a region not noted for prosperous farming (Fig. 8-C). It is also an area of unique lifestyle in a region that contains a variety of unusual ones. The striking dichotomy of the area is that the plain people have become a major tourist attraction, which is a great annoyance to them[b] but an economic boon to many of their less conservative neighbors. Their self-imposed isolation has persisted for two and a half centuries but is becoming increasingly difficult to maintain.

TLM

[a] James T. Lemon, "The Agricultural Practices of National Groups in Eighteenth-Century Southeastern Pennsylvania," *Geographical Review*, 56 (October 1966), 467–496.

[b] The plain people believe that photographs in which individuals can be recognized violate the biblical commandment against graven images; hence, they are unalterably opposed to being photographed. Tens of thousands of camera-carrying tourists, on the other hand, find the plain people to be irresistibly photogenic. These incompatible viewpoints bring about frequent confrontations—always nonviolent.

FIGURE 8-23 A successful fishing trip for menhaden. The hold of the boat is full, and fish are piled more than a foot deep on the deck. The port is Gloucester, Massachusetts (TLM photo).

FIGURE 8-24 A crowded beach on the sheltered bay side of Cape Cod in Midsummer (Dennis MacDonald/PhotoEdit).

region. Local people spend much of their holiday or vacation time within the region; in addition, a great many visitors come from other regions and overseas. Tourist interest focuses mainly on three general categories of attractions: cities and their points of interest, coastal areas, and historical sites.

Without doubt it is the urban attractions of Megalopolis that are most beguiling to resident and visitor alike. New York City is the number-one tourist goal on the subcontinent, and Washington is not far behind. Here, as in other large cities of Megalopolis, there are a host of things to see and do, although famous landmarks, such as the Empire State Building, United Nations, National Capitol, and the White House, seem to rank first in popularity.

The long seacoast and numerous islands of Megalopolis are important summer playgrounds, but their recreational significance in winter is limited. There are scores of beach resorts, the more famous being Cape Cod (Fig. 8-24), Nantucket Island, Martha's Vineyard, Asbury Park, Atlantic City, Ocean City, and Virginia Beach.

The almost-continuous beaches of southern New Jersey attract the greatest flood of patronage; some 45 million people (including repeats) are estimated to visit this area each summer. The resort towns, from Point Pleasant in the north to Cape May in the south, are built right along the beach, usually separated from the sand by only a boardwalk. Real estate prices continue to soar, and construction of motels, condominiums, town houses, and other types of housing units flourish, catering to the immense nearby markets of Philadelphia, New Jersey, and New York.

A special attraction of coastal New Jersey is Atlantic City, once the nation's most storied beach resort but now the leading gambling town and the single most visited travel resort in the country (Fig. 8-25). The first casinos were opened in 1978, and by 1984 Atlantic City had surpassed Las Vegas in both visitors and money wagered.

The historical attractions of Megalopolis may occupy the visitor for a shorter period of time than the city or shore, but the great number of historic sites and the frequently intriguing nature of their presented interpretation make many of them hard to pass up. Most states of the region have capitalized on "selling" history to tourists; but Virginia (with hundreds of permanent historical markers along its highways), Pennsylvania, and Massachusetts have done the most thorough jobs. Battlefield sites of the Revolutionary and Civil wars are particularly notable. An interesting and popular trend is the integrated restoration of early settlements on a large-scale basis—for example, the Pilgrim village of Plymouth, Massachusetts; the restored seaport of Mystic, Connecticut; and the colonial town of Williamsburg, Virginia.

THE ROLE OF THE REGION

The significance of Megalopolis is much greater than its area or even its population would indicate. It is not enough to say that within the region are one-sixth of the people, one-fifth of the factory output, one-third of the high-rise office space, and one-fourth of all wholesale sales of the nation. It must be reiterated that Megalopolis is also the financial heart of the country: six of the eight largest banks in the nation are in New York City. Of even greater importance, Megalopolis is

FIGURE 8-25   Atlantic City's renowned boardwalk separates casinos and shops on the right from beach and ocean on the left (courtesy Atlantic City Convention and Visitors Authority).

It is impossible to quantify the magnitude of the decision-making role of Megalopolis, but some indication of this factor can be seen by tabulating the locations of the headquarters of the major business corporations of the nation. According to the annual *Fortune* magazine survey for 1995, one-third of the largest industrial corporations in the nation were headquartered in Megalopolis. The region also housed headquarters for 25 percent of the largest retailing companies, 46 percent of the largest life insurance companies, and 60 percent of the largest diversified financial companies. Add to this the Megalopolitan location of both the seat of government of the United States and the United Nations, as well as a host of state, city, and local government centers, and clearly much of the economy and politics of the nation is guided from Megalopolis—for better or worse.

Another way in which the importance of Megalopolis transcends its size is in its manner of coping with urban problems. This is an urban region with an unmatched intensity of living. Its urban problems, too, are unmatched in their complexity and magnitude. Equally abundant within the region are workers, brainpower, skills, and technology with which to face the problems. Jean Gottmann called Megalopolis the "cradle of the future," implying its role, both actual and potential, in meeting the challenge of an urbanized world.

There is yet another way in which Megalopolis has a role to play. What can it show the world in the matter of despoliation or protection of the environment? As cities grow and green spaces shrink, how does our relationship to nature change and how does this change affect the quality of life for people and all other living things? Must air pollution continue to worsen, for there are increasing numbers of machines and people to expel pollutants into the atmosphere? Will New York Harbor eventually be so choked with garbage scows that there will be no room for ocean liners? Must Chesapeake Bay become as sterile as Newark Bay, where in some reaches even coliform (intestinal) bacteria cannot survive? The inhabitants, abruptly and increasingly conscious of ecological relationships, watch with interest and apprehension.

the brain and nerve center of the nation. A large share of the decisions that shape the economy and government of the United States, thus significantly influencing economic and political decisions and events over most of the world, are made here.

SELECTED BIBLIOGRAPHY

ALEXANDER, LEWIS M. *The Northeastern United States* (2d ed.). New York: D. Van Nostrand Company, 1976.

ALFORD, JOHN J., "The Chesapeake Oyster Fishery," *Annals,* Association of American Geographers, 65 (1975), 229–239.

BERGMAN, EDWARD F., AND THOMAS W. POHL, *A Geography of the New York Metropolitan Region.* Dubuque, IA: Kendall/Hunt Publishing Company, 1975.

BORCHERT, JOHN R., *Megalopolis: Washington, D.C., to Boston.* New Brunswick, NJ: Rutgers University Press, 1992.

CAREY, GEORGE W., *A Vignette of the New York—New Jersey Metropolitan Region.* Cambridge, MA: Ballinger Publishing Company, 1976.

CONZEN, MICHAEL P., AND GEORGE K. LEWIS, *Boston: A Geographical Portrait.* Cambridge, MA: Ballinger Publishing Company, 1976.

DE SOUZA, ANTHONY, ED., *The Capital Region: Day Trips in Maryland, Virginia, Pennsylvania, and Washington, D.C.* New Brunswick, NJ: Rutgers University Press, 1992.

DILISIO, J. E., *Maryland.* Boulder, CO: Westview Press 1983.

GOTTMANN, JEAN, *Megalopolis: The Urbanized Northeastern Seaboard of the United States.* New York: The Twentieth Century Fund, 1961.

———, *Megalopolis Revisited: 25 Years Later.* College Park: University of Maryland, Institute for Urban Studies Monograph Series No. 6, 1987.

HART, JOHN FRASER, "The Perimetropolitan Bow Wave," *Geographical Review*, 81 (1991), 35–51.

KANTROWITZ, N., *Ethnic and Racial Segregation in the New York Metropolis: Residential Patterns among White Ethnic Groups, Blacks and Puerto Ricans.* New York: Praeger Publications, 1973.

KIERAN, JOHN, *A Natural History of New York City.* New York: Fordham University Press, 1982.

LIPPSON, ALICE JANE, *The Chesapeake Bay in Maryland: An Atlas of Natural Resources.* Baltimore: Johns Hopkins University Press, 1973.

MCMANIS, DOUGLAS R., *Colonial New England: A Historical Geography.* New York: Oxford University Press, 1975.

MCPHEE, JOHN A., *The Pine Barrens.* New York: Farrar, Straus & Giroux, Inc., 1968.

MEYER, DAVID R., *From Farm to Factory to Urban Pastoralism: Urban Change in Central Connecticut.* Cambridge, MA: Ballinger Publishing Company, 1976.

MULLER, PETER O., KENNETH C. MEYER, AND ROMAN A. CYBRIWSKY, *Metropolitan Philadelphia: A Study of Conflict and Social Cleavages.* Cambridge, MA: Ballinger Publishing Company, 1976.

OLSON, SHERRY, *Baltimore.* Cambridge, MA: Ballinger Publishing Company, 1976.

———, *Baltimore: The Building of an American City.* Baltimore: Johns Hopkins University Press, 1980.

PROCTER, MARY, AND BILL MATUSZESKI, *Gritty Cities: A Second Look at Allentown, Bethlehem, Bridgeport, Hoboken, Lancaster, Norwich, Paterson, Reading, Trenton, Troy, Waterbury, Wilmington.* Philadelphia: American University Press Services, 1978.

ROBICHAUD, BERYL, AND MURRAY F. BUELL, *Vegetation of New Jersey: A Study of Landscape Diversity.* New Brunswick, NJ: Rutgers University Press, 1973.

SAXENIAN, ANNALEE, *Regional Advantage: Culture and Competition in Silicon Valley and Route 128.* Cambrige, MA: Harvard University Press, 1994.

SHANKLAND, GRAEME, "Boston—The Unlikely City," *Geographical Magazine,* 53 (February 1981), 323–327.

STANSFIELD, CHARLES A., *New Jersey.* Boulder, CO: Westview Press, 1983.

THOMPSON, DEREK, ED., *Atlas of Maryland.* College Park: University of Maryland, Department of Geography, 1977.

THOMAS, JEAN-CLAUDE MARCEAU, "Washington," in *Contemporary Metropolitan America,* Vol. 4, *Twentieth Century Cities,* John S. Adams, ed., pp. 297–344. Cambridge, MA: Ballinger Publishing Company, 1976.

WARF, BARNEY, "The People and Landscapes of Brooklyn, New York," *Focus,* 40 (1990), 6–11.

———, "The Port Authority of New York–New Jersey," *Professional Geographer,* 40 (August 1988), 288–296.

WHEELER, JAMES O., "Corporate Role of New York City in the Metropolitan Hierarchy," *Geographical Review,* 80 (1990), 370–381.

The Appalachians and the Ozarks

9

The hill-and-mountain country of eastern North America is noncontiguous. The highlands of the Appalachians are separated from the uplands of the Ozark–Ouachita section by more than 300 miles (480 km) of exceedingly flat terrain. Yet the physical and cultural characteristics of these disconnected slopeland areas are generally similar, so it is logical to include them within the same broad region despite their separated positions.

The Appalachian highlands extend from central New York State, south of the Mohawk Valley, to central Alabama, where they terminate at the edge of the Gulf coastal plain. The eastern boundary of this subregion follows the topographic distinction between the Appalachian Piedmont and the Blue Ridge–Great Smoky Mountains from Alabama to southernmost Pennsylvania (Fig. 9-1). From there the boundary is considered to follow the northeasterly trend of Blue Mountain across Pennsylvania to the northwestern corner of New Jersey, where the name changes to Kittatinny Mountain and into New York state, where it is known as the Shawangunk Mountains.

The western boundary of the Appalachian subregion is more indistinct physically but is conceptually situated along the western edge of the Appalachian "plateaus."

The uplands of the Ozark–Ouachita subregion extend in a general southwestward direction from southeastern Missouri through northwestern Arkansas into southeastern Oklahoma. The boundary of this section on all sides is determined primarily by the topographic pattern: the prevalence of slopelands within and the prevalence of relatively flat land without.

THE REGIONAL CHARACTER

As with any large region, generalizations about predominant characteristics abound with exceptions. There are many differences in the various parts of the region. Some authorities would say that the northern Appalachians (roughly from the Kanawha River northward) are quite distinct from the southern Appalachians; others would note that the main differences are between coal-mining areas and non-coal-mining areas; and still others would point up differences between the Appalachians and the Ozarks. Throughout the region there are also pronounced variations in lifestyle among the citizens: between the African American factory worker in Chattanooga and the white subsistence farmer on Hickory Ridge, between the hardscrabble coal miner in West Virginia and the externally oriented resort operator on Lake of the Ozarks.

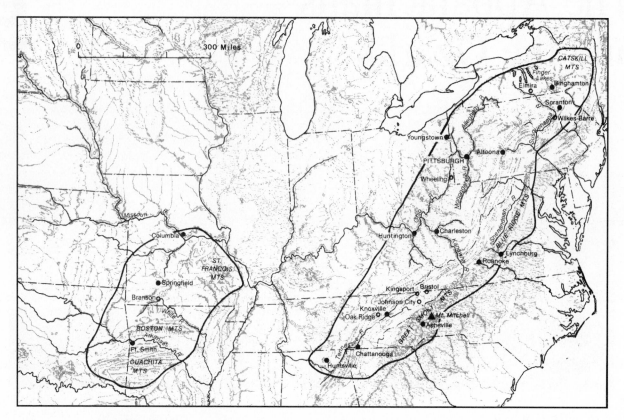

FIGURE 9-1   The Appalachians and Ozarks Region (base map copyright A. K. Lobeck; reprinted by permission of Hammond, Inc.).

It has been tempting to paint the region as a hill people haven, replete with colorful speech, a charming folk culture, and hidden moonshine stills. This exaggerated image belongs to another era if, indeed, it ever pertained. Hill people and mountain folk certainly exist today in various parts of the region, but with few exceptions their isolation and distinctiveness are gone forever.

Certain elements contribute to geographical generalizations about the region. Most of the land is in slope, and the slopes are often steep. As a result, life is focused in the valleys; settlements, transportation routes, and industrial developments compete with river or stream for the limited amount of flat land on the valley floor. Another pervasive aspect of the region is forest; almost all the area was originally forest covered and most of it is still clothed with virgin or second-growth trees.

Another significant facet of regional character is recent population trends. During the middle decades of the twentieth century—the 1940s and 1950s—there was an actual population decline in most parts of the region, a situation that pertained to no other populous portion of North America. Appalachian birthrates, long among the highest on the continent, declined rapidly. More significant, however, was the impact of out-migration, predominantly teenagers and young adults, that characterized most of the region until the mid-1960s. Beginning in the late 1960s, however, more people moved into the region than left. This pattern accelerated in the 1970s, and the traditionally declining areas of southern Appalachia and the Ozarks–Ouachitas are now experiencing a population resurgence that is not restricted only to urban areas but also includes a great many rural counties.

The population growth is indicative of greater economic opportunities in the region, but it does not necessarily indicate a reversal of the long-standing economic plight of the people of Appalachia and Ozarkia. The 1930 Census of Agriculture showed that the Appalachians encompassed the highest proportion of low-income farms in the country. Later the significant decline in coal-mining employment added another dimension to economic difficulty, for farming and coal-mining were long the principal occupations of the people of Appalachia; thus this region came to be recognized as the number-one long-run problem area (in a geographical sense) in the nation's economy. Any discussion of the region's geography must consider the low incomes and restricted economic opportunities of a large proportion of the population.

THE ENVIRONMENT

Although the Appalachian and Ozark portions of the region are separated from one another by several hundred miles, their geologic and geomorphic affinities are so marked that there is general agreement that Appalachian lithology and structure continue westward beneath coastal-plain sediments to reappear at the surface in the Ozark–Ouachita subregions.[1] The result is notable similarity in rock types, structure, and landform patterns, which give the Appalachian and Ozark Region topographic distinctiveness that provides a convenient rationale for delineating physical subregions. The prominent subregions are

1. The Blue Ridge–Great Smoky Mountains
2. The Ridge and Valley section
3. The Appalachian Plateaus, often further subdivided into the Allegheny Plateau in the north and the Cumberland Plateau in the south
4. The Ouachita Mountains and Valleys
5. The Ozark Plateaus (Fig. 9-2)

The Blue Ridge–Great Smoky Mountains

The Blue Ridge consists largely of crystalline rocks of igneous and metamorphic origin: granites, gneisses, schists, diorites, and slates (Fig. 9-3). Extending from Pennsylvania to Georgia, the Blue Ridge exceeds all other mountains in the East in altitude. North of Roanoke it consists of a narrow

---

[1] William D. Thornbury, *Regional Geomorphology of the United States* (New York: John Wiley & Sons, 1965), p. 262.

ridge cut by numerous gaps; south of Roanoke it spreads out to form a tangled mass of mountains and valleys more than 100 miles (160 km) wide. The mountains are steep, rocky, and forest covered. The highest peak is Mount Mitchell in North Carolina (6684 feet or 2005 m), in the range known as the Great Smoky Mountains.

Because of heavy rainfall, the Blue Ridge was covered with magnificent forests, originally consisting of hardwood trees, especially oak, chestnut, and hickory (Fig. 9-4). The greater part of the original forest was logged years ago, part of it being converted into charcoal for iron furnaces in the nineteenth century. Much of the area has been cut repeatedly and a great deal has been burned; however, most of the subregion is still cloaked with extensive forest.

The Ridge and Valley

This subregion consists of a complex folded area of parallel ridges and valleys.

The Great Valley This nature-chiseled groove that trends northeast–southwest from the Hudson Valley of New York to central Alabama is one of the world's longest mountain valleys. Its flattish floor is divided into separate sections by minor cross ridges that serve as watershed separations for the several streams that drain the valley. Different names are applied to various segments of the Great Valley. Near the Delaware it is called the Lehigh Valley; north of the Susquehanna, the Lebanon Valley; south of the Susquehanna, the Cumberland Valley; in northern Virginia, the Shenandoah; in Virginia as a whole, the Valley of Virginia; and in Tennessee, the Valley of East Tennessee.

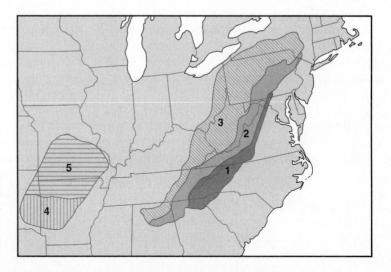

FIGURE 9-2 Topographic subdivisions of the Appalachians and Ozarks Region: (1) the Blue Ridge–Great Smoky Mountains, (2) the Ridge and Valley section, (3) the Appalachian Plateaus, (4) the Ouachita Mountains and Valleys, and (5) the Ozark Plateaus.

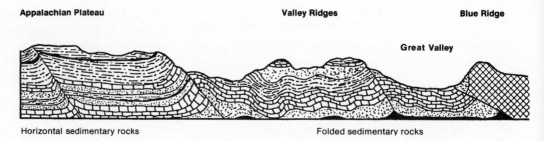

FIGURE 9-3   Generalized geologic cross section of the Appalachian subregion.

FIGURE 9-4   A view of the Blue Ridge in Virginia's Shenandoah National Park (TLM photo).

The Great Valley has long been the north–south highway in the Appalachian region as well as one of the most productive agricultural areas in the East. It has never been of outstanding industrial importance, although several of its cities (particularly in Pennsylvania and Tennessee) have important manufacturing establishments.

The eastern and western confines of the Great Valley are definite: the knobby wooded crest of the Blue Ridge towers above the valley floor on the east and the wild, rugged, though less imposing, Appalachian ridges bound it on the west.

The Ridge and Valley Section   West of the Great Valley is a broad series of roughly parallel ridges and valleys that developed along parallel folds in Paleozoic sediments.

Topographical development results primarily from differential erosion, with the configuration based on differences in the resistance of the bedrock. The ridgemakers are mostly hard sandstones, and the valleys usually indicate outcrops of softer limestones and shales. The ridges vary in size and shape, but most valleys are narrow, flattish, and cleared for agriculture. From the air the forested ridges show up as dark parallel bands and the cleared valleys as light ones.

This area gets less rainfall than other parts of the Appalachians—about 40 inches (102 cm). This difference is ascribed to the rain shadow effect of the Blue Ridge on the one side and the Allegheny–Cumberland Escarpment on the other. The seasonal distribution is quite even, although autumn is somewhat more dry than spring. About half the precipitation falls during the growing season, whose average length

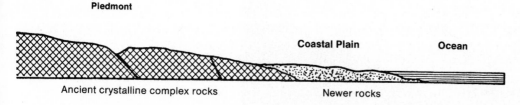

varies from 176 days in the north to more than 200 days in the south.

The natural vegetation of the valley lands consisted primarily of oak and hickory, with sycamore, elm, and willow near the streams. On the ridges it was similar to that found on the Blue Ridge Mountains to the east. Some of this magnificent forest was destroyed by the pioneers who settled in the area, but the ridges are still tree covered, even though much of the cover is second and third growth. There were also areas of grassland in the broader river valleys—the result of burning by the Native Americans.

The Appalachian Plateaus

The western division of the Appalachian Highlands is a broad belt of land known as the Appalachian Plateaus. Along its eastern edge it has a high, bold escarpment, the Allegheny Front, that is so steep that roads and railroads ascend it with difficulty. Most of the layers of rocks that form the plateaus lie flat, one on the other. This subregion extends from the Catskill Mountains to north-central Alabama and from the Allegheny Front to the Interior Lowland. The term *plateau* is applicable in a structural sense, but the present topography is chiefly hill country, with only accordant summits as reminders of any previous plateau surface. The area can be subdivided into the Allegheny Plateau (glaciated and nonglaciated sections) and the Cumberland Plateau.

The Glaciated Allegheny Plateau   Rounded topography typifies most of this area, which on the north is bounded by the Mohawk Valley and the Ontario Plain. The nearly flat-lying sandstones and shales are much dissected by streams that have cut down and back into the plateau. In the northern part—the Finger Lakes country—six slender lakes, trending north–south, occupy the valleys of preglacial streams that were modified by ice erosion and blocked by ice deposition. Here is an area of rolling terrain. In northeastern Ohio and adjacent northwestern Pennsylvania the plateau was modified by the ice, and the relief is gentle, with broad

divides. Northeastern Pennsylvania consists of a hilly upland with numerous streams, lakes, and swamps.

The Unglaciated Allegheny Plateau   Topographically this plateau is more rugged than its glaciated neighbor to the north. Most of the area might properly be regarded as hill country, for the plateau has been maturely dissected. In the Kanawha Valley, the streams lie 1000 to 1500 feet (300 to 500 m) below the plateau surface. Some valleys are so narrow and canyonlike as to be uninhabited; others have inadequate room even for a railroad or highway.

The Cumberland Plateau   This "plateau" is mostly rugged hill country that is so maturely dissected that practically none of its former plateau characteristics remains except locally, as in parts of Tennessee. No sharp boundary separates it from the Allegheny Plateau to the north; thus the Cumberland Plateau is regarded here as beginning in southern Kentucky (the upper reaches of the Kentucky River) and extending to the Gulf Coast Plain. It includes parts of southeastern Kentucky, eastern Tennessee, and northern Alabama.

The Ozark–Ouachita Uplands

The Ozark–Ouachita subregion consists of three major divisions:

1. The Ozark section, consisting mainly of eroded plateaus, such as the Salem and Springfield plateaus and two hilly areas, the St. Francois Mountains in Missouri and the Boston Mountains in Arkansas (Fig. 9-5).
2. The broad structural trough of the Arkansas River valley.
3. The Ouachita Mountains, whose strongly folded and faulted structures result in ridge and valley parallelism similar to the Ridge and Valley section of the Appalachians. The Ouachita Mountains reach their highest elevation—about 2800 feet (840 m) above sea level—near the Arkansas–Oklahoma border.

FIGURE 9-5 The heavily forested Ozark Mountains in northern Arkansas (courtesy Arkansas Department of Parks and Tourism).

SETTLEMENT OF THE APPALACHIAN HIGHLANDS

It was a century and a half after the founding of Atlantic seaboard colonies before settlement began to push into the hill country of the Appalachians. Welsh Quakers and Scotch-Irish Presbyterians were in the vanguard that funneled through William Penn's Philadelphia and spread out in southeastern Pennsylvania. It was German Protestants, however, who were the bulk of the settlers, attracted by both economic opportunity and the promise of religious freedom. By 1750 half of the population of the Pennsylvania colony consisted of immigrant Germans.[2]

From the Pennsylvania Piedmont through the Triassic Lowland was the only easy route to the Great Valley of the Appalachians. Into this valley in the early part of the eighteenth century came pioneer settlers from Pennsylvania; later they were joined by a trickle and then a flood of people moving more directly west from tidewater Virginia and Maryland. A road was finally cut through the western mountains via the Cumberland Gap, providing access to the interior. This "Wilderness Road" furnished the only connection to the infant settlements in Kentucky for several decades, later being supplemented by the old national road west from Baltimore.

In the latter part of the eighteenth century and the early part of the nineteenth the coves and valleys of the Appalachians began to be occupied. "Many of the early settlers were hardy Ulster Scots, descendants of Scots who had been settled in the ancient Irish province of Ulster more than a century before they had moved to the New World."[3] People of English and German stock were also numerous, and there was a minority of French Huguenot and Highland Scots as well. The Great Valley, with its prime agricultural land, was fairly densely occupied before the settlers began to push into the mountains in large numbers. By the 1830s following the eviction of Cherokees from their homes in northern Georgia, that area was occupied by whites, and the settlement of the Appalachians was more or less complete.

For many years most of the Appalachians south of Pennsylvania were rather like landlocked island, isolated from the rest of the country. Most rivers were too swift to be used for transportation, and the rugged terrain inhibited railroad and highway construction. Even today most of the railways are branch lines built solely to exploit the coal and timber resources.

When the attractions of urban living became known to the hill people, particularly during World War II, an out-migration began. The availability of industrial employment and urban amenities drew tens of thousands of inhabitants from the Appalachians eastward to Washington, southward to Atlanta and Birmingham, and especially westward and northward to Nashville, Louisville, Cincinnati, Pittsburgh, Cleveland, Detroit, and Chicago. There they often settled in relatively close-knit neighborhood communities, maintaining a strong flavor of hill country living in the midst of the city until time eventually wore away the traditions.

The Great Valley did not significantly share in the egress; rather, its attractions of cities, industry, and better

[2] Ezra Bowen and Editors of Time-Life Books, *The Middle Atlantic States* (New York: Time-Life Books, 1968), p. 35.

[3] John Fraser Hart, *The South* (Princeton, NJ: D. Van Nostrand Company, 1976), p. 52.

farming areas produced a population increase. Indeed, most parts of the subregion experienced population expansion during the past decade, a striking turnaround from the declining trend that prevailed for more than half a century.

The present population of the Appalachians is preponderantly native-born and white. African Americans constitute a significant minority in the Pittsburgh area, in some of the other Pennsylvania and Ohio industrial towns, and in parts of the Piedmont; elsewhere they live mostly in Chattanooga and Charleston.

SETTLEMENT OF THE
OZARK—OUACHITA UPLANDS

The earliest white settlements in the Ozark–Ouachita Uplands were those of the French along the northeastern border of Missouri. Such attempts were feeble and were based on the presence of minerals, especially lead and salt. Because silver, the one metal they wanted, was lacking, the French did not explore systematically or try to develop the subregion. The first recorded land grant was made in 1723. The French, who never penetrated far into the Upland, were reduced to a minority group by English-speaking colonists toward the close of the eighteenth century. After the purchase of this territory in 1803 by the United States, settlement proceeded rapidly.

After the initial occupance of the Missouri, Mississippi, and Arkansas River valleys, succeeding pioneers entered the rougher and more remote sections of the Upland and remained there. The hilly, forested habitat, which provided so amply for the needs of the pioneers, later retarded their development by isolating them from the progress of the prairies.

World War I was an important factor in the breakdown of isolation in this subregion. The draft and the appeal for volunteers caused many ridge dwellers to leave the hills to join the armed forces. High wages in the cities during the war also enticed many from their mountain fastnesses. After the war, those who returned brought back new ideas. But the Great Depression of the 1930s had a regressive effect, for many who went to the cities lost their jobs and returned to the Upland, and a large part of the plateau population went on relief when various social service agencies were developed under the New Deal.

World War II repeated the effect of the earlier war. Selective Service and high wages, particularly in munitions and aircraft factories in St. Louis, Kansas City, Tulsa, Dallas, and Fort Worth, attracted many younger persons away from the hills.

Modern highways opened most of the subregion to much greater interaction with the surrounding lowlands. In the last few years the long-term trend of out-migration has been reversed. Increasing numbers of people are settling in the Ozarks and Ouachitas, not only in the urban areas but also throughout the rural counties. The attraction is partly the availability of jobs in a more diversified industrial scene, but, more importantly, it is based on tourism and retirement, reflecting the recreational, climatic, and scenic amenities of the subregion.

AGRICULTURE

Most early settlers of the Appalachian and Ozark Region depended on crop growing and livestock raising for their livelihood, and farming continues as a prominent factor throughout most of the region today. In general, however, agriculture has not been a very prosperous occupation. Scarcity of flat land and fertile soils combined with a lack of accessible markets to circumscribe the agricultural opportunities.

Historically farms were small and cropped acreage per farm was limited. Even in recent decades of increasing average farm size elsewhere, many parts of Appalachia have experienced an opposite trend. Farm mechanization also lags behind the rest of the country. Generally farm income and living standards continue to be the lowest in the nation.

A common generalization is that the characteristic Appalachian land-use pattern is one of forested slopes and cleared valley bottoms. Such a pattern clearly prevails in many of the more productive areas, such as the ridge-and-valley portion of Pennsylvania. For most of the region, however, this generalization is invalid; the countryside appears disorganized and patternless, with a hodgepodge of land use on both slopeland and bottomland. One astute student of Appalachia described this situation as follows:

> As you travel through the core of Appalachia . . . you quite literally do not know what to expect when you turn the next corner or go over the crest of the next hill . . . Field, forest, and pasture are scattered across flat land and hillside alike, with no apparent logic, and the slope of the land seems to have scant power to predict how man will use it. Steep slopes are cultivated but level land is wooded. The tiny tobacco patches stick to the more or less level land, but rows of scraggly cornstalks march up some treacherously steep hillsides, often just across the fence from stands of equally scraggly trees. On the far side of the woods, as like as not, and on the selfsame slope, a herd of scrawny cattle mopes through a gulley-scored "pasture" choked with unpalatable

grasses, unclipped weeds, blackberry briers, sumac bushes, and sprouts of sassafras, persimmon, thornapple, and locust.[4]

The seemingly random and nondescript character of land use can be explained in part by the small size of many holdings and the relatively high rate of farm abandonment that prevailed for some time. More fundamental, however, is the more or less inadvertent cycle of land rotation that is traditional in much of the region: land is cleared and cultivated as cropland for some years and then allowed to lapse into a state of less intensive use (pasture) or disuse (woodland) for a period. This cycle is usually lengthy, the farmer clearing, or abandoning, a particular piece of land only once in a working lifetime (perhaps without realizing that he is reclearing a parcel that had previously been cleared by his father or grandfather). The long-range nature of such a cycle makes it difficult to delineate and comprehend.

Crops and livestock are varied over this extensive region. It is primarily a general farming area, and most farms yield a variety of products, generally in small quantities. Corn is the most common row crop, although it is usually not grown for commercial purposes. Hay growing is widespread and occupies the greatest acreage of cropland. Tobacco is the typical cash crop (Kentucky, with one-fourth of the national

[4] John Fraser Hart, "Land Rotation in Appalachia," *Geographical Review,* 67 (April 1977), 148.

crop, ranks second only to North Carolina in tobacco production, although much of the output is west of the Appalachians), but average acreages are quite small. Livestock are significant and usually provide the greater part of farm income (Fig. 9-6).

Product specialization is notable in some areas, the most widespread specialties being dairying and apple growing (Fig. 9-7). Dairy farming is particularly notable in the northern part of the region (New York and northern Pennsylvania) and in the Springfield Plateau of the Ozark section. Apple orchards are most prominent in and near the Great Valley, throughout most of its length. Five states of the region (New York, Pennsylvania, Virginia, West Virginia, and North Carolina) are among the leaders in apple production, and most of their output is associated with the Great Valley or with other ridge-and-valley locations. A notable specialty in New York's Finger Lakes district is the growing of wine grapes.

FOREST INDUSTRIES

Almost all of this upland region was originally forested and supported logging or lumbering activities at one time or another. Well over half the area is now tree covered; most of it is second growth. Relatively valuable hardwoods constitute most of the total stand, sometimes in relatively pure situations but often intermingled with softwoods.

FIGURE 9-6   A Shenandoah Valley farm scene in Rockingham County, Virginia. Hereford cattle are in the foreground, with large chicken houses behind (TLM photo).

FIGURE 9-7 An apple orchard scene in northwestern Arkansas (John D. Cunningham/Visuals Unlimited).

A large proportion of total forested acreage is within the boundaries of national forests. In some districts, however, extensive tracts are owned by major forest products companies and there are a great many small private holdings, particularly in Pennsylvania and Virginia.

In recent years there has been a modest resurgence of forest industries in the Appalachians. With increasing farmland abandonment, it is obvious that woodland acreage is expanding. The replacement process has been described as follows:

> A hard-scrabble, hillside farm is finally abandoned. The first summer, weeds quickly take over. The next summer some grass gets a foothold under the weeds, and blackberry seedlings make their appearance. After several more years, clusters of trees push up above the brambles. The trees may be gray birch from wind-borne seed, and Eastern red cedar trees from seeds dropped by birds that had dined on red cedar berries. Ultimately maples and oaks crowd the birches and cedar for sunlight. By and by the oaks and maples predominate.[5]

Forestry is generally minor in the total economy of the region, but almost every county has at least a little of it, and its local significance may be great even though it is usually characterized by low wages and temporary employment. Lumber is the principal product in most areas, but there has been a continued increase in pulping operations and hardwood products (furniture, flooring, etc.). No part of the region is a major lumber producer, but taken as a whole, about one-third of the nation's hardwood lumber comes from the Appalachians and Ozarks.

MINERAL INDUSTRIES OF THE APPALACHIANS

Various ores and other economic minerals are produced in the Appalachians, but coal is much more important than all the rest combined. No other activity has made such an imprint on the subregion and is so intimately associated with the "problem" of Appalachia.

Bituminous Coal

A large proportion of the Appalachian subregion is underlain by seams of bituminous coal that occur in remarkable abundance. This area has been the world's most prolific source of good-quality coal and still yields about two-thirds of the nation's total output. Every state in the region except New York and North Carolina is a producer of bituminous coal.

The history of the industry is one of ups and downs, but the energy crisis of the 1970s demonstrated conclusively that the United States must rely heavily on its domestic resources, and its reserves of coal are far larger than its supplies of all other practical energy sources combined.

Coal mining had been declining from the end of World War II until the early 1960s, losing its two biggest markets, railway steam engines and domestic heating, and experiencing a national employment decrease from 400,000 to 150,000. Increasing use of coal in electric power generation halted the decline in the early 1960s, but there was only a slow growth in output and almost no increase in employment until the accelerated demand of the early 1970s. By 1975 domestic bituminous coal production had surpassed the previous record-year output (1947), and a new peak was reached every year until 1982. After a brief decline, output again began to rise, and by the early 1990s new production records were being set, with a billion tons of coal being mined each year. At the time of this writing, coal generates more than 55 percent of America's electric power.

There are more than 5000 coal mines in the Appalachian and Ozark Region, employing about 150,000 miners. Many mines are small operations ("dogholes") with only a few workers, but most of the output comes from large mines owned by only a dozen or so companies, most of which are subsidiaries of giant corporations.

About 60 percent of the coal is extracted from strip mines, in which gigantic power shovels, the largest machines ever to move on land, exploit seams that are close to the surface (Fig. 9-8). Mechanization is virtually complete in the

[5] Evan B. Alderfer, "A Jogtrot through Penn's Woods," *Business Review,* Federal Reserve Bank of Philadelphia (February 1969), p. 11.

FIGURE 9-8   Removing a mountain top for strip mining of coal in West Virginia (Terry Wild Studios).

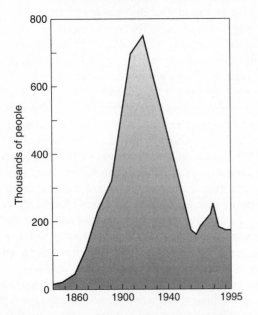

FIGURE 9-9   Historical fluctuations in coal-mining employment in the United States. The half-century downtrend has now ceased, but the resurgence has been erratic.

larger underground shaft mines as well. Mechanization has meant a dramatic increase in productivity per miner but a significant decrease in the number of miners needed (Fig. 9-9).

The strongest production record has been in the southern Appalachians, partly because of the higher proportion of more desirable coals with low sulfur content. Wyoming is now the national leader in coal production, but the next three leading states are Appalachian—West Virginia, Kentucky, and Pennsylvania (Fig. 9-10).

Coal has been a mixed blessing for Appalachia in the past; the same holds true today. Relatively minor market fluctuations can have a significant impact in the coal counties. The long-range prognosis however, is more favorable than in decades. With current technology and at contemporary levels of production, more than three centuries' worth of coal is still in the ground.

Anthracite Coal

In the northern end of the ridge-and-valley country in northeastern Pennsylvania lies the once important anthracite coal field. Anthracite, a very high quality coal, is quite expensive to mine because of its narrow and highly folded seams. For over a century this was a prominent mining activity, employing nearly 180,000 miners in the peak year of 1914. Competition with lower-cost bituminous coal and other fuels overwhelmed the anthracite industry. Despite massive reserves, fewer than 1000 miners are currently employed, and production is negligible.

Petroleum

Petroleum has long been important in Pennsylvania, Kentucky, and New York. The nation's first oil well was put down near Titusville in northwestern Pennsylvania in 1859, and this state led all others in the production of crude oil until 1895. Although still productive, the area is outstanding more for the high quality of the lubricants derived from the oil than for the quantity of production. Today less than 2 per-

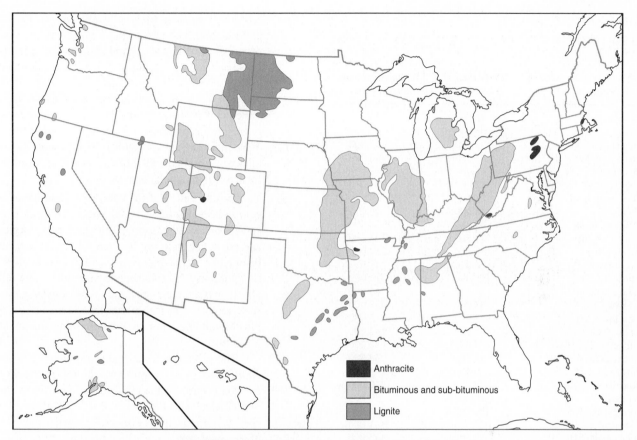

FIGURE 9-10   Principal coal deposits of the United States.

cent of the country's petroleum is produced in the Appalachians and Ozarks, and only one of the 60 leading oil fields (the Bradford–Allegheny field in Pennsylvania and New York) lies in the region.

Copper and Zinc

Numerous deposits of metallic ores yielding copper, zinc, and small quantities of other metals are worked along the western flanks of the Great Smoky Mountains in Tennessee. This state is the leading zinc producer of the nation, with nine mines in the area just east of Knoxville. In the extreme southeastern corner of Tennessee are five underground copper mines, the only significant source of copper in the eastern United States.

MINERAL INDUSTRIES OF THE OZARK–OUACHITA UPLANDS

For an upland area, the Ozark–Ouachita subregion is poor in minerals—with two exceptions. Southeastern Missouri is one of the oldest mining areas of the United States, having produced almost continuously since 1725. It is the principal

source of lead ore in the nation, with much of the output from recently developed or expanded mines. The Tri-State District where the Missouri, Kansas, and Oklahoma borders meet is another underground mining district that has been prominent during the past half-century. A leading zinc producer in the past, the Tri-State District now yields chiefly lead. Combined output from mines in these two districts makes Missouri the leading lead-producing state, normally with more than 90 percent of national output. Missouri ranks fourth among the states in zinc production.

CITIES AND INDUSTRIES

The Appalachian and Ozark Region is not highly urbanized. Urbanization is a much more recent phenomenon here than in most of the rest of the nation, and in many extensive areas rural dwellers are in the majority (see Table 9-1 for the region's largest urban places).

Pittsburgh is the only major metropolis in the entire region (Fig. 9-11). Including associated industrial satellites in the valleys of the lower Monongahela and Allegheny Rivers and the upper valley of the Ohio and the nearby complexes of

TABLE 9-1

Largest Urban Places of the Appalachians and Ozarks Region, 1995

Name	Population of Principal City	Population of Metropolitan Area
Altoona, PA	52,500	133,000
Asheville, NC	66,500	220,000
Binghamton, NY	53,000	263,000
Charleston, WV	52,000	256,000
Chattanooga, TN	161,000	442,000
Columbia, MO	67,000	121,000
Cumberland, MD	23,500	103,000
Elmira, NY	33,000	98,000
Fayetteville, AR	45,000	230,000
Fort Smith, AR	76,500	185,000
Hickory, NC	35,000	310,000
Huntington, WV	54,000	316,000
Huntsville, AL	165,000	318,000
Jamestown, NY	35,000	145,000
Johnson City, TN	47,000	455,000
Johnstown, PA	29,000	245,000
Joplin, MO	44,000	143,000
Knoxville, TN	180,000	640,000
Parkersburg, WV	37,000	152,000
Pittsburgh, PA	390,000	2,403,000
Roanoke, VA	99,500	240,000
Scranton, PA	79,000	639,000
Springfield, MO	155,000	296,000
State College, PA	47,500	141,000
Steubenville, OH	22,000	143,000
Wheeling, WV	37,000	155,000
Williamsport, PA	34,000	125,000
Youngstown, OH	97,000	604,000

Youngstown and *Wheeling*, the Pittsburgh district was one of the great metal-manufacturing areas of the world. Its long-time economic focus was on the production of primary iron and steel and on further processing in fabricated metals and machinery industries. However, in the 1980s the American steel industry virtually died, and the Pittsburgh area probably suffered more than any other, losing some 150,000 high-paying manufacturing jobs. It is no longer among the leading two dozen manufacturing cities of the nation. The city responded with an imaginative program called Renaissance II, which has revived the urban economy with downtown renovation, the attraction of high-tech businesses, and neighborhood revitalization.

Besides this notable urban agglomeration, there are three general areas in which loose clusters of medium-sized cities occur.

1. In northeastern Pennsylvania and adjacent parts of New York state are several cities whose economy significantly depended on coal mining or special-ized manufacturing for several generations. The decline of both anthracite coal mining and the once prominent textile and apparel manufacturing industries has been spectacular. Long-run population decline continues. The principal Pennsylvania urban places are *Scranton* and *Wilkes-Barre* in the upper valley of the Susquehanna River. *Binghamton* and *Elmira* are separate urban nodes in southern New York that have fairly well balanced economies and have absorbed considerable in-migration from the Pennsylvania anthracite area.

2. The middle valley of the Ohio River and the adjacent valley of the lower Kanawha have long been centers of heavy industry, particularly metallurgical and chemical factories. The lure of firm power has been particularly important as an industrial attraction; large nearby coal supplies are much more reliable than hydroelectricity with its seasonal vagaries. Inexpensive river transportation and a nearby surplus of suitable labor are other recognizable assets for industry.

    The principal urban nodes are the tri-cities of *Huntington* (West Virginia)–*Ashland* (Kentucky)–*Ironton* (Ohio) on the Ohio River, and *Charleston* on the Kanawha (Fig. 9-12). But many of the most spectacular industrial developments in this area took place away from the older centers, often in the splendid isolation of a completely rural riverside setting some distance from any urban area.

3. The valleys of eastern Tennessee have had considerable urban and industrial growth, much of it associated with TVA power development. The principal nodes are *Chattanooga, Knoxville, Johnson City, Kingsport,* and *Bristol.*

Two very atypical cities are also found in the region. *Huntsville*, Alabama, is the location of Redstone Arsenal, the rocket and guided-missile center of the U.S. Army, which has generated much industrial and service support and has stimulated rapid urban growth. *Oak Ridge*, Tennessee, is the site of Oak Ridge National Laboratory, which is the nation's largest nuclear energy research and development center and which conducts a variety of other programs in the general fields of energy production and conservation.

RESORTS AND RECREATION

The highlands of the northeastern Appalachians are not particularly spectacular or unusually scenic; they consist of pleasant, forested hills, with many rushing streams and a

FIGURE 9-11 The "golden triangle" of rejuvenated downtown Pittsburgh. The Monongahela River (coming in from the right) joins the Allegheny River (coming in from the left) to form the Ohio River (courtesy Greater Pittsburgh Convention and Visitors Bureau).

number of lakes. They also contain, particularly in the Catskill and Pocono mountains, one of the densest concentrations of hotels, inns, summer camps, and resorts to be found on the continent. The great advantage is that they are literally next door to Megalopolis and provide a relatively cool summer green space for urban millions to visit.

The principal tourist attractions of the Pennsylvania Piedmont and the Shenandoah Valley are historical. This was an area of almost continual conflict during the Civil War; in few other areas are historical episodes presented so clearly and accurately to the visitor, most notably at Gettysburg.

Farther south the national parks attract large numbers of tourists. Shenandoah National Park, with its beautifully timbered slopes and valleys, is well known for picturesque Skyline Drive. Great Smoky Mountains National Park is a broad area of lofty (by eastern standards) mountains clothed with dense forests of pine, spruce, fir, and hardwoods. In most years it receives more visitors than any other national park on the continent.

The recreational possibilities of the Ozark–Ouachita Uplands were recognized by Congress as early as 1832,[6] but the resort industry as it exists today is a recent development. Although Hot Springs and other centers became important locally in the 1890s, the present development had to wait until better railroads and highways were built into the mountains

of the area and until the urban centers in surrounding regions attained sufficient size to support a large nearby resort industry. Both goals have been achieved and today the Ozark–Ouachita Uplands occupy the unique position of being the only hilly or mountainous area within a few hours' drive of such populous urban centers as Kansas City, St. Louis, Memphis, Little Rock, Dallas, Fort Worth, Oklahoma City, and Tulsa.

As in other parts of the Southeast and Gulf Southwest, some of the most successful recreational areas developed around the large, branching reservoirs that were constructed in various river valleys. Water sports, in the form of boating, fishing, swimming, and skiing, are now very much a part of holiday living for hundreds of thousands of families in an area where natural lakes are almost nonexistent. Most important as a recreational center is Lake of the Ozarks in Missouri but also notable are Lake O' the Cherokees in Oklahoma, Lake Ouachita in Arkansas, and Bull Shoals Reservoir on the Arkansas–Missouri border.

In the Ozarks subregion the single largest tourist attraction, by far, is the small town of Branson, Missouri, which has become a country-and-western music entertainment center that rivals fabled Nashville.

THE OUTLOOK

The plight of the Appalachian and Ozark Region is celebrated in song and story. It has been recognized as a major nega-

---

[6] An area around Hot Springs was set aside as a federal preserve some four decades before Yellowstone, normally considered to be the first of the national parks, was established.

A CLOSER LOOK   River-Basin "Development"

The "taming" or "harnessing" of a river has become one of the favorite tools for local or regional economic development in the twentieth century. All across the United States, from the Penobscot to the Sacramento, rushing rivers have been turned into quiet reservoirs by straightening channels, constructing levees, and, particularly, building dams. In most cases, such development has meant multiple use because of the advantages of flood control, decreased erosion, improved navigation, expanded hydroelectricity potential, increased water availability, and broadened recreational opportunities. Opponents of river development schemes point out that the economic logic of such projects is nearly always suspect at best and that ecological problems often arise.

The controversy between proponents and opponents of river development is an old one. Meanwhile, most rivers of the nation have experienced some sort of development, and plans are in process for the rest. In the Appalachian and Ozark Region lie major portions of two of the most integrated and best known of all river-basin development schemes: the Tennessee Valley Authority (TVA) and the Arkansas River Waterway.

THE TENNESSEE VALLEY AUTHORITY

To date, the TVA is perhaps the greatest experiment in regional socioeconomic planning and development carried out by the federal government. It was created in the 1930s to aid in controlling,

conserving, and using water resources of the area. It deals with such diverse matters as flood control, power development and distribution, navigation, fertilizer manufacturing, agriculture, afforestation, soil erosion, land planning, housing, and manufacturing.

The area of the Tennessee Valley Authority encompasses the watershed of the Tennessee River and its tributaries—more than 40,000 square miles (104,000 km^2) in parts of seven states (Fig. 9-A). This area was selected because it was the most poverty-stricken major river basin in the country. Except that it is a drainage basin, it is not a unified region because land utilization, agriculture, manufacturing, transportation, and the distribution of power all cut across the drainage boundary.

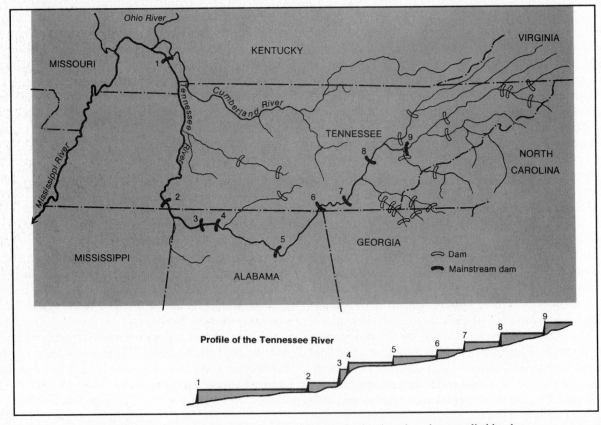

FIGURE 9-A   The dams of the Tennessee Valley Authority (based on data supplied by the Tennessee Valley Authority).

There are no longer any free-flowing stretches of the river except in its upper headwaters; it has been dammed into a series of quiet lakes for its 650-mile (1040-km) length from Knoxville to its confluence with the Ohio River near Paducah. Besides 9 mainstream dams, there are more than 24 others on tributaries. Hydroelectricity is generated at each dam, but most of the TVA's power output now emanates from a dozen thermal electric plants that it controls. This comprehensive electric system was a great boon to the valley, but it is also the most controversial feature of the TVA's operation. Private power companies contend that their very existence is threatened by the cheaper TVA power, part of the cost of which is federally subsidized. Furthermore, many complaints about destructive strip mining in the region are aimed at the TVA, which is the single largest purchaser of coal.

A major initial purpose of the TVA was to control flooding in the valley, a sporadic, major hazard. This goal was largely achieved, but some critics point out that flood damage was prevented by permanently flooding much of the best valley land.

Other benefits attributable to the TVA include attraction of industry to the region, alleviation of much accelerated soil erosion, provision of new water recreational areas, and maintenance of a permanent 9-foot (3-m) navigation channel. The benefits that resulted are unquestionable, but there is considerable difference of opinion about costs in relation to benefits. Critics claim that the money could have been used more effectively in other ways.

The controversy will doubtless continue; nevertheless, many lessons can be learned from the experiment. The TVA functions at a level between centralized federal government and fragmented local authorities and yet on an interstate basis. It is blessed by some and cursed by others but is often referred to, particularly by foreigners, as a comprehensive and relatively successful example of functional regional planning. On the other hand, many residents of the region consider the TVA to be little more than a large electric power utility.

THE ARKANSAS RIVER NAVIGATION SYSTEM

On a somewhat smaller scale, but not at a significantly lower cost, is the Arkansas River Navigation System, dedicated in 1971 after nearly two decades of construction. Its purpose was to produce a 440-mile (700-km) navigable channel (9 feet or 3 m in depth) up the Arkansas River from its confluence with the Mississippi, via Pine Bluff, Little Rock, Fort Smith, and Muskogee, to the head of navigation near Tulsa. Eighteen dams have "stabilized" the river, thus adding flood control and hydroelectricity production to the project's benefits (Fig. 9-b).

There is no doubt that the local areas have benefited from cheaper transportation, power generation, and industrial attraction; however, the expense of the undertaking, which was approximately equal to the amount spent on construction of the Great Lakes–St. Lawrence Seaway, makes it one of the most costly public works projects in the history of the nation. Critics who contend that "it would be cheaper to pave it" are speaking only partly in jest.

TLM

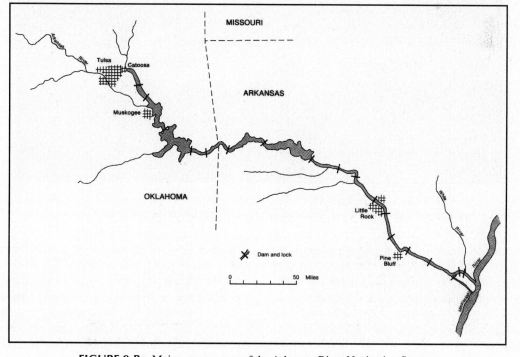

FIGURE 9-B   Major components of the Arkansas River Navigation System.

FIGURE 9-12   Large factories, especially chemical plants, crowd the valley of the Kanawha River around Charleston, West Virginia (courtesy Charleston Chamber of Commerce).

tive economic anomaly, an extensive region of poverty in the heart of the richest nation on earth. Its way of life has been called a "culture of despair." The reasons underlying such a situation are complex and imperfectly understood but certainly include a litany of environmental difficulties and a variety of questionable economic approaches and negative human attitudes.

Into a region of small, hill-country farmers came three waves of economic development, each largely financed by "outside" entrepreneurs, each largely sending the profits out-

side the region, and each despoiling the environment to a notable (sometimes disastrous) degree. The story of logging and mining, the first two waves, is well known. Some local people made money from sale of land or resources and many jobs were provided; however, many sales were at relatively low prices and most jobs were low paying, part time, or both. Recreation and tourism, the third wave, is more recent but is nearly as massive and sudden as the other two. Developers, usually corporate and often from outside the region, purchased large acreages of high-amenity (scenic or waterside) land on which to build massive recreational and housing projects. Lack of integrated planning and zoning allows development that often does not conform to the landscape.

Federal and state governments attempted to alleviate the situation with massive infusions of capital and ambitious development programs. Such efforts, most notably represented by the TVA, the Arkansas River Waterway, and the Appalachian Regional Commission, provided many advantages to the region but usually at a cost-benefit ratio of depressing proportions.

Until fairly recently the popular solution was graphically shown by migration statistics; most parts of the region experienced a massive and continuous out-migration and population decline for several decades. Beginning in the late 1960s, however, and continuing to the present, new demographic trends have appeared. Many sections, particularly in southwestern Appalachia (Kentucky, Tennessee, and Alabama) and the Ozarks of Arkansas and Missouri, have experienced an upsurge of population growth. This growth has been based partly on expanded employment opportunities in manufacturing (new factories) and recreation (mostly reservoir-related services) that stemmed the prevailing out-migration of working-age people and partly on an influx of older people who have opted for retirement in the pleasant rural surroundings offered by the hill country.

Some parts of the region have functioned for a considerable period as pockets of prosperity—for example, the Poconos and much of the Great Valley. Economic stimulation has been provided by long-term government installations at such places as Oak Ridge, Tennessee, and Huntsville, Alabama. And many larger cities, such as Pittsburgh, Chattanooga, Knoxville, and Springfield, continued to experience "normal" urban growth patterns based on their diversified economies.

The renewed importance of coal in the national energy scene augurs well for the economy of most of the coal-mining areas despite the production and employment fluctuations of the 1980s.

It seems likely that the Appalachians and Ozarks are partly free of the traditional syndrome of regional poverty and depression. Agricultural specialization, especially in beef cattle and poultry, will result in more efficient and profitable output in many farming areas. Industrial and recreational developments will continue to make contributions to economic diversification. Population statistics will be watched with interest to see if the recent short-term growth in rural areas develops into a long-term trend.

Many areas, however, did not escape the patterns of the past. There are still too many areas of poor farms, eroded soil, and marginal mines with underpaid workers interspersed among the districts of improving conditions.

SELECTED BIBLIOGRAPHY

BATTEAU, ALLEN, ED., *Appalachia and America: Autonomy and Regional Dependence*. Lexington: University Press of Kentucky, 1983.

BENHART, JOHN E., AND MARJORIE E. DUNLAP, "The Iron and Steel Industry of Pennsylvania: Spatial Change and Economic Evolution," *Journal of Geography,* 88 (September–October 1989), 173–184.

BINGHAM, E., "Appalachia: Underdeveloped, Overdeveloped, or Wrongly Developed?" *Virginia Geographer,* 7 (Fall–Winter 1972), 9–12.

CUFF, DAVID J., ET AL., *The Atlas of Pennsylvania*. Philadelphia: Temple University Press, 1989.

HART, JOHN FRASER, "Land Rotation in Appalachia," *Geographical Review,* 67 (1977), 148–166.

KARAN, P. P., AND COTTON MATHER, EDS., *Atlas of Kentucky*. Lexington: University of Kentucky Press, 1977.

KARAN, P. P., AND WILFORD A. BLADEN, *Across the Appalachians: Washington, D.C., to Lake Michigan*. New Brunswick, NJ: Rutgers University Press, 1992.

MARSH, BEN, "Continuity and Decline in the Anthracite Towns of Pennsylvania," *Annals,* Association of American Geographers, 77 (September 1987), 337–352.

———, "Environment and Change in the Ridge and Valley Region of Pennsylvania," *Journal of Geography,* 88 (September–October 1989), 162–166.

MILLER, E. JOAN WILSON, "Ozark Superstitions as Geographic Documentation," *Professional Geographer,* 24 (1972), 223–226.

MILLER, E. WILLARD, "The Anthracite Region of Northeastern Pennsylvania: An Economy in Transition," *Journal of Geography,* 88 (September–October 1989), 167–172.

RAFFERTY, MILTON D., *Missouri: A Geography*. Boulder, CO: Westview Press, 1983.

———, *The Ozarks: Lands and Life*. Norman: University of Oklahoma Press, 1980.

RAITZ, KARL B., RICHARD ULACK, AND THOMAS LEINBACH, *Appalachia: A Regional Geography*. Boulder, CO: Westview Press, 1983.

VERNON, PHILIP H., AND OSWALD SCHMIDT, "Metropolitan Pittsburgh: Old Trends and New Directions," in *Contemporary Metropolitan America*. Vol. 3, *Nineteenth Century Inland Centers and Ports,* ed. John S. Adams, pp. 1–59. Cambridge, MA: Ballinger Publishing Company, 1976.

WALLACH, BRET, "The Slighted Mountains of Upper East Tennessee," *Annals,* Association of American Geographers, 71 (September 1981), 359–373.

ZELINSKY, WILBUR, "The Pennsylvania Town: An Overdue Geographical Account," *Geographical Review,* 67 (1977), 127–147.

The Inland South

10

For many generations "the South" has been an evocative term that applies to a distinctive section of the continent, roughly the southeastern quarter of the United States. Although its boundaries were never easily delimited, it had certain cultural (that is, economic, demographic, political, and social) characteristics that tended to set it apart. Referred to in the past as "Cotton Belt" or "Old South," this broad section of the nation has experienced notable changes in many aspects of its geography in recent decades and is now more properly conceptualized as the "New South" or "Eastern Sun Belt."

In this text, the South, like New England, is depicted more accurately as two regions rather than as one. There are important differences, both physical and cultural, between the interior and coastal portions. The actual boundary between the two regions—Inland South and Southeastern Coast—is determined primarily on the basis of topography, hydrography, and land use. As delineated in Figure 10-1, the Inland South Region occupies most of the broad coastal plain southeast, south, and southwest of the Appalachian–Ozark Region. The immediate littoral zone, including much of Louisiana and all of Florida, is excluded from the Inland South Region.

The northern boundary of the Inland South is demarcated by the southern margin of the two upland areas (Appalachians and Ozark–Ouachitas). Between and on either side of the uplands—in the Mississippi River lowland, in Virginia, and in Oklahoma–Texas—the extent of the Inland South is determined by variation in crop patterns. In each of these three areas there is a transition from a southern pattern featuring cotton to some other pattern: a wheat belt pattern in Oklahoma–Texas, a corn belt pattern in the Mississippi Valley, and a less distinctive general farming pattern in Virginia.

The southern and eastern boundaries of the Inland South are in a transition zone that roughly parallels the Gulf and South Atlantic coasts, some 75 to 100 miles (120 to 160 km) inland. Dense forest, poor drainage, and spotty agriculture predominate coastward; more open woodland, better drainage, and less discontinuous agriculture are found interiorward. The littoral zone of the southeastern states is prominently oriented toward the sea; its port activities, industrial development, and seaside recreation set it apart from the Inland South.

The western boundary of the region is less clear cut than the others. Differences in land use seem to be the most important criteria. In south central Texas the change from flat land to the short but steep slopes of the "hill country" is accompanied by a change from crop farming to pastoralism. North of the "hill country," however, there is only an indefinite transition from relatively small general farms eastward to

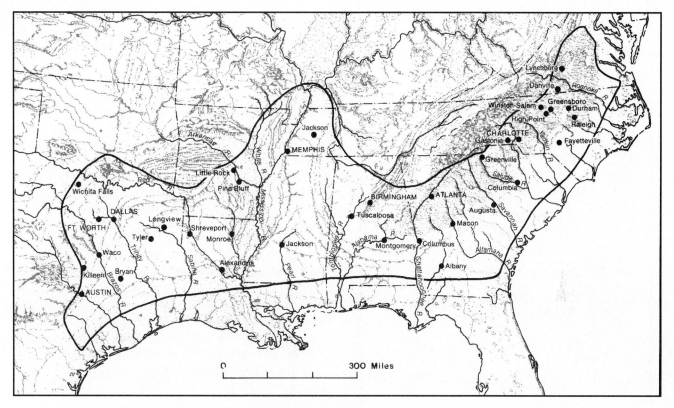

FIGURE 10-1   The Inland South Region (base map copyright A. K. Lobeck; reprinted by permission of Hammond, Inc.).

somewhat larger specialty farms (grain sorghums, wheat, irrigated cotton, and cattle) westward.

THE PHYSICAL ENVIRONMENT

There is widespread uniformity of physical attributes in the Inland South, several of which provide conspicuous elements of unity for the region. The land surface is quite flat in most places, the drainage pattern is broadly centrifugal and functionally simple, summer and winter climatic characteristics are grossly uniform, and red and yellow soils of only moderate fertility predominate. It is, however, the prevalence of forest and woodland over almost all of the region that is the most noticeable feature of the regional landscape.

The Face of the Land

The Inland South is primarily a coastal plain region; the flat or gently undulating land surface is only occasionally interrupted by small hills or long, low ridges. The underlying rocks consist of relatively thick beds of unconsolidated sediments of Tertiary or Cretaceous age. These beds have a monoclinal dip that is gently seaward, so that progressively older rocks are exposed with increasing distance from the coast.

Along the interior margin of the region in the east is an extensive section of older crystalline rocks that form the Southern Appalachian Piedmont. The topography here is somewhat more irregular than in the coastal plain, although slopes are not steep and the landscape is best described as gently rolling. Between the rocks of the Piedmont and those of the coastal plain is a continuation of the Fall Line. The Southern Piedmont extends in an open arc from central Virginia to northeastern Alabama.

That portion of the Inland South Region that is north of the Gulf of Mexico exemplifies a pattern of landform development known as a belted coastal plain. The sedimentary beds outcrop in successive belts that are arranged roughly parallel to the coast. Some of these strata are less susceptible to erosion than others and thus stand slightly above the general level of the land as long, narrow, resistant ridges called *cuestas* (Fig. 10-2). The overall pattern is a series of broad lowlands developed on weaker limestones and shales, which

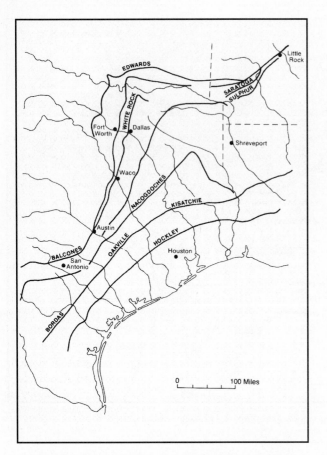

FIGURE 10-2 Principal cuestas making up the belted coastal plain of the western Gulf area (from N. M. Fenneman, *Physiography of Eastern United States*. New York: McGraw-Hill Book Company, 1938; reprinted by permission).

are bounded on the seaward side by the low but abrupt scarps that mark the inward edge of the resistant sandstone cuestas (Fig. 10-3). In some cases, the cuestas extend for hundreds of miles. They are more numerous in the west Gulf Coast Plain of Texas and Louisiana but are slightly bolder and more conspicuous in the east Gulf Coast Plain of Mississippi and Alabama.

Between the two belted zones of the coastal plain is the broad north–south alluvial plain of the Mississippi Valley. This alluvial lowland varies in width from 25 to 125 miles (40 to 200 km) and is bounded on the eastern and western sides by prominent bluffs that rise as much as 200 feet (60 m) above the lowland. Although a few residual upland ridges interrupt the valley floor, most of the lowland is extraordinarily flat and has a southward slope averaging only about 8 inches (20 cm) per mile. With such a gentle gradient, the Mississippi River and its low tributaries meander broadly and produce many oxbow lakes and winding scars.

The general drainage pattern of the region consists of a number of long, relatively straight rivers that flow sluggishly toward the ocean from interior upland areas (Appalachians, Ozarks, Ouachitas, and high plains of West Texas). The normally dendritic pattern of their tributaries is interrupted in several places by the cuestas, resulting in right-angle bends and trellising. Very few natural lakes, other than small oxbows, are found along the rivers, but swamps, and bayous, indicative of poor drainage, are widespread.

Climate

The climate of the Inland South is sometimes described as humid subtropical, but such terminology is belied by occasional severe winter cold spells. Nevertheless, summer is clearly the dominant season in this region. It is a long period of generally high humidity that is hot by day and warm by night. Summer is also the time of maximum precipitation, with most rainfall coming in brief convective downpours.

Winter is a relatively short season, but it is punctuated by sporadic sweeps of continental polar air across the region that push the normally mild temperatures well below the freezing point. In no part of the region is snow unknown, and most sections can anticipate one or more snowfalls each win-

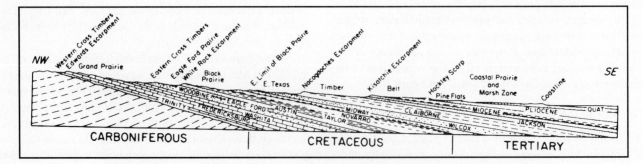

FIGURE 10-3 Generalized geologic cross section of the western Gulf coastal plain, extending northwest–southeast across Texas (from Fenneman, *Physiography of Eastern United States*; reprinted by permission).

ter, although the length of time of snow cover is usually measured in hours rather than days. Winter is only slightly less moist than summer, and most winter precipitation falls in protracted drizzles rather than brief showers. Total annual rainfall in the region varies from 55 inches (140 cm) in the east to 20 inches (50 cm) on the western margin.

Spring and fall are relatively long transition seasons, marked by pleasant temperatures. The former is a notably windy time of the year and, in the western part of the region, is often characterized by major dust storms carried on westerly winds.

Soils

The most widespread soils of the Inland South are Ultisols. They are primarily red or yellowish gray in color, indicating a considerable degree of leaching and the subsequent concentration of insoluble iron and aluminum as well as accumulation of a clay horizon. With careful management, they can be agriculturally productive soils; unfortunately, careful management has often been lacking and some of the nation's worst examples of accelerated soil erosion can be found in the region.

In the Black Belt of Alabama–Mississippi and the Black Waxy Prairie of central Texas are extensive areas of soils derived from underlying limestone and marl. Classed as Mollisols, they are among the most naturally fertile soils anywhere; but they, too, have been subjected to severe erosion and in many localities the black topsoil has been stripped away, revealing the lighter-colored subsoil.

The rich alluvial soils of the Mississippi Valley are also dark in color, rich in organic matter, and highly productive. They are mostly classed as Inceptisols and Alfisols.

Natural Vegetation

The Inland South was originally a timbered region (Fig. 10-4) characterized by southern yellow pines on most of the interfluves and southern hardwoods (gums, oaks, cypress) in the stream valleys, with a proportion roughly half pine and half hardwood. There are 11 species in the group called southern yellow pines, of which 7 are prominent in the Inland South. The most widespread pine species in the region are loblolly, shortleaf, and longleaf.[1] The principal concentrations of hardwoods are in the Mississippi River lowland and scattered widely over the northern portion of the state of Mississippi. In the natural state, three parts of the region apparently were relatively treeless. The Black Belt and Black Waxy Prairie were probably covered with prairie grasses[2] and the extreme western part of the region had a mixed cover of grassland, low open woodland, and scrubby brush.

PEOPLING AND PEOPLE OF THE INLAND SOUTH

The aboriginal inhabitants of the Inland South consisted of several strong and well-organized Native American tribes and a number of minor ones. Most important were the "Five Civilized Tribes"—Cherokee, Choctaw, Chickasaw, Creek, and Seminole—that originally occupied most of the area between the Mississippi River and the Atlantic Coast. West of the Mississippi, the Caddo, Osage, and Apache were important; later the Comanche moved down from the northern plains and dominated the western frontier of the region for several decades.

During the early years of European contact conflict with the Native Americans was relatively limited, primarily because profitable trading relationships were established between tribesmen and coastal merchants and because pressure for European settlement in the region was slow in building up. Once the Europeans began to move inland significantly, however, the days of the Native Americans were numbered regardless of the treaties that were frequently promulgated and just as frequently dishonored by the colonial and federal governments.

Indian wars became commonplace in the early 1800s, and before long most of the recalcitrant tribes had either been wiped out or shifted to new homes west of the Mississippi. An acerbic contemporary critic described the process in which "the most grasping nation on the globe" would take the Native Americans "by the hand in brotherly fashion and lead them away to die far from the land of their fathers ... with wonderful ease, quietly, legally, and philanthropically. ... It is impossible to destroy men with more respect for the laws of humanity."[3]

European settlement in the eastern part of the Inland South was little more than a tiny trickle until the eighteenth century, and the occupance of most of the region dates from the early nineteenth century. Five generalized tides of settlement can be discerned:

[1] Elbert L. Little, Jr., and William B. Critchfield, *Subdivisions of the Genus Pinus (Pines)* U.S. Department of Agriculture Miscellaneous Publication no. 1144 (Washington, DC: Government Printing Office, 1969).

[2] As in the case with many seemingly "natural" grasslands, there is considerable debate as to whether the prairie association of the Black Belt was indeed natural or had been induced by repeated burning by Native Americans. See, for example, Erhard Rostlund, "The Myth of a Natural Prairie Belt in Alabama: An Interpretation of Historical Records," *Annals,* Association of American Geographers, 47 (December 1957), 392–411.

[3] Alexis de Tocqueville, *Democracy in America,* trans. George Lawrence (New York: Harper & Row, 1966), p. 312.

A CLOSER LOOK   The Irresistible Kudzu

One of the most prominent features in the vegetational landscape of the Inland South is kudzu (*Pueraria lobata*) a leguminous, broad-leaved, climbing vine with clusters of purple flowers (Fig. 10-A). A native of East Asia, it was imported from Japan in 1876 as an ornamental decoration. It soon became popular as a shade plant, and then it was discovered to have value as a forage crop for cattle, pigs, and goats. It became even more popular in the 1930s, when the Soil Conservation Service became its advocate as a means of controlling soil erosion. Indeed, kudzu is probably the only weed pest that can boast its own fan club: In 1943 the Kudzu Club of America was formed and soon had 20,000 members.

Kudzu flourished remarkably in its adopted environment in the southeastern states. Despite its undeniable usefulness as forage and as soil holder, it is now considered a nuisance in many localities, where it has aggressively invaded and overwhelmed the natural vegetation. It grows with astonishing speed—it can spread up to 12 inches (30 cm) in 24 hours—hence its nickname, the "mile-a-minute vine." Kudzu rapidly grows over anything that stands in its way, mercilessly smothering trees, climbing up utility poles and highway signs, taking over agricultural fields, even inundating buildings. Indeed, it is sometimes referred to as "the plant that ate the South" (Fig. 10-B).

Control programs vary, involving selective herbicides, physical removal, prescribed burning (sometimes with flamethrowers), and grazing by sheep, goats, and hogs. Control is feasible, if costly, whereas eradication probably is impossible. Kudzu is found as far north as Pennsylvania, but cold winters inhibit its growth. However, throughout the Inland South, from Virginia to east Texas, it is an implacable presence in the landscape.

TLM

FIGURE 10-A   A close-up view of the kudzu vine (TLM photo).

FIGURE 10-B   The relentless spread of the kudzu vine has overwhelmed this landscape in central Mississippi. It is kept from encroaching onto the highway by the use of chemical pesticides and flamethrowers (TLM photo).

FIGURE 10-4 A typical landscape in the Inland South Region is forested, as in this view near Hattiesburg, Mississippi (TLM photo).

1. From the early colonial coastal settlements of Virginia and Maryland freemen moved west and southwest, joined by settlers coming directly from Europe but funneled through the Chesapeake Bay ports; this stream was augmented by a flow up the Shenandoah Valley into eastern Tennessee and beyond.

2. From the South Atlantic coastal ports, particularly Charleston and Savannah, more European settlers were channeled into the interior.

3. A third route of settler flow was southward from the Midwest and the upper South (Kentucky and Tennessee) and, in part, down the Mississippi Valley.

4. A fourth stream moved northward through the Gulf Coast ports of New Orleans and Mobile.

5. There was also an early movement of settlers of Spanish ancestry northeastward from Mexico into central and eastern Texas; this flow was circumscribed first by contact with the French in Louisiana and later by the persistent movement of Anglos into Texas from the East and Northeast.

Except for those coming from Mexico, nearly all settlers of the region were northwest European in origin. It is probable that most initial settlers were American-born, but many came directly from Europe, and certainly the great majority were of recent European ancestry. British people (English, Welsh, Scots, and Scotch-Irish) were the most numerous, but Germans, French, Swiss, and Irish were also significant.

Cotton as a Settlement Catalyst

By the 1790s most of the Carolina portion of the Inland South and part of eastern Georgia had been settled and there were

---

other small settled districts in the lower Mississippi Valley and central Texas. In 1793 young Eli Whitney, visiting on a Georgia plantation near Savannah, developed a vastly improved cotton gin. It was the first practical machine for separating "green seed" cotton lint from seed. With remarkable suddenness, cotton was adopted as the commercial crop of settlement, for it could be produced on small farm or large plantation, provided that cheap labor was available. Thus cotton production boomed, settlement spread, and slave importation expanded.

By 1820 eastern and central Georgia were fully occupied, the good lands of central and western Alabama were settled, much of the Mississippi Valley was being farmed, and settlements in eastern Texas were expanding. Within another three decades nearly all the Inland South was under settlement except for much of the alluvial Mississippi valley. Plantations, with large slaveholdings, were widespread. In addition, many small farmers were engaged in limited commercial enterprises with only a few slaves.

Yeoman farmers, who operated without labor assistance, were commonplace, especially in areas of less productive soil or steeper slope. A few freed slaves were farming in Virginia and North Carolina. The total slave population amounted to more than 3 million in 1850, compared with a white population of twice that number.

The Contemporary Population

The present population of the Inland South is more homogeneous than that of most regions, although this homogeneity is rapidly diminishing due to a rising tide of immigration from such places as Mexico, China, and Korea (which are, for example, the three leading sources of immigrants to Atlanta). Basically, the region's population has two principal components: some 30 million whites, who are mainly of Anglo-Saxon ancestry and Protestant religious affiliation, and 12 million African Americans, who are also mostly Protestant (Fig. 10-5). The largest denominational affiliations in the region are Baptist and Methodist, although various evangelical Protestant groups are strongly entrenched and are experiencing a rapid growth in membership.

Compared with the total population of the region, adherents of Roman Catholicism are relatively few in number except among the large Hispanic minorities of several Texas cities. Jews constitute an extremely small proportion of the population but have a disproportionately large impact because of their prominence in economic and civic affairs.

In terms of population mobility in the region, there are three recognizable trends in partial opposition to one another:

1. A large share of the people who live in the Inland South were born there. For example, more than

A CLOSER LOOK   The Fire Ant—An Exotic Scourge

The conspicuous presence of an exotic plant, the kudzu, is more than matched by the insidious invasion of an exotic animal, the South American fire ant, which has become a scourge throughout the southeastern United States over the last half century. A native of the floodplain of the Paraguay River, the "invincible" fire ant (*Solenopsis invicta*) was introduced inadvertently via cargo ship to the port of Mobile, Alabama, in the 1930s. From that foothold it has spread to the farthest reaches of the Southeast, presumably being halted only by cold on the north and dryness on the west.

These tiny (one-eighth inch or 3.5 mm long) reddish brown or black creatures congregate in colonies that have up to 200 queens and half a million worker ants. Each colony lives in a largely subterranean mound, and an acre of infested land may hold up to 400 mounds (Fig. 10-C).

The virtues of fire ants have yet to be discovered, but the problems they cause are legion. Their durable mounds damage plows and other farm machinery. They nibble on the insulation of utility cables, causing power outages and other breakdowns in services. But their most daunting impact is the result of their voraciously carnivorous appetites. They possess formidable mandibles and an acutely venomous sting. They prey particularly on other invertebrates, but will attack larger creatures if they can get to them. Reptiles, rodents, and birds are particularly susceptible, but such larger animals as raccoons, deer, goats, and cattle are recurrent victims. Humans, too, are attacked with frequency.

FIGURE 10-C Fire ant mounds in a Georgia pasture (William J. Weber/Visuals Unlimited).

Only a handful of human deaths have been ascribed to fire-ant attacks, but thousands of people have suffered from their stings, and in areas of major infestation humans modify their behavior to avoid fire-ant locales.

Chemical warfare has been waged against the invading fire ants for more than three decades, at a cost of hundreds of millions of dollars. Such deadly chemicals as dieldrin, heptachlor, chlordane, and mirex are effective pesticides, but they wipe out many components of the ecosystem in addition to the fire ants. When these ecologically unsound chemicals are prohibited, the fire ants rebound quickly.

The outlook for comprehensive fire-ant control is dismal. Decades of research have produced no feasible answers. It is likely that we will just have to learn to live with them as we have done with such unpleasant but more tolerable pests as cockroaches and mosquitoes. The discouraging prognosis is summed up in the words of a Texas entomologist: "Basic research may eventually give us some answers. We may come pretty close to at least understanding why we can't control them."

TLM

---

80 percent of the population of Mississippi, Alabama, Georgia, and South Carolina are still living in the state of their birth in contrast to the national average of less than 70 percent.

2.   The urban areas, particularly the larger ones, are significant foci of in-migration, particularly of people from outside the region. Newly arrived whites tend to settle in the suburbs, whereas African American newcomers mainly go to the central cities.

3.   The long-run high rate of net out-migration of African Americans from the Inland South has now been reversed. Today there is a significant net inflow of African Americans to the region from the Northeast and north-central states (but not from the West). Even so, nearly 90 percent of all migrants into or out of the Inland South are nonblacks, and the net impact of migration on the racial composition of the region's population is to make it whiter rather than blacker.

FIGURE 10-5   The Inland South is sometimes referred to as the "Bible Belt" because of numerous and enthusiastic adherents of various forms of Protestantism (TLM photo).

THE CHANGING IMAGE OF THE INLAND SOUTH

Perhaps in no other part of North America is the sense of regional identity so pervasive as in the Inland South. The southern way of life is recognized by southerner and non-southerner alike as being regionally distinctive, partially in tangible ways and partially as a state of mind. For many people, this regional identity arouses strong emotions, ranging from reverence to abhorrence. To some, the southern way of life is "genteel"; to others, "decadent." The strongly entrenched and nondiversified economic base of the southern past is part of the image, but more a part of contemporary consciousness are feelings about social conditions, particularly in regard to the African American minority of the population.

The origin of the southern way of life is complex and beyond the scope of this presentation. Briefly, a generation or two ago there was a regional character to the Inland South that was simple of generalization even if imperfect of image: the region was economically depressed and socially divided. Agriculture, the traditional basis of the southern economy, was straitjacketed by corn for subsistence and by cotton and tobacco for cash. The unholy duo of tenancy and soil erosion had a stranglehold on rural life. Industry was present, but it was undiversified and paid low wages. Per capita income was well below the national average, and poverty was relatively widespread. Average educational attainment was low, and illiteracy was a problem. Practically speaking, only one political party was extant. Class distinctions were strong and the

African American was universally accorded—by white and African American alike— the bottom rung of the social and economic ladder. There was more to the image—veneration of womanhood, religious piety, economic pluck, and patriotic valor; only inadequacies have been emphasized. The point is that the Inland South displayed certain significant disadvantages and faced certain significant handicaps.

What of its regional character today? Certainly many of the "old" elements of regional distinctiveness still prevail. Regional speech characteristics, for example, show no significant change; the southern drawl is as distinctive today as it ever was despite population diversification and the influence of mass media. Dietary preferences have been modified, particularly in the large cities, but there is still a pronounced regional "flavor" to both choice of food and method of cooking; corn bread, hominy grits, hush puppies, pot likker, black-eyed peas, turnip greens, and other delicacies are distinctly southern. Protestant religious fundamentalism continues to be strong, and the term "Bible Belt" is still quite apt. And "southern hospitality" is more than a cliche; there is a regional claim to personal warmth and friendliness that is readily discernible.

Race relations in the Inland South, as in most of the nation, are far from tranquil, although racist feelings of the white majority have considerably softened and overt discrimination is either eliminated or notably reduced. Public as well as most private facilities are desegregated, and African Americans are accorded the rights and responsibilities of full citizenship in most places and activities. Black legislators have been elected in Georgia, black mayors in Mississippi, and black sheriffs in Alabama. Perhaps more important, state and local administrations in such key states as North Carolina and such major cities as Atlanta, Birmingham, and Dallas have become more cognizant of and more responsive to the needs of their African American citizens.

The economy of the Inland South has undergone expansion, diversification, and substantial strengthening. Industrial growth is notable and widespread. In every state of the region, value added by manufacturing far exceeds the value of agricultural production. As recently as 1947 Atlanta was the only city in the region that ranked among the 50 leading industrial centers of the nation, whereas in the early l990s the region had four urban areas (Dallas, Atlanta, Charlotte, and Greensboro–Winston-Salem) among the top 20 industrial cities. The tertiary and quaternary components of the economy also expanded and diversified, with greatly increased employment opportunities in trade, services, finance, construction, transportation, and government.

Perhaps, however, the changing character of the South is best exemplified by the agricultural scene. The cotton-and-corn, plantation-oriented duoculture (never merely a cotton monoculture) that characterized primary production in the region for several generations gave way to a diversification of significant proportions. This diversification is most conspicuously demonstrated by the dethroning of "King Cotton."

THE RISE AND DECLINE OF COTTON IN THE INLAND SOUTH

When the American Southeast was first settled by Europeans, the colonists felt that whites could do little sustained physical labor in its tropical summers. Consequently, Charleston and Savannah became great slave-importing cities. The first crops raised with the help of slave labor were rice, indigo, sugarcane, tobacco, and some cotton. In 1786 long-fibered Sea Island cotton was introduced and successfully grown along the coastal lowlands.

In 1793 Whitney's invention of the cotton gin revolutionized the cotton industry. European demand expanded, settlement spread westward, cotton acreage enlarged, and

many more slaves were brought in to work the fields. By the middle of the nineteenth century the American South was supplying 80 percent of the world's export cotton.

The Civil War and emancipation of the slaves were big setbacks for the cotton industry, as was the spread of the boll weevil across the South in the early 1900s. In both cases, however, the industry rebounded and grew to new heights. By the late 1920s the United States was growing well over half of all the world's cotton and exporting an all-time high of more than 11 million bales annually (Fig. 10-6).

With the onset of the economic depression of the 1930s, however, cotton production in the Inland South went into a steady decline. Cotton acreage diminished from a peak of about 17 million acres to a nadir of about 3 million. By the mid-1980s the old Cotton Belt was growing only about 40 percent of all U.S. cotton; most production had shifted to irrigated lands farther west, primarily in west Texas, Arizona, and California. Texas, for decades the undisputed leader in cotton output, was surpassed by California in 1982. Moreover, four-fifths of the Texas production is west of the borders of the Inland South Region.

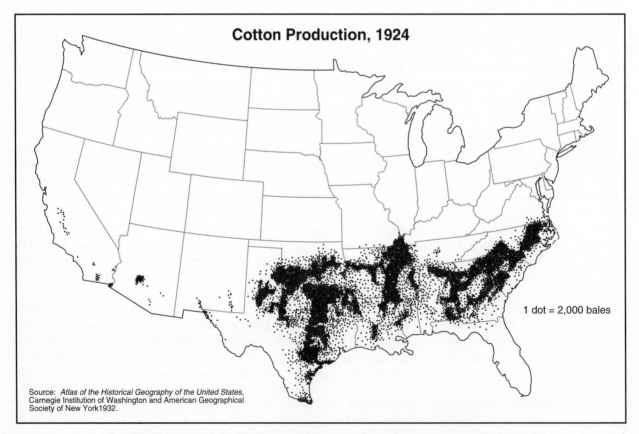

Source: *Atlas of the Historical Geography of the United States,* Carnegie Institution of Washington and American Geographical Society of New York1932.

FIGURE 10-6   Three quarters of a century ago cotton production was prominent in almost all parts of the Inland South Region.

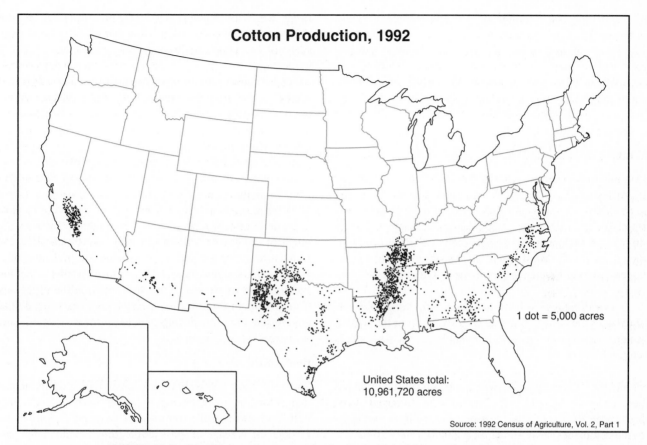

FIGURE 10-7   Distribution of cotton production in the United States. The only part of the old Cotton Belt that is still a major producing area is the middle Mississippi Valley.

In the late 1980s, however, there was a cotton renaissance in the region, occasioned by a variety of factors, but particularly by a dramatic rise in demand for cotton products and a federal farm program that bolstered market prices. By the early 1990s Texas was again the leading producer, with one-fourth of the national crop (though mostly in west Texas). Mississippi ranked third (after California), followed by Arkansas and Louisiana (Fig. 10-7).

FARMING IN THE INLAND SOUTH: PRODUCTIVE DIVERSITY

For several decades social scientists often referred to the "changing South," listing tendencies and prognosticating trends in economic and social matters that will make a "New South" out of the "Old South." Many predicted changes have long since materialized, and in no field have they been more striking than in agriculture.

Crop rotation, improved soil management and fertilization, supplemental irrigation, variation in field crops, and in-creased mechanization are all part of the scheme, but the most pronounced aspect is the shift toward mixed farming, with pastures for cattle as the dominant feature of the farm landscape. Cropped acreage has decreased, and many eroded fields have been restored to useful production by the planting of forage and pasture.

With the decline of cotton as undisputed king, many attendant evils have also diminished. Soil erosion and fertility loss have decreased. The established and practical but often corrosive pattern of absentee ownership and farm tenancy has been broken—presumably forever.

If southern agriculture has diversified and cotton acreage has declined, what has replaced cotton and what are the elements of diversification? In terms of gross acreage, the answer is pasture and woodland.[4] A great deal of former

---

[4] For the region as a whole, forest and pasture occupy a much greater acreage than cropland does. One result is that various types of wildlife are now found in much greater numbers than at any time in the past century. In Georgia, for example, there are more than 3 million white-tailed deer, and the legal bag limit is 5 deer per hunter. In Alabama the legal bag limit under certain circumstances is 170 deer per hunter per year!

cropland, especially cotton land, has been returned to a grass or tree cover and is now used mainly for grazing cattle and, to a lesser extent, as a source of pulpwood. The major component of agricultural diversification has been a shifting emphasis from crops to livestock, particularly beef and dairy cattle and, locally very important, poultry. The most notable cropping diversification has been the expanded cultivation of soybeans.

Crops

Cotton farmers were attracted by the adaptability of *soybeans* to mechanized farming and by their value as a cash crop, and in the 1940s soybean cultivation spread throughout the region (Fig. 10-8). In most parts of the Inland South there was a significant retrenchment of soybean production in the 1980s, although the middle Mississippi Valley continues as the second largest soybean-producing area in the country (after the Corn Belt).

Corn acreage has been on a decreasing trend for several decades, but it is still a widely grown crop in the region. Another grain crop that is widespread in the Inland South is *winter wheat*.

Tobacco is the outstanding specialty crop in the northeastern part of the region. Although tobacco consumption has been declining steadily in the United States in recent years, production has experienced less of a decline because the tobacco companies have shifted their emphasis toward the export market. More than one-third of the national output comes from eastern and central North Carolina and adjacent parts of southern Virginia and northern South Carolina (Fig. 10-9). Most tobacco is grown on small- to medium-sized farms, which are distinguished by the presence of an array of shiny new metal bulk-curing barns that have replaced thousands of antiquated traditional flue-curing barns. Tobacco harvesting extends over a period of 4 to 8 weeks during which the pickers comb the fields about once a week, selecting ripe leaves from the bottom of the stalks. Mechanization has become prominent in both harvesting and curing, with the result that the previous high-density farm population has diminished significantly since 1970.

Various other crops are grown in quantity in the region. Nearly one-half of the national production of *rice* comes from the bottomlands of eastern Arkansas and adjacent parts of Mississippi. More than half the national output of *peanuts* is in southwestern Georgia and southeastern Alabama and much of the rest is from northeastern North Carolina and southeastern Virginia and from scattered localities in central Texas and Oklahoma (Fig. 10-10). Three-fourths of the national output of *pecans* comes from this region, especially Georgia. South Carolina and Georgia rank second and third among the states producing *peaches*. Other vegetable and fruit crops are widespread.

Beef Cattle

Since the early 1950s, livestock have yielded more income than cotton, or any other crop, to farmers of the Inland South, with beef cattle in the forefront. In association with cattle raising, the acreage of grass and legume meadows for pasturage has greatly increased; however, a considerable amount of grazing is carried on in the fairly open forest that characterizes more than half the region. It is estimated that five-sixths of the forested area is grazed.

FIGURE 10-8 A field of soybeans near Columbia, South Carolina (TLM photo).

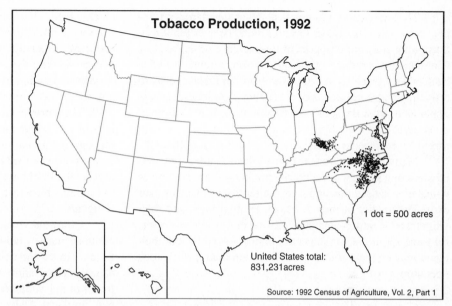

FIGURE 10-9 Tobacco is widely grown on the Piedmont and Inner Coastal Plain of the Carolinas and Virginia.

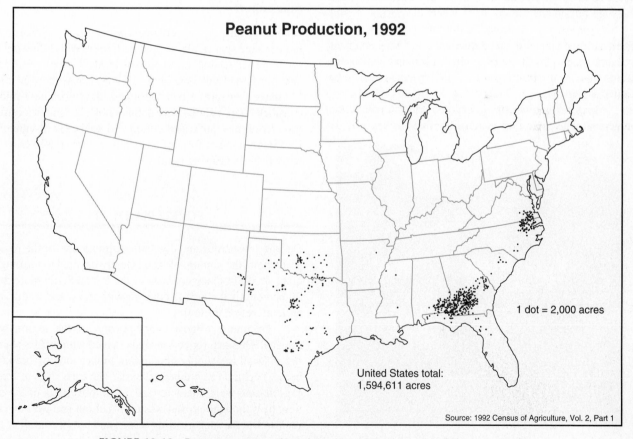

FIGURE 10-10   Peanuts are grown throughout most of the Inland South Region, with the major concentration in southwestern Georgia and southeastern Alabama.

The American market for beef has changed from an emphasis on steaks and roasts toward ground beef (hamburger), thus quantity is more important than texture or tenderness. Consequently, producers have shifted from purebred English beef breeds (such as Angus, Hereford, and Shorthorn) to crossbreeds that contain Brahma (Zebu) bloodlines (such as Santa Gertrudis, Brangus, and Beefmaster). It is a common practice to run a Brahma or Santa Gertrudis bull in with mixed-breed cows to produce mixed offspring (Fig. 10-11).

Many cattle enterprises of the Inland South are operated more or less as an avocation by urban business executives who derive satisfaction and a tax shelter from a ranch of their own. The quality of their livestock is often high, and their willingness to experiment and innovate makes them a powerful force for upgrading the entire industry even though their ranch may be more for outdoor recreation than for livestock operation.

Poultry

In the past, poultry raising, particularly chickens, was of importance for home consumption and local market only; today there are certain areas in the South where the raising of poultry for commercial consumption dominates farm activities. There is some emphasis on producing eggs, frying chickens, and turkeys, but the principal product is broilers, which constitute more than two-thirds of all poultry consumed in the United States.

Much of this industry is handled under a contractual agreement whereby a feed merchant or meatpacker provides chicks, feed, vitamins, medicines, scientific counsel, and a market, and the farmer supplies only housing and labor. The contractee is benefited by needing little capital and having an assured market; the contractor, by having an assured supply of dependable quality; and the consuming public, by lower poultry prices.

The scale of these contract farming operations often is immense. The eggs may be incubated in gigantic hatcheries that turn out a million chicks a week. The day-old chicks are taken to nearby feeding farms, where chicken houses as big as a football field may hold 50,000 birds. Seven weeks later the mature birds are trucked to the processing plants for slaughter.

Despite skyrocketing poultry consumption, an insecure national market, labor disputes at packing plants, and cutthroat competition combine to inhibit stability in the industry. Nevertheless, major centers of commercial broiler raising are found in the Southern Piedmont of Georgia and the Carolinas, northern Alabama, central Mississippi, southwestern Arkansas, and easternmost Texas (Fig. 10-12).

Aquaculture

A notable "farming" development within the last two decades has been the spectacular rise of the farm-raised catfish industry in the Mississippi Delta country of Mississippi, Arkansas, and northern Louisiana (Fig. 10-13). Long shunned as a "trash fish" by most urban consumers, the catfish now enjoys considerable popularity. In a number of Delta counties catfish production has surpassed cotton and soybeans in value. In Louisiana, another form of aquaculture, farm-raised crayfish (crawfish), is growing rapidly.

FOREST PRODUCTS

Despite prosperous and diversified agriculture for the region as a whole, the amount of land actually devoted to farming is quite small. In many counties of the Inland South more than 90 percent of the area is covered with trees, and traditional agriculture is hard to find.

Originally almost every portion of the region was forested. Demand for cotton land coupled with reckless burning reduced the supply of standing timber long before lumbering began on a commercial basis. The early days of forest exploitation were characterized, as in most parts of the nation, by a thoughtless and wasteful "cut out and get out" approach. But the long growing season and heavy rainfall of the region produce rapid tree growth, and forest industries have become well established in every state. A forest planted with pine seedlings can be thinned for pulpwood after only 12 to

FIGURE 10-11 The raising of beef cattle is a major enterprise throughout the Inland South. This purebred Brahma herd is near San Augustine, Texas (TLM photo).

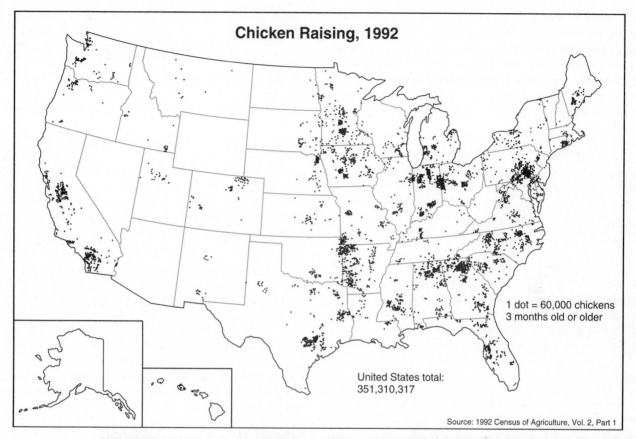

Chicken Raising, 1992

1 dot = 60,000 chickens
3 months old or older

United States total:
351,310,317

Source: 1992 Census of Agriculture, Vol. 2, Part 1

FIGURE 10-12   Chickens are raised commercially in most parts of the United States. One of the principal concentrations is in northern Georgia.

15 years, and a mature crop of sawtimber can be harvested within four decades.

Over the last several years there has been continually increasing logging in the region, with the result that timber, especially softwood timber, is now being harvested faster than it is being grown. Moreover, a growing deficiency in pine regeneration is resulting in substantial replacement of pines by hardwoods. These trends produce concerns about

long-range productivity, which have partly been allayed by the development in the late 1980s of the federal Conservation Reserve Program, in which landowners receive an annual payment for idling farmland on a long-term basis. Much of this idled land is being planted to trees.

Softwood Lumbering

Cutting southern yellow pines for lumber was the first forest industry to be widespread in the region. Small operations, feeding small sawmills, were characteristic. Prior to the 1960s there had been a long-term decline in regional lumber output. The trend, however, was reversed in that decade and production has increased, partly because of tree farming and partly because of increased scale of operations and the establishment of larger sawmills. The region yields about one-third of the nation's softwood lumber and the same proportion of its plywood.

Hardwood Industries

Hardwood logging is much more localized. The greatest concentration is in the Mississippi River lowland, where cutting

FIGURE 10-13   A catfish farm along the Mississippi River in southwestern Mississippi (courtesy Mississippi Department of Economic and Community Development).

is a seasonal operation because of the boggy conditions caused by winter rains. Hardwood output is also notable in the Carolina Piedmont, where it is carried on year-round. Southern hardwoods—oak, hickory, cypress, gums—are in great demand for furniture, veneers, and shingles.

Pulp and Paper Industry

This branch of forest industries is the leading source of income in the region, having grown rapidly in the last few decades. The wood from southern pine has long fibers, which are preferred for making heavy duty paper and paperboard. The region yields nearly two-thirds of the total pulpwood cut; Georgia is the national leader in production of pulpwood, pulp, and paper, with Alabama a close second in pulpwood output.

Throughout the Inland South, a conspicuous feature of the landscape is a clearing along a railway or highway that serves as a depot (called a *woodyard*) for stacking pulpwood cut in nearby areas (Fig. 10-14). From the woodyard the pulpwood is shifted, usually by rail, to a pulp mill. Many mills, especially the larger ones, are located outside the Inland South in the Southeastern Coast Region, often at a seaport.

MINERALS AND MINING

Apart from hydrocarbons and bauxite, the region is poor in economic mineral resources. Moreover, the abrupt decline of the national petroleum industry in the 1980s impacted the western part of the Inland South cataclysmically.

Petroleum

The great oil boom of the region dates from 1930, when the East Texas Field, the most prolific in North American history, was brought in. At maximum extent, the field was 42 miles (67 km) wide and 9 miles (14 km) long. Within an area of about 300 square miles (777 km^2) were more than 27,000 producing wells. This field yielded more than 5 percent of all oil ever produced in the United States, nearly three times as much as any other field.

The western states of the Inland South are all among the national leaders in petroleum production: Texas, first; Louisiana, third; Oklahoma, fifth. Other than the East Texas Field, there are three principal areas of production: north-central Texas and south-central Oklahoma, centering on Wichita Falls and Ardmore; the northwestern corner of Louisiana and the southwestern corner of Arkansas, with adjacent portions of northeastern Texas, centering on Shreveport and El Dorado; and northeastern Louisiana and adjacent portions of Mississippi.

The drastic decline in demand for domestic petroleum during the 1980s was a severe blow to the economy of the oil-producing states. Unemployment rates soared, and Louisiana even experienced a population decrease, unprecedented in the Sunbelt states. As of this writing, recovery is still slow and erratic.

Natural Gas

Production and consumption of natural gas increased dramatically in North America over the last three decades. Proved reserves are now at critically low levels, however, and production trends continue downward despite accelerated explo-

FIGURE 10-14 A typical pulp log woodyard in southwestern Mississippi (TLM photo).

ration. Texas and Louisiana are by far the leading producers of natural gas, although most of their production is outside the Inland South Region. In recent years natural gas production from these two states amounted to about three-fourths of total national output.

Bauxite

Although seven-eighths of the bauxite ores used in the United States are imported (largely from the Caribbean area), domestic production is also important. All domestic supplies now being worked lie within the boundaries of the Inland South Region: three surface mines in Alabama and Georgia. Arkansas was long the principal source of domestic bauxite, but its production has now ceased.

URBAN-INDUSTRIAL DYNAMISM

An important part of the new look of the Old South is furnished by the dynamic urban and industrial growth of the region (see Table 10-1 for the region's largest urban places). From an area of sparse and specialized manufacturing in small cities it has developed into a region of notable industrial diversity and strength in booming metropolises.

While the Northeast experienced a significant reduction in manufacturing employment in recent years, the Southeast had a large increase in its industrial base. The traditional "southern" industries in the Inland South Region grew only slowly; it is the "northern" industries of the region that expanded rapidly. Thus textiles, apparel, and tobacco products had an inconspicuous growth; major expansion was focused in such durable goods categories as metal fabrication, transportation equipment, and electronics-electrical equipment. As a result, there is a convergence of the industrial structure of region and nation; that is, the Inland South is becoming less distinctive economically and is developing a balanced industrial structure that is reflective of the national pattern. Despite continuing absolute growth, manufacturing in the Inland South, as in the nation as a whole, is becoming relatively less important in the economy because of the rapid expansion of the tertiary sector.

It should be noted that much of the industrial growth of the region has taken place in nonmetropolitan areas. Cities are no longer the favored locales for many manufacturing firms. Instead, it is commonplace for the firm to purchase a rural tract and build single-story plants with large parking lots to accommodate cars of the workers who commute from as far as 50 or 100 miles (80 or 160 km) away. Such developments are particularly notable on the Carolina piedmont.

TABLE 10-1

Largest Urban Places of the Inland South Region, 1995

Name	Population of Principal City	Population of Metropolitan Area
Albany, GA	84,000	117,000
Alexandria, LA	49,000	132,000
Anniston, AL	28,000	121,000
Athens, GA	42,000	132,000
Atlanta, GA	405,000	3,335,000
Augusta, GA	50,500	451,000
Austin, TX	564,000	1,007,000
Birmingham, AL	282,000	878,000
Bryan, TX	71,000	139,000
Charlotte, NC	446,000	1,284,000
Charlottesvile, VA	44,000	148,000
Columbia, SC	102,500	492,000
Columbus, GA	207,000	282,000
Dallas, TX	1,046,000	2,944,000
Danville, VA	54,500	112,000
Decatur, AL	49,500	142,000
Durham, NC	152,000	
Fayetteville, NC	86,000	289,000
Florence, AL	35,000	137,000
Fort Worth, TX	464,000	1,490,000
Gadsden, AL	42,500	103,000
Gastonia, NC	56,000	
Greensboro, NC	198,000	1,122,000
Greenville, SC	64,000	884,000
High Point, NC	71,500	
Jackson, MS	198,000	417,000
Jackson, TN	54,000	80,500
Killeen, TX	71,000	301,000
Little Rock, AR	181,000	543,000
Longview, TX	78,000	210,000
Lynchburg, VA	79,000	210,000
Macon, GA	109,000	306,000
Memphis, TN	615,000	1,068,000
Monroe, LA	55,000	146,000
Montgomery, AL	203,000	317,000
Pine Bluff, AR	61,000	89,000
Raleigh, NC	239,000	998,000
Sherman, TX	30,500	96,000
Shreveport, LA	198,000	362,000
Texarkana, TX	35,000	125,000
Tuscaloosa, AL	76,000	164,000
Tyler, TX	76,000	157,000
Waco, TX	108,000	197,000
Wichita Falls, TX	99,000	133,000
Winston-Salem, NC	151,000	

Primacy of the Gateway Cities

At the top of the urban hierarchy of the Inland South are two dominant metropolises that epitomize the concept of the gateway city. *Atlanta* in the east and *Dallas* in the west serve as regional capitals in terms of commerce, finance, transporta-

A CLOSER LOOK   Sunbelt vs. Snowbelt

The term "Sunbelt" was coined by journalists in the late 1970s to refer to the Florida–Texas–California axis of mild-winter states within which there had been, and presumably would continue to be, population and economic growth considerably above the norms for the United States as a whole, thus implying a hitherto unprecedented concentration of political power as well. The appropriateness of this concept and the catchiness of its appellation were affirmed by the rapidity with which both idea and name were adopted by scholars and laypersons alike.

No precise boundaries have been attached to the Sunbelt, although most references seem to accept the parallel of 37° north latitude (roughly the northern border of several states from North Carolina to Arizona) as approximately its northern limit. Nor does the Sunbelt have clear-cut attributes. Despite the name, its principal environmental earmark is not the presence of sunshine but the relative absence of cold weather. Concomitant characteristics include significant population growth and economic well-being.

The Sunbelt concept goes hand in hand with its antithesis—a Snowbelt or Frostbelt—which vaguely refers to the northeastern quarter of the United States. This was the "core" of the nation and its economic heartland throughout history. Although there have been precursors of significant economic/population/political reorientation for years, it was not until the 1970s that the trends became sufficiently persistent and vigorous to induce the widespread acceptance of the Sunbelt idea.

There is no question that the established patterns have experienced significant change. The Sunbelt has been gaining, both absolutely and relatively, over the Snowbelt in terms of both population growth and economic prosperity. The causes of the changes are complex and variable, but they include at least four factors: (1) lower production costs in the Sunbelt; (2) favorable developments in transportation, communication, and industrial technology in the Sunbelt; (3) a decrease in the significance of many of the cost efficiencies once enjoyed in the Snowbelt; and (4) the increasing importance of amenity factors (including mild winters) as an attraction for choosing a place of residence.

Thus the rise of the Sunbelt is very real. It should be cautioned, however, that the Sunbelt has never in actuality been a monolithic area of growth and prosperity; it had (and continues to have) significant areas of economic stress and population decline. Moreover, economic advantages are relative things, and in the late 1980s there emerged prominent areas of economic progress in the Snowbelt (witness the resurgence of the Boston area), whereas some of the prodigy areas of the Sunbelt have suffered severely (as the "oil patch" of Texas/Louisiana/Oklahoma). Still, the Sunbelt concept is a notion whose time has come, and it is now well established in public perception.

TLM

tion, and other economic aspects. These two cities are the principal funnels and nerve centers through which extraregional goods, services, ideas, and people are channeled into the Inland South, and to a lesser extent, they accommodate the reverse flow of regional output to the nation. This gateway function is best shown by the magnitude of wholesale sales, the concentration of financial institutions, especially banks and insurance companies, the large number of national and regional corporate headquarters, and the daily passenger flow through the respective airports, which were the second and third busiest airports in the world in the mid-1990s (exceeded only by Chicago's O'Hare Field).

Although approximately the same in size and function, the two cities have quite different origins and histories. Industrial growth in recent decades was spectacular in both urban nodes, particularly Dallas, which now ranks as the tenth largest industrial center in the nation (Fig. 10-15). In both metropolises most industrial expansion occurred in the related fields of aerospace and electronics, although their industrial structures are generally well diversified. The economic function of these gateway cities, however, is much more significantly commercial than industrial, and the impressive skyline of Dallas and the progressive atmosphere of Atlanta suggest the improving economic and cultural image of the region (Fig. 10-16).

Secondary Regional Centers

One step lower in the regional hierarchy are the medium-sized cities of Memphis, Fort Worth, and Birmingham, each of which is also growing rapidly. *Memphis* is the traditional river city of the middle Mississippi basin, dominating the area between the spheres of New Orleans and St. Louis.

Fort Worth, although near enough to have a twin-city relationship with Dallas, is remarkably different from its neighbor.[5] Whereas Dallas's eastward orientation emphasizes commerce, finance, oil, and electronics, Fort Worth is clearly oriented to the west, focusing more on cattle, railways, and agricultural processing.

[5] For statistical purposes, the Bureau of the Census has declared Dallas and Fort Worth to be part of the same Consolidated Metropolitan Statistical Area (CSMA), which encompasses a dozen small north Texas counties. The local media coined the inelegant term "Metroplex" to refer to this conurbation.

FIGURE 10-15 Most of the major cities of the Inland South have become prominent industrial centers. This extensive electronics plant is a few miles north of Dallas (courtesy Texas Instruments, Incorporated).

The story of *Birmingham* is unusual; it has long been the only heavy-industry center in the South. Local deposits of iron ore and coal made it the least expensive place in the nation to manufacture steel. In common with all other North American steel centers, it suffered severe decline in the 1980s, but in the last few years has experienced an economic renaissance into a growing high-tech center specializing in medical research and medical services.

The Carolina Urban-Industrial Complex

Although the Carolinas lack a major metropolis, they have a zone of notable urban and industrial development centered on the Piedmont of North Carolina (Fig. 10-17). This zone extends northward into southern Virginia, southwestward across South Carolina into northern Georgia, and eastward onto the North Carolina coastal plain. The heart of the district is the North Carolina Piedmont, with its medium-sized cities of *Charlotte, Winston-Salem,* and *Greensboro.* It is a highly industrialized district; many factories are located in smaller towns, such as Reidsville and Shelby, and in nonurban areas, especially in Gaston and Cabarrus counties. New industries tended to locate in rural areas or in small towns rather than in cities, partly because of a tight labor market. Manufacturing is not diversified in the district, being mainly tobacco, textiles, chemicals, and furniture. The area contains the greatest concentrations of cigarette, cotton-textile, and furniture factories to be found in the nation.

The importance of manufacturing to the local economy is shown by the fact that North and South Carolina have greater proportions of their labor forces employed in factories (more than 30 percent) than do any other states.

This Piedmont industrial district contains about half as many industrial facilities and employees as the six New Eng-

FIGURE 10-16 Skyline scene in Atlanta.

land states combined. Moreover, three of the urban nodes, centered on Greensboro, Greenville, and Charlotte, are among the three dozen leading industrial centers in the nation despite their relatively small populations.

Other facets of urbanism are also notable in the district. For example, in the area encompassing *Raleigh, Durham,* and *Chapel Hill* is the "Research Triangle," probably the most successful corporate research enterprise in the nation. In connection with the three nearby major universities, it is oriented toward industrial and environmental research and has spurred considerable economic growth in the area.

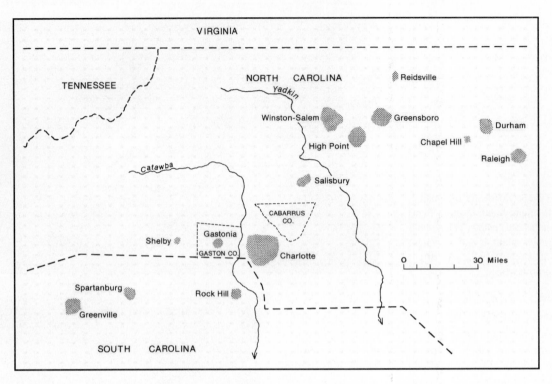

FIGURE 10-17 The urban-industrial complex of the Carolinas.

THE OUTLOOK

The Inland South is a region with a distinctive cultural heritage, but the only really unifying physical phenomena are climate and forest. It is a region that has traditionally produced a few staple commercial crops but where an agricultural revolution has taken place. It is a region that has been traditionally rural and agrarian, but it has become significantly urbanized. It is a region of slow population growth and out-migration that has become the nation's principal focus of interregional in-migration.

Livestock raising, particularly beef cattle and chickens, probably will persist as dominating features of the agricultural scene, despite the recent resurgence of cotton growing and continuing crop diversification. Supplemental irrigation will become commonplace to carry crops vigorously through short dry spells and to increase overall yields.

The continued decrease of rural population, both white and African American, will probably be a blessing rather than a curse to the region. The displaced marginal agriculturalists will probably find more satisfactory urban employment, both within and without the Inland South; and either machinery or migrant workers will take up the labor slack on the fewer but larger remaining farms.

Production from timberlands will become increasingly important for both lumber and pulp. The trend toward clearcutting will probably spread throughout most of the larger pine holdings, although its efficiency as a management technique will be partly counterbalanced by environmental and aesthetic objections. The inadequate replacement of pines could become a serious problem before the turn of the century.

The economic outlook in nonagricultural fields is generally promising, particularly in the western part of the region. The attraction of the Inland South to manufacturing industries is now solidly based, and industrial investment will probably continue to be strong in the near future, even though the relative role of manufacturing in the economy is now diminished. Low-skilled and low-paid industries will continue to decline, thus shrinking the region's competitive advantage, but worker retraining and entrepreneurial leadership should overcome this problem.

Urban growth and associated commercial expansion seem ensured. Some of the nation's fastest-growing medium-sized cities are in this region, and the two gateway metropolises continue as leaders of expansion. At the other end of the urban spectrum, however, many smaller towns seem destined to wither, and there are many parts of the "Old South" (primarily in rural areas) that have seen and can expect to see little change.

The population mix will probably become more diverse with continued influx of people from other regions. The rate of in-migration may slow down, but it is unlikely to diminish significantly. The concentration of African Americans in the larger cities, particularly the inner-city districts of Atlanta, Birmingham, and Memphis, will provide increased political power to African American voters.

All things considered, the passage of time will continue to blur the image of the Inland South as a distinctive region even as the region's economic and social role in the nation becomes more significant.

SELECTED BIBLIOGRAPHY

ADKINS, HOWARD G., "The Imported Fire Ant in the Southern United States," *Annals,* Association of American Geographers, 60 (1970), 578–592.

CLAY, JAMES, W., DOUGLAS M. ORR, JR., AND ALFRED W. STUART, EDS., *North Carolina Atlas: Portrait of a Changing Southern State.* Chapel Hill: University of North Carolina Press, 1976.

CONWAY, DENNIS, ET AL., "The Dallas–Fort Worth Region," in *Contemporary Metropolitan America,* Vol. 4, *Twentieth Century Cities,* ed. John S. Adams, pp. 1–37. Cambridge, MA: Ballinger Publishing Company, 1976.

CROMARTIE, JOHN, AND CAROL B. STACK, "Reinterpretation of Black Return and Nonreturn Migration to the South, 1975–1980," *Geographical Review,* 79 (July 1989), 297–310.

CROSS, R. D., ET AL., *Atlas of Mississippi.* Jackson: University of Mississippi Press, 1974.

DE VORSEY, LOUIS, JR., AND MARION J. RICE, *The Plantation South: Atlanta to Savannah and Charleston.* New Brunswick, NJ: Rutgers University Press, 1992.

DOSTER, JAMES F., AND DAVID C. WEAVER, *Tenn-Tom Country.* Tuscaloosa: University of Alabama Press, 1987.

FALK, WILLIAM W., AND THOMAS A. LYSON, *High Tech, Low Tech, No Tech: Recent Industrial and Occupational Change in the South.* Albany: State University of New York Press, 1988.

FOSCUE, EDWIN J., "East Texas: A Timbered Empire," *Journal of the Graduate Research Center,* Southern Methodist University, 28 (1960), 1–60.

HAMLEY, W., "Research Triangle Park: North Carolina," *Geography,* 67 (January 1982), 59–62.

HART, JOHN FRASER, "Land Use Change in a Piedmont County," *Annals,* Association of American Geographers, 70 (December 1980), 492–527.

———, "Migration to the Blacktop: Population Redistribution in the South," *Landscape,* 25 (1981), 15–19.

———, "The Demise of King Cotton," *Annals,* Association of American Geographers, 67 (1977), 307–322.

HARTSHORN, TRUMAN A., ET AL., "Metropolis in Georgia: Atlanta's Rise as a Major Transaction Center," in *Contemporary Metropolitan America.* Vol. 4, *Twentieth Century Cities,* ed. John S. Adams, pp. 151–225. Cambridge, MA: Ballinger Publishing Company, 1976.

HAYES, CHARLES R., *The Dispersed City: The Case of Piedmont North Carolina.* Chicago: University of Chicago, Department of Geography, 1976.

HEATWOLE, CHARLES A., "The Bible Belt: A Problem in Regional Definition," *Journal of Geography,* 77 (February 1978), 50–55.

HILLIARD, SAM B., *The South Revisited: Forty Years of Change.* New Brunswick, NJ: 1992.

HODLER, THOMAS W., AND HOWARD A. SCHRETTER, EDS., *The Atlas of Georgia.* Athens: Institute of Community and Area Development of the University of Georgia, 1986.

JORDAN, TERRY G., JOHN L. BEAN, AND WILLIAM M. HOLMES, *Texas.* Boulder, CO: Westview Press, 1984.

KOVACIK, CHARLES F., AND JOHN J. WINBERRY, *South Carolina: The Making of a Landscape.* Columbia: University of South Carolina Press, 1989.

MEINIG, DONALD W., *Imperial Texas: An Interpretive Essay in Cultural Geography.* Austin: University of Texas Press, 1969.

PRUNTY, MERLE C., AND CHARLES S. AIKEN, "The Demise of the Piedmont Cotton Region," *Annals,* Association of American Geographers, 62 (1972), 283–306.

REED, JOHN SHELTON, *One South: An Ethnic Approach to Regional Culture.* Baton Rouge: Louisiana State University Press, 1982.

SINIARD, L. ARNOLD, "Dominance of Soybean Cropping in the Lower Mississippi River Valley," *Southeastern Geographer,* 15 (May 1975), 17–32.

TRIMBLE, STANLEY W., *Man-Induced Soil Erosion on the Southern Piedmont, 1700–1970.* Ankeny, IA: Soil Conservation Society of America, 1974.

WINBERRY, JOHN J., AND DAVID M. JONES, "Rise and Decline of the 'Miracle Vine': Kudzu in the Southern Landscape," *Southeastern Geographer,* 13 (November 1973), 61–70.

The Southeastern Coast

11

Attenuated along the "subtropical" shores of eastern United States is the Southeastern Coast Region. It is long and narrow in shape, dynamic in recent population and economic development, and represents a curious mixture of the primitive and the modern. It is a region that contains some of the most pristine natural areas as well as some of the most completely unnatural areas to be found anywhere on the continent. Burgeoning cities and busy factories nestle side by side with quiet backwaters where fin and feather and fur dominate life, a juxtaposition restricted to places where a low-lying coastal environment permits an infinite variety of land-and-water relationships. Sophisticated urbanites live close to backwoods trappers, and some of the nation's most scientific and mechanized agriculture is carried on only a stone's throw from areas of hardscrabble subsistence farms.

The region extends from the Dismal Swamp of southeasternmost Virginia along the littoral zone of the Atlantic Ocean and Gulf of Mexico to the international border at the Rio Grande; thus is abuts the Inland South Region along most of its interior margin, brushing the Great Plains Region in southern Texas and touching the Megalopolis Region in southern Virginia (Fig. 11-1).

The inland margin of the Southeastern Coast Region is not clearly defined. Physically it is marked by a transition zone that separates the poorly drained coastal lands from the better-drained country interiorward. Culturally it can be considered to be inland of the seacoast cities and their immediate city-serving farmlands. As thus defined, this region encompasses a relatively narrow coastal strip of Virginia, the Carolinas, Georgia, Alabama, Mississippi, and Texas; it also includes about one-third of Louisiana and all of Florida.

THE PHYSICAL SETTING

Relief of the Land

Although almost uniformly characterized by flatness and impeded drainage, the regional topography is varied in detail and primarily reflects variations in geomorphic history. Except in much of Florida, where Tertiary formations predominate, almost the entire region is underlain by relatively unconsolidated Quaternary sediments of marine origin. These beds extend seaward to constitute also the continental shelf. The shelf offshore of this region is quite broad—generally more than 75 miles (120 km) in width—everywhere except off the southeastern corner of Florida and near the delta of the Mississippi River. Along almost all the coast the ocean is very shallow for a considerable distance offshore; in many

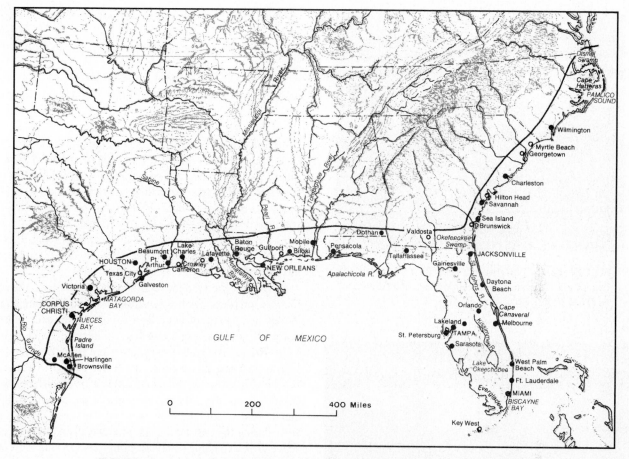

FIGURE 11-1   The Southeastern Coast Region (base map copyright A. K. Lobeck; reprinted by permission of Hammond, Inc.).

localities a swimmer must wade out tens or even hundreds of yards to find water deep enough to swim in.

Low-lying, sandy islands occur off the coast of most of the region except parts of Louisiana and the Gulf of Mexico side of peninsular Florida. The almost continuous chains of islands off the North Carolina and Texas coasts represent barrier sandbars at an extreme stage of lengthwise development; they are long, narrow, and low and encompass lagoons and sounds that are in the process of being filled in as the coastline prograes outward. Other offshore islands (along the Georgia and Mississippi coasts, for example) are more varied in origin, ranging from erosional remnants to old beach ridges, with gradations in between. The Florida Keys are totally different in formation. The eastern keys are exposed remnants of an elongated fossil coral reef cut by tidal channels into a series of separate islands, whereas the western keys represent limestone shoals upraised from an earlier sea bottom.

With the important exceptions of peninsular Florida and southeastern Louisiana, the topography of the mainland is generally uniform throughout the region. The land is exceedingly flat, sloping gently toward the sea. A great many rivers meander across this lowland. The lower reaches of most streams were drowned by coastal subsidence; thus the coastline is studded with estuaries and bays in addition to the lagoons and sounds previously mentioned. Inadequate drainage is, therefore, common and there are extensive stretches of swampland as well as some large marshes (Fig. 11-2). The two largest and most famous of these poorly drained lands are Okefenokee Swamp in Georgia–Florida (Fig. 11-3) and Dismal Swamp on the Virginia–North Carolina border.

The topography of peninsular Florida is distinctive from that of the rest of the region. The peninsula is a recently emerged mass of carbonate rocks, largely limestone, characterized by karst features in the north and immense areas of in-

FIGURE 11-2 Tall grass marshes are widespread in this region. The scene is near Hilton Head, South Carolina (TLM photo).

FIGURE 11-3 Southeastern swamplands contain a great variety of vegetation. This is the Okefenokee Swamp near Waycross, Georgia (TLM photo).

adequate drainage in the south. The caves and sinks of typical karst regions are modified in Florida by the uniformly high water table; most caverns are water-filled, and most sinks have become small lakes. There are so many sinkhole lakes in central Florida that this area is referred to as the "lakes district."

Associated with the karst is the most extensive artesian system in North America. Some components of the system

are deep and discharge on the sea bottom many miles offshore. Shallower components often discharge as artesian springs. Florida has the greatest concentration in the country, and many serve as the source of short rivers. Silver Springs is the largest known of the artesian outflows, with an average daily discharge of nearly half a billion gallons (1.9 gigaliters) of water. [1]

The exceedingly flat terrain of southern Florida has a natural drainage system that is both complex and delicate. Lake Okeechobee, with a surface area of about 750 square miles (1900 km^2) and a maximum depth of less than 20 feet (6 m), is fed in part by overflow from the Kissimmee Lakes to the north. Lake Okeechobee, in turn, overflows broadly southward with a 50-mile (80-km)-wide channel that maintains the natural water supply of the Everglades and Big Cypress Swamp. Human interference with this drainage system has caused major problems.

The area around the mouth of the Mississippi, the continent's mightiest river, is generally flat and ill-drained, and its detailed pattern of landforms is notably complex. Essentially all southeastern Louisiana within the Southeastern Coast Region is a part of the Mississippi River deltaic plain. During the past 20 centuries the lower course of the river shifted several times, producing at least seven different subdeltas that are the principal elements of the present complicated "bird's foot" delta of the river. The main flow of the river during the past 600 years or so has been along its present course extending southeast from New Orleans. This portion of the delta was built out into the Gulf at a rate of more than 6 miles (9.6 km) per century in that period.

The Atchafalaya distributary is the principal secondary channel of the river. It carries about 25 percent of the Mississippi's water; and under the normal pattern of deltaic fluctuation the main flow of the river would now be shifting to the shorter and slightly steeper channel of the Atchafalya. An enormous flood-control structure was erected above Baton Rouge, however, in an effort (thus far successful) to prevent the Mississippi from abandoning its present channel and delta.

The terrain of the Mississippi deltaic plain consists of many bayous and swamps, with marshes occupying all the immediate coastal zone. In addition to the natural bayous, the coastal marshlands are crisscrossed by a maze of artificial waterways, mostly shallow, narrow ditches (called *trainasse*) that were crudely excavated by local people to provide smallboat access. Many of these simple canals are now established elements of the drainage systems, as are more recent canals

[1] William D. Thornbury, *Regional Geomorphology of the United States* (New York: John Wiley & Sons, 1965), p. 47.

dredged to provide boat access to oil and gas wells. The only slightly elevated land is along natural levees, which parallel the natural drainage channels, and on low sandy ridges (*cheniers*) that roughly parallel the coast of southwestern Louisiana.

Climate

This region is typical of the warmer, more humid phase of the humid subtropical climate. It is characterized by a heavy rainfall and a long growing season (from 240 days in the north to almost frostless areas of southern Florida). The total annual precipitation decreases from more than 60 inches (152 cm) in southern Florida and along the eastern Gulf Coast to less than 30 inches (76 cm) in the southwest. Over most of the region the rainfall is evenly distributed throughout the year, but the maximum comes during the summer and early autumn, which are thunderstorm and hurricane seasons. The torrential rains (most stations in this region have experienced more than 10 inches (25 cm) of rain in a 24-hour period) coupled with high-velocity winds have caused considerable damage to crops and wrought great destruction to many coastal communities at some time or other.

From a psychological standpoint mild, sunny winters largely offset the hurricane menace and thus enable this region to capitalize on climate in its agriculture and resort business. But severe, killing frosts affect part of the region almost every year.

Soils

The region has a considerable variety of soil types, but most are characterized by an excess of moisture, quartz particles, or clay. Most widespread are hydromorphic (poorly drained) varieties, but sandy soils are also common. From a taxonomic standpoint the soils of the Southeastern Coast Region are the most complex of the continent. All soil orders except Aridisols and Andisols are extensively represented, often in a complicated patchwork pattern.

Natural Vegetation

Because of heavier rainfall, the eastern part of the region is largely covered with forests, mainly yellow pines and oaks on the better-drained, sandy lands, with cypress and other hardwoods dominating the swamps and other poorly drained areas. Hardwoods were probably more widespread long ago, before the Native Americans began burning the woodlands in winter to aid their hunting. Periodic burning favored the pines and oaks, for they are more resistant to fire than other trees and shrubs. Until fairly recently burning was commonplace in the region, generally done on a casual and indiscriminate basis in order to "improve" grazing for scrub cattle. A common feature of the region is the so-called Spanish moss, which festoons live oaks and cypress but is much less abundant on pines except in swampy areas.

Some extensive grasslands, usually marshy, are found along the coast. Dotted irregularly through these marshlands are bits of elevated ground, usually called *islands* or, in Louisiana, *cheniers,* which are generally covered with big trees laden with moss. From southwestern Louisiana westward most of the natural vegetative cover is coastal prairie grassland; it is partially replaced by scrubby brush country south of Corpus Christi.

Most of Florida south of Lake Okeechobee is part of the Everglades (literally "sea of grass") where tall sawgrass dominates the landscape (Fig. 11-4). Patches of woodland,

FIGURE 11-4   A view of the Everglades in a wet year (TLM photo).

A CLOSER LOOK   *Major Wetland Problem Areas*

Coastal wetlands are among the most critical and most fragile environments. Where they are subjected to rapid human population growth and pell-mell development, the ecological ramifications may shortly become overwhelming. Much of the Southeastern Coast Region is susceptible to such problems, but two major areas are of particular concern.

THE EVERGLADES

The semitropical flatlands of southern Florida contain a unique ecosystem that depends on a delicate balance of natural factors, the cornerstone of which is an unusual natural drainage system that includes nearly one-fifth of the state's total area. Human disruption of this drainage system is the root of the problem.

Under normal circumstances there is a steady and reliable flow of water from the Kissimmee chain of lakes into Lake Okeechobee, from which the natural overflow sends a very broad and very shallow sheet of water southward to sustain Big Cypress Swamp and the Everglades before eventually draining into the Gulf of Mexico (Fig. 11-A). As more and more settlers were attracted to southern Florida, farms and cities were established and grew apace. The relatively frequent minor, and occasional major, natural floods brought hardships to farmers and urbanites alike, and a number of artificial drainage canals were constructed for flood-control purposes and to "improve" the land for agricultural and urban expansion.

As metropolitan Miami burgeoned, specialty farmers of southern Florida were displaced and the only direction they could move was westward, into the 'Glades. Two highways, the Tamiami Trail and Alligator Alley, were built east–west across the area. Further land development, primarily for residential and second-home purposes, was undertaken in Big Cypress Swamp, as a sort of subsidiary of the west coast urban clus-

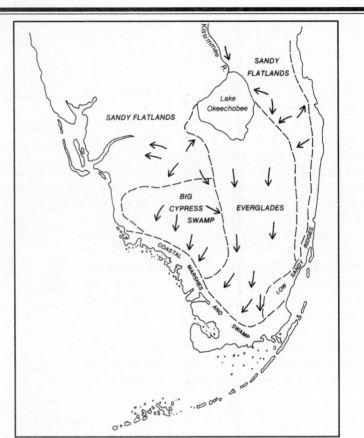

FIGURE 11-A   The natural drainage pattern of southern Florida is generally southwestward from Lake Okeechobee through the Everglades (after G. G. Parker, U.S. Geological Survey Water Supply Paper 1255).

ters of Fort Myers and Naples. Even the Kissimmee River was channeled into a straight-line, stagnant canal.

These varied developments were increasingly opposed by individuals and groups from many parts of the country and the "conservation coalition," gaining in strength and political muscle, succeeded at some points. Everglades National Park was established in 1947 and in 1974 the federal and state governments authorized the purchase of

more than half a million acres (200,000 ha) of Big Cypress Swamp to stop piecemeal development and to set up a watershed reserve.

The ecological threat, however, persists. Much of the land north and east of Everglades National Park and the Big Cypress Reserve is in private ownership, and pressures for development have not abated. They are mostly "upstream" areas in the watershed, and any disruptions in the normal drainage pattern may have serious repercussions in the downsteam sections. Moreover, most original drainage canals are still functioning, thus cutting off much of the natural southward flow, which is particularly serious in minimum-rainfall years. Only one-fifth of the water that used to reach the ecosystem at the turn of the century is getting there, and only 5 percent of the wading birds that used to nest in the wetlands are still doing so.

The problem is to maintain a flow of water throughout the Everglades–Big Cypress area from the Kissimmee drainage southward. Dechannelization of the Kissimmee River has been "authorized," but no funds are available to implement the project. Meanwhile, most of the normal southward flow from Lake Okeechobee can be (and usually is) legally diverted.

In 1990, for the first time, the Army Corps of Engineers (which built most of the flood-control and drainage structures) and the South Florida Water Management District (which operates the drainage system) began planning a comprehensive project to resolve the major problems by restoring water to the Everglades in a systematic fashion. The plan is very controversial in the state, and its ultimate price tag would be hundreds of millions of dollars, so the salvation of the 'Glades is far from assured.

THE MISSISSIPPI DELTA

The problem is more complex in southeasternmost Louisiana but the basic cause—human disruption of the natural drainage system—is the same. In order to provide a dry surface for human settlement and activities, the land was increasingly drained and the rivers channeled for two centuries, a process that was greatly accelerated in recent decades. Artificial levees and flood-control structures keep the Mississippi and its distributaries in relatively narrow channels, thus denying both silt and freshwater to the marshes. In addition, a maze of canals (extending for an estimated 10,000 miles [16,000 km]), dredged mostly to gain access to marshland oil wells, accelerates erosion and allows saltwater to encroach.

The result is a continuing diminishment of land that is unprecedented on the North American coastline (Fig. 11-B). The delta is both washing away and sinking. The amount of marsh continues to decrease, and the amount of open water in the delta continues to expand. Since the 1950s the amount of land loss has averaged about 40 square miles (104 km^2) annually, or an acre every 15 minutes. The freshwater and brackish ecosystems suffer, and some of the higher land containing human settlements is sinking.

The relevant authorities agree that large-scale introduction of sediment-laden river water is urgently needed to assuage the problem and that canal dredging should be strictly curtailed. But the economic imperative of more shipping and more oil development continue to cloud the issue. Indeed, there are plans to dredge the river 15 feet (4.5 m) deeper to accommodate larger bulk carriers.

TLM

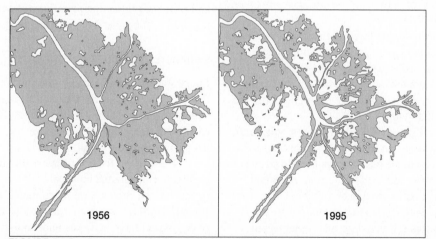

FIGURE 11-B   Land loss in the lower Mississippi delta. Much of the marshland shown in the left diagram (1956) has sunk or been washed away by the intruding ocean in the right diagram (1995). (*From Physical Geography: A Landscape Appreciation, Tom L. McKnight, Prentice Hall. 5th ed., 1996, p. 469.*)

FIGURE 11-5   In Florida a walking catfish seems to have lost its shoes (courtesy Charles N. Trainor from Omikron/Photo Researchers, Inc.).

slightly elevated above the surrounding Everglades marsh, are called *hammocks;* they are the favorite haunts of Florida's most vicious wildlife, mosquitos. The littoral zone of the coastal Everglades has North America's only extensive growth of tangled mangroves.

Wildlife

The Southeastern Coast Region contains the only part of the conterminous states that is a major wintering ground for migratory birds, particularly waterfowl. From Florida to Texas coastal marshes and swamps provide a last stronghold for the wintering (for example, whooping cranes) or breeding (various egrets and spoonbills) of numerous rare and endangered birds as well as seasonal or permanent homes for many other avian species.

The region's poorly drained wild areas also provide an important habitat for a number of native quadruped species. Mustelids, such as mink, otter, and skunk, are prominent, but muskrats and raccoons are the most numerous of the region's native furbearers. Here, too, is the last stronghold for the cougar in eastern United States.

The American alligator, North America's only large reptile, is found exclusively in the Southeastern Coast Region. Heavily persecuted in the past, it responded to stringent protection in the 1960s and is once again abundant in Louisiana and Florida and increasing in suitable areas from the Big Thicket of Texas to the Dismal Swamp of Virginia.

The vast expanses of marshes and other types of wilderness and the relatively mild winters in this region make the Southeastern Coast an attractive habitat for a host of exotic animals that were introduced either deliberately or inadvertently. It is in Florida, where natural ecosystems have been most disturbed and where winter weather is most permissive, that the greatest variety of alien species has become established. Siamese walking catfish (Fig. 11-5), Mexican armadillos, African cattle egrets, Indian rhesus monkeys, South American giant toads, and Australian parakeets represent only a sampling of the international Noah's ark that Florida is becoming.

The exotic species that became most conspicuous in the regional biota, however, is the South American nutria (*Myocastor coypu*). It was deliberately introduced to Louisiana in the 1930s for a dual-purpose: to provide another source of furs for trappers of the area and to help in controlling the rapid expansion of water hyacinth (another exotic from South America) in the state's bayous and lakes. The nutria is larger than the muskrat and its fur is richer, more akin to beaver fur; so nutria pelts are considerably more valuable than muskrat. The first nutria brought to the United States were kept on fur farms, and today there are several hundred operating nutria farms in Louisiana. In the 1940s, however, nutria were released or escaped from farms and are now very numerous in Louisiana and adjacent parts of Texas and Mississippi. The animal has now become well established, is a major quarry for trappers, and has found many items that it prefers to eat other than water hyacinths.

PRIMARY INDUSTRIES

In a region that is close to nature numerous primary economic activities are likely to be carried on, provided that basic resources are present. The Southeastern Coast is such a region, and primary production significantly contributes to both the regional economy and the regional image.

Commercial Fishing

Every state in the region has a prominent commercial fishing fleet, and the total catch from South Atlantic–Gulf of Mexico waters is exceeded only by the harvest from Alaskan waters. Menhaden are by far the bulk of the catch, although their value is comparatively small.

The shrimp fishery is easily the most important of the region. Shrimp are the most valuable variety of ocean product in the United States, and more than half the total national shrimp landings are made at Southeastern Coast Region ports, particularly in Texas and Louisiana. Shrimping is so widespread in this region that almost every estuary and bay has at least one little fishing hamlet and a small complement of shrimp boats (Fig. 11-6). Larger port towns and cities, such as Corpus Christi, Pascagoula, and Savannah, are likely to have a more numerous shrimping fleet. Shrimping is, however, a wasteful fishery. The shrimp trawlers take an average of 10 pounds (5 kg) of bycatch (unwanted species) for every pound (0.5 kg) of shrimp caught.

FIGURE 11-6 There are dozens of small fishing ports in the region, and almost every one has its shrimping fleet. This harbor scene is at Pass Christian, Mississippi (TLM photo).

FIGURE 11-7 A crab packing plant at Brunswick, Georgia (TLM photo).

Louisiana is second only to Alaska in volume of seafood landings. In terms of value of catch, Louisiana also ranks second, with Florida and Texas in fourth and fifth places. Half of the leading fishing ports of the nation are located along the Gulf of Mexico or South Atlantic coasts (Fig. 11-7), particularly in Louisiana, where such tiny towns as Cameron, Empire, Venice, Dulac, and Chauvin are notable.

Lumbering and Forest Products Industries

Except for the prairie sections of Texas and Louisiana and the Florida Everglades, most of this region was originally forested. Consequently, such forest products as lumber and pulp-wood were important. The picture in this region is similar to that in the Inland South, with the additional fact that some larger pulp mills have tidewater sites, as at Mobile, Savannah, and Houston.

Another forest product that is of only minor significance in the Inland South but of major importance in the Southeastern Coast Region is naval stores. These are the tar and pitch that were used before the advent of metal ships to caulk seams and preserve ropes of wooden vessels. When it was discovered that turpentine and resin could be distilled from the gum of southern yellow pine, many new uses for these substances were found in such products as paints, soaps, shoe polish, and medicines.

The normal method of obtaining resin is to slash a tree and let the gum ooze into a detachable cup that can be emptied from time to time. A more recently developed and popular method is to shred and grind stumps and branches (formerly considered as waste) and put them through a steam-distillation process. A considerable amount of naval stores is also obtained as byproduct from the sulfate process of paper making.

Extensive forests and high-yielding species of trees permit this region to produce more than half the world's resin and turpentine. The principal area of production is around Valdosta in southern Georgia and in adjacent portions of northern Florida.

Agriculture

Farming is by no means continuous throughout the region. Large expanses have little arable land, particularly such poorly drained areas as the Dismal and Okefenokee Swamps, the Everglades, the coastal flatwoods zone of western Florida and adjacent Alabama and Mississippi, the coastal marshlands of Louisiana, and the drier coastal country between Corpus Christi and the Lower Rio Grande Valley. Although terrain hindrances are nil, agriculture is frequently handicapped by drainage problems. Also, soils tend to be infertile; considerable fertilizer, strongly laced with trace elements, must be applied, especially in Florida.

Regional agriculture faces many problems: insect pests thrive in the subtropical environment, frost damage is sometimes heavy, marketing is often complex, and overproduction sometimes occurs. A long growing season, adequate moisture, and relatively mild winters compensate, with the result that a considerable quantity of high-quality, and often high-cost, crops is grown. The grains that are the staple crops of most North American farming areas are virtually absent in this region. Instead, farmers concentrate on growing special-

ty crops of various kinds. The region's major agricultural role is the output of subtropical and off-season specialties.

Citrus Fruits If any crop typifies both the image and the actuality of specialty production in the Southeastern Coast Region, it is probably citrus (Fig. 11-8). Central Florida and the Lower Rio Grande Valley of Texas, two of the four principal citrus areas of the country, are in this region (Fig. 11-9).

Florida is the outstanding citrus producer, supplying about 70 percent of the national orange and tangerine crops, about 80 percent of the grapefruit, and even larger shares of limes and tangelos. Commercial groves have always been concentrated in the gently rolling, sandy-soiled lakes district of central Florida (Fig. 11-10), but in the last few years there has been a distinct shift southward in order to avoid the devastating effects of freezes that were so prominent during the 1980s. Indeed, the "big freeze" of 1985 killed more than 200,000

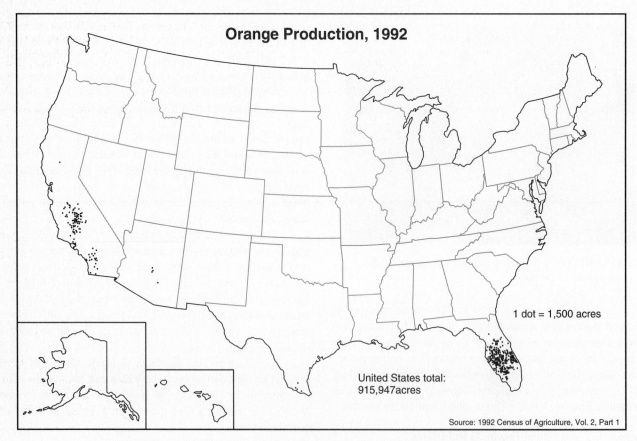

FIGURE 11-8 Florida is the principal source, by far, of oranges in the United States.

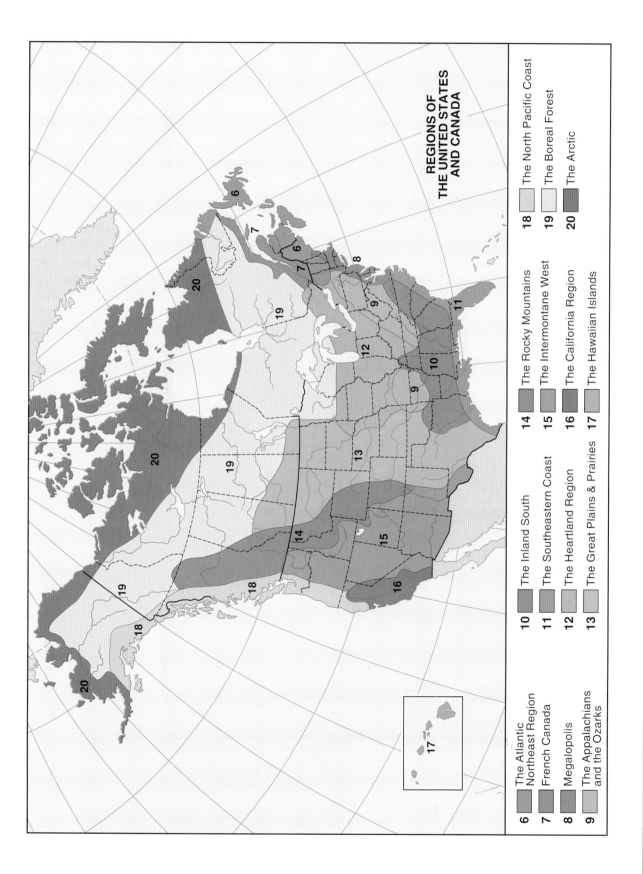

REGIONS OF
THE UNITED STATES
AND CANADA

6 The Atlantic Northeast Region
7 French Canada
8 Megalopolis
9 The Appalachians and the Ozarks
10 The Inland South
11 The Southeastern Coast
12 The Heartland Region
13 The Great Plains & Prairies
14 The Rocky Mountains
15 The Intermontane West
16 The California Region
17 The Hawaiian Islands
18 The North Pacific Coast
19 The Boreal Forest
20 The Arctic

Plate 6-1
Many lighthouses are operating along the irregular coastline of the Atlantic Northeast. This is the Portland Head Light, at the southern tip of the region *(TLM photo)*.

Plate 6-2 ▶
There are numerous building stone quarries in northern New England. The most famous is the Rock of Ages granite quarry near Barre, Vermont, shown here *(TLM photo)*.

◀ **Plate 6-3**
Rolls of paper produced in a gigantic paper mill at Corner Brook, Newfoundland *(TLM photo)*.

Plate 6-4 ▶
A view across the St. John River at Fredericton, New Brunswick, showing a massive log boom (holding area) *(TLM photo)*.

Plate 7-1 ▶

An impressive olympic village was built to house the athletes who participated in the 1976 Olympiad in Montreal. After the games, the complex was converted to public housing *(TLM photo)*.

◀ **Plate 7-2**

The small town of La Baie is built around a huge aluminum smelter and pulp mill alongside the Saguenay River near Chicoutimi, Quebec.
Inexpensive power is the primary lure for this industrial development *(TLM photo)*.

Plate 7-3 ▶

Quebec City is the only walled city existing in North America. Shown here is one of the half dozen apertures in the wall *(TLM photo)*.

Plate 8-1
The canyons of Manhattan. This is Fifth Avenue *(courtesy New York Convention and Visitors Bureau)*.

Plate 8-2
The nation's capital contains a great host of monuments and memorials. One of the most heavily visited is the Vietnam War Memorial *(TLM photo)*.

Plate 8-3
The Philadelphia skyline at night *(A. Gurmankin/Visuals Unlimited)*.

Plate 8-4
The Mall in Washington, showing the Lincoln Memorial (in foreground), Washington Monument, and Capitol (in distance) *(John Sohlden/ Visuals Unlimited)*.

Plate 9-1 ▶
A water gap in the ridge-and-valley section of the Appalachians. This scene is near Warm Springs, Virginia *(TLM photo)*.

◀ **Plate 9-2**
Tobacco is still a major crop in the hill country of eastern Kentucky, as in this scene near Morehead *(TLM photo)*.

Plate 9-3 ▶
Several large reservoirs have been created in the Ozarks subregion. This sunset scene shows Lake of the Ozarks in southwestern Missouri *(TLM photo)*.

◀ **Plate 9-4**
Gatlinburg, Tennessee, the gateway to Great Smoky Mountains National Park, at night *(Gary W. Carter/Visuals Unlimited)*.

◀ **Plate 10-1**
Cotton growing has made a comeback in the last few years in the Inland South, and tobacco persists in several areas. Often these two traditional crops are grown together in the Carolinas. This scene is near Roanoke Rapids, North Carolina *(TLM photo)*.

Plate 10-2 ▶
Beef cattle cooling off in a farm pond near Augusta, Georgia *(TLM photo)*.

◀ **Plate 10-3**
A 1996 Olympic scene at Centennial Olympic Park in Atlanta *(Mark Farmer/Aristock)*.

Plate 11-1 ▶
Florida has more miles of beaches, by far, than any other state. This scene is at Miami Beach *(TLM photo)*.

◀ **Plate 11-2**
The introduction of Brahma cattle has been widespread in the hot, humid parts of the country. This scene is near Kissimmee, Florida *(TLM photo)*.

Plate 11-3 ▶
Tobacco mostly is sold by auction in large sheds, as in this scene at Waycross in southern Georgia *(TLM photo)*.

◀ **Plate 11-4**
Recreational fishing is a popular avocation all along the Gulf coast. These successful fishermen show off their catch at Cocodrie, Louisiana, southwest of New Orleans *(TLM photo)*.

Plate 11-5 ▶
Grand Casino in Gulfport, Mississippi *(courtesy Grand Casino, Gulfport, Mississippi)*.

◀ Plate 12-1
The verdant green of early summer in a representative Corn Belt farm in southern Wisconsin *(TLM photo).*

Plate 12-2 ▶
Canada's impressive Parliament building in Ottawa *(TLM photo).*

Plate 12-3 ▲
Gleaming skyscrapers characterize the Renaissance Center in Detroit *(David R. Frazier/Photo Researchers. Inc.).*

Plate 12-4 ▶
The Sears Tower looms over other skyscrapers in downtown Chicago. The Chicago River is in the foreground *(Porterfield/ Chickering/Photo Researchers. Inc.).*

Plate 13-1 ▶
A six-row cotton cultivator at work on the flat lands of the Llano Estacado near Lamesa in west Texas *(TLM photo).*

Plate 13-2 ◀
Louis Riel, who led "Riel's Rebellion" in western Canada, was a hero to the Metis, but was finally apprehended and executed. His grave is in the St. Boniface suburb of Winnipeg, Manitoba *(TLM photo).*

Plate 13-3 ▶
A view across the downtown section of Edmonton, the second largest city in western Canada *(TLM photo).*

Plate 13-4 ◀
Beef cattle at a feedlot near Greeley in northeastern Colorado *(TLM photo).*

Plate 14-1
The town of Ouray, which occupies a small valley surrounded by the San Juan Mountains in southwestern Colorado, is a very popular tourist spot *(TLM photo)*.

Plate 14-2 ▶
Irrigated crops are grown sporadically in the Rocky Mountains Region. Near Kamloops, British Columbia, a specialty is hops *(TLM photo)*.

Plate 14-3
The region is replete with spectacular scenery. This hiker is headed for Kokanee Lake in Kokanee Glacier Provincial Park near Nelson, British Columbia *(TLM photo)*.

Plate 14-4 ▶
Mt. Sarbach towers above the silt-laden North Saskatchewan River in Alberta's Banff National Park *(TLM photo)*.

Plate 15-1 ▶
Apple orchards along the Columbia
River near Wenatchee, Washington
(TLM photo).

◀ **Plate 15-2**
The largest dam on the lower Colorado
River is Hoover Dam near Las Vegas,
Nevada, which was built primarily to
generate electric power *(TLM photo).*

Plate 15-3 ▶
A sheep round-up on a cold November
morning in the high plateaus of central
Utah, near Tooele *(TLM photo).*

◀ **Plate 15-4**
Corn depending on flood-water
farming in the Navajo Reservation's
Monument Valley of northeastern
Arizona *(TLM photo).*

◀ **Plate 16-1**
Citrus orchards in a "cove" or side valley of the San Joaquin Valley. This scene is near the town of Lemoncove *(TLM photo)*.

Plate 16-2 ▶
A Sequoia "sapling" has fallen across the road in Sequoia National Park *(TLM photo)*.

◀ **Plate 16-3**
The spectacular scenery of California's central coast is epitomized by The Pinnacle at Point Lobos *(TLM photo)*.

Plate 16-4 ▶
Tomato harvesting in the San Joaquin Valley near Modesto (TLM photo).

Plate 17-1 ▶
Sunrise over Diamond Head, an
enduring image of Waikiki *(TLM photo).*

◀ **Plate 17-2**
Cattle ranching is by far the most
widespread land use in the region.
This Santa Gertrudis bull is on the
island of Maui *(TLM photo).*

Plate 17-3 ▶
The silversword is a rare and endangered
plant species that grows only on Maui's
Haleakala volcano *(TLM photo).*

◀ **Plate 17-4**
The growing skyline of Honolulu, as
seen from the east *(TLM photo).*

Plate 18-1
One of the deepest and most attractive lakes in the world is Crater Lake in south-central Oregon. The lake occupies an enormous volcanic caldera, in which the small volcano called Wizard Island has emerged *(TLM photo).*

Plate 18-2 ▶
Small boat harbors are widespread in the region. Shown here is Friday Harbor in Washington's San Juan Islands. A Washington State ferry approaches in the background *(TLM photo).*

◀ **Plate 18-3**
A flourishing pulp mill at Gold River on the west coast of British Columbia's Vancouver Island *(TLM photo).*

Plate 18-4 ▶
A mountain goat surveys the high country of the Olympic Mountains in Washington's Olympic National Park *(TLM photo).*

FIGURE 11-10 The lake-and-orchard district of central Florida. The orderly geometry of citrus groves is interrupted by the patchy pattern of shallow lakes (courtesy Florida Citrus Commission).

oils that characterize much of the state. Vegetable growing in the Lower Rio Grande Valley is largely confined to the alluvial lands of the delta.

The climate in most of the region is suited to the production of early vegetables. In the southern parts of Florida and Texas some winters have no killing frosts; however, even here an occasional "norther" may sweep down from the interior and kill the sensitive vegetable crops.

Southern Florida is a major producer of tender vegetables—for example, 66 percent of U.S. radishes, 40 percent of eggplant, 29 percent of green peppers, 24 percent of squash, 22 percent of celery; and the proportion of national *winter* vegetable production is much higher. Nevertheless, the largest acreage and most valuable crop is tomatoes.

Sugarcane Sugarcane growing in the delta country of Louisiana began in 1751 when Jesuits introduced the crop from Santo Domingo. Most Louisiana cane production today is on alluvial soil or drained land west of New Orleans, where there is a shorter growing season than in the other domestic sugarcane states; consequently, both yields per acre and average sugar content are relatively low. Production and harvesting are almost entirely mechanized.

Florida has been a sugarcane producer since 1931, but the output was relatively limited until the 1960s, when the

United States ceased buying Cuban sugar. Production is concentrated on the organic soils just south of Lake Okeechobee where very high yields are attained. These soils, however, oxidize and "evaporate" when exposed to the atmosphere, which causes subsidence at a rate of about 12 inches (30 cm) per decade. This factor creates major water-control problems, and the expanding industry must move to sandy soils farther from the lake. Even so, Florida cane planting has been on a steady upward trend for several years and now provides nearly half the national cane crop (as compared with less than 25 percent from Louisiana).

Some 400,000 acres (160,000 ha) are now in sugarcane [compared with 300,000 acres (120,000 ha) in vegetables] on very large farms. There are fewer than 150 cane farms, most of which contain tens of thousands of acres. Labor costs are the most expensive phase of the operation; the muck soils are too soft for efficient mechanical cutting, so cane must be hand cut. About 10,000 cane cutters, mostly from Jamaica, are flown in to work during the November–April cutting season (Fig. 11-11).

The newest sugarcane area is the Lower Rio Grande Valley of Texas, which had almost a century of commercial cane production until its total demise in the 1920s. The rise in sugar prices, however, tempted valley farmers to begin planting cane again in 1973. Results were encouraging and

Plate 19-1 ▶
Loggers use long-handled peaveys to guide floating logs to an outlet of Lake of the Woods near Kenora in western Ontario *(TLM photo).*

◀ **Plate 19-2**
A typical spruce forest scene near Fairbanks in central Alaska *(TLM photo).*

Plate 19-3 ▶
Alluvial gold mining, often with huge dredges, occurs in many Alaskan valleys. This is the Chatanika Valley east of Fairbanks *(TLM photo).*

◀ **Plate 19-4**
The only urban place of any size in the Yukon Territory is Whitehorse *(Jeff Greenberg/Visuals Unlimited).*

◀ **Plate 20-1**
Jagged peaks in Auyuittuq National
Park on Canada's Baffin Island
(Mike Beedell/Corel Corporation).

Plate 20-2 ▶
Hanging whitefish out to dry
in a Yukon River delta village
(Mike Beedell/Corel Corporation).

◀ **Plate 20-3**
Camping in the continent's farthest
north national park, Ellesmere Island
(Mike Beedell/Corel Corporation).

Plate 20-4 ▶
A polar bear sow and cub near the
Hudson Bay coast in northeastern Manitoba
(Mike Beedell/Corel Corporation).

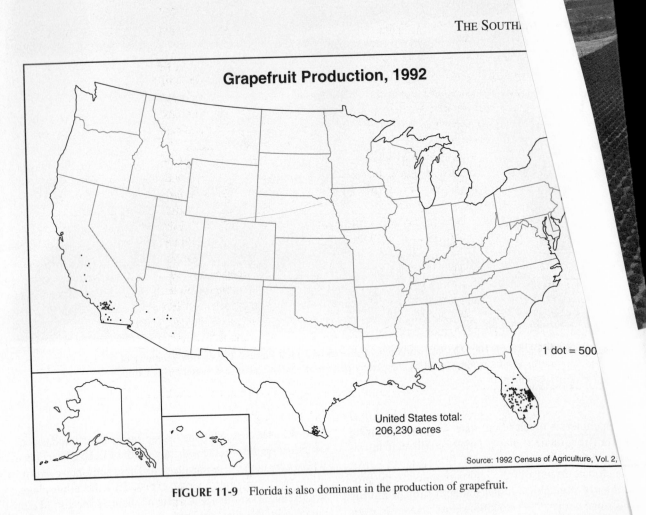

Grapefruit Production, 1992

1 dot = 500

United States total:
206,230 acres

Source: 1992 Census of Agriculture, Vol. 2,

FIGURE 11-9   Florida is also dominant in the production of grapefruit.

acres (81,000 ha) of citrus in central Florida. Frost damage, along with vigorous supply competition from Brazil, has led to a one-third decline in Florida citrus acreage since 1970.

The exacting requirements of modern citrus production make it increasingly practical for farmers, particularly if their holdings are not large, to contract all operations to a production company. Such farmers may never set foot on the land; indeed, many live tens or hundreds of miles away from their orchards. Many growers are members of cooperatives, which usually also have their own production crews and equipment and operate their own processing plants for citrus concentrate.

Marketing arrangements are often complicated. More than half of all citrus harvested in Florida is sold as processed products, particularly frozen concentrated juice. Citrus waste (pulp and peeling) is generally fed to cattle.

The Texas citrus area is located on the terraces of the Lower Rio Grande delta. Almost all production is based on

irrigation water diverted from the river. Severe frost sionally damages the groves, sometimes so badly that trees must be planted. On such occasions, the horticultu usually raise cotton or vegetables for a few years until groves become reestablished. Seedless pink grapefruit is specialty of the valley.

Truck Farming   With increasing demands for fresh vegetables, the Southeastern Coast has become one of the major regions for early truck products of the continent. Aside from parts of southern California and the gardens under glass in the North, most of the nation's winter vegetables grown for sale come from this region, particularly from Florida and the Lower Rio Grande Valley.

Climate is the dominant factor affecting the growth of early vegetables, and soils help to determine the specific locations. In Florida the best truck crops are produced on muck or other lands having a higher organic content than the sandy

Plate 19-1 ▷
Loggers use long-handled peaveys to guide floating logs to an outlet of Lake of the Woods near Kenora in western Ontario *(TLM photo)*.

◁ **Plate 19-2**
A typical spruce forest scene near Fairbanks in central Alaska *(TLM photo)*.

Plate 19-3 ▷
Alluvial gold mining, often with huge dredges, occurs in many Alaskan valleys. This is the Chatanika Valley east of Fairbanks *(TLM photo)*.

◁ **Plate 19-4**
The only urban place of any size in the Yukon Territory is Whitehorse *(Jeff Greenberg/Visuals Unlimited)*.

◀ **Plate 20-1**
Jagged peaks in Auyuittuq National
Park on Canada's Baffin Island
(Mike Beedell/Corel Corporation).

Plate 20-2 ▶
Hanging whitefish out to dry
in a Yukon River delta village
(Mike Beedell/Corel Corporation).

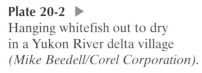

◀ **Plate 20-3**
Camping in the continent's farthest
north national park, Ellesmere Island
(Mike Beedell/Corel Corporation).

Plate 20-4 ▶
A polar bear sow and cub near the
Hudson Bay coast in northeastern Manitoba
(Mike Beedell/Corel Corporation).

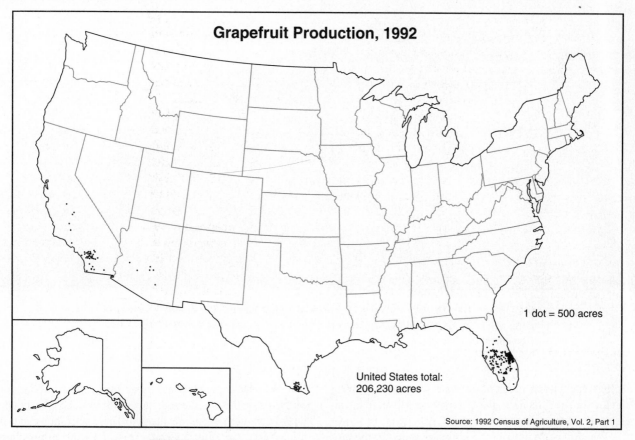

Grapefruit Production, 1992

1 dot = 500 acres

United States total:
206,230 acres

Source: 1992 Census of Agriculture, Vol. 2, Part 1

FIGURE 11-9   Florida is also dominant in the production of grapefruit.

acres (81,000 ha) of citrus in central Florida. Frost damage, along with vigorous supply competition from Brazil, has led to a one-third decline in Florida citrus acreage since 1970.

The exacting requirements of modern citrus production make it increasingly practical for farmers, particularly if their holdings are not large, to contract all operations to a production company. Such farmers may never set foot on the land; indeed, many live tens or hundreds of miles away from their orchards. Many growers are members of cooperatives, which usually also have their own production crews and equipment and operate their own processing plants for citrus concentrate.

Marketing arrangements are often complicated. More than half of all citrus harvested in Florida is sold as processed products, particularly frozen concentrated juice. Citrus waste (pulp and peeling) is generally fed to cattle.

The Texas citrus area is located on the terraces of the Lower Rio Grande delta. Almost all production is based on

irrigation water diverted from the river. Severe frost occasionally damages the groves, sometimes so badly that new trees must be planted. On such occasions, the horticulturists usually raise cotton or vegetables for a few years until the groves become reestablished. Seedless pink grapefruit is the specialty of the valley.

Truck Farming   With increasing demands for fresh vegetables, the Southeastern Coast has become one of the major regions for early truck products of the continent. Aside from parts of southern California and the gardens under glass in the North, most of the nation's winter vegetables grown for sale come from this region, particularly from Florida and the Lower Rio Grande Valley.

Climate is the dominant factor affecting the growth of early vegetables, and soils help to determine the specific locations. In Florida the best truck crops are produced on muck or other lands having a higher organic content than the sandy

FIGURE 11-10 The lake-and-orchard district of central Florida. The orderly geometry of citrus groves is interrupted by the patchy pattern of shallow lakes (courtesy Florida Citrus Commission).

soils that characterize much of the state. Vegetable growing in the Lower Rio Grande Valley is largely confined to the alluvial lands of the delta.

The climate in most of the region is suited to the production of early vegetables. In the southern parts of Florida and Texas some winters have no killing frosts; however, even here an occasional "norther" may sweep down from the interior and kill the sensitive vegetable crops.

Southern Florida is a major producer of tender vegetables—for example, 66 percent of U.S. radishes, 40 percent of eggplant, 29 percent of green peppers, 24 percent of squash, 22 percent of celery; and the proportion of national *winter* vegetable production is much higher. Nevertheless, the largest acreage and most valuable crop is tomatoes.

Sugarcane Sugarcane growing in the delta country of Louisiana began in 1751 when Jesuits introduced the crop from Santo Domingo. Most Louisiana cane production today is on alluvial soil or drained land west of New Orleans, where there is a shorter growing season than in the other domestic sugarcane states; consequently, both yields per acre and average sugar content are relatively low. Production and harvesting are almost entirely mechanized.

Florida has been a sugarcane producer since 1931, but the output was relatively limited until the 1960s, when the

United States ceased buying Cuban sugar. Production is concentrated on the organic soils just south of Lake Okeechobee where very high yields are attained. These soils, however, oxidize and "evaporate" when exposed to the atmosphere, which causes subsidence at a rate of about 12 inches (30 cm) per decade. This factor creates major water-control problems, and the expanding industry must move to sandy soils farther from the lake. Even so, Florida cane planting has been on a steady upward trend for several years and now provides nearly half the national cane crop (as compared with less than 25 percent from Louisiana).

Some 400,000 acres (160,000 ha) are now in sugarcane [compared with 300,000 acres (120,000 ha) in vegetables] on very large farms. There are fewer than 150 cane farms, most of which contain tens of thousands of acres. Labor costs are the most expensive phase of the operation; the muck soils are too soft for efficient mechanical cutting, so cane must be hand cut. About 10,000 cane cutters, mostly from Jamaica, are flown in to work during the November–April cutting season (Fig. 11-11).

The newest sugarcane area is the Lower Rio Grande Valley of Texas, which had almost a century of commercial cane production until its total demise in the 1920s. The rise in sugar prices, however, tempted valley farmers to begin planting cane again in 1973. Results were encouraging and

FIGURE 11-11   Some sugar cane harvesting in Florida is done by hand (Dennis McDonald/PhotoEdit).

now about 5 percent of the national output comes from "the Valley" (Fig. 11-12).

Rice   In colonial times the coastal areas of South Carolina and Georgia produced large quantities of rice, but it did not become an important commercial crop in Louisiana until about a century ago when the introduction of harvesting machinery permitted large-scale farming on the prairies in the southwestern part of that state and adjacent areas in Texas.

Mills, concentrated mainly in Crowley, Lake Charles, and Beaumont, are equipped with complicated machinery for drying, cleaning, and polishing the rice and for utilizing its byproducts. Favorable geographical conditions and complete mechanization enable this region to grow rice at a low per-acre cost.

Rice production in the United States has been on an upward trend for more than two decades. This expansion is predicated mostly on overseas markets; normally the country

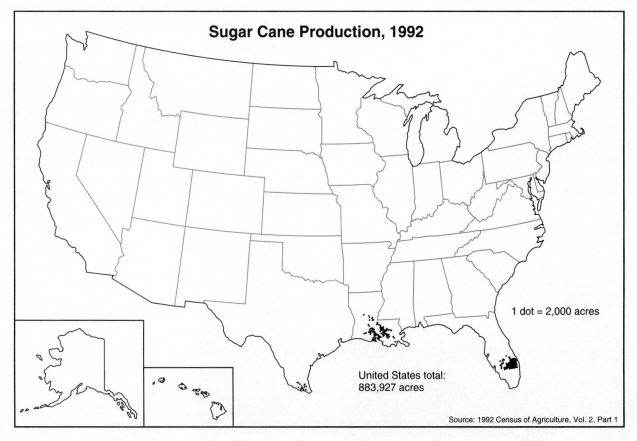

Sugar Cane Production, 1992

1 dot = 2,000 acres

United States total:
883,927 acres

Source: 1992 Census of Agriculture, Vol. 2, Part 1

FIGURE 11-12   All sugarcane in the conterminous states is grown in the Southeastern Coast Region—in southern Florida, southern Louisiana, and the Lower Rio Grande Valley of Texas.

grows less than 2 percent of the world rice crop but supplies more than 20 percent of total world rice exports. Output increased more in Arkansas and California than in this region, however. About one-third of the U.S. rice crop is grown here, divided about equally between Texas and Louisiana.

Other Crops Other prominent crops in the Southeastern Coast Region include cotton and grain sorghums in the Lower Rio Grande Valley and the Coastal Bend area (around Corpus Christi) of Texas, tobacco in southern Georgia and northern Florida, and peanuts in southwestern Georgia–southeastern Alabama.

The Livestock Industry

The Southeastern Coast Region has been an important producer of beef cattle since French and Spanish colonial times. But after the Great Plains were opened to grazing in the 1870s, the poorer pastures of the Gulf Coast fell into disfavor. For several decades cattle raising was all but abandoned except in the Acadian French country of southwestern Louisiana, the coastal grasslands of southwestern Texas, and the open-range cattle country of northern and central Florida. Through the introduction of new breeds, particularly the Brahman, the cattle industry has expanded in recent years to other parts of the coastal region (Fig. 11-13).

Florida is a major cattle-raising state, although it has only one large meatpacking plant. Most of the cattle are shipped out of the state for fattening and slaughter. Typically they are sold as yearlings to feeder lots in the Midwest or other places, including California. Some 600,000 head are sent out from Florida annually.

Brahman, or Zebu, cattle *(Bos indicus)* were brought from India to the Carolina coastal country more than a century ago because cattlemen thought they would be better adapted to the hot, humid, insect-ridden area than the English breeds common in the rest of the country. The oversized, hump-shouldered, lop-eared, slant-eyed, flappy-brisketed newcomer has been a thorough success and is often crossbred to produce special-purpose hybrids.

One of these hybrids, the Santa Gertrudis, meticulously developed on Texas's gigantic King Ranch, is considered to be the first "true" cattle breed ever developed in North America. It is five-eighths Shorthorn and three-eighths Brahman. Although there are many purebred Brahmans, as well as Herefords, Santa Gertrudis, and other breeds in the region, most of the beef cattle are of mixed ancestry, and their physiognomy usually reveals the presence of some Brahman inheritance.

Mineral Industries

The economically useful minerals of the region are few in number but occur in enormous quantities.

Phosphate Rock The United States produces more than one-third of the world's phosphate rock, and about 75 percent of the output is from central Florida. A little hard-rock phosphate is dug, but most production comes from unconsolidated landpebble phosphate deposits east of Tampa Bay. There are mines, all open pit, in three counties (Fig. 11-14). Large holes (often ponded because of the high water table), spoils banks, and considerable smoke from the processing plants (a council on air-pollution control has been established) are characteristic landscape features in the mining

FIGURE 11-13 A typical mixed-breed herd of Florida cattle, near Winter Haven (TLM photo).

FIGURE 11-14   A phosphate quarry near Lakeland in central Florida (TLM photo).

area. Phosphate mining began in 1888 in Florida, and abundant reserves still exist. Most of the output is used in fertilizer manufacture, but there is considerable export, especially to Japan.

Salt   The entire region from Alabama westward is dotted with deposits of rock salt. The deposits occur mostly in salt domes of almost pure sodium chloride (Fig. 11-15). Mining is limited to Louisiana and Texas, the two leading salt-mining states in the nation. The reserves are enormous. Although only a few domes are mined in the region at present, they yield great quantities of salt, and none is approaching exhaustion.

The mine usually lies at the top of the salt dome, 600 feet (180 m) or more below the surface. A shaft is driven down, and large chambers are excavated. Mine props need not be used because large supporting columns of salt are left in place. The chambers are sometimes more than 100 feet (30 m) high. Some salt is extracted in brine solution by a modified Frasch process. Although the Gulf Coast has no monopoly in salt production, it has sufficient reserves to make it an important producer for a long time. There are many known salt domes (some of them are now producing petroleum or sulfur) where salt can be obtained when needed. Moreover, half a dozen of the domes have been hollowed out (by solution mining) to serve as enormous storage reservoirs for crude oil. Some 750 million barrels of oil were stockpiled here during the 1980s as a cushion against a future world oil shortage.

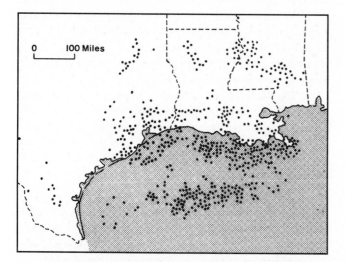

FIGURE 11-15   Distribution of salt domes on and near the Gulf Coast Plain (after William D. Thornbury, *Regional Geomorphology of the United States* [New York: John Wiley & Sons, Inc., 1965], p. 67; reprinted by permission).

Sulfur   Coastal Louisiana and Texas produce about two-thirds of the continent's sulfur. Deposits are found in the caprock overlying certain salt domes, where extraction is accomplished by the Frasch process. Wells are drilled into the deposits on top of the salt plugs (Fig. 11-16). Superheated water is pumped into the deposit and the sulfur is converted into a liquid. Compressed air is used to force the molten material to the surface. The pipes are heated to prevent the sulfur from solidifying until it reaches huge temporary vats on the

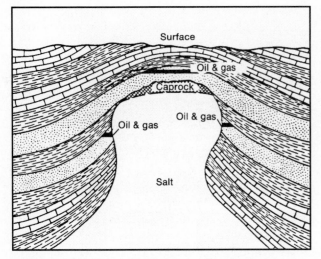

FIGURE 11-16   Generalized geologic cross section of a salt dome, showing typical reservoirs of oil and gas.

surface. From the vats the molten sulfur is transported by pipeline or allowed to harden and then broken into fragments for shipping. Frasch production declined during the 1970s and 1980s but increased significantly in the 1990s, although only a handful of Frasch mines are in operation.

Petroleum and Natural Gas   The Southeastern Coast Region encompasses most of the area known as the Gulf Coast Petroleum Province, one of the continent's major oil producers. It normally yields about one-fifth of the U.S.

FIGURE 11-17   Mountains of oyster shells that have been dredged from Matagorda Bay along the Texas coast (TLM photo).

total flow. The oil fields, almost exclusively in association with salt domes, are scattered along the west Gulf Coast Plain from the mouth of the Mississippi to the mouth of the Rio Grande.

Because much crude oil is transported by pipelines today, the Gulf Coast port cities have become the termini of most pipelines from the Gulf Coast and midcontinent oil provinces. These pipelines have permitted the rapid development of the refining industry along the west Gulf Coast between Baton Rouge and Corpus Christi. Enormous quantities of both crude and refined petroleum are also shipped by tankers to the North Atlantic seaboard.

The continental shelf off the Gulf Coast is broad and shallow and underlain by geologic structures of the same types as those of the coastal plain. Consequently, salt domes, with their associated petroleum, natural gas, and sulfur, are widespread in the "tidelands," as the shelf has come to be called. Many favorable structures have been discovered geophysically and numerous wells have been sunk.

Offshore drilling has been extended farther and farther out into the Gulf. More than 25,000 offshore wells have been drilled, the vast majority directly south of Louisiana, but some as far east as offshore Alabama. These operations are extremely expensive, primarily because of the elaborate, self-contained drilling platforms used. As many as 60 wells can be drilled from a single platform, using whipstocking techniques to angle the wells in various directions.

In the mid-1990s the Gulf waters experienced a major oil rush, made possible by drilling platforms the size of football fields that drill for oil at previously unimaginable depths and supercomputers that locate once-invisible reserves. Major oil companies poured billions of dollars into new exploration and drilling in waters once considered to be too deep for development and in regions where oil was hidden beneath massive layers of salt. By 1996 three dozen confirmed deepwater discoveries had been made.

The Southeastern Coast Region is also a major source of natural gas, which is distributed by pipeline to consumers in many parts of the country. Reserves of gas in the tidelands area are thought to be gigantic, but as yet there is little production of natural gas from the continental shelf because of the high cost of drilling.

Shells   Oyster and clam shells have been used for many years as a road-building material, but the big development in this industry did not begin until the heavy-chemical industry started its phenomenal growth in the 1940s. Today one sees large mounds of gray, gravelly material piled up at many industrial establishments; they are oyster shells that have been dredged from the shallow bays along the Gulf Coast (Fig. 11-17). With the nearest source of limestone in

Texas more than 200 miles (320 km) from the coast, oyster shells from dead reefs provide industry with a cheap and abundant supply of lime.

The extraction operation is totally mechanized. Dredges of various sizes suck up the dead oyster shell, mostly from the Virginia oyster (*Crassostrea virginica*), clean it, and expel it onto large barges floating alongside, which, in turn, transport it to the numerous markets in the coastal area. The devastating nature of a dredging operation causes significant controversy in an area where a number of competing activities are relatively incompatible. Commercial fishermen, sports fishermen, tourists, and nature lovers frequent the same bays as the shell dredgers; petroleum extraction, commercial shipping, and pleasure boating add to the activities in these shallow waters.

MANUFACTURING

For several decades the Southeastern Coast Region has experienced booming industrial growth. Dozens of new factories, many of them large and highly mechanized, have been opened, particularly along the western coast. The region offers many attractions to industrial firms, but the principal locational advantages are

1. Large supplies of basic industrial materials, such as petroleum, natural gas, sulfur, salt, and lime (oyster shells).
2. An abundance of natural gas for fuel.
3. A tidewater location providing cheap water transport rates for bulky manufactured products.

Other factors, such as available labor, a growing local market, and local venture capital, are also important but of less overall significance in attracting major industries.

In most parts of the region the industrial sector is specialized rather than diversified, on the basis of specific resources or other locational attractions. The principal types of manufacturing are limited in variety, although often enormous in magnitude.

The basic specialized manufacturing industry of the region is petroleum refining. From Pascagoula to Corpus Christi there are numerous nodes of refineries, inevitably connected by a network of pipelines to other industrial facilities as well as to storage tanks and shipping terminals (Fig. 11-18). Much of the crude oil is obtained within the region, but a considerable proportion of the refinery input comes from oil fields to the north of the coastal zone and from imported oil. The entire region contains approximately 40 per-

cent of the nation's refinery capacity. The principal refining centers are Houston–Texas City and Beaumont–Port Arthur in southeastern Texas.

In conjunction with refining, a vast complex of petrochemical industries have developed. These are manufacturing operations in which various "fractions" (components) are separated from petroleum or natural gas and used to make other marketable products, ranging from butane (for heating gas), butadiene (for synthetic rubber), and toluene (for explosives) to more sophisticated industrial hydrocarbons.

Petrochemical plants are usually located near refineries. Often a complex pattern of pipelines conducts liquid or gaseous products from one plant to another, the finished product of one factory serving as the raw material ("feedstock") for another. Petrochemical plants are scattered along the coast from Mobile to Brownsville, but the main concentrations are in the major refining areas: along the Houston Ship Canal, around Texas City, in the vicinity of Sabine Lake, around Lake Charles, and along the Mississippi River between Baton Rouge and New Orleans.

Other prominent specialized manufacturing industries of the region include plastics (particularly polyethylene and styrene), inorganic chemicals (most notably sulfuric acid and soda ash), alumina (more than 90 percent of domestic alumina output comes from coastal plants in Texas and Louisiana), pulp and paper, rice mills, sugar refineries, and seafood processing.

In consonance with the situation all across the country, manufacturing has declined in importance in this region since the early 1980s. However, it continues to be a significant and distinctive component of the regional economy.

URBAN BOOM IN THE SPACE AGE

Few parts of North America have experienced such notable urban growth in recent years as the Southeastern Coast Region (see Table 11-1 for the region's largest urban places). A relative abundance of employment opportunities combines with the attraction of mild winters and a coastal environment to make this a region of rapid, continuing in-migration from other parts of the country. The location of a number of military bases as well as several major space-related facilities (in Texas, Florida, Louisiana, and Mississippi) have contributed significantly to the boom in the region.

Gulf Coast Metropolises: Houston and New Orleans

Houston and New Orleans are the well-established metropolitan centers of the region, although the former long since surpassed the latter in terms of both absolute and relative

FIGURE 11-18   A massive refinery complex at Baton Rouge, Louisiana, situated on the east bank of the Mississippi River. Most of the refinery's output moves by water transport on the river, both by inland waterway barges and by ocean-going tankers (courtesy Humble Oil and Refining Company).

growth. Houston, perhaps more than any other city east of the Rockies, epitomizes the story of North American urban expansion in recent decades, whereas New Orleans is an older city that has grown with unspectacular vigor.

Houston is the largest city in the South, and no city in its size-category has grown at a faster rate in the last 30 years (Fig. 11-19). Its image is one of a brash, sprawling, oil-oriented boom town. It is noted for low taxes, no zoning, a high crime rate, a very unsatisfactory municipal transportation system, and unfettered economic boosterism. It is the classic American example of the application of the private enterprise ethic to a metropolis.

Houston is one of the busiest ports in the nation, although its traffic is largely coastal shipment of bulk materials. Its artificial harbor is located at the end of a narrow, dredged, 52-mile (83-km) channel, which is one of the most dynamic and most dangerous (owing to the great quantity of volatile materials being shipped on a cramped waterway)

canals in the world. As the site of the Manned Spacecraft Center of the National Aeronautics and Space Administration (NASA), Houston's preeminent place in the space age is ensured. Nearby elements in the Houston conurbation include residential Pasadena and Baytown, industrial Texas City, and the island port and beach resort of Galveston.

New Orleans is one of those uncommon American cities that is distinguished by uniqueness of character. Its wretched and crowded site between the Mississippi River and Lake Pontchartrain,[2] its splendid location near the mouth of the continent's mightiest river, its remarkable past that encompasses the full span of southern history, and its unusual French cultural flavor that is unparalleled south of Montreal

---

[2] See Peirce Lewis's list of the 10 major environmental handicaps that give New Orleans what is probably the most miserable site of any North American metropolis in *New Orleans: The Making of an Urban Landscape* (Cambridge, MA: Ballinger Publishing Co., 1976), pp. 31–32.

TABLE 11-1

Largest Urban Places of the Southeastern Coast Region, 1995

Name	Population of Principal City	Population of Metropolitan Area
Baton Rouge, LA	246,000	563,000
Beaumont, TX	117,000	375,000
Biloxi, MS	52,000	348,000
Brazoria, TX	46,000	212,000
Brownsville, TX	116,000	310,000
Charleston, SC	82,000	525,000
Corpus Christi, TX	280,000	386,000
Daytona Beach, FL	72,000	448,000
Dothan, AL	55,000	136,000
Florence, SC	33,000	121,000
Fort Lauderdale, FL	153,000	1,402,000
Fort Myers, FL	49,000	375,000
Fort Pierce, FL	48,000	275,000
Fort Walton Beach, FL	24,000	157,000
Gainesville, FL	87,000	195,000
Galveston, TX	58,000	239,000
Harlingen, TX	59,000	
Houma, LA	34,000	190,000
Houston, TX	1,780,000	3,715,000
Jacksonville, FL	905,000	983,000
Lafayette, LA	89,000	364,000
Lake Charles, LA	73,000	175,000
Lakeland, FL	67,000	438,000
McAllen, TX	94,000	480,000
Melbourne, FL	64,000	452,000
Miami, FL	379,000	2,047,000
Miami Beach, FL	95,000	
Mobile, AL	211,000	518,000
Naples, FL	27,000	181,000
New Orleans, LA	502,000	1,315,000
Ocala, FL	49,500	229,000
Orlando, FL	196,000	1,390,000
Panama City, FL	37,000	140,000
Pensacola, FL	65,000	378,000
Port Arthur, TX	62,000	
St. Petersburg, FL	236,000	
Sarasota, FL	59,000	510,000
Savannah, GA	144,000	280,000
Tallahassee, FL	139,000	256,000
Tampa, FL	293,000	2,178,000
Victoria, TX	57,000	79,000
West Palm Beach, FL	77,000	987,000
Wilmington, NC	61,000	191,000

combine to set New Orleans apart. Its busy port depends heavily on income from ocean-borne commerce (rather than on coastwise shipping). The industrial base is limited; proportional employment in manufacturing is only half the average for large cities. Very important to the economy of New

Orleans is the tourist industry; it is one of the major tourist destinations in the country.

The Boom Cities of Florida

Florida has experienced several periods of rapid population and economic expansion in its history, but the growth trends of recent years are unparalleled in continuity and stability. This growth was predominantly in urban areas. During the 1980–1990 decade 10 of the 20 fastest-growing metropolitan areas in the nation were in Florida.

The state's permier metropolis, *Miami*, now sprawls widely from its focus on Biscayne Bay, and the resort suburb of Miami Beach merges imperceptibly northward with Fort Lauderdale. The urbanized zone extends almost uninterruptedly to Palm Beach, which is 50 miles (80 km) north of the Miami central business district. Miami's rapid growth makes it the second largest city in the region and fourth largest (after Houston, Dallas, and Atlanta) in the South. Its economic orientation is essentially commercial and recreational, but its attraction as a tourist center is waning, presumably because Miami's large urban area and unusual ethnic mix engendered a host of social problems (for example, the highest rate of major crimes in the nation), and the city has lost some of its charm as a tourist goal. Construction of new hotels was virtually nonexistent for more than a decade, and attempts to restore the image of a swinging resort by obtaining legalized casino gambling failed.

Manufacturing is of relatively minor importance in the economy, and maritime commerce is relatively limited, although increasing. Nevertheless, the city's role as gateway to Latin America is stronger than ever. Its airport is one of the nation's busiest for both domestic and international flights, and Miami has displaced New York as the leading U.S. port of departure for international sea passengers (with nearby Port Everglades ranking third). A significant "hidden" component of the local economy is drug smuggling; it is estimated that the total value of drugs brought into the United States annually through Dade County (the county in which Miami is located) is only slightly less than the county income generated by tourism. This factor presumably explains why the Dade County branch of the Federal Reserve Bank has a greater cash surplus than any other branch in the country.

The great influx of refugees from the Caribbean (chiefly Cubans and Haitians) in the 1970s–1980s placed an enormous strain on the economic and political infrastructure of the metropolitan area and required major social adjustments that significantly heightened ethnic tensions. Hispanics now constitute a plurality of greater Miami's population. Most of this has come about recently and abruptly; no other

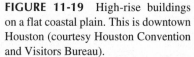

FIGURE 11-19 High-rise buildings on a flat coastal plain. This is downtown Houston (courtesy Houston Convention and Visitors Bureau).

American city has ever experienced such a massive and sudden change in its ethnic composition (Fig. 11-20).

Florida's other principal urban areas are also growing rapidly, although the rate of increase is greater in the cities of central Florida than in those of the northern part of the state. The *Tampa–St. Petersburg* metropolitan area now has a population of 2.2 million; its diversified economy and busy phosphate port continue to attract many newcomers. *Jacksonville* is one of Florida's oldest cities, but its recent growth pattern has been more erratic than that of cities farther south. The burgeoning *Orlando–Winter Park* area of central Florida is the state's major inland metropolis. It has shared significantly in the space boom because of its relative proximity to the Kennedy Space Center on Cape Canaveral, and it is the nearest city to North America's leading theme amusement park, the Walt Disney World–Epcot Center, which records more than 25 million visitors annually.[3] The *Lakeland–Winter Haven* area, between Tampa and Orlando, is also growing rapidly; a midstate megalopolis is in the making here, stretching from St. Petersburg to Orlando.

Smaller Industrial Centers

Several small- to medium-sized cities are highly industrialized and represent, in microcosm, the recent economic history of the region. All are ports and all depend heavily on one or more aspects of the petroleum industry.

Baton Rouge, situated on the Mississippi River more than 150 miles (240 km) from the Gulf of Mexico is a beehive of petrochemical activity and oil refining. *Corpus Christi* is the busy port of the central Texas coast. As such, its commercial function is more notable than its manufacturing one, although it is the site of a significant concentration of factories, many of which are petroleum oriented. In the extreme southeastern corner of Texas are the Sabine Lake cities of *Beaumont,* noted for oil refining, rice milling, and wood products manufacturing; *Port Arthur,* petrochemicals and synthetic rubber; and *Orange,* shipbuilding.

The Older Port Cities

Several of the region's older ports followed the trend of economic and population expansion at a somewhat slower rate. *Charleston* and *Savannah* had glory days during colonial and Confederate times but experienced limited and erratic growth during most of the past century (Fig. 11-21). Today their river-mouth harbors are busy again, their fishing fleets are active, and their rate of industrial expansion (pulp and paper, shipbuilding, and food processing) is impressive.

As the only major Gulf Coast port serving much of the Southeast, *Mobile* enjoys a privileged position. It is the ore-import port for the Birmingham steel industry and serves as the funnel for goods moving in and out of Alabama's inland waterway system. A few major industrial plants—ship-building, paper making, alumina reduction—are notable. Several small cities serve as central places in the Lower Rio Grande Valley, but only *Brownsville* has a deep-water harbor and effectively functions as a port. Urban growth in the valley is related primarily to agriculture, fishing, tourism, and services rather than to manufacturing or port activities.

INLAND WATERWAYS

Inland waterway freight transportation is notable in the Southeastern Coast Region. Most of the traffic is by means of

[3] This figure represents 10 million more people than the second most heavily visited theme park (Disneyland in California).

A CLOSER LOOK   Florida's Spreading Cities

Florida now ranks among the nation's most highly urbanized states. Although Florida exhibits a far-flung network of cities and suburbs—stretching from the far west of the Panhandle (where Pensacola is actually closer in road mileage to Chicago than to Miami) to the Upper Keys off the southern tip of the Peninsula—the Sunshine State's metropolitan areas share many features. First, the cities are marked by landscapes of newness and constant change, testimony to a continuing episode of prodigious growth that began in the 1970s and is certain to spill over into the 21st century. Second, they are home to a diverse, affluent population with few ties to Florida's past, especially its older cities. And third, they are dominated as nowhere else by the highway culture spawned by the ubiquitous use of the automobile.

These forces have combined to shape urban spatial patterns marked by *deconcentration*, accelerated in the 1980s and 1990s by the continuing availability of developable land, expanding infrastructures, and comparatively few restrictions on new development. Thus with few incentives to cluster people and activities, the settlement fabric keeps spreading outward from the older cores, with urban space used as lavishly in Florida as anyplace else on Earth. Judging by automobile usage, commuting, and other travel behavior, distances between trip origins and destinations are of little concern to most urban Floridians. (Is it therefore any wonder that such an on-the-go lifestyle has given the Miami–Fort Lauderdale area the nation's highest per capita rate of cellular telephone use?)

The internal spatial structure of the Florida metropolis reflects all the characteristics just discussed. With the exception of Jacksonville (which annexed its surrounding county), the central cities tend to be geographically compact, mostly built-out, and increasingly underbounded with respect to their mushrooming suburban rings. These older cores, therefore, contain only a modest proportion—usually less than one-third—of their metropolitan area's population and jobs. Despite a superficial veneer of relatively new skyscrapers, the central-city downtowns hark back to the days before Florida became a megastate in the 1970s. Most are small, haphazardly planned (especially Tampa's), and lose most of their street life after their office workforces disperse at the end of the business day.

Not surprisingly, urban Florida's leading business centers are more likely to be located in the suburbs of large cities, such as Coral Gables outside Miami, Maitland just north of Orlando, and the Butler Boulevard complex several miles south of Jacksonville's CBD. One of the best examples of such a suburban downtown is West Shore, located on Tampa's western edge adjacent to its ultramodern airport. As far back as the mid-1980s, so much commercial development had agglomerated here that West Shore became (and has remained) the state's single largest concentration of office space. It first surpassed downtown Tampa in 1984 and a decade later contained more than twice the workforce and office space of the CBD as West Shore completed its rise to become the premier business location in the Tampa Bay metropolis.

Residential landscapes in urban Florida have also responded to the strong centrifugal drift. The suburbs of large metropolitan areas exhibit a sprawling mosaic of discrete, low-density communities. Closer inspection of this mosaic reveals that each of its tiles caters to a homogeneous, specialized population. The overall result, however, is a particularly segregated urban society—not just in terms of race and income level, but also according to lifestyle and age (the retirement community, of course, has reached its apogee in Florida).

Given the forces turned loose on Florida's cityscape for the past quarter-century, we should not expect aesthetics to have played much of a role. To be sure, rapidity of development did not help matters (between 1980 and 1990 8 of the 10 fastest-growing metropolitan areas in the nation were located in the Sunshine State). But Floridians as a whole apparently cared so little about their immediate environment that they permitted a planner's nightmare of sprawl, strip development, and near universal visual blight to materialize during the 1970s and 1980s without even token coordinated resistance at the county and state levels. Moreover, tourism booms were allowed to reinforce the headlong rush toward landscape messiness. The gaudy strips of motels and fast-food facilities (such as Orlando's appalling International Drive) notwithstanding, Florida in recent years has also contributed some innovations to the U.S. urban scene. The Gulf Coast resort of Seaside in the far northwest reintroduced notions inspired by nineteenth-century houses and communities, concepts that have taken root and been expanded in certain suburbs around the country. Within Florida itself, the Disney Corporation in the late 1990s pioneered such a community (to be called Celebration) a few miles to the south of the Magic Kingdom. A less pleasant innovation of the 1970s was the nation's first entirely fenced-in municipality (Atlantis near West Palm Beach)—in a state that has since built thousands of miles of private walls, berms, and street barriers.

Florida is often called a microcosm of the United States, and this profile of its current metropolitan trends describes an urban direction not unlike that being taken by other heavily populated parts of the country. Quite clearly, the American city is today turning inside out. In many parts of Florida, however, things are moving at fast-forward speed, and this bears watching because it is here that the full consequences of this national transformation are likely to first come to light in the years immediately ahead.

Professor Peter O. Muller
University of Miami
Coral Gables, Florida

FIGURE 11-20   There are now many ethnic neighborhoods in the cities of Florida. Cuban districts are particularly extensive in greater Miami (courtesy James R. Curtis).

FIGURE 11-21   Old waterfront mansions in the High Battery section of Charleston, South Carolina (courtesy Charleston Convention and Visitors Bureau).

of interconnected channels exists. Most of the region's important ports are set inland some distance from the open sea and can function as ports only through channels that were dredged along stream courses in the flat coastal plain. Houston, with its dredged Buffalo Bayou, is the prime example, but the same process has been carried out elsewhere. The port of New Orleans, for example, had a second channel cut directly east from the city, thus shortening the distance to the Gulf considerably.

The most remarkable of the region's water routes is the Gulf Intracoastal Waterway, which extends from Brownsville to the Florida panhandle just inland of the coast, using lagoons behind barrier islands where possible and cutting through the coastal plain where necessary. For most of its 1100-mile (1760-km) length it is only 12 feet (3.6 km) deep; thus it is designed primarily for barge traffic (Fig. 11-22). It

FIGURE 11-22   Barge tows moving on a busy stretch of the Gulf Intracoastal Waterway near Morgan City Louisiana (courtesy U.S. Army Corps of Engineers, New Orleans District).

barges, flat-bottomed craft that usually are connected in tandem and pushed or pulled by tugs. Barges provide the least expensive form of transport, but they can be used only in quiet waters because of their susceptibility to swamping or capsizing.

The region is particularly well suited to barge traffic because of the large number of available canals and other dredged interior channels and because of its direct connection with the Mississippi River waterway system. Flat land and a high water table make the construction of inland waterways in the region relatively inexpensive; thus an extensive system was only partially completed by World War II but was particularly useful during that conflict because of the activities of German submarines in the Gulf of Mexico.

The waterway was finally completed in 1949. It is most heavily used in its central portion; two-thirds of total traffic is in the section between Houston and the Mississippi River. Sulfur, cotton, grain, and other bulk goods are important commodities shipped, but the most important are petroleum and petroleum products. Heavy use of the waterway is shown by the fact that 9 of the 15 largest U.S. seaports (in tonnage) are located along its course.

Of major significance to the Gulf Intracoastal Waterway is its connection with the Mississippi River system. Barge traffic from the waterway can move into the Mississippi via two major channels in southeastern Louisiana, which means that barges from the Gulf Coast can be towed to the far-flung extremities of the Mississippi navigation system— up the Tennessee, Ohio, or Missouri Rivers and even into the Great Lakes via the Mississippi and Illinois Rivers.

The Alabama, Tombigbee, and Black Warrior Rivers of the state of Alabama were also dredged and stabilized to provide navigation systems for barges. A modest amount of traffic traverses these rivers, mainly between Birmingham and Mobile. Much more ambitious is the Tennessee–Tombigbee Waterway (called Tenn-Tom), which connects the previously dredged Tombigbee River with the previously dammed Tennessee River, thus providing a barge route directly north from Mobile to the Tennessee River system. This, the nation's largest and costliest ($1.8 billion) navigation project was opened for business in 1985, to disappointingly low traffic usage.

Along the Atlantic coast is the Atlantic Intracoastal Waterway, which is theoretically analogous to the Gulf Intracoastal Waterway. The former, however, does not have the advantage of tapping a major region of bulk production or of connecting to the Mississippi navigation system. Consequently, commercial traffic on the Atlantic Intracoastal Waterway is exceedingly sparse except in a few short reaches. It is used much more extensively for pleasure boating.

A Cross-Florida Canal has long been proposed to join the two intracoastal waterways. Construction actually began in the late 1960s but halted in 1970 because of anticipated ecological problems and the questionable economics of the project.

RECREATION

The recreation and tourist industry of the Southeastern Coast Region is a major segment of the regional economy. The mild winter climate is part of the attraction, as are an abundance of shows, amusement parks, and historical sites. It is the coastline and its beaches, however, that draw visitors to the region. There is a greater total mileage of usable beaches in this region than in the rest of North America combined. And except in the vicinity of the largest cities or principal resorts, the beaches are likely to be both clean and uncrowded—even on a hot Sunday afternoon in August.

The coast of Florida is most heavily used, of course (Fig. 11-23). The Atlantic margin of that state consist of an almost-continuous beach from the St. Johns estuary in the north to Biscayne Bay in the south. Beaches are spaced more irregularly on Florida's Gulf Coast but are splendid in quality.

The Texas coast also contains hundreds of miles of beaches, although most are on offshore islands and not all are readily accessible. The principal beach resort is Galveston Island, near the Houston conurbation. The Corpus Christi area also has fine beaches, and Padre Island, extending for 100 miles (160 km) south from Corpus Christi, has the longest stretch of undeveloped beach in the nation.

The Mississippi coastline, from Pascagoula to Bay St. Louis, is one continuous beach and has an interesting pattern of development. In the area from Pass Christian to Ocean Springs, a stretch of some 35 miles (56 km), there is an almost unbroken line of lovely old homes set on the beach ridge, a few feet higher than the magnificent stretch of white sand, with the busy coastal highway intervening. Wherever this pattern is interrupted, the string of big homes is replaced by a newer resort or commercial development, significantly stimulated by newly legalized gambling casinos.

Beaches are intermittent and usually located on offshore islands along the South Atlantic coast north of Florida. The most famous resort area is near Brunswick, where each

FIGURE 11-23 Hundreds of hotels and apartment buildings line the ocean front almost continuously from Miami Beach to Fort Lauderdale (courtesy Miami Beach Tourist Development Authority).

A CLOSER LOOK   The Dynamic Demographics of Florida

Although most parts of the Southeastern Coast Region experienced significant population changes of various kinds in recent years, the state of Florida is particularly notable for its dynamic demographics. Some characteristics of its population are unusual, some are unique, and some probably represent portents of the future for larger areas of the United States.

For the state as a whole, the population increase of the last few decades, particularly the last few years, has been remarkable.[a] At midyear of the twentieth century Florida ranked 20th among the states, with a population of 2.7 million. By 1960 it ranked 10th, with 4.9 million people. Its 1980 population of 9.7 million gave it 7th place among the states. With 15 million people in 1995, Florida had become the 4th most populous state.

[a] Much of the data presented here is based on research by John W. Stafford of the University of South Florida and was kindly provided by him.

During the most recent census decade, the 1980s, Florida was the only state that ranked among the leaders in both total and relative growth. Its total decennial increase was exceeded only by those of California and Texas, and its proportional increase also gave it third ranking, behind only Nevada and Arizona.

The most striking aspect of Florida's population expansion is the fact that almost all of it is due to net in-migration rather than to natural increase. In terms of natural increase (excess of births over deaths), Florida has almost reached the zero threshold; its annual natural increase rate is 0.2 percent, the lowest among the 50 states. The components of this equation are simple—Florida has the highest death rate of all states and it ranks 39th in birthrate.

Thus Florida has been nurtured by a sustained high rate of in-migration for many years, particularly during the 1970s, when 92 percent of the state's total population increase was the result of the excess of immigration over emigration. This represents a long-continu-

ing 2 percent annual population growth due to net migration.

This population expansion has extended over most of the state, although it is most prominent in the central and southern parts and least notable in the northern panhandle. During the 1970s three-fourths of Florida's counties experienced "fast growth" (22 percent or more) and the remainder were in the "moderate-" or "slow-" growth category; none had a population decline. In no other southern state did as many as one-third of the counties experience fast growth.

People who move to Florida come from many different places and represent a great diversity of backgrounds and characteristics. Florida has a positive migration balance with almost every state, but the great bulk of its domestic immigrants come from Megalopolis and from the Heartland. Only a relatively small proportion of these immigrants are nonwhites; of the 2.7 million net immigrants of the 1970s, it is estimated that only 100,000 were African American (which nevertheless represented the first decade in a century

---

of the three nearby offshore islands has a different use pattern. Sea Island is the classic resort island; its relatively small acreage is devoted to pretentious summer homes, many of which are mansions. Just to the south is St. Simons Island. This is a much larger landmass that shows evidence of sporadic development over a long period of time, ranging from summer homes to modest motels. Of more recent vintage is the development of Jekyll Island; most of this island is maintained as uncrowded stretches of white sand, but in the center is a prominent concentration of modern motels and convention facilities. There are similar developments on Hilton Head Island in South Carolina and the 60-mile (96-km) beach centering at Myrtle Beach (South Carolina) is one of the most heavily used "camping" beaches in the country, with over 11,000 developed campsites. Much of the North Carolina mainland shore is without beaches, but projecting as a curved crescent into the Atlantic are the remarkable sandbar islands of Cape Lookout and Cape Hatteras, with a continuous oceanside beach.

The region has many places of historical interest, but the most notable are the four urban centers that have pre-

served the architectural flavor of yesteryear. The Vieux Carré (French Quarter) of New Orleans is the preeminent attraction of this type and sets the theme—with its Royal Street shops and Bourbon Street nightspots—for one of North America's most distinctive tourist centers (Fig. 11-24). Both Savannah and Charleston are cities of similar and unusual historic interest. Their extensive areas of eighteenth- and nineteenth-century architecture are unmatched elsewhere and are nicely counterpointed by Charleston's waterfront Battery area and Savannah's delightful system of city squares. In northeastern Florida the Jacksonville–St. Augustine area has maintained a smaller but more varied sampling of historic architecture.

In a region where recreation and tourism are big business it is not surprising that there has been heavy capital investment in constructed or "modified" attractions to which the public is invited for an admission charge. Florida, again, is the leader in such development. Most early endeavors were associated with some sort of water show featuring swimmers, skiers, or fish at one of the numerous artesian springs or lakes. Then the scope of the attractions broadened and now that Florida experienced net African American immigration).

Probably the most conspicuous feature of the immigrant cohort is its relatively advanced age. Fully 20 percent of the immigrants are retirees, compared with a national average of about 5 percent. Thus the continuing flow of immigrants contributes particularly to the "graying" of the state's total population. More than 18 percent of Florida's population is now 65 or over, which is half again greater than the national average. The Florida population is easily the oldest of any state, which primarily explains the state's low birthrate and high death rate.

From an ethnic standpoint Florida's population becomes proportionally whiter year by year. In 1940 some 27 percent of the population was African American; by 1990 the proportion had declined to 14 percent despite the fact that a record number of African Americans (1.8 million) were residents of the state. It should be noted that the African American component of the population is a major contributor to the natural in-

crease rate; nearly three-fourths of the state's natural increase during the past two decades was accounted for by African Americans.

The "internationalization" of Florida's population since the 1960s has received much attention, as well it should. It is epitomized by the flood of Cuban immigration to southern Florida; nearly 60 percent of the people of Cuban ancestry in the United States live in Florida, and the majority are concentrated in metropolitan Miami. There have also been notable influxes from Haiti, Jamaica, and other parts of Latin America. Even so, the magnitude of the domestic migration to Florida is such that the proportion of Hispanics in the state population is only slightly higher than the national average.

Prospects for the future look like more of the same. The ambience and reputation of Florida as a Sunbelt refuge for snow-weary Northerners are clearly well established, and Florida seems to be both functionally and psychologically more accessible to the emigrant states of the Northeast than are the states of

the Texas-to-California axis. Moreover, Florida attracts people of working age because it has an expanding job market; it attracts business people and retirees because it has no state income tax (only five other states have no income tax), and its sales tax is still relatively moderate.

Planners in other states are watching with interest as Florida copes with its dynamic population growth and unusual population mix. The age structure of Florida's population in the 1980s is likely to be replicated in the nation as a whole in the 1990s as our total population ages. Will an elderly population support capital expenditures for roads, sewers, schools, and other infrastructural elements that will be needed in the future? And in greater Miami there is the conspicuous example of another language finding a semiequal footing with English in a situation where the second language is not simply that of an ethnic enclave excluded from the power structure but is actually a language of big business and international trade.

TLM

FIGURE 11-24 New Orleans' *Vieux Carré* (French Quarter) is an extremely popular tourist destination. This is a view of Bourbon Street (TLM photo).

A CLOSER LOOK   Sequential Changes on America's Hippest Beach

In the winter of 1940 *Time* magazine declared Miami Beach "a booming pleasure dome . . . like no other city in the U.S., or in the world." What was startling about this declaration is that the country was still in the grip of the Great Depression! But there were no soup lines here, people were not jumping out of windows because the stock market had collapsed, men were not "takin' to the rails" in search of a job. No, in Miami Beach, life was sweet. The economy was embarrassingly robust. Gangsters, Hollywood stars, sports celebrities, and lots of ordinary people were flocking to the city's fabled shores, its glamorous nightclubs, its smart shops. To accommodate them, more than 250 hotels and 400 apartments had sprung up in the last seven years alone. Nowhere was this spirit of prosperity and good times more evident than on South Beach. A tightly-compacted square-mile hugging the island's lovely southeastern coast, South Beach was fringed by powdery white sands and palm tree-studded parks, and washed by the crystal clear, blue-green waters of the Atlantic (Fig. 11-C).

As a subtropical seaside resort, South Beach was symbolic of freshness, vitali-ty, and invigoration. For a beleaguered nation, it offered the tantalizing fantasy of escape. In spite of the Depression (or perhaps because of it), this was an era in American life and culture that was self-consciously modern, optimistic, and ut-terly enthralled by science, technology, and industry. The future held great promise. Machines were revered. Speed and efficiency were the cries of the era. Responding to this fascination, the art and architecture of the day became dominated by not only the functional-ism and aesthetics of cars, trains, ships, and airplanes, but also by rockets, washing machines, and vacuum clean-ers! After all, these machines and appli-ances were thoroughly modern. But more importantly, they offered hope—hope that life would be not only better in the future, but more exciting too.

On South Beach, the machine aes-thetic found its best expression in Art Deco architecture, a twentieth century invention that had recently come from Europe. It was a playful, whimsical style, ideally suited to a place devoted to the pursuit of fantasy and leisure. The many small, rather modest hotels—typi-cally with less than 100 rooms and catering to middle-class visitors who made up the bulk of South Beach's tourist trade—were designed in a streamlined form that was meant to con-vey a feeling of smooth, continuous, machine-like movement. Here, of course, life was supposed to be smooth, so why not smooth out the buildings? To add a theatrical dimension, some of the hotels were shaped like cruiseships, complete with "portholes," "hatches," and "masts." Others, with their aerody-namic designs, fins, spires and anten-nas, resembled a Buck Rogers rocket ready to blast off for Mars. Befitting the location, the structures were painted in soft, dreamy pastels and decorated with tropical motifs—flamingos and palms, dolphins and waves. Yes, South Beach was garish and flamboyant, a place that traded shamelessly on illusion and fan-tasy. But it also was a fun, hip place where it seemed as if the good times would never end (Fig. 11-D).

In the fall of 1982 South Beach had a reputation that was more infamous than famous: It ranked among Miami's mean-est, most poverty-stricken, and crime-ridden neighorhoods. It also was one of the city's most neglected and controver-sial communities. Art Deco had long since fallen out of fashion. In the minds

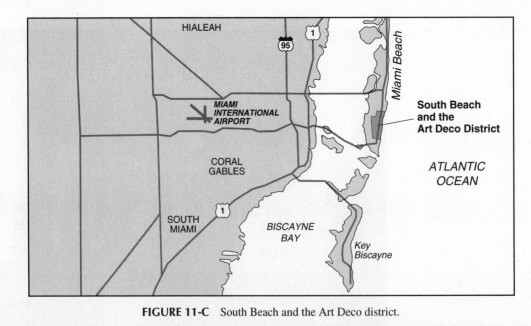

FIGURE 11-C   South Beach and the Art Deco district.

FIGURE 11-D   South Beach in its glory years of the 1930s. Looking south along Lummus Park and Ocean Drive (courtesy Miami-Dade Public Library, Romer Collection).

of most, the style now seemed silly at best, an eyesore at worst. The city fathers and the developers, each for their own reasons, were grasping for an urban plan that was contemporary, and progressive, and that most of all would make lots of money and propel the community out of this choking malaise. An ambitious plan was embraced that called for dredging a series of Venice-like canals, lined with upscale marinas, fancy restaurants, boutiques, and towering condos. While awaiting financing and the approval of numerous government agencies, a building moratorium was put into effect. It dragged on for several years. Meanwhile, the not-really-so-old hotels either lapsed into disrepair, stood empty and abandoned, or had been razed by the wrecker's iron ball. It was a sad, tragic sight. Prodevelopment demonstrators marched on city hall carrying placards that read, "Scrap This Junk," "New Is Beautiful," "Deco Schmeco."

The broken down hotels were populated by disadvantaged people, a poor and powerless urban underclass. The most numerous and visible, at least during the day when they sat quietly in rows in the shade of the hotel porches, were Jewish retirees. Most were born in Eastern Europe; many spoke Yiddish. Perhaps remembering the South Beach of their youth, they had journeyed from New York and other big eastern cities to spend their final days. Now old and frail, for most of them, their dreams of a happy and safe retirement in the south Florida sun and surf had collapsed before their very eyes into a kind of cruel nightmare. Increasingly they shared turf with drug addicts and homeless people, but especially with another poor, desperate group: Caribbean refugees, mainly from Cuba. Two years earlier, in 1980, more than 125,000 Cubans had sailed from the port of Mariel to Key West, Florida, mariners aboard the so called "Freedom Flotilla." Several hundred of the least fortunate and dangerous among them, including former mental patients and hardened criminals, ultimately found shelter in the decaying morass of South Beach.

Amid this depressing portrait of community decline, there were glimmers of hope. In the late 1970s most of South Beach had been officially designated an historic district and listed in the *National Register of Historic Places*. Although such recognition did not in itself insure that the buildings would be preserved, it did suggest that Art Deco architecture was considered, at least by some, to be both historically significant and worth saving. If only experimentally, several young Deco supporters from the local arts community had already pooled their resources, bought a couple of hotels, and had begun the restoration process. Most observers laughed and thought it was a waste of money and talent.

In the spring of 1996 South Beach had become one of the most celebrated, talked about, and photographed ocean-front resorts in the world, a new American Riviera. The current darling of the international jet set, the rich, the beautiful, the hip, South Beach is a glittering glamour capital and arts colony. Madonna hangs out here; so too does Sylvester Stallone and Cindy Crawford. It is equally famous for its exotic, non-stop nightlife, its topless and gay beaches, its chic sidewalk cafés, its galleries, its alternative music venues, and especially for its supermodels who have made it one of the premier centers of the fashion world. (On any given morning, when the light is just right, there may be as many as four or five fashion "shoots," all drawing their share of onlookers.) And, of course, it is famous for its signature feature: Tropical Art Deco architecture.

More than 600 of the Art Deco buildings—the world's greatest concentration—have been or are currently being restored. Structures that sold for $200,000 in 1982, now command five million dollars or more. Investors have flocked in from Europe, Canada, Japan, and Latin America. So too have the tourists. On a Saturday night there may be 75,000 people, searching for a good time, an escape. Looking west across Ocean Drive from Lummis Park at the continuous row of freshly painted, pastel colored Deco beauties, is like looking at a dreamscape: They seem unreal, like an illusion. But then again, this place was born of illusions and fantasies; it has only come full-circle.

Professor James R. Curtis
California State University
Long Beach

runs the gamut from specialized commercial museums (circus, vintage car, and so on) to the variety of Walt Disney World and its associated Epcot (Experimental Prototype City of Tomorrow) Center.

Florida is the most popular tourist destination in North America and its tourist industry is a major source of income. In the southern half of the state it flourishes primarily in the winter, with a secondary smaller peak in summer; in the northern half business is greatest in the summer. Visitors come from almost every state and many foreign countries, but the majority are urban dwellers from east of the Mississippi and north of the Ohio and Potomac Rivers. Avoidance of winter is one of the major stimuli for coming to Florida, in a pattern that began nearly a century ago.

The tourist industry in Florida has experienced more than one major setback and is still overextended at times. But today it is solidly based and the most important revenue-producing activity in the state. The principal concentration of resorts is along the southeastern coast, the best known being the extravagant and yet decaying hostelries of Miami Beach. Peripheral attractions include Everglades National Park, the various islands of the Keys, and tourist-oriented Seminole Indian settlements.

THE OUTLOOK

The Southeastern Coast Region remained one of the most backward parts of North America until almost the beginning of the present century. Prior to 1900 its chief activities were forest exploitation, agriculture, and ranching. The few run-down ports were in need of modernization.

With the discovery of oil at Spindletop and the development of salt, phosphate, lime (oyster shells), and sulfur, the coastal region began to attract industry even though no major development took place until the 1940s. In this region industry has found a favorable habitat; indeed, few parts of North America present a brighter outlook for manufacturing than this area that is the world's most extensively industrialized subtropical region.

Petroleum overshadows all other factors in the economy of the western part of the region, and its serious decline in the 1980s was a staggering blow. Texas experienced net out-migration for a couple of years in the mid-1980s, and Louisiana actually lost population for the first time in several decades. In the mid-1990s the oil industry started to make a slow recovery, although it is still far from robust, and Louisiana, in particular, continues to exhibit a fragile economy.

Conventional port business is expanding throughout the region, notably led by the export of bulk foodstuffs (particularly soybeans) through New Orleans and manufactured goods destined for Latin America through Miami. Port facilities are being enlarged significantly at Houston, New Orleans, Mobile, and Charleston.

Specialized farming will continue to occupy an important place in the economy, led by citrus, sugarcane, and vegetables. Intensification of beef cattle raising is probable.

The space boom of the 1960s has significantly ebbed, but the period of hardship resulting from the sharp decline seems to have passed and a steadier but less spectacular era of space-related activities will probably become established at the prominent centers of Cape Canaveral–Titusville in Florida, Houston, and the Bogalusa–Picayune border area of Louisiana–Mississippi.

The attraction of mild winters and outdoor living remains compelling for Northeasterners; thus people are likely to continue to pour into the region at a rapid rate to visit, work, or retire, especially in Florida and the Lower Rio Grande Valley. Rapid population growth and a continually expanding tourist trade will be major elements in the economic and social geography of the region for some time to come. Increasingly too, Florida serves as a way station for tourists on the way to the Bahamas or the Caribbean.

The striking juxtaposition of natural areas and constructed landscapes in this region makes it a prominent place for ecological confrontations. Several major battles have already occurred: the Everglades jetport has been delayed if not canceled; the Cross-Florida Canal has at least been postponed; "development" plans for Padre Island and Cape Hatteras are being contested; the ramifications of pesticides are being debated heatedly in southern Louisiana; the maintenance of a water supply for the Everglades remains an unresolved issue. In this region the "developers" and the "preservationists" are in serious conflict, and with the passage of time the arena of combat is certain to widen as the need for rational land-use plans becomes increasingly pressing.

A CLOSER LOOK   Winter Texans

When winter comes to North America, there is an annual migration of hundreds of thousands of people to the warm parts of the continent. These are people who are sufficiently affluent and/or sufficiently footloose to leave their normal domiciles and live for a few days or a few weeks or a few months in a temporary residence in a milder climate. The vast majority are senior citizens, and mostly they travel in motor homes or recreational vehicles. When the snow is deep and the wind is cold in such places as New York, Ohio, Iowa, and Saskatchewan, the "snow birds" pack up and head south.

Some go to southern California, where it is common for San Diego motels to fly the Canadian flag in winter. Some are attracted to the lower Col-

orado River valley, where mild temperatures and inexpensive lodging at such places as Bullhead City and Lake Havasu City (AZ) are close to the West's most rapidly expanding gambling center, Laughlin (NV). Some go to the long-established winter resort cities of Phoenix and Tucson. The largest movement of snow birds is from Megalopolis and the Midwest to Florida.

Another prominent movement is from the north-central states to the Lower Rio Grande Valley of Texas, focussed on the city of Brownsville, but spreading over four counties to include Harlingen, McAllen, and various smaller urban places (Fig. 11-E). The lure of "The Valley" to northern snowbirds is primarily mild weather and inexpensive living conditions, but there are other attrac-

tions as well. The Hispanic culture and exotic shopping of Mexico is just across the river. The pristine beaches of South Padre Island are close at hand. Numerous wildlife refuges offer excellent possibilities for bird watching.

The local communities provide many programs for "winter Texans," and the economy is particularly geared to their needs. About 90,000 people come to the Valley each winter, many of them repeat visitors year after year. Most of them arrive in recreational vehicles and stay in one of the dozens of RV parks that are available. The average age of the winter Texans is 68, and their average length of stay is 25 days. They mostly come from the Midwest, especially Iowa and Minnesota.

TLM

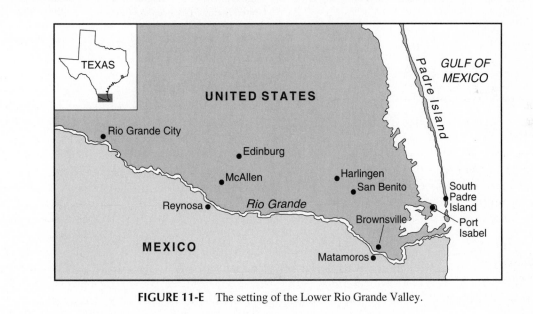

FIGURE 11-E   The setting of the Lower Rio Grande Valley.

SELECTED BIBLIOGRAPHY

BUCZACKI, STEFAN T., "Florida for Fruit and Vegetables," *Geographical Magazine,* 51 (November 1978), 100–102.

CARTER, L. J., *The Florida Experience: Land and Water Use Policy in a Growth State.* Baltimore: Johns Hopkins University Press, 1975.

DANIELSON, MICHAEL N., *Profit and Politics in Paradise: The Development of Hilton Head Island.* Columbia: University of South Carolina Press, 1995.

DAVIS, DONALD W., "Trainasse," *Annals,* Association of American Geographers, 66 (1976), 349–359.

FERNALD, EDWARD A., AND ELIZABETH D. PURDUM, EDS., *Atlas of Florida*. Gainesville: University of Florida Press, 1992.

HART, JOHN FRASER, AND ENNIS L. CHESTANG, "Rural Revolution in East Carolina," *Geographical Review,* 68 (October 1978), 435–458.

HENRY, JAMES A., KENNETH M. PORTIER, AND JAN COYNE, *The Climate and Weather of Florida*. Sarasota, FL: Pineapple Press, 1995.

JORDAN, TERRY G., JOHN L. BEAN, AND WILLIAM M. HOLMES, *Texas*. Boulder, CO: Westview Press, 1984.

KNIFFEN, FRED B., *Louisiana, Its Land and People*. Baton Rouge: Louisiana State University Press, 1968.

LEE, DAVID, "Black Districts in Southeastern Florida," *Geographical Review*, 82 (October 1992), 375–387.

LEWIS, PEIRCE F., *New Orleans: The Making of an Urban Landscape*. Cambridge, MA: Ballinger Publishing Company, 1976.

LONGBRAKE, DAVID B., AND WOODROW W. NICHOLS, JR., *Sunshine and Shadows in Metropolitan Miami*. Cambridge, MA: Ballinger Publishing Company, 1976.

MARCUS, ROBERT B., AND EDWARD A. FERNALD, *Florida: A Geographical Approach*. Dubuque, IA: Kendall/Hunt Publishing Company, 1974.

NEWTON, MILTON B., JR., *Louisiana: A Geographical Portrait* (2d ed.). Baton Rouge, LA: Geoforensics, 1987.

PALMER, MARTHA E., AND MARJORIE N. RUSH, "Houston," in *Contemporary Metropolitan America*. Vol. 4: *Twentieth Century Cities,* ed. John S. Adams, pp. 107–149. Cambridge, MA: Ballinger Publishing Company, 1976.

RANDALL, DUNCAN, P., "Wilmington, North Carolina: The Historical Development of a Port City," *Annals,* Association of American Geographers, 58 (1968), 441–451.

SHELTON, M. L., "Surface-Water Flow to Everglades National Park," *Geographical Review*, 80 (October 1990), 355–369.

STANSFIELD, CHARLES A., JR., "Changes in the Geography of Passenger Liner Ports: The Rise of the Southeastern Florida Ports," *Southeastern Geographer,* 17 (May 1977), 25–32.

TONER, M. F., "Farming the Everglades," *National Parks and Conservation Magazine,* 50 (August 1976), 4–9.

TURNER, R. EUGENE, AND EDWARD MALTBY, "Louisiana Is the Wetland State," *Geographical Magazine,* 55 (February 1983), 92–97.

WINSBERG, MORTON D., *Florida Weather*. Orlando: University of Central Florida Press, 1990.

ZIEGLER, JOHN M., "Origin of the Sea Islands of the Southeastern United States," *Geographical Review,* 49 (1959), 222–237.

The Heartland

12

The Heartland is a broad region of moderately high population density and enormous economic productivity. It includes all of 3 states (Indiana, Illinois, Iowa) and parts of 12 others as well as a significant portion of the province of Ontario. For comparative purposes, it can be noted that the regional population is only two-thirds that of Britain and France combined, but its gross output (industrial, agricultural, and mineral) is more than twice as great.

It is a region with a remarkably favorable combination of physical factors for the development of agriculture. These factors together with intelligent farm management and the considerable application of inanimate energy make it the largest area of highly productive farmland in North America, if not the world. The Heartland Region is also the industrial core of the continent. Although containing less than one-third of North America's population, it has over half the manufactural output of both countries.

This widespread and well-balanced regional economy provides much of the stability and diversity that permit the people of the United States and Canada to enjoy one of the world's highest standards of living. And yet it is not economic muscle alone that gives this region its "heartland" appellation. This is in many ways the "core" region of North American society. Here are exemplified the ideas, attitudes,

and institutions that are most representative of the way of life in North America. Here the population amalgam is most thoroughly distilled and the "average" American or Canadian (not, however, the average French Canadian) is most likely to be found. The region possesses an inordinate amount of political power, partly because the American political system has given rural people a greater proportional representation than urbanites in legislative bodies, and the rural population is larger here than in other parts of the continent.

The Heartland is the transportation and communication hub of the continent. Its productive lands and relatively level terrain stimulate and make feasible a dense network of highways and railroads. Waterway traffic on the lower Great Lakes and the major midwestern rivers—Ohio, Mississippi, Missouri, and Illinois—continues to increase. Air transportation among the cities of the region is remarkably dense. Indeed, Chicago, in addition to being one of the world's leading railway centers, is served by the busiest airport in the world.

Several important parts of the two countries are not located in the Heartland: the great decision-making centers of New York and Washington; the rapidly growing cities of the West and South; the totality of French Canada; and many others. Nevertheless, the role of the Heartland Region in the life of North America is critically important. In this regard

Robert McLaughlin noted, "Had any of America's other regions been detached over the course of history, the United States would not be the nation as we know it—but it would still exist. Without the Heartland, however, the United States would be inconceivable."[1] Wilbur Zelinsky declared the region to be "justly regarded as the most modal, the section most nearly representative of the national average."[2]

The Heartland encompasses a smaller portion of Canada, but that portion (southernmost Ontario) contains such a concentration of wealth, power, and population that it is often regarded as the "norm" of Canadian life and has had a more important policymaking role than any other part of the country.[3] The "Americanism" of southern Ontario has attracted much attention from scholars, as has its cultural affinity with the U.S. Midwest.[4]

The Heartland, then, is a broad region that includes what is generally referred to as the Midwest. It covers, however, somewhat more than the cultural Midwest: western Kentucky and the Nashville Basin are more properly "southern" in culture, western New York is only marginally midwestern, and the term is rarely used in southern Ontario.

EXTENT OF THE REGION

The Heartland occupies only part of the vast interior plain of North America; it is set off from adjoining regions, particularly to the west and north, by imprecise and transitional boundaries (Fig. 12-1).

[1] Robert McLaughlin and the Editors of Time-Life Books, *The Heartland* (New York: Time Incorporated, 1967), p. 16.

[2] Wilbur Zelinsky, *The Cultural Geography of the United States* (Englewood Cliffs, NJ: Prentice-Hall, 1973), p. 128.

[3] John Warkentin, "Southern Ontario: A View from the West," *Canadian Geographer,* 10 (1966), 157.

[4] See, for example, Andrew H. Clark, "Geographical Diversity and the Personality of Canada," in *Readings in Canadian Geography,* ed. Robert M. Irving (Toronto: Holt, Rinehart & Winston of Canada, 1972), pp. 9–10; and Zelinsky, *Cultural Geography of the United States,* p. 128.

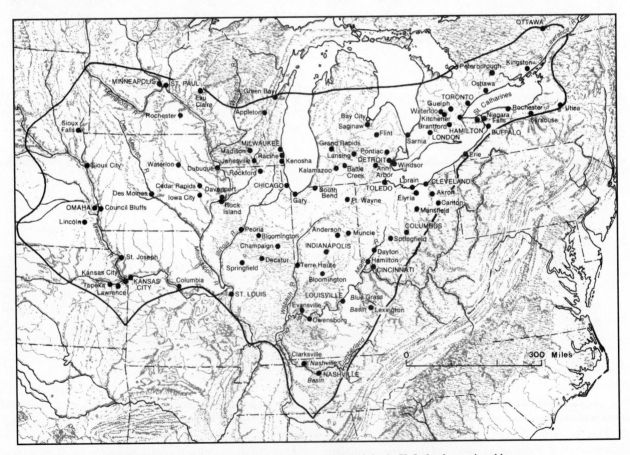

FIGURE 12-1   The Heartland Region (base map copyright A. K. Lobeck; reprinted by permission of Hammond, Inc.).

The eastern and southern margins of the region are relatively definite because of land-use contrasts that are associated with physiographic differences. The interior lowlands support more intensive, diversified, and prosperous agricultural activities, as well as more and bigger cities, than do the slopelands of the Appalachians and Ozarks; thus the regional boundary generally follows the topographic trend from northernmost New York State south and west to the Missouri–Kansas border, with a southerly salient to include parts of western Kentucky and Tennessee.

The northern boundary of the Heartland Region approximately follows the southern edge of the Canadian Shield, where differences in bedrock geology are accompanied by related vegetation and agricultural variations. On the west the Heartland merges with the Great Plains; scanty rainfall separates the prairie margin from the short-grass country of the plains at about the 98th meridian. Here precipitation is inadequate for the profitable production of unirrigated corn and this crop tends to be replaced by those that are more drought resistant. The western boundary of the Heartland Region is generalized as the transition from corn-dominated farming on the east to wheat-grain sorghum-pasture-dominated land use on the west.

The Heartland is by no means the largest region in North America; several exceed it in areal extent. It has, however, the greatest population of any major region as well as the largest economic output. Despite the magnitude of these factors, it is difficult to subdivide the region in a satisfactory manner, for there is an essential homogeneity of pattern and interdependence of relationship of most elements of geographical significance. It is, however, possible to delineate meaningful subregions on the basis of limited factors—for example, agriculture.

The Corn Belt is a widely accepted term applied to the core of the Heartland; it is an east–west band extending from central Ohio to eastern Nebraska in which the predominance of corn in the agricultural scene has long been recognized. North of the Corn Belt, and most prominent in Minnesota–Wisconsin–Michigan, is the Dairy Belt. Southeast of the Corn Belt is a tobacco and general farming area. But if subdivision by agricultural pattern is meaningful, on a more general geographical basis it is not unless a relatively large number of subregions is recognized. Accordingly, we will not be concerned with delimiting subregions.

THE LOOK OF THE LANDSCAPE

The landscape of the Heartland has a certain similarity of appearance throughout its length and breadth, which is another element of unity in the region. The conspicuous environmental features are absence of slopeland, prevalence of trees, and abundance of small bodies of water. The terrain is flat to undulating throughout and extends to a relatively featureless horizon in all directions. Despite the extensive acreage of cropland and pasture, trees are present in many portions of the region: wide-branching oaks, maples, and other hardwoods in the forests and woodlots, and tall cottonwoods and poplars along the stream courses. Large lakes, with the exception of the Great Lakes, are uncommon, but thousands of small marshes and ponds dot the landscape everywhere north of the Ohio and Missouri Rivers.

No other human activity is nearly so conspicuous as agriculture; crops and pastures cover more of the land than does anything else. Appearance, of course, varies greatly with the season. In spring, for example, deep green fields of winter wheat, pale green pastures, and still paler green blocks of oats are just breaking through the earth, and black, brown, and tan, depending on the type of soil, squares of land are plowed and ready to be planted to corn and soybeans.

Of great prominence in the landscape, particularly if viewed from the air, is the rectangularity of areal patterns. Primarily as a result of the systematic rectangular land survey of the eighteenth century, landholdings in most of the region have right-angled boundaries. Thus fields and farms appear as a gigantic checkerboard intersected by a gridiron pattern of roads, which chiefly cross at 1-mile (1.6-km) intervals. Modern interstate highways may appear as diagonal scars superimposed on the grid, but the functional roadway network of the rural Heartland is almost as straight and angular as ever.

Conspicuous farmsteads dot the patterned rectangles of the farmlands. Although large and pleasant, the farmhouses (Fig. 12-2) may suffer in comparison with the architectural gems of New England, New York, eastern Pennsylvania, and Maryland, but the total farmstead is usually large and impressive. The two-story farmhouse, normally ringed with trees, is dwarfed by imposing and generally well kept outbuildings, such as huge barns, corncribs, silos, machine sheds, dairy buildings, and chicken houses.

Villages and small towns occur at regular intervals over the land, delicately sprawling around a road junction or alongside a railway line. They lie flat against the earth, with only a water tower or grain elevator rising above the leafy green trees. Businesses are clustered along one or two main streets, but commercial bustle is usually lacking, for most small towns are stagnating or withering anachronisms in an era of metropolitan expansion and transportation ease. The residential sections are characterized by big old homes, sometimes frame and sometimes brick, whose unused front porches look across broad lawns to tree-lined streets (Fig. 12-3).

FIGURE 12-2   Large well-kept farmsteads are typical of the Heartland, as in this example from southern Ontario, near Waterloo (TLM photo).

FIGURE 12-3   The residential tranquility of a small Heartland town, as represented by Fairfield in eastern Iowa (TLM photo).

Large and small cities are spaced more irregularly across the region. In contrast to the serenity of small towns, they are places of constant movement. These are representative North American cities, sprawling outward around the periphery and rising upward in the center. Above all, the urban centers are foci of activity: hubs of transport routes that con-verge from all directions with a steady stream of incoming and outgoing traffic.

THE PHYSICAL SETTING

Few, if any, large regions of the world have a more favorable combination of climate, terrain, and soils for agriculture. J. Russell Smith's classic accolade, "The Corn Belt is a gift of the gods,"[5] also almost equally applies to other parts of the Heartland Region.

Terrain

Almost all the Heartland Region is in the vast central lowland of North America. Throughout the lowland the land is mostly level to gently undulating, with occasional steeper slopes marking low hills, ridges, or escarpments (Fig. 12-4). The entire region is underlain by relatively horizontal sedimentary strata, one of the most extensive expanses of such bedrock to be found on Earth. These various sedimentaries—mainly limestone, sandstone, shale, and dolomite—are relatively old, originating primarily in Paleozoic time. Their structural arrangement is generally subdued, consisting of broad shallow basins and low domes.

The limited relief and gentle slopes of the region are partly ascribed to the underlying structure, but are predomi-

[5] J. Russell Smith and M. O. Phillips, *North America* (New York: Harcourt, Brace and Co., 1942), p. 360.

FIGURE 12-4 The terrain of most of the Heartland is flat to gently rolling, as in this typical scene from southeastern Michigan near Detroit (TLM photo).

nantly the result of glacial action during Pleistocene time. The several ice advances of the last million years had lasting effects, for they leveled the topography from its preglacial profile. There was some planing of hilltops and gouging of valleys, but generally ice action in this region was depositional rather than erosional in nature; thus it was the filling of valleys rather than the wearing down of hills that produced the present land surface. Over most of the region the bedrock has been buried many tens or hundreds of feet by glacial debris; for the Corn Belt section glacial drift is thought to average more than 100 feet in depth. The greater part of the Heartland is, therefore, indebted in no small measure to the Ice Age for its flat lands and productive soils.

Impaired drainage and loess are two other significant legacies of the Pleistocene in the region. The areas of more recent glaciation, the so-called Wisconsin stage, contain many marshes, bogs, ponds, and lakes because "normal" drainage patterns have not as yet had time to develop since the most recent ice recession (about 8000 years ago). Beyond (south of) the margin of Wisconsin glaciation are extensive areas that are mantled with deep deposits of loess, which is fine-textured windblown silt, believed to have originated from the grinding action of ice on rock, which often produces fertile soils.

There are several relatively discrete topographic sections in the Heartland Region, with characteristics sufficiently distinctive to warrent separate mention (Fig. 12-5).

The northeastern part of the region, from the Upper St. Lawrence Valley in Ontario–New York to south-central Wisconsin, is the *Great Lakes* (1) section, where large and small lakes dominate the landscape. Two of the Great Lakes are entirely and two are partially within this section, as are thou-

sands of smaller bodies of water. There are many prominent examples of glacial or glaciofluvial deposition: drumlins, eskers, outwash plains, and, particularly, the long irregular ridges of terminal moraines. Several significant scarps occur in this section, the most conspicuous of which is the Niagara Escarpment. Its abrupt cliffs of gray dolomite overlook the gentler glaciated plain between Lakes Erie and Ontario, then arc around the north side of lakes Huron and Michigan, and reappear in Wisconsin.

In southwestern Wisconsin and adjacent Minnesota, Iowa, and Illinois lies the *Driftless Area* (2), an unglaciated island in a sea of glacial drift. Missed by the continental ice sheets, it differs topographically and economically from the surrounding territory. Its landscape, consisting mostly of low but steep-sided hills, is similar to the one that existed before the glaciers came.

The central portion of the Heartland, from central Ohio to eastern Iowa, is the *Till Plain* (3) section. The terrain varies from extremely flat to gently rolling, the remarkable lack of relief presumably resulting from cumulative deposition of at least three ice advances. Drainage is much better integrated here than in the Great Lakes section and there are fewer lakes and marshes.

The *Dissected Till Plain (4)* occupies the northwestern portion of the region, extending from Minnesota to Missouri. It is covered with till from earlier glacial advances and there has been more time for stream erosion to modify the surface; hence there is a greater degree of dissection and a general absence of lakes as well as a lack of terminal moraines.

In the southwestern corner of the Heartland is the *Osage Plains* (5) section that extends from Missouri southwestward into Texas. This is an area that was essentially

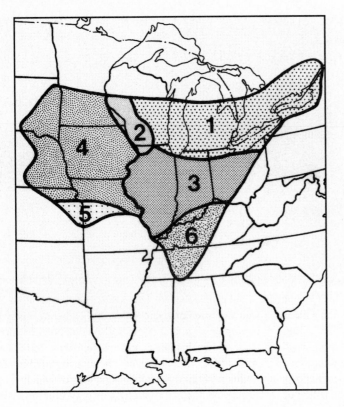

FIGURE 12-5 Topographic subdivisions of the Heartland: (1) Great Lakes, (2) Driftless Area, (3) Till Plain, (4) Dissected Till Plain, (5) Osage Plains, and (6) Interior Low Plateaus.

unaffected by glaciation and is represented by a nearly featureless plain developed on horizontal sedimentary beds.

In the southeastern portion of the Heartland Region is a section known as the *Interior Low Plateaus* (6), although most of the terrain has the appearance of low hill country. This area is structurally and topographically complex, with a series of cuestas, escarpments, basins, major fault lines, and some intrusive vulcanism. Much of the area is referred to as the Highland Rim; it consists mostly of low hills and scarp ridges and has a great many karst features, particularly sinkholes and caverns. There are two famous and fertile basins in this section, the Nashville Basin and the Bluegrass Basin, that are flat lands surrounded mostly by infacing escarpments.

Climate

The region's interior location in the eastern part of the continent results in a humid continental climate that is noted for significant seasonal and abrupt day-to-day changes in weather conditions. Climatic patterns are broad and transitional across the region, for there is relatively uniform relief and no significant topographic barriers. Thus temperatures and length of growing season increase more or less uniformly from north to south and precipitation generally decreases from east to west.

Summer is a time of hot days and relatively warm nights. Humidity is generally high, and well over half the total annual precipitation falls during this season. Thunderstorms are common in summer and tornadoes are frequent.

Although summer conditions are quite benign for crop growing, winter is generally severe. The cold weather is not continuous, for there are periodic spells of relative warmth with the passing of low-pressure centers. But cold fronts are frequent, often accompanied by blizzards and followed by anticyclonic spells of clear and very cold weather. Snow is commonplace throughout the region, although only in the far north does it remain unmelted on the ground for long periods (Fig. 12-6).

The transitional seasons, spring and fall, are usually brief but dramatic, with rapid shifts of temperature and abrupt periods of storminess. Flooding is a perennial natural hazard in spring and early summer, sometimes resulting from heavy spring rains falling on frozen ground but often simply the result of prolonged rain with rapid runoff. In any event, the densely populated floodplains of the region's major rivers are frequently threatened with inundation.

FIGURE 12-6 Winter is a prominent season throughout the Heartland Region. This farm scene is in southeastern Wisconsin, near Racine (TLM photo).

Natural Vegetation

The region's original vegetation consisted of forest and grass. The eastern part was forested. Ohio, Indiana, southern Michigan, southern Wisconsin, and southern Illinois all were a part of the oak–hickory southern hardwood forest, and the Kentucky Bluegrass and the Nashville Basin were a part of the chesnut–oak–yellow poplar southern hardwood forest. The forest near the northern boundary of the region was characterized by conifers on the sandy soils and by magnificent stands of hardwoods on the clay lands. Elsewhere the forest consisted wholly of hardwoods.

Southern Minnesota, all of Iowa, central Illinois, northern Missouri, and eastern Nebraska and Kansas formed the prairie, a vast billowy sea of virgin grass without timber except along the streams. It was tall grass with long blades and stiff stems, growing to a height of 1 to 3 feet (0.3 to 0.9 m) and sometimes 6 to 8 feet (1.8 to 2.4 m). Trees growing along streams were chiefly cottonwoods, oaks, and elms in the western portion, with occasional sycamores and walnuts farther east.

The true prairie extended from Illinois (small patches existed in western Ohio and northern Indiana) to about the 98th meridian, where it was gradually replaced by the short grass of the steppe. The boundaries of the prairie were never sharply defined. They were not the meeting place of two contrasted vegetation belts; rather they were broad mobile zones that moved with pronounced changes in precipitation. Many interesting theories for the origin of the prairie were advanced, but none as yet is wholly acceptable to botanists, plant ecologists, and plant geographers.

Soils

Nowhere else on the continent is there such a large area that combines generally fertile soils with a humid climate. This combination is at its best in the region's core, the Corn Belt. The Bluegrass area and the Nashville Basin are also highly productive. Some soils, however, as in central Michigan and central Wisconsin, in the Driftless Area, and in the Highland Rim of Kentucky and Tennessee, are far from rich.

Most of the region is characterized by Alfisols and Mollisols. Although the former develop under deciduous forest in the milder of the humid continental climates, the fact that they are forest soils means that, in general, they are less productive than the dark brown to black Mollisols of the prairie. Still, some forest soils, such as those in western Ohio and north-central Indiana, yield about as well per acre as prairie soils. It may be said that forest soils that develop on calcareous till or on limestone, granite, gneiss, and schist are superior to those that evolve from shale and sandstone. All forest soils, however, are permanently leached, acid in reaction, and poor in humus.

The true prairie soils are generally fertile. They develop in cool, moderately humid climates under the influence of grass vegetation and are characterized by a dark brown to black topsoil underlain by well-oxidized subsoils. Being relatively well supplied with moisture, they are moderately leached and acid in reaction and they lack a zone of lime accumulation. They are primarily silt loams and clay loams in texture and are derived largely from glacial till.

In summary, much of the eastern portion of the Heartland is dominated by Alfisols, which are gray-brown in color and have subsurface clay accumulation. In the west there is a preponderance of Mollisols, which tend to be black in color and rich in organic matter.

HUMAN OCCUPANCE OF THE HEARTLAND

The Heartland has always been a productive region for goods that were valuable in each period of its history: game, furs, crops, minerals, and factory output. Thus it has been a region that was coveted and struggled over by a diversity of peoples who learned of its riches. Three great nations, as well as a dozen major Native American tribes, fought for supremacy here, and the early history of the region is punctuated by battles, massacres, wars, and alliances.

Aboriginal Occupance

Although relatively little is known about the early aboriginal inhabitants of the region, the antiquity of human presence is well established, dating from about 10,000 B.C. Over much of the Heartland these prehistoric people are generally referred to as "Mound Builders" and just about the only landscape evidence of their presence is the large number of burial mounds and other scattered earthen structures.

At the time of European contact many well-organized tribes existed in what is now the Heartland Region. They were chiefly forest dwellers of the Algonquian linguistic group. There were, however, Iroquoian tribes on the northeastern fringe, including the important Hurons in southern Ontario and Siouan tribes (especially the Sioux and the Osage) on the western prairie margin. The forest Native Americans were semisedentary in pattern, their economy combining hunting with farming (corn, beans, squash, and tobacco). In some areas, as with the Hurons north of Lake Erie, a considerable section of forest was cleared for agriculture, although in most cases such cleared land had again reverted to woodland between the times the Native Americans were expelled and the white settlers arrived in any numbers.

French Exploration and Settlement

Most early explorers and pioneering fur traders in the region were French. During the seventeenth and the early part of the eighteenth centuries various French individuals and expeditions explored most of the major waterways of the Heartland.

They were primarily interested in fur and wanted to monopolize the fur trade; consequently, they helped various Native American tribes to keep colonial settlers east of the Appalachians.

The French made little attempt at colonization, but eventually they founded a number of settlements that were the first towns of the region. Most started out as trading posts, forts, or missions and were located along the Mississippi River, the Wabash River, or near the Great Lakes. Cahokia and Kaskaskia, in what is now Illinois, date from 1699 and became thriving wilderness towns in a short while. Vincennes, on the Wabash River, was founded soon afterward. In 1701 the French established a fort where Detroit now stands, but it was not incorporated as a village for another century. The only original French settlement in southern Ontario that still exists was also on the Detroit River. St. Louis, founded in 1764, was another French trading post; it soon became the principal settlement in the Upper Mississippi Basin.

The Opening of the Midwest to Settlement

Britain gained title to most of the present Heartland Region in 1763 by overwhelming the French in Canada. A Native American alliance, led by the Ottawa tribe, was immediately formed to keep the British out of the region. The tribes destroyed eight British forts and laid siege to Detroit. Eventually the siege was raised, but the British government agreed to reserve that part of the continent between the Appalachians and the Mississippi solely for Native American occupance— a highly impractical resolution, considering the sentiment and politics of the time. Within two decades the American colonies successfully revolted against the Crown, and the new nation inherited control of the trans-Appalachian Midwest by virtue of the surrender of land claims by the states of the eastern seaboard.

The first Congress of the United States drew up ordinances in 1785 and 1787 to provide for the systematic survey and disposition of lands in the "Northwest Territories"—the territory northwest of the Ohio River—that proved to be some of the most enduring legislation ever promulgated. A grid system, based on principal meridians and baseline parallels, was staked out to divide the entire area into a township-and-range pattern, a township to consist of 36 sections of 640 acres (1 square mile) each (Fig. 12-7). Thus the land could be accurately surveyed and realistically sold on a sight-unseen basis. The result of this surveying system can be seen in the field, land ownership, settlement, and road patterns of most of the Heartland today.

Provision was soon made for a territory to be admitted to the Union as a state, in all respects equal to the original state, as soon as its population reached 60,000. Ohio was admitted under this provision in 1803, although both Kentucky and Tennessee had already become states in the 1790s.

Prior to the American Revolution, white settlement had begun on a small scale in the Bluegrass portion of Kentucky and parts of Tennessee; these areas attracted a small flood of settlers immediately after the Revolution. There was also a considerable influx into Upper Canada (southern Ontario) at this time, mostly people whose property had been confiscated in New York and other revolutionary colonies; some 10,000 such Tories settled along the upper St. Lawrence and at either end of Lake Erie by 1783.

The Ohio River was a major artery of movement during this period. Fort Duquesne had already evolved into Pittsburgh. Louisville was founded at the falls of the Ohio in 1779. Cincinnati began a decade later and soon became the principal river town.

Ohio did not attract many settlers until the recalcitrant Native Americans had been dealt with. During the early 1790s several thousand soldiers fought a series of battles with the "lords of the forest," finally achieving a decisive defeat of the Miamis and their allies in 1794. This action opened a floodgate of immigration and within less than a decade Ohio had enough inhabitants to become a state. Cleveland was founded in 1796. By the War of 1812 there were more than a quarter of a million people in Ohio, although the only places of significant settlement farther west were in southern Indiana and the Mississippi Valley below St. Louis.

Westward Expansion

During the War of 1812 Tecumseh, the Shawnee chief, tried to organize a Great Lakes-to-Gulf Native American confederation to fight the United States. He was only partly successful in his mission but did persuade many tribes, from the Creeks in Alabama to the Chippewas in the Lake Superior country, to join the alliance. The venture ended in 1813 in a battle in Ontario; Tecumseh was killed and the confederation collapsed. After the war most Heartland Native Americans

were deported west of the Mississippi, where they were promised that the land would be theirs forever.

Settlement in the Heartland basically flowed from three fountainheads: (1) New England, whose Puritans came by

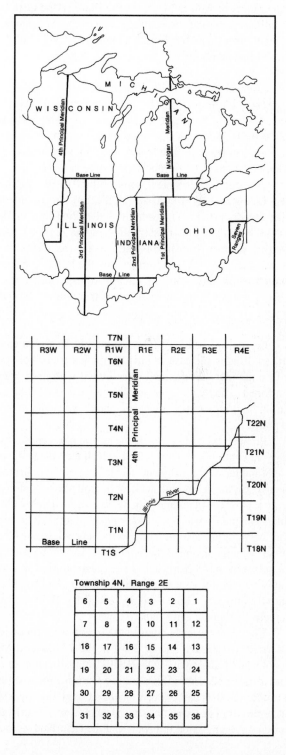

FIGURE 12-7   A range-and-township pattern of land survey was begun in the lands west of the Appalachians in 1785. A rectangular grid system, surveyed from principal meridians and base line parallels, was started in seven rows of townships ("the seven ranges") in eastern Ohio and gradually extended westward. All the basic survey of the area shown here was completed within half a century. Detail of the range-and-township pattern in a portion of western Illinois is shown in the second drawing as an example of the whole. The third drawing shows the 6-mile-by-6-mile grid of a typical township, with the sequential numbering of the sections.

way of the Mohawk Valley; (2) the South, whose frontiersmen broke through the mountains of Kentucky and Tennessee via the Cumberland Gap; and (3) Pennsylvania, whose Scotch-Irish and Germans came via Pittsburgh and the Ohio River country as well as by way of the Cumberland Gap. Only 1 million Americans were living west of the Appalachians in 1800 (fewer than 50,000 of this total in Ontario), but their numbers had increased to 2.5 million by 1820 and to 3.5 million by 1830.

After the War of 1812, settlement rapidly expanded in the forested portions of Indiana and Illinois; these two states were admitted to the Union in 1816 and 1818, respectively. The Driftless Area of Wisconsin attracted settlers because of discoveries of lead ore. The Missouri Valley attracted enough settlement so that Missouri became a state in 1821.

The tide of settlement soon shifted to the southern Great Lakes area. Fort Dearborn (Chicago) was founded in 1816; Toledo in 1817. The opening of the Erie Canal in 1825 was the harbinger of a reorientation of the regional transportation pattern from north–south along the rivers to an east–west axis, which was furthered by the building of more canals and the beginning of railway transportation in the 1840s. The major flow of settlers was then via the Mohawk corridor (Erie Canal route) from New York and New England.

On reaching the forest–prairie margin, in such areas as northern Illinois and eastern Iowa, migrant settlers were puzzled by the fact that the land was clothed with grass rather than forest. They were even suspicious, reasoning that soils bearing no timber must be inferior, and they usually avoided the prairie.

Pioneers who did settle on the prairie often chose tracts that were contiguous to forest land, which they made the real base of the farm establishment. The taming of the prairie was not easy, for the heavy soil stuck to the iron plows then in use until the plow could not move in the furrow, and many plows broke. The prairie was not really conquered until 1837, when John Deere, a blacksmith living in the tiny village of Grand Detour, Illinois, invented the steel plow. It soon became apparent that the good crop yields of the deep prairie soil would readily pay for the cost of breaking the sod. And so Illinois and Missouri was almost totally settled by 1850, as were southern Wisconsin and southern Iowa.

The more westerly cities in the region were founded at this time: Milwaukee and St. Paul in the 1840s, and Minneapolis and Kansas City in the 1850s. Intensification of settlement proceeded rapidly; southern Ontario, for example, had more than 40 percent of Canada's population by 1861. But as the frontier moved westward beyond the Heartland, a continual out-migration followed it; if the Heartland was easy

to move into, it was also easy to move out of. This was particularly true of the Ontario portion of the region.

In the second half of the nineteenth century, Chicago emerged as the primate city of the Heartland. Its preeminent location at the south end of lake Michigan gave it a crossroads position that allowed it to dominate the fertile agricultural hinterland that extended outward in most directions from the city. Chicago developed into the nation's dominant railway hub at this time. Twenty-six mainline tracks radiated from the Chicago hub, each with Chicago as its terminus. Literally you could get *to* Chicago from anywhere, but you could not go *through* Chicago on any railway; all lines terminated there. By 1890 1 million people lived in Chicago.

Immigrants to the Cities

After about 1880 there was a greatly increased flow of European immigrants to the Heartland, which continued through the early decades of the twentieth century. By 1920 one-sixth of the region's population was foreign-born. Generally, these immigrants settled in urban areas, often giving them a distinctive ethnic flavor, such as Germans in Milwaukee or Scandinavians in Minneapolis.

In many ways, the Heartland has become the true melting pot of North America. The midwestern blend of culture, speech, and lifestyle is the one most readily identified as the standard for the United States; similarly, southern Ontario represents Anglo-Canada.

THE INCREDIBLE OPULENCE OF HEARTLAND AGRICULTURE

Although the economy of the region is not primarily agricultural and the regional population is not predominantly rural, the Heartland is easily the preeminent producer of crops and livestock on the continent. Its large expanse of productive agricultural land is unparalleled in North America (Fig. 12-8); and despite the continuing downtrend in farm numbers and farm population, the level of farm output continues to rise. By far the greatest concentration of agricultural counties in the United States is found in this region; this is less so for Canada.

Throughout the greater part of this region the country appears to be under almost complete cultivation, with four or five farmsteads to each square mile in the eastern portion and two or three in the western portion. The farms are based on the subdivision of sections into half-sections, quarter-sections, and 40-acre (16-ha) plots. Corn, winter wheat, soybeans, oats, and hay are almost universal crops. Tobacco and fruits are locally important and much land is in pasture. The

FIGURE 12-8 Environmental conditions are favorable for agriculture in most parts of the Heartland. Fertile floodplain soils, as represented here along the Illinois River, are particularly productive (TLM photo).

region not only grows tilled crops but also supports the densest population of cattle and swine in North America.

FARM OPERATIONS

The typical Heartland farm has been a family-operated enterprise of modest size (a few hundred acres) that is highly mechanized, highly productive, and yields a variety of staple commodities in a system referred to as "mixed" (i.e., crops and livestock) farming. Grains, particularly corn, have been the dominant crops, with much of the output being fed to livestock in a relatively small feedlot that is the cornerstone of the operation.

Many of these characteristics still pertain; however, some notable changes have occurred in the last few years. Mixed farming, for example, is much less prominent. Cash-grain farming is now the principal farming system, and many farms no longer are fenced because there are no livestock to exclude from the cropped fields.

The total area in farms has been on a declining trend for decades in both Canadian and U.S. portions of the region. The number of farms is also decreasing. Average farm size, on the other hand, has doubled in the past third of a century, as the more successful (or more dedicated) farmers buy and especially rent more land. Family farms continue to dominate the region's agrarian scene, although the formation of family-held corporations is increasingly common. About half the nation's employment in agriculture occurs in the Heartland. This proportion has been on an increasing trend primarily because the number of Heartland farmers has been declining at a slower rate than in most other parts of the country. Moreover, more than half of all Heartland farmers now have their primary employment off the farm; i.e., they have part-time (or sometimes full-time) work in nearby towns or cities.

Changing technology made crop diversification and rotation less necessary in the region. The trend to specialize more and diversify less is especially pronounced on the better lands, involving, in particular, a concentration on corn and soybeans at the expense of hay crops and small grains.

In common with agriculturalists elsewhere on the continent, Heartland farmers fell on hard times during the 1980s, owing to a variety of factors, many of them international in scope. Low incomes, declining asset values, and bankruptcy became widespread. Like it or not, North American farmers are relatively high-cost producers by world standards, and the income derived from sale of their commodities partly reflects this fact. Conditions were improving for Heartland farmers in the early 1990s, but only slowly.

Crops

Corn Corn (maize) thrives under the favorable conditions of hot, humid summer weather; fertile, well-drained, loamy soils; and level to rolling terrain. No other country has this favorable combination of growing conditions over such a wide territory; thus the United States produces nearly half the world's corn, most of which is grown in this region.

Genetic and agronomic technological improvements have made crop rotation and diversification less necessary, and today corn is seldom grown in a rotational cycle in this region. The seeds, furthermore, are planted in much greater numbers in more closely spaced rows, and fertilizer and herbicide chemicals are heavily used. The resulting yields in many areas are more than 150 bushels per acre; for the Corn Belt as a whole, the average yield in most years is nearly 100 bushels per acre.

The principal corn-growing areas are still in the heart of the Corn Belt, from central Indiana to eastern Nebraska; this section also achieves the highest average yields. Corn continues as the dominant crop in most of the Heartland; it occupies nearly twice as much acreage as any other crop in the Corn Belt (Fig. 12-9). About 75 percent of all grain corn grown in the United States is in the Heartland; the comparable figure for Canada is about 80 percent.

Soybeans   This shallow-rooted legume is popular because it yields a heavy crop of beans; is valuable for meal and oil; makes good hay, silage, and pasturage; has few diseases; and is not attacked by pests (Fig. 12-10).

No crop in twentieth-century North America has experienced such expanded production as the soybean. The United States is now the world's leading producer (75 percent) and leading exporter (90 percent). Soybeans are second only to corn as a source of cash farm income to American farmers and are usually the leading agricultural export of this country.

Three-quarters of the national output of soybeans emanate from this region, and the six leading states are all Heartland states (Fig. 12-11). Nearly all of Canada's soybeans are grown in the extreme southern part of Ontario, just north of Lake Erie.

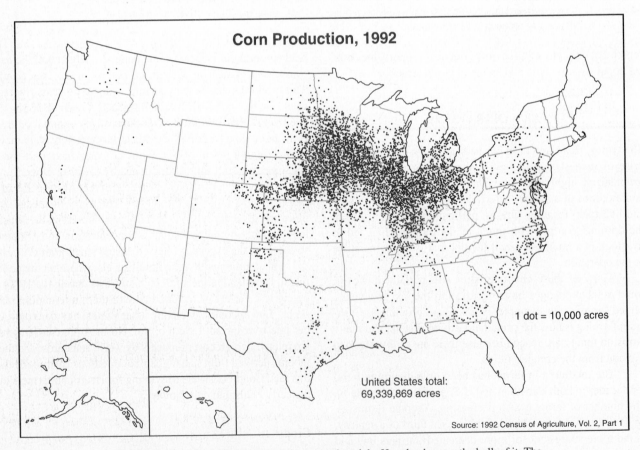

Corn Production, 1992

1 dot = 10,000 acres

United States total:
69,339,869 acres

Source: 1992 Census of Agriculture, Vol. 2, Part 1

FIGURE 12-9   Corn is the great American cereal, and the Heartland grows the bulk of it. The Corn Belt subregion alone produces half the national total.

FIGURE 12-10 A field of soybeans in southeastern Wisconsin, near Sturtevant (TLM photo).

Alfalfa The legume alfalfa is well adapted to the region, especially to the prairie portion where winter rainfall is less abundant and the soils less leached and hence higher in calcium. It thrives best on soils rich in lime. The crop has greatly increased in importance in the Corn Belt and the Dairy Belt since 1920, even in the eastern part of the Heartland.

Because alfalfa is harvested several times each season and recovers quickly after cutting, the per-acre yield exceeds that of any other hay crop. In the short-summer dairy portion of the region alfalfa becomes a very significant crop.

Wisconsin, for example, grows almost as much alfalfa as corn. In most years Wisconsin vies with California as the leading grower of alfalfa.

Other Hay Crops Numerous other crops are also planted for hay production—the so-called tame hays. None approaches alfalfa in acreage or output, but their combined total in the region is almost as great as that of alfalfa. *Clover, timothy,* and *clotim mixtures* are the most widespread types of tame hay after alfalfa. Emphasis on nonalfalfa hay production is greatest in the southern part of the region, especially Missouri and Kentucky.

Tobacco Kentucky is second only to North Carolina in tobacco growing, and output is also significant in and around the Nashville Basin in Tennessee. Both burley and dark-fired tobacco are produced. Although generally tobacco is not considered an important Canadian crop, it is the leading one in several counties bordering on Lake Erie, particularly around Norfolk. Tobacco is the most valuable cash crop in Ontario.

Fruit Commercial growing of fruit is not a widely distributed enterprise but is concentrated in definite localities. Tree fruits, more exacting in climate than in soil requirements, frequently suffer from extremes of temperature; so the best suited areas are those with a minimum of danger from late spring and early fall frosts, notably peninsulas, hillsides, and leeward sides of lakes. This tempering effect of a large body of water gave rise to a fruit belt just east of Lake Michi-

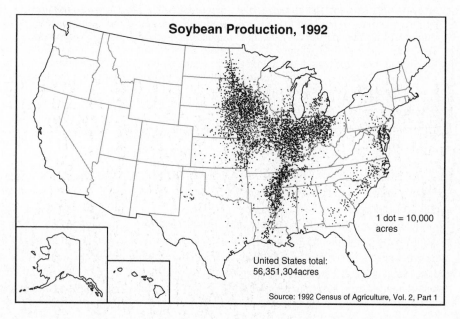

FIGURE 12-11 The United States is the world's leading producer of soybeans, and the Heartland is its outstanding growing region.

Soybean Production, 1992

1 dot = 10,000 acres

United States total:
56,351,304acres

Source: 1992 Census of Agriculture, Vol. 2, Part 1

A CLOSER LOOK   Preserving Canada's Fruitlands—The Niagara Fruit Belt and the Okanagan Valley

SOFT-FRUIT-GROWING AREAS IN CANADA ARE LIMITED

The areas in Canada that can grow grapes and soft fruits such as apricots, peaches, and sweet cherries are severely limited by climatic conditions.

The Niagara Fruit Belt, a narrow strip of land along Lake Ontario, stretching from Hamilton to the Niagara River, lays claim to fame as the best grape and soft-fruit-growing region in all of Canada. The water of Lake Ontario retains its warmth well into the middle of the winter and moderates the cold air masses that sweep in from the north. In spring, the lake is slow to warm up and thus keeps the weather cool enough to retard the opening of fruit blossoms until after most risk of frost is over. The slope of the land provides air drainage (cold air runs downhill just like water), which further protects against frost damage on clear, cold nights. It may surprise you to learn that Niagara has less risk of spring frost damage to peaches than has the "peach" state of Georgia.

The Okanagan Valley, tucked in the mountain-and-plateau country of south-central British Columbia, rates second only to the Niagara Fruit Belt for the growing of grapes and soft fruits. It is also internationally renowned for its production of apples, particularly the red Delicious variety. The winters of the Okanagan are moderated by a great deal of cloud cover, and the sloping sites of the broad terraces above the val-ley floor provide excellent air drainage to protect against spring frosts.

URBAN SPRAWL IN CANADA'S FRUIT BELTS

In Canada, cities are located on the country's best agricultural land. In fact, 50 percent of Canada's population lives on the best 5 percent of its farmland. To make matters worse, urban growth has occurred in a low-density urban sprawl pattern that consumes, and sterilizes for agricultural production, much more land than would be required with more orderly and compact development. Nowhere is this low-density urban sprawl more evident than in the two major soft-fruit-growing regions.

In the Niagara Peninsula, several major cities and numerous towns are located on the prime fruit-growing strip along Lake Ontario. The Queen Elizabeth Way (QEW) not only takes up valuable land but also has facilitated the sprawling of urban land uses right across the fruitland. The greatest density of urban uses is found on the welldrained sandy soils that are required for crops such as peaches and sweet cherries and vinifera and hybrid grapes. Several studies have shown that by directing urban growth into non-fruit-growing areas and requiring contiguous compact urban development, there is room in the Niagara Peninsula for a great amount of urban development without completely destroying the land resource for fruit growing.

The degree and pattern of urban sprawl in the Okanagan Valley is strikingly similar to that of the Niagara region. Considering the fact that the Okanagan has a population of only about one-fifth that of Niagara, the degree of low-density sprawl is relatively greater in the Okanagan than in the Niagara case. As in Niagara, most of the urban sprawl is occurring on the best fruitlands. Also, as in Niagara, a major highway goes right through the best fruitlands, thus facilitating urban sprawl on those lands. However, unlike in Niagara, in the Okanagan the terrain does not provide for feasible alternative locations for a highway and urban development.

Visible urban expansion can be mapped; the indirect effects on the fruit-growing industry are more difficult to assess. Although we cannot measure quantitatively the indirect impact of urbanization, interviews with hundreds of Niagara and Okanagan fruit growers yield the following list: vandalism, crop pilfering, trespass, complaints from nonfarm residents about spraying and noisemakers (to scare off birds), high land prices that preclude farm expansion, and high land taxes to pay for urban type services.

Besides the problems associated with urbanization, the fruit-growing industry faces many other hazards and problems: vicissitudes of weather (frost, hail, poor weather for bee pollination at blossom time, droughts, windstorms, rain at harvest time), disease, insects, and lack of skilled workers. All of the above are se-

---

gan (particularly apples and cherries), just south of Lake Erie (primarily grapes), and just south of Lake Ontario (especially apples and cherries).

The Niagara Peninsula in Canada, benefiting from the same climatic principle, is famous for fruits, particularly grapes and peaches. The narrow lake plain between Hamilton and the Niagara River is one of only two major areas in Canada where tender fruit crops can be produced.

The Livestock Industry

Livestock raising and feedlot operations are widespread throughout the region (Fig. 12-12). As a generalization, however, it can be stated that livestock feeding dominates in the west, particularly in Iowa, where both beef cattle and hog feedlots are ubiquitous. Feedlots are also notable in the east, especially in Indiana where hogs dominate the scene. In the rious but they are usually taken in stride by the growers; they are part of the business of farming.

All fruit growers agree that their most serious problem is the ever-tightening cost-price squeeze. Costs of things such as energy, fertilizer, chemicals, and equipment have skyrocketed in the last several decades. On the other hand, prices received for fruit have been kept low by competition from foreign countries where costs are lower and the agricultural industry is more heavily subsidized. The General Agreement on Tariffs and Trade (GATT) prevents Canada from protecting the fruit-growing industry with trade tariffs. The Canada–U.S. Free Trade Agreement (1989) and the North American Free Trade Agreement (1992) have resulted in even stiffer competition for Canadian fruit growers.

When growers are not receiving a reasonable return on their investment of capital, management, and labor, it is not surprising that many of them welcome the opportunity of selling their land at high prices for urban purposes. By 1996, over half of the Niagara and Okanagan fruitlands had been ruined for agricultural purposes by urban development.

If the fruit-growing and related processing and wine industries were to collapse, there would be serious economic dislocation in the Niagara Peninsula and the Okanagan Valley. Conservationists are even more concerned that the collapse of the fruit-growing industry would result in the destruction of scarce and irreplaceable land resources that can produce the widest range of crops in Canada.

FRUITLAND PRESERVATION MEASURES

In Canada, provincial governments are responsible for resource management and land-use planning. Since the late 1970s, the province of Ontario has had a policy that states that development must not occur on prime farmland if any other reasonable alternative is available. The Regional Municipality of Niagara (similar to a county government) implements this policy through its Official Plan and Zoning Bylaws. By the early 1990s, fruit growers were experiencing serious financial losses and began pressuring the regional and provincial government to relax the land preservation policies so that they could profit from selling their holdings for urban development. The provincial government of that time (the New Democratic party) responded by initiating a land conservation easement program that would pay the farmers for "development rights" and thus permanently preserve the land for agriculture.

In 1995, before the conservation easement program came into effect, a new Conservative government was elected, and it promptly canceled the program. To date, the provincial and regional farmland preservation policies remain in force, but the fruit growers have renewed their demand to be permitted to sell farmland for development.

In British Columbia, the provincial government passed strong legislation to preserve agricultural land with its *British Columbia Land Commission Act* (1973). The act resulted in the establishment of Agricultural Land Reserves in which all good agricultural land is preserved for farming and related uses. The *Land Commission Act* was accompanied by the *Farm Income Assurance Act,* which provides for the payment of indemnities to farmers when the price of produce falls below the cost of production.

Despite continued opposition by developers and some fruit growers, and despite several changes in provincial government, the Agricultural Land Preserve legislation has succeeded in greatly reducing the urbanization of fruitland in the Okanagan Valley.

Although conservationists have won some major battles in the struggle to preserve the Niagara and Okanagan fruitlands, it is not safe to assume that these resources have been preserved in perpetuity. Governments, policies, and legislation can change. The long-term preservation of these and other renewable resources depends on continuously informed and alert citizens who must insist that governments implement the necessary conservation measures.

Professor Ralph R. Krueger
University of Waterloo
Waterloo, Ontario

central part of the region (Illinois and vicinity) livestock feeding is less important than cash-grain farming.

Beef Cattle   Beef cattle are most numerous in the western part of the region—the old prairie portion—for, unlike swine, they are essentially grass eaters. Contrary to the common notion that the range states provide only the grazing and breeding lands and the Corn Belt the fattening areas, the fact is that about two-thirds of the animals slaughtered within the Corn Belt are bred in it.

Feeding is a major agricultural enterprise in the western part of the region, the animals being carried through the winter on hay and other home-grown feeds and fattened on corn. Most cattle feedlots in this region are relatively small, fattening less than 100 animals annually. This is in marked contrast to the situation in other regions, particularly the

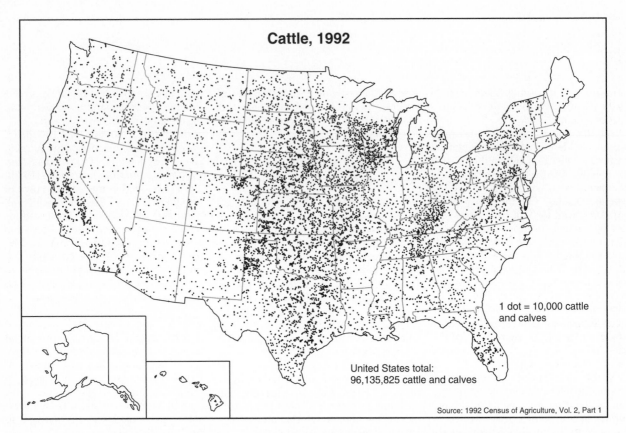

FIGURE 12-12   Although cattle are widely distributed over the nation, the greatest concentrations occur where there is an overlap of Corn Belt beef cattle with milk cattle of the Hay and Dairy belt.

West, where there is a strong trend toward ever-larger feedlots.

Swine   Of all large domesticated animals, swine most efficiently and rapidly convert corn into meat. Spring shoats, for example, are ready for market in approximately 8 months. During the feeding period they gain in weight from 1 to $1\frac{1}{4}$ pounds (0.45 to 0.56 kg) per day. Two-thirds of the swine in the United States are raised in this region.

The generally smaller farms in the eastern part of the region often emphasize pork rather than beef production, for swine require less space than cattle (Fig. 12-13). Even so, the largest numbers of swine are found where there is the greatest production of corn; thus Iowa has more than twice as many hogs as the second-ranking state, Illinois (Fig. 12-14).

Dairy Cattle   Dairying is widespread in the Heartland but is dominant only in the northern part of the region and around the major cities (Fig. 12-15). In Wisconsin and

FIGURE 12-13   A typical hog lot in central Iowa (Sylvan Wittwer/Visuals Unlimited photo).

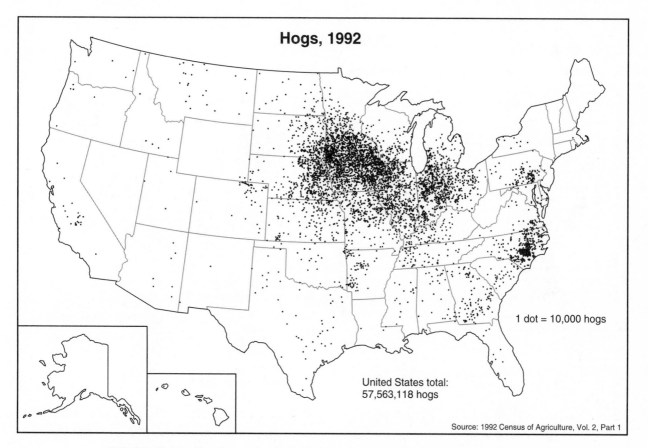

FIGURE 12-14   Corn is a major feedstuff for swine, so the distribution pattern of hogs is similar to that of corn.

Minnesota the principal products are manufactured items—butter, cheese, evaporated milk, dried milk—to combat the high cost of shipping fluid milk. Dairy farms tend to be relatively small and often involve marginal land that is not well suited for crop farming but is satisfactory for pasture. Many dairy farmers now also raise and fatten dairy animals for sale as beef; this activity has to some extent replaced the secondary raising of hogs on dairy farms, which was a traditional activity.

The National Government and the Farmer

Today in both Canada and the United States agriculture, that "last stronghold of free enterprise," is ever more dependent on the government. Within the last six decades a bureaucracy of astounding magnitude has evolved, doing many worthwhile things for the farmer but enmeshing agriculture in an endless series of quotas, regulations, and artificial conditions.

Direct government influence on farming occurs in many ways and differs somewhat in the two countries. The geography of agriculture, however, is most likely to be affected by the following practices.

Cost Sharing of Conservation Practices.   Interested farmers can obtain advice, plans, and technical assistance for soil conservation from the government. Also, the government pays approximately half the cost of establishing certain conservation practices on farmers' lands. In addition, farmers can take some acres out of cultivation and receive direct payments for not growing crops. All these measures are beneficial for overused and eroded land. But at the same time the high productivity of intensified farming on the better-cultivated land offsets the tendency toward decline in total production that would be an anticipated effect of conservation practices.

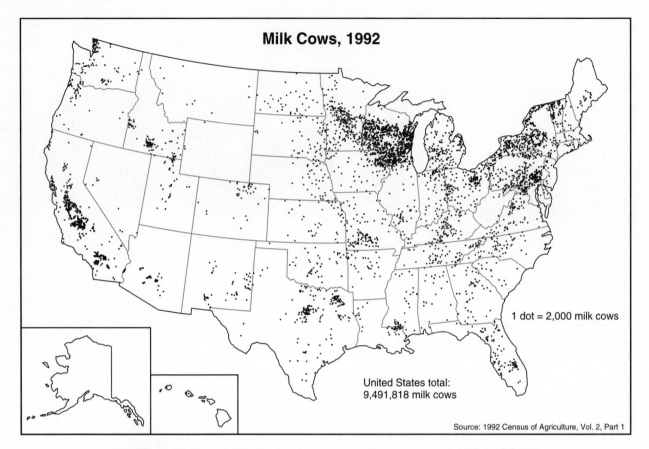

Milk Cows, 1992

1 dot = 2,000 milk cows

United States total:
9,491,818 milk cows

Source: 1992 Census of Agriculture, Vol. 2, Part 1

FIGURE 12-15   Dairy cattle are particularly concentrated in a northern belt from Minnesota to New England, with the largest numbers in the western lakes states of Wisconsin and Minnesota.

Farm Credit.   U.S. and Canadian governments make loans to their qualifying farmers who cannot obtain credit from conventional sources.

Crop Insurance.   Both governments set up crop insurance agencies to indemnify farmers for crop losses due to "acts of God."

Acreage Allotments and Marketing Quotas.   Most major crops are now grown under some sort of allotment and quota system whereby maximal acreages are designated for each year's harvest.

Price Supports.   Several basic agricultural commodities in both countries achieve what amounts to a guaranteed sale on the basis of federal price supports, regardless of national or international market conditions. In the early 1980s the U.S. government embarked on an innovative but costly program designed to make agriculture less reliant on govern-

ment intervention and more market oriented. Instead of making direct payments to farmers, the government provides grain (corn, wheat, sorghum, or rice) or cotton from government surplus stocks to farmers, who reduce their own plantings of these crops. This process is referred to as "payment in kind" or PIK. The farmer is then able to sell these commodities on the free market. More than one-third of all eligible farmers opted to participate in this program in its first year of operation (1983). Government-owned crop surpluses diminished rapidly and cash outlays were reduced, but foreign competitors complained vociferously about commodity "dumping," and overall costs to the U.S. government actually increased.

Export Controls.   In a highly fluctuating pattern the two governments sporadically impose export controls on certain basic commodities. An aggressive program to restimulate export demand for farm products was started in the United

States, primarily through "blended credit," which has the effect of reducing the interest cost component of credit sales.

What to do about the complex federal farm programs (which are grossly oversimplified in this brief discussion) is enigmatic. Friend and foe alike agree that they are too big, too all-pervading, and too expensive. And yet neither Republicans nor Democrats nor Conservatives nor Liberals nor Social Crediters nor Independents have been able to devise a scheme for getting government out of agriculture without wrecking the farm economy.

As the leading producer, consumer, and exporter of grain in the world, the United States exerts an inordinate influence on international grain trade. Its long-established philosophy of management by exception (that is, intervening in response to perceived needs) never worked too satisfactorily and the likelihood of a satisfactory program is increasingly remote as the agricultural sector of the economy becomes ever more complex and heterogenous.

MINERALS

Although relatively inconspicuous in this region of remarkable agricultural and industrial output, mining is also an important enterprise in many localities.

Coal

Coal deposits are widely distributed, and one of the nation's leading provinces, the Eastern Interior, is in this region. The Eastern Interior Province is second (although distantly) only to the Appalachian Province as a coal producer. Illinois is the fourth-ranking coal state, with about 10 percent of total national output. Mines in western Kentucky and southern Indiana together yield slightly more coal than Illinois, and their output is increasing. In most years Muhlenberg County in Kentucky produces nearly twice as much bituminous coal as any other county in the nation. Most of the coal in the Eastern Interior Province is obtained from strip mines.

Petroleum and Natural Gas

Important oil and gas reserves have been tapped in Michigan, Ohio, Indiana, Kentucky, and Illinois. Production is scattered, however, and the region as a whole yields only 2 percent of total U.S. output; about half comes from Illinois.

The only oil and gas of any consequence in eastern Canada are found in this region, somewhat scattered in extreme southern Ontario. Oil output amounts to less than 1 percent of the national total, but further modest discoveries

are anticipated. The proportion of the national output of natural gas is somewhat higher.

Limestone

In south-central Indiana, in the Bedford–Bloomington area, are the famous limestone quarries that supply a superior limestone used in buildings throughout the East and Midwest. About three-fifths of the dimension (block) limestone of the country comes from these quarries. The building-stone business is declining, however, for it suffers from competition with cheaper concrete, brick, and lumber.

Salt

Salt occurs widely in North America, but major deposits are in the vicinity of the southern end of Lake Huron. Michigan is one of the leading states and Ontario is the leading province in salt output of their respective countries. The salt is obtained in solid form by underground digging and in the dissolved state by the modified Frasch process. There are several mines in Michigan, but the principal output is actually beneath metropolitan Detroit. The Ontario mines are at Goderich, Watford, Sarnia, and Windsor.

HEARTLAND MANUFACTURING

Much of the character, prestige, and reputation of the Heartland Region was derived from two outstanding components of its economy—agriculture and manufacturing—that have been in decline for some years. Of the two, manufacturing has suffered the most severe and long-lasting deterioration. The Heartland was the preeminent industrial region of the continent for many decades, but both the relative significance and absolute significance of manufacturing have plummeted. Industrial output is still of great importance to the regional economy, but, as is true in most regions, its dominant role has been superseded by the tertiary sector (services and trade).

For more than a century the Heartland was recognized as one of the great industrial domains of the world. Its metropolises—Chicago, Toronto, Detroit, Cleveland, St. Louis, Milwaukee—were the classic industrial cities of the subcontinent, blue-collar urban nodes with a multiplicity of workshops large and small (Fig. 12-16).

By the 1980s, however, the situation had changed and is continuing to change. The industrial dominance of the Heartland is waning rapidly, challenged by vigorous industrial growth in the South and West and by the subcontinent-

FIGURE 12-16 Great Lakes Steel Corporation mill on the bank of the Detroit River (John Sohlden/Visuals Unlimited photo).

FIGURE 12-17 An automobile assembly line in Sterling Heights, Michigan (courtesy Chrysler Corporation).

wide decline of manufacturing in favor of service industries. The bellwether industries of the region—steel and automobiles—fell on hard times, experiencing reduced demand that resulted in many shutdowns and layoffs. Most other types of manufacturing did not suffer such sharp declines, but almost all were affected negatively.

The development of high-technology manufacturing in other localities was particularly significant because these industries (electronics, computers, aerospace, instrumentation, pharmaceuticals, etc.) had been flourishing at twice the rate of that of manufacturing as a whole. Although high-technology industries expanded somewhat in the region, the major high-tech growth areas are in California, Texas, Florida, and other places far removed from the industrial heartland.

Meanwhile, the "core" industries of the region are severely depleted. Most notable has been the decline of steel production. Because of low-cost foreign competition, high-cost output, diminished domestic demand, and other factors, steel production in the United States and Canada nosedived. By the mid-1990s, employment in U.S. and Canadian steel mills was less than one-third what it was three decades earlier and one-half the total of just 15 years previously. In the United States more than a quarter of a million steel industry jobs vanished. In Canada the leading steel center, Hamilton, lost 25,000 jobs.

The automotive industry also suffered notably. Massive layoffs took place throughout the prime manufacturing areas, particulartly in Michigan, Ontario, and Ohio. New auto manufacturing plants were established in such southern states as Tennessee, Kentucky, and Georgia, largely in response to lower labor costs.

Leading Industrial Centers

Only 7 Heartland metropolitan areas rank among the top 20 industrial centers of the continent.

Chicago surpassed New York in the last half decade to become the largest manufacturing center of the subcontinent. Heavy industry predominates. The fountainhead of all manufacturing was the great primary iron and steel industry at the southern tip of Lake Michigan. This was the meeting place of iron ore brought by lake carrier from the Minnesota iron ranges and coal railed north from West Virginia, Kentucky, and southern Illinois. As the domestic steel industry faltered in recent years, Chicago was less severely affected than most other centers.

Although justly famous for its primary steel industry now and its meatpacking industry in the past, its prominent types of manufacturing today are machinery output and metal fabrication. Chicago is the subcontinent's leading center in the fabrication of metals, that relatively prosaic heavy industry in which primary metal (mostly sheets, bars, and rods of steel) is shaped and fashioned into pipes, screws, wire, beams, cans, and other products of specific use. However, Chicago's leading type of manufacturing is machinery.

Despite a significant decline in recent years, *Detroit* still ranks as the fourth largest industrial center of the subcontinent (exceeded only by Chicago, Los Angeles, and New York). Detroit became the first great automotive production center because several pioneers of the industry—notably Henry Ford and Ransom Olds—got their start there. Ford was particularly instrumental in the rise of the industry in Detroit: He developed an automobile cheap enough for the average family; he adapted the assembly line to the industry; and he introduced standardization and interchangeable parts, thereby making mass production possible. A number of nearby cities, such as Pontiac and Ypsilanti in Michigan and Windsor in Ontario have parts manufacturing and assembly plants that buttress Detroit's position (Fig. 12-17).

The twin cities of *Minneapolis–St. Paul* have experienced less industrial decline than most urban areas and now rank as the sixth largest manufacturing center of the subcontinent. Minneapolis was long noted for its flour milling and farm machinery manufacture, and St. Paul for its meatpacking. In recent years a variety of other industries have brought much diversity to the industrial structure of the Twin Cities.

Cleveland is another heavy manufacturing center that has experienced industrial decline but still ranks among the 10 leaders of the continent. Both steel and automobiles are prominently produced in Cleveland, but its leading industry is machinery production.

Canada's largest industrial center is *Toronto*, which ranks about twelfth among the subcontinent's manufacturing cities. Overall, Toronto has one of the most diversified industrial structures in North America.

St. Louis also has a well-diversified industrial structure, with an emphasis on aircraft and automotive production.

Milwaukee is primarily a heavy manufacturing center, but its industrial reputation comes particularly from its brewing industry.

Other Heartland cities that rank among the 40 leading industrial centers of the subcontinent include *Grand Rapids, Cincinnati, Indianapolis, Rochester, Dayton*, and *Hamilton*.

TRANSPORTATION

The extensive and busy Heartland Region is well served by transportation facilities. Gentle relief and the lack of topographic barriers made the construction of surface transport lines relatively easy, partly accounting for the dense networks of roads, railways, and pipelines in the region. In addition, the two outstanding natural inland waterway systems of North America, the Great Lakes and the Mississippi River drainage, are largely within the region, and numerous canals have at one time or another helped to augment the waterway transport system.

Railways

The first railroads in the Midwest were built not as competitors to navigable waterways but as links connecting them. The rapid extension and improvement of railway facilities after 1850, however, profoundly changed agriculture and revolutionized the whole course of internal trade. By 1860 railroads had triumphed over inland waterways and since then port rivalries have been expressed in the competition of the railroads serving them. Railroads, by spanning the great interior with a network of steel rails, also stimulated the growth of cities and the development of manufacturing.

The present railway network is dense, but its major flow is east–west (Fig. 12-18). The main lines principally connect the eastern metropolises of New York, Philadelphia, and Toronto–Montreal with the major midwestern hubs of Chicago and St. Louis. The principal north–south traffic is associated with the Mississippi Valley axis of Chicago–St. Louis–New Orleans.

Since World War II there has been a long-term downtrend in railway usage in the Heartland, as over most of North

FIGURE 12-18 A major railway bridge crosses the Mississippi River just below a dam near St. Louis (TLM photo).

America. Passenger traffic is limited. Many short lines have been abandoned and trackage has been considerably consolidated. The role of the railroad today, more than ever before, is to take bulk commodities, particularly mineral and agricultural products, on long hauls. Most of the short-haul freight business, as well as a considerable amount of long-haul traffic, has been lost to truckers.

Roads

The truck and the automobile revolutionized transportation in the twentieth century almost as much as the railway did in the nineteenth. Motor trucking operations benefit from low capacities, high speed, and flexibility of route; therefore they can provide frequent shipments at relatively low cost. Their principal advantage over railroads is on short hauls, although they can often compete on intermediate and even long hauls.

Most parts of the Heartland are well served by highways and roads, for both long-distance and local travel. Several major toll roads were built in this region in the 1940s and 1950s in an effort to speed cross-country traffic and alleviate congestion around cities. Later, however, the idea of toll roads was virtually abandoned because of the construction of the national interstate highway system, which was essentially completed in 1978 [42,000 miles (67,000 km) at a cost of $70 billion]. All large and most medium-sized cities in the nation are connected by this system, normally with a four-lane divided controlled-access highway. A large portion of this roadway system is located in the Heartland Region.

Inland Waterways

River, lakes, and canals have been and, in many cases, continue to be of considerable importance for transportation in the Heartland Region.

Rivers   Rivers, the principal routeways of pioneer days, were used whenever possible in preference to hard and slow overland routes. Their chief advantages as highways were low cost and convenience. Many a stream that now seems too small or shallow for transportation was extensively used, and many a settlement would have died out had there been no stream over which to float products to market. Almost all large western communities in the period from 1800 to 1850 were located on the Ohio or the Mississippi.

The Ohio River   The channel of the Ohio River was not navigable during the droughts of late summer until the federal government established a permanent 9-foot (2.7-m) stage with a system of four dozen dams that back the water into a succession of lakes deep enough for navigation. Locks permit boats to get around the dams.

The Ohio thus has become one of the subcontinent's leading carriers of waterborne freight. Thousands of commodious barges, shackled in tows, are propelled over the river at all seasons except for several days in spring when the water is too high or in winter when ice is hazardous. The tonnage is several times greater now than it was at the height of the steamboat period. Bulk products constitute 95 percent of the total freight: coal, coke, ore, sand and gravel, stone, grain, pig iron, and steel.

The dams are several decades old and their locking systems are slow and cumbersome. Replacement construction has begun, however, according to a master plan that calls for a 60 percent reduction in the number of dams and locks on the river, which will result in longer pools and considerable time saving for the carriers.

The Mississippi and Missouri Rivers   Barge traffic is also significant on these two rivers, located in the western Heartland. Minneapolis is the head of navigation on the Mississippi, and Sioux City is the effective head of navigation on the Missouri. There are more than 50 flood-control and power-generation dams on the two rivers, with locks to let the barges pass through. Most of these dams or locks are also antiquated; during the busy shipping season bottlenecks create significant traffic jams.

Canals   During the nineteenth century literally dozens of canals were in use in the Heartland. Most were small and short, serving as connectors between rivers or lakes. A few, however, were of considerable length. Most have long since been abandoned, but a handful still play a vital role in freight transportation.

The longest of the Heartland canals is still in use. The Erie Canal crossed New York state from east to west to connect Lake Erie with the Hudson River, following the only practical low-level route through the Appalachian barrier. It was opened in 1825, became the busiest inland waterway in the world, and was a major factor in making New York City the dominant Atlantic port. Railway competition eventually caused it to decline in importance. The entire waterway was reconstructed as the New York State Barge Canal, which was completed in 1918. Traffic today is only moderate.

THE URBAN SYSTEM OF THE HEARTLAND

In no other major region of North America is there such regular development of an urban hierarchy as in the Heartland (see Table 12-1 for a listing of the region's largest urban places). The concept of city and hinterland is prominently

FIGURE 12-19   A lift bridge on the Welland Canal, near Niagara Falls, Ontario (courtesy Ontario Ministry of Industry and Tourism).

displayed in most parts of the region, and there is a "finested hierarchy" of at least half a dozen levels of magnitude from metropolis to village. The almost classical regularity of the pattern in the Corn Belt and the western margins of the region is interrupted by three major factors that counterpoint the scheme:

1. The Great Lakes, which constrict the areal pattern, provide an important functional connection with overseas regions and stimulate urban development on lakeshores to an unusual degree.

2. Major deposits of economic mineral resources (coal and petroleum) distort the pattern with regard to the distribution of smaller cities and towns.

3. The presence of an important international boundary essentially divides the pattern into two separate systems, one of which has a significant influence on the other.

It would require a detailed analysis of the economic base and function of each individual city to determine its exact position in the hierarchy; even then the position would depend on the stated limits of the scheme. Here we can note only some major components and guess about the position of individual cities.

1. Although the economic influence of New York City is felt to some extent throughout the region, particularly in the eastern portion, *Chicago* is clearly the dominant metropolis. Its position as the wholesaling center and transportation hub is an obvious indicator of its primacy in any hierarchical urban system for the Midwest.

2. Subregional metropolises, at the second level of magnitude, would probably be represented by Toronto, Detroit, and St. Louis. *Toronto's* sphere is essentially limited to the Ontario portion of the Heartland, but as one of the two primate cities of

TABLE 12-1

Largest Urban Places of the Heartland Region, 1995

Name	Population of Principal City	Population of Metropolitan Area	Name	Population of Principal City	Population of Metropolitan Area
Akron, OH	227,000	681,000	La Crosse, WI	49,000	125,000
Anderson, IN	62,000	135,000	Lafayette, IN	47,000	171,000
Ann Arbor, MI	111,000	512,000	Lansing, MI	129,000	436,000
Appleton, WI	71,000	342,000	Lawrence, KS	64,000	90,000
Battle Creek, MI	57,000		Lexington, XY	239,000	436,000
Benton Harbor, MI	15,000	165,000	Lima, OH	47,000	158,000
Bloomington, IL	51,000	139,000	Lincoln, NE	207,000	240,000
Bloomington, IN	57,000	119,000	London, ON	280,000	419,000
Buffalo, NY	326,000	1,190,000	Lorain, OH	73,000	
Canton, OH	88,000	406,000	Louisville, KY	281,000	988,000
Cedar Rapids, IA	117,000	176,000	Madison, WI	211,000	394,000
Champaign, IL	63,000	179,000	Mansfield, OH	55,000	181,000
Chicago, IL	2,770,000	7,715,000	Milwaukee, WI	620,000	1,458,000
Cincinnati, OH	368,000	1,590,000	Minneapolis, MN	366,000	2,711,000
Clarksville, TN	75,000	182,000	Muncie, IN	74,000	123,000
Cleveland, OH	508,000	2,225,000	Nashville, TN	510,000	1,095,000
Columbia, MO	79,000	122,000	Niagara Falls, NY	61,000	
Columbus, OH	675,000	1,440,000	Niagara Falls, ON	74,000	
Council Bluffs, IA	58,000		Omaha, NE	346,000	668,000
Davenport, IA	98,000	364,000	Oshawa, ON	135,000	275,000
Dayton, OH	186,000	856,000	Ottawa, ON	310,000	1,024,000
Decatur, IL	86,000	124,000	Owensboro, KY	59,000	91,000
Des Moines, IA	201,000	399,000	Peoria, IL	111,000	345,000
Detroit, MI	1,005,000	4,311,000	Peterborough, ON	67,000	
Dubuque, IA	58,000	89,000	Pontiac, MI	72,000	
Eau Claire, WI	56,000	145,000	Racine, WI	85,000	186,000
Elyria, OH	54,000		Rochester, MN	67,000	121,000
Elkhart, IN	47,000	163,000	Rochester, NY	237,000	1,090,000
Erie, PA	110,000	280,000	Rockford, IL	147,000	349,000
Evansville, IN	133,000	289,000	Saginaw, MI	71,000	406,000
Flint, MI	139,000	439,000	St. Catharines, ON	134,000	391,000
Fort Wayne, IN	180,000	472,000	St. Joseph, MO	76,000	103,000
Gary, IN	117,000	623,000	St. Louis, MO	380,000	2,544,000
Grand Rapids, MI	198,000	995,000	St. Paul, MN	270,000	
Green Bay, WI	108,000	215,000	Sarnia, ON	52,000	
Guelph, ON	103,000		Sheboygan, WI	48,000	107,000
Hamilton, OH	71,000	315,000	Sioux City, IA	81,000	119,000
Hamilton, ON	321,000	647,000	Sioux Falls, SD	111,000	152,000
Independence, MO	120,000		South Bend, IN	108,000	256,000
Indianapolis, IN	781,000	1,481,000	Springfield, IL	110,000	195,000
Iowa City, IA	54,000	106,000	Springfield, OH	72,000	
Jackson, MI	39,000	154,000	Syracuse, NY	163,000	756,000
Janesville, WI	53,000	148,000	Terre Haute, IN	61,000	156,000
Kalamazoo, MI	78,000	443,000	Toledo, OH	330,000	614,000
Kansas City, KS	148,000		Topeka, KS	123,000	165,000
Kansas City, MO	436,000	1,660,000	Toronto, ON	645,000	4,306,000
Kenosha, WI	79,000	148,000	Utica, NY	68,000	323,000
Kingston, ON	63,000		Waterloo, IA	69,000	130,000
Kitchener, ON	165,000	397,000	Waterloo, ON	63,000	
Kokomo, IN	45,000	101,000	Windsor, ON	195,000	287,000

FIGURE 12-20   High-rise buildings in the heart of Toronto's central business district (TLM photo).

Canada, it serves many of the same functions as Chicago (Fig. 12-20). *Detroit* would probably be a major subregional metropolis even without the automobile industry because of its outstanding situation on a strategic isthmus alongside the principal waterway of the subcontinent; the addition of the automotive industry adds another major dimension to its significance. *St. Louis* is a more straightforward example of the hierarchical pattern, serving as the principal gateway city for much of the midsubcontinent (Fig. 12-21).

3. The third level of the hierarchy should probably include such major industrial cities and Great Lakes ports as *Cleveland, Buffalo,* and *Milwaukee,* as well as such sectional gateway cities as *Cincinnati, Minneapolis–St. Paul,* and *Kansas City.*

4. The fourth level should probably include cities of major intrastate influence, such as *Ottawa, Hamilton, Rochester, Columbus, Dayton, Indianapolis, Louisville, Nashville,* and *Omaha.*

5. Succeeding levels in the hierarchy would successively enumerate smaller urban centers with successively less extensive fields of influence.

FIGURE 12-21   The Gateway Arch overlooks downtown St. Louis (courtesy St. Louis Visitors Commission).

THE OUTLOOK

The importance of the Heartland to North America cannot be overstated. It includes a large share of the population and economic power of the United States and it includes that relatively small part of Ontario that provides much of the economic and political leadership of Canada. Thanks to benign

nature and enterprising people, the Heartland generally has had a prosperous past; its near future is less sanguine, but the long-run advantages remain.

Although the farmers of the region occasionally suffer from the caprices of nature—drought, flood, tornado, hail,

A CLOSER LOOK   The Twin Cities

The Twin Cities of Minneapolis and St. Paul are at the confluence where the Minnesota River, which drains the prairies to the southwest, joins the Mississippi, which flows south from the boreal forest. The river was the lifeline that first brought people and goods into the area, but its deep trenchlike valley, a glacial spillway, was and still remains a major barrier to movement overland. Sheer bluffs 100 to 200 feet (30 to 60 m) high frame a milewide marshy bottomland that floods when heavy spring rains coincide with rising temperatures that melt the winter's accumulation of snow.

At the end of the glacial epoch the Mississippi River tumbled over the brink of the bluff in a magnificent waterfall that had eroded its way 7 miles (11 km) upstream to the present site of downtown Minneapolis by the time the first white explorers entered the area. The Falls of St. Anthony represented the largest waterpower site west of Niagara. Downstream from the falls the river flows through a deep gorge, but just upstream it is divided into two shallow channels by Nicollet Island, where the first bridge ever built across the Mississippi anywhere was opened in 1855.

Downtown St. Paul is perched on the bluffs overlooking the floodplain of the Mississippi. For all practical purposes it is the head of navigation on the river. Just to the east a broad tributary valley, now an industrial area, gives easy access to the level uplands back of the bluffs. St. Paul was the focus of the lucrative fur trade over the ox cart trail that ran northwestward along the edge of the boreal forest to the Red River valley and Winnipeg, and in 1858 St. Paul was made the state capital of Minnesota.

The railroads that began pushing into the upper Midwest after the Civil War could ignore neither the power site and bridging point nor the capital and head of navigation, but the two places were 9 miles (4 km) apart, too far to be served from a single central station in the horse-and-buggy era. The railroads built complete sets of facilities at each location, and they were largely responsible for the development of two separate, almost equal, and fiercely competing central cities, which the automobile still has not succeeded in blending into one. Minneapolis claims to be more progressive, Scandinavian, and Lutheran, whereas St. Paul is more traditional, German and Irish, and Catholic.

In 1848 the Falls of St. Anthony were harnessed to saw logs floated downstream from the northern pineries, and sawmilling flourished for 60 years, but the last sawmill was closed in 1919 because the forest had all been depleted. The wheat boom on the prairies after the Civil War sparked the growth of flour milling, which dominated the economy for half a century, but after World War I it too began a gradual decline when wheat production moved westward. Fortunes from the milling business were invested in the growth industry of the times, namely, the manufacture of electrical controls, which laid a foundation for the development of the contemporary electronics and computer industries.

Today no single firm or type of activity dominates the economy of the Twin Cities, which have become a diversified regional capital comparable to Kansas City, Dallas, or Atlanta, the "front office" for a large but sparsely populated hinterland that stretches westward to the Montana Rockies. The population of the Twin Cities is predominantly white, middle-class, and northwest European. The only real cluster of people of southern or eastern European ancestry is in the industrial area of northeast Minneapolis.

The streets of the two central business districts often seem deserted, because in cold or hot weather most pedestrians use the glass-enclosed "skyways" that lace together office buildings, stores, and hotels at the second-story level. Perhaps also for climatic reasons, the Twin Cities were a national leader in developing large enclosed shopping malls in suburban areas. Narrow commercial strips follow former streetcar lines in the central cities and arterial highways in the suburbs. The commercial strip along the interstate highway west of the airport has become a serious rival to the two downtown areas.

The first industrial areas were near the Mississippi River for power and cheap transportation. Hulking flour mills and great batteries of grain elevators still line the river downstream from the Falls, but most are now derelict or have been converted to other uses. Later industrial areas spread along railroad lines, and the newest planned industrial districts, with pleasant structures on landscaped lots, are near major highways; a nice example is the 3-M Company complex north of the interstate highway east of St. Paul. The Twin Cities have no heavy "smokestack" industries, and the principal "nuisance"

---

freezes, insect pests—agricultural problems are much more associated with marketing than with production. Free-market prices tend to be soft and erratic and only the continuance of considerable government support, as distasteful as it is to all concerned, is likely to keep the farm economy viable. Current trends of fewer farms, larger farms, decreasing acreage, and increasing yields will probably continue—at least in the short run. The levels of accumulated crop surpluses may fluctuate from time to time, partly as a result of the international market and partly with changes in federal agricultural policies. Without strict government controls, however, increasing yields and falling prices would be a predictable scenario.

Despite the prominence of the agricultural sector and the drift of population away from the metropolises, the tempo of the Heartland is geared to the city, not to rural areas. Metropolitan expansion, although slowed, is the norm in the United States and Canada, and this region reflects the pattern.

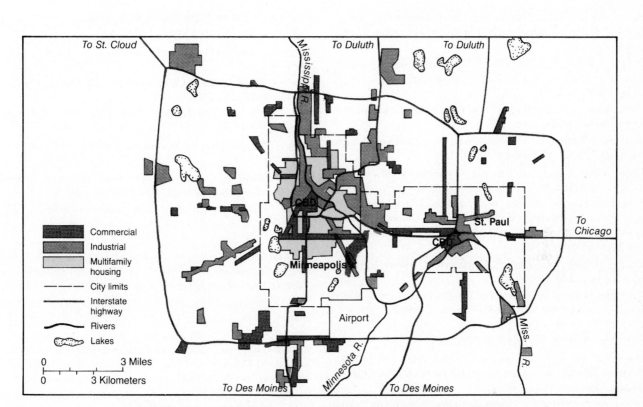

FIGURE 12-A   Generalized land use of the Twin Cities (drawn by Don Pirius).

industries are the steel mill, former stockyards, chemical plants, and oil refineries well downriver from St. Paul.

The area around downtown Minneapolis is the only part of the Twin Cities that is densely built-up, with significant numbers of multifamily dwellings, but apartment buildings also buffer some shopping strips (Fig. 12-A). The rest of the built-up area has single-family homes on individual lots that are interspersed with schools, churches, parks, lakes, and streams. The flat, sandy outwash plains have large tracts of inexpensive houses, and the more expensive houses are in the rolling glacial moraines, especially near parks and water bodies. The southern part of the metropolitan area tends toward white-collar workers and office jobs, whereas most blue-collar workers live closer to the industrial areas in the north.

Before 1940 the built-up area was fairly well contained within the two central cities, but since World War II it has spread well beyond the encircling interstate highway bypass. The Twin Cities have enjoyed most of their growth during the automobile era and it is impossible to design any sensible public transit system to serve them. Their built-up area is more than half as large as the built-up area of Los Angeles, but it has less than one-fifth as many people; it sprawls three times as much as the classic city of sprawl.

Professor John Fraser Hart
University of Minnesota
Minneapolis

Economic indicators in recent years have been distressing. In the 1980s, for the first time on record, the Heartland fell below the national income average. Still, the economic significance of the Heartland cannot be denied.

Not that stagnation, poverty, and other problems will not occur from place to place and from time to time. Any contemporary discussion of "distressed cities" would certainly feature Detroit, Cleveland, St. Louis, Akron, Dayton, Hamilton, and other Heartland examples. A depressed steel industry and a struggling automotive industry undermine important pillars of the regional economy. Yet the region's automotive industry already appears to be making a comeback; for example, 60 percent of the nation's active auto assembly plants are now in the Heartland. Adjustment to the various economic problems and situations may not come easily, but it is difficult to imagine that the inherent environmental and societal advantages of the region will not prevail in the long run.

SELECTED BIBLIOGRAPHY

ABLER, RONALD, JOHN S. ADAMS, AND JOHN R. BORCHERT, *The Twin Cities of St. Paul and Minneapolis*. Cambridge, MA: Ballinger Publishing Co., 1976.

ADAMS, JOHN S., AND BARBARA J. VAN DRASEK, *Minneapolis–St. Paul: People, Place, and Public Life*. Minneapolis: University of Minnesota Press, 1993.

AKIN, WALLACE E., *The North Central United States*. Princeton, NJ: D. Van Nostrand Company, 1968.

BAERWALD, THOMAS J., "The Twin Cities: A Metropolis of Multiple Identities," *Focus*, 36 (Spring 1986), 10–15.

BERRY, BRIAN J. L., ET AL., *Chicago: Transformation of an Urban System*. Cambridge, MA: Ballinger Publishing Company, 1976.

BORCHERT, JOHN R., AND NEIL C. GUSTAFSON, *Atlas of Minnesota: Resources and Settlement* (3d ed.). Minneapolis: University of Minnesota and the Minnesota State Planning Agency, Center for Urban and Regional Affairs, 1980.

BURNS, NOEL M., *Erie: The Lake That Survived*. Totowa, NJ: Rowman and Allanheld, 1985.

CHAPMAN, L. J., AND D. F. PUTNAM, *The Physiography of Southern Ontario*. Toronto: University of Toronto Press, 1966.

CRONON, WILLIAM, *Nature's Metropolis: Chicago and the Great West*. New York: W. W. Norton and Company, 1991.

DARDEN, JOE T., RICHARD CHILD HILL, JUNE THOMAS, AND RICHARD THOMAS, *Detroit: Race and Uneven Development*. Philadelphia: Temple University Press, 1987.

DAVIS, ANTHONY M., "The Prairie–Deciduous Forest Ecotone in the Upper Middle West," *Annals*, Association of American Geographers, 67 (1977), 204–213.

DEAR, M. J., J. J. DRAKE, AND L. G. REEDS, EDS., *Steel City: Hamilton and Region*. Toronto: University of Toronto Press, 1987.

EHRHARDT, DENNIS K., "The St. Louis Daily Urban System," in *Contemporary Metropolitan American*. Vol. 3, *Nineteenth Century Inland Centers and Ports*, ed. John S. Adams, pp. 61–107. Cambridge, MA: Ballinger Publishing Company, 1976.

EICHENLAUB, VAL, *Weather and Climate of the Great Lakes Region*. Notre Dame, IN: University of Notre Dame Press, 1979.

FRAMPTON, ALYSE, "Toronto's Harbourfront: An Exciting Blueprint for Urban Renewal," *Canadian Geographic*, 104 (December 1984–January 1985), 62–69.

FULLERTON, DOUGLAS, "Whither the Capital?," *Canadian Geographic*, 107 (December 1987–January 1988), 8–19.

GENTILCORE, LOUIS, ED., *Ontario*. Toronto: University of Toronto Press, 1972.

GORRIE, PETER, "Tobacco Alternatives: New Crops Offer Hope for Hard-Pressed Ontario Growers," *Canadian Geographic*, 108 (June–July 1988), 58–65.

———, "Great Lakes Clean-up at Critical Turning Point," *Canadian Geographer*, 110 (December 1990–January 1991), 44–57.

HARRIS, RICHARD, "Chicago's Other Suburbs," *Geographical Review*, 84 (October 1994), 394–410.

HART, JOHN FRASER, "The Middle West," *Annals*, Association of American Geographers, 62 (1972), 258–282.

———, "Change in the Cornbelt," *Geographical Review*, 76 (January 1986), 51–72.

———, "Small Towns and Manufacturing," *Geographical Review*, 78 (July 1988), 272–287.

———, *The Land That Feeds Us*. New York: W.W. Norton and Company, 1991.

———, "Nonfarm Farms," *Geographical Review*, 82 (April 1992), 166–179.

HUDSON, JOHN C., *Crossing the Heartland: Chicago to Denver*. New Brunswick, NJ: Rutgers University Press, 1992

———, *Making the Corn Belt: A Geographical History of Middle-Western Agriculture*. Bloomington: University of Indiana Press, 1994.

JOHNSON, HILDEGARD BINDER, *Order upon the Land: The U.S. Rectangular Land Survey and the Upper Mississippi Country*. New York: Oxford University Press, 1976.

KEATING, MICHAEL, "Fruitlands in Peril: We're Covering Them with Houses, Factories, and Asphalt," *Canadian Geographic*, 106 (October–November 1986), 26–35.

KRUEGER, RALPH R., "The Struggle to Preserve Specialty Cropland in the Rural-Urban Fringe of the Niagara Peninsula of Ontario," *Environments*, 14 (1982), 1–10.

———, "Urbanization of the Niagara Fruit Belt," *Canadian Geographer*, 22 (Fall 1978), 179–194.

MALCOMSON, ROGER, "The Niagara River in Crisis," *Canadian Geographic*, 107 (October–November 1987), 10–19.

MARTIN, VIRGIL, *Changing Landscapes of Southern Ontario*. Toronto: Boston Mills Press, 1989.

MATHER, COTTON, ET AL., *Upper Coulee Country*. Prescott, WI: Trimbelle Press, 1975.

MATTHEW, MALCOLM R., "The Suburbanization of Toronto Office," *Canadian Geographer*, 37 (Winter 1993), 293–306.

MAYER, HAROLD M., AND THOMAS CORSI, "The Northeastern Ohio Urban Complex," in *Contemporary Metropolitan American*. Vol. 3, *Nineteenth Century Inland Centers and Ports*, ed. John S. Adams, pp. 109–179. Cambridge, MA: Ballinger Publishing Company, 1976.

RAITZ, KARL B., *The Kentucky Bluegrass: A Regional Profile and Guide*. Chapel Hill: University of North Carolina, Department of Geography, 1980.

———, "Kentucky Bluegrass," *Focus*, 37 (Fall 1987), 6–11.

RUBENSTEIN, JAMES R., *The Changing U.S. Auto Industry: A Geographical Analysis*. New York: Routledge, 1992.

SANTER, RICHARD, A., *Michigan: Heart of the Great Lakes*. Dubuque, IA: Kendall/Hunt Publishing Company, 1977.

SHORTRIDGE, JAMES R., *The Middle West: Its Meaning in American Culture*. Lawrence: University Press of Kansas, 1989.

SIMMONS, JIM, "Toronto's Changing Commercial Structure," *The Operational Geographer*, 9 (December 1991), 5–9.

SINCLAIR, ROBERT, AND BRYAN THOMPSON, "Detroit," in *Contemporary Metropolitan America.* Vol. 3, *Nineteenth Century Inland Centers and Ports,* ed. John S. Adams, pp. 285–354. Cambridge, MA: Ballinger Publishing Company, 1976.

SOMMERS, LAWRENCE M., JOE T. DARDEN, JAY R. HARMAN, AND LAURIE K. SOMMERS, *Michigan: A Geography.* Boulder, CO: Westview Press, 1984.

SPELT, JACOB, *Toronto.* Don Mills, Ont.: Collier-MacMillan Canada, 1974.

———, *Urban Development in South-Central Ontario.* Toronto: McClelland & Stewart, 1972.

SQUIRES, GREGORY, LARRY BENNETT, KATHLEEN MCCOURT, AND PHILIP NYDEN, *Chicago: Race, Class and the Response to Urban Decline.* Philadelphia: Temple University Press, 1987.

STANLEY, JAMES, "Salmon Revival in Lake Ontario," *Canadian Geographer,* 101 (August–September 1981), 46–51.

VOGELER, INGOLF, ED., *Wisconsin: A Geography.* Boulder, CO: Westview Press, 1986.

WELLER, PHIL, *Fresh Water Seas: Saving the Great Lakes.* Toronto: Between the Lines, 1990.

YEATES, MAURICE, *Main Street: Windsor to Quebec City.* Toronto: MacMillan Company of Canada, 1975.

———, "The Extent of Urban Development in the Windsor–Quebec City Axis," *Canadian Geographer,* 31 (Spring 1987), 64–69.

The Great Plains and Prairies

13

Cutting through the center of North America in a south–north orientation is one of the most distinctive and widely recognized regions of the continent—the Great Plains and Prairies. As identified in this text, the region corresponds roughly with the Great Plains physiographic province that has been described by many geomorphologists and geographers.[1] In the United States the region is universally referred to as "the Great Plains." In Canada, however, that term is seldom used; instead the Canadian portion of the region is called "the prairies" or occasionally "the interior plains."[2] To accommodate both local usages, "The Great Plains and Prairies" is the name chosen for use here.

The extent of the region as recognized here is broader than the extent of the physiographic province. On the basis of land use and cropping patterns, the delimitation of the region has been pushed somewhat to the south to include ranching and irrigated farming country in south-central Texas and to the northeast to encompass the agricultural area of the Red River valley (Fig. 13-1). The entire eastern boundary of the region varies somewhat from the physiographic boundary, again on the basis of land-use patterns.

The problem of delimitation of the regional boundaries of the Great Plains was discussed in some detail in Chapter 5. In summary, the western boundary is marked fairly abruptly by the rise of the frontal ranges of the Rocky Mountains; the northern boundary represents the southern margin of the boreal forest; and the eastern boundary is transitional between the extensive wheat-farming systems to the west and corn-and-general-farming systems to the east.

THE CHANGING REGIONAL IMAGE

The dramatic weather and unpredictable climate of the vast interior plains region have defied accurate assessment by its occupants. As a result, fluctuating patterns of land use and economic well-being occurred but never changed the region's basic role as quantity producers of selected agricultural and mineral resources for the continent.

[1] For example, Wallace W. Atwood, *The Physiographic Provinces of North America* (Boston: Ginn & Company, 1940); Nevin N. Fenneman, "Physiographic Divisions of the United States," *Annals,* Association of American Geographers, 18 (1928), 261–353; Charles B. Hunt, *Natural Regions of the United States and Canada* (San Francisco: W. H. Freeman & Company, 1973); and William D. Thornbury, *Regional Geomorphology of the United States* (New York: John Wiley & Sons, 1965).

[2] Note J. Brian Bird, *The Natural Landscapes of Canada* (Toronto: Wiley Publishers of Canada, 1972); P. J. Smith, ed., *The Prairie Provinces* (Toronto: University of Toronto Press, 1972).

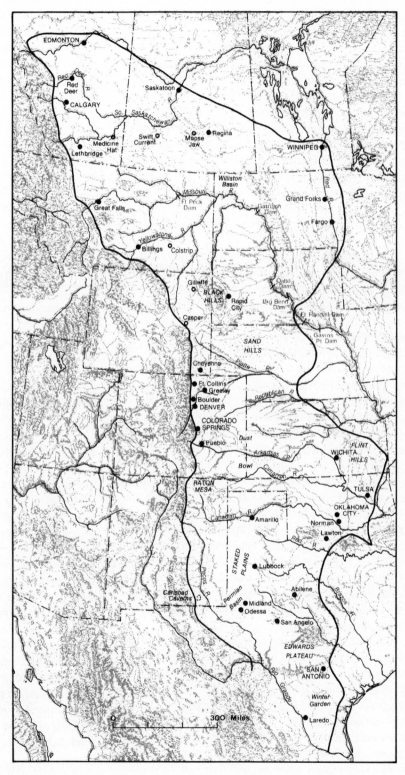

FIGURE 13-1 The Great Plains and Prairies Region (base map copyright A. K. Lobeck; reprinted by permission of Hammond, Inc.).

The Great Plains and Prairies Region contains some of the best soils and potentially most productive farmlands of North America; yet crop failures have alternated with crop surpluses, and accelerated soil erosion is commonplace. Relatively deep, dark-colored soils, which contain a considerable amount of organic matter and lime, are widespread. These soils, low relief, and much summer sunshine combine to provide several of the necessary ingredients for productive agriculture, but erratic precipitation, sometimes ill-advised farming practices, and the vagaries of the marketplace are inhibitory factors. Average annual precipitation is on the minimal side for crop production, and its usefulness is further limited by considerable fluctuations from the average in any given years, by the cloudburst nature of much of the rain, and by spring floods caused by rapid snowmelt and heavy showers.

An expanding international market for wheat, and occasionally for other crops, in the past persuaded farmers to attempt cultivation of land that should not have been plowed. Crop failures and soil abuse resulted, most notably in dust bowl conditions. In such a subhumid-to-semiarid region the dangers of accelerated soil erosion are great, and wind erosion, in particular, has left its mark on extensive areas. The history of the region is thus marked by occasional monumental crop failures, although in many years bumper crops resulted in stupendous surpluses.

Relatively few people within the region are engaged in the primary production of relatively few products, but their output is tremendous. The total farm and ranch population of the region is only a small percentage of the North American total; yet well over half the continent's annual output of wheat, grain sorghums, barley, rye, flax, canola, mohair, and potash originates here. In addition, there is notable production of coal, cotton, cattle, wool, petroleum, and natural gas. Gathering, storing, transporting, and sometimes processing these products are major activities that employ a large number of people, although there are marked seasonal fluctuations. Ultimately the great majority of the output leaves the region for most of its processing, as well as for final disposition to consumers.

Although cities are growing, there is a strong rural orientation to life. As in the rest of the continent, increasing urbanization is characteristic of the Great Plains and Prairies. Many small towns and villages, however, do not show marked growth tendencies (Fig. 13-2); the larger cities are the growth centers. Overall, even though the majority of the populace is urban, primary production has always been such a significant backbone of the economy that rurally oriented points of view, values, opinions, and judgments are often prevalent.

The character of the region encompasses a curious mixture of the drab and the grandiose. Flat land, endless horizons, blowing dust, colorless vistas, withering towns, and workaday tasks emphasize the former. But many facets of the regional scene are on a heroic scale: sweeping views, dramatic weather, natural calamities, stupendous production, staggering problems, and immense distances.

In the past the Great Plains stereotype was a vast land of dry, treeless plains, sparsely settled and given over primarily to raising cattle. But technological change came to this region as it has to most of North America. Improved dry-farming techniques, development of large-scale irrigation enterprises, establishment of state-of-the-art livestock feedlots, expanded output of energy minerals, and rapid growth of significant urban-industrial nodes combined to reshape the distinctive image of the region.

THE PHYSICAL SETTING

It is convenient to think of the physical geography of the plains–prairies as being uniform, the flat land engendering basic homogeneity in other physical aspects. There is some validity to this concept of broad regional unity in terms of gross patterns; in any sort of detailed consideration of the region, however, there is notable variety of contrasts—in physical as well as cultural geography.

Terrain

The region has the basic topographic unity of an extensive plains area (Fig. 13-3), but in detail the plains character is only true in the broadest sense, owing to significant variations from area to area. The underlying structure is a broad geosyncline composed of several basins separated by gentle arches, the surface bedrock consisting mostly of gently dipping sedimentary strata of Cretaceous and Tertiary age. The surface expression is an extensive plain that is highest in the west and gradually descends to the east at an average regional slope of about 10 feet per mile (1.9 m per km). Near the western margin the plain is more than 6000 feet (1800 m) above sea level in some places; at the eastern edge the average altitude is less than 1500 feet (450 m). There are great aprons of alluvial deposits near the Rocky Mountain front and along the major river valleys, and glacial deposits thinly cover the surface north and east of the Missouri River.

Several prominent physiographic subdivisions can be recognized on the basis of their landform associations. From south to north, they are as follows.

1. The Rio Grande Plain is flattish throughout, with some incised river valleys.

FIGURE 13-2   The characteristic regional landscape consists of extensive cropland (strip-cropped wheat here) and pasture, dotted with small towns that are marked by tall grain elevators. This is Cowley, Alberta (Alberta Government photo).

2. The Central Texas Hill Country consists of a broad crescent of low but steep-sided hills that form the dissected margin of the Edwards Plateau on its eastern and southern sides and somewhat to the north. Associated features include an eroded dome of Precambrian rocks and a number of large fault-line springs that discharge around the edge of the hills.

3. The High Plains section of the Great Plains occupies most of the area from the Edwards Plateau northward to Nebraska. Much of it is extraordinarily flat except where crossed by one of the major eastward-flowing rivers. The surface rock is chiefly a thick mantle of Tertiary sediments. The extreme flatness is partly a result of a concentration of carbonates (caliche) in a "caprock" layer that resists erosion and is partly due to surface formations that are sandy and thus highly porous. In both cases, water erosion is at a minimum except along the escarpment-like edges of the caprock and where certain rivers, particularly the Canadian and Red, have cut down through the resistant surface. In west Texas and eastern New Mexico is the Llano Estacado (Staked Plains) where there are some 30,000 square miles (78,000 km^2) of almost perfect flatness, essentially unmarked by stream erosion. The surface of the Llano is pockmarked by some 30,000 small playas that collect water briefly after the scarce rains, most of which soon evaporates. The Edwards Plateau, extending southeastward from the Llano Estacado, is geologically different but topographically identical.

4. The longitudinal valley of the Pecos River is a gentle trough lying below the level of the Llano Estaca-

FIGURE 13-3   The image of the Great Plains is one of flatness, and reality is much like the image in many parts of the region. This scene is in southeastern Alberta, just north of the international border (TLM photo).

do; it is characterized by karst features and gravel-capped terraces.

5. The Raton Mesa section along the New Mexico–Colorado border consists of a series of mesas and buttes supported by basalt flows, along with a few cinder cone volcanoes.

6. The Colorado Piedmont is an irregularly shaped zone extending along the Rocky Mountain front, from the Arkansas Valley to the Platte Valley, where much of the Tertiary alluvium has been eroded, causing the surface to be lower than the High Plains to the east. Topography here is strongly controlled by stream dissection.

7. The Nebraska Sand Hills cover much of the western and central portions of that state. The area is a maze of sand dunes and ridges that rise to several hundred feet in height and are separated by numerous small basins.

8. The Unglaciated Missouri Plateau occupies most of the northern Great Plains north of Nebraska and south of the Missouri River. There is considerable variety to the topography, but most is gently undulating. There are conspicuous badlands in South Dakota, North Dakota, and Montana as well as notable outliers of the Rocky Mountains.

9. The Glaciated Missouri Plateau section, north and east of the Missouri River, demonstrates many features of glacial origin on its surface, particularly moraines and ponds.

10. The Lake Agassiz Basin encompasses the valley of the Red River of the North as well as much of southern Manitoba and eastern Saskatchewan. Lake Agassiz was the largest of the late Pleistocene ice marginal lakes, and its ancient lake bed is extremely flat and deeply floored with silty clay. Several dozen beach lines are identifiable.

In the northwestern portion of the region several isolated ranges are offset from the Rocky Mountains. Although most are topograpically and geologically related to the Rockies, their outlying position makes them a part of the Great Plains and Prairies Region. The largest and most conspicuous of the outliers is the Black Hills; others are shown in Fig. 13-4.

Drainage and Hydrography

There is generally good drainage throughout the region, with some significant exceptions. Much of the area north and east of the Missouri River is dotted with small lakes and marshes, which are chiefly the result of Pleistocene glacial deposition. Drainage in the Nebraska Sand Hill area is also irregular, with many small basins and pockets that do not have exterior drainage outlets.

The basic stream-flow pattern of the region is from west to east; the rivers rise in the Rocky Mountains and flow down the regional slope to join the Mackenzie, Hudson Bay, Mississippi, or Gulf of Mexico systems (Fig. 13-5). The only two significant variants from this pattern are the Red River of the North, which flows northward into Lake Winnipeg, and the Pecos River, which flows generally southward to become a tributary of the Rio Grande. Most of the principal river valleys are conspicuous as narrow strings of irrigated agriculture, denser rural settlement, and urban clusters.

Large or medium-sized lakes are virtually unknown in the region. A number of large reservoirs were constructed, however, and more are planned. The most prominent are those along the Missouri River, where there is now little free-flowing water from eastern Montana to northeastern Nebraska.

Climate

Such a latitudinally extended region has considerable variation in climate, particularly in temperature. The essential characteristics, however, are clear-cut: moisture conditions are subhumid to semiarid, with evaporation usually exceeding precipitation; there are pronounced seasonal extremes;

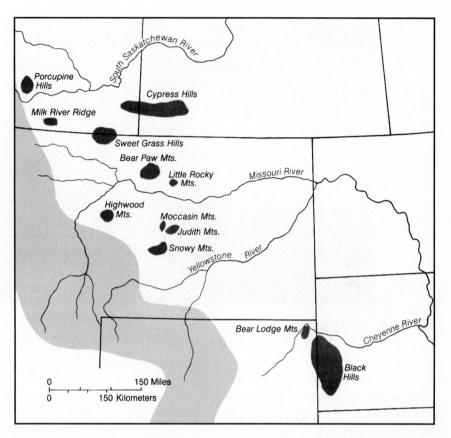

FIGURE 13-4 Principal mountain outliers in the northern Great Plains and Prairies. The extensive shaded area on the left represents the Rocky Mountains.

FIGURE 13-5 Late spring and early summer constitute a time of flooding for many of the rivers of the region. This is a June scene on the Arikaree River near Haigler in central Nebraska (TLM photo).

and much drama and violence occur in day-to-day weather conditions (Fig. 13-6).

The climate of the plains and prairies is continental; precipitation ranges from 15 inches (38 cm) in the northwest to 35 (89 cm) in the southeast and varies greatly from year to year. There are periods of dry years when the westerly margins become almost desertic. The growing season varies from 120 days in the north to 300 days in the south. Summers are

FIGURE 13-6   Dramatic weather is almost commonplace in the Great Plains and Prairies Region. Here a thunderstorm moves in on Gillette, Wyoming (TLM photo).

normally very hot, although the duration of high temperatures is much shorter in the northern part of the region than the southern.

Winters are bitterly cold[3] and dry and therefore very hard on such perennials as cultivated hay and fruit trees. The differences in temperature between winter and summer are so great as to give this region the distinction of having one of the greatest seasonal ranges of any region.

Summer winds are so hot and dry and those of winter so biting and cold that most farms have trees planted as a windbreak to reduce surface wind velocities. In the western part of the region, however, from Colorado northward, the winter weather is sporadically ameliorated by *chinook* winds. These warming, drying, downslope winds from the Rockies bring periods of relative mildness that are a welcome relief to both people and lifestock.

The Great Plains and Prairies experience the highest incidence of hail of any region in North America. The crop-destroying nature of the ice pellets is so intense that hail insurance is important to most farmers, particularly in the northern portion.

Above all, this is a region of violent weather conditions and abrupt day-to-day or even hour-to-hour weather changes. The horizon may be flat and dull in the Great Plains and Prairies, but the skies are often turbulent and exciting. Cold fronts, warm fronts, tornadoes, thunderstorms, blizzards, heat waves, hailstorms, and dust storms are all part of the annual pageant of weather in this region.

Soils

The soils of the wheat belts, among the most fertile in North America, are mostly Mollisols. They have a lime zone, a layer of calcium carbonate a few inches or a few feet beneath the surface within reach of plant roots. Because of the scanty rainfall, these soils have not had the lime leached from them. Their fertility—when combined with greater rainfall—makes them the most productive, broadly distributed soils in the world, although there is less humus than in the grassland soils to the east. They are characterized by being dark colored, rich in organic matter, well supplied with chemical bases, and usually containing a subsurface accumulation of carbonates, salts, and clay.

In drier localities Entisols and Aridisols are dominant, particularly in eastern Colorado, Wyoming, and Montana and in western Nebraska. These soils contain little organic matter

---

[3] Among world cities of over 500,000 population, Winnipeg has the coldest mid-winter temperatures. Thus it serves as the locus on many folk tales about winter temperature extremes. A long-standing weather legend is that the coldest street corner in Canada is at Portage Avenue and Main Street in downtown Winnipeg. Vilhjalmur Stefansson, the last of the great Arctic explorers, has been quoted as saying that if you can live in Winnipeg, you can live anywhere in the Arctic, as regards winter discomfort.

and are either dry and clayey or dry and sandy, although their level of natural fertility is generally high.

Natural Vegetation

Between the forests on the east and the mountains on the west lie the prairies and the steppe. The prairie, whose grasses usually attain a height of 1 to 3 feet (0.3 to 1 m), characterizes areas with 20 to 25 inches (50 to 64 cm) of precipitation in the north and 35 to 40 inches (89 to 102 cm) in the south. Merging with the prairie on the semiarid fringe to the west is the steppe, whose grasses are of low stature and where rainfall is less than 20 inches (50 cm).

The native vegetation of the semiarid grazing portion of the Great Plains is primarily short grass, with grama and buffalo grasses most conspicuous. Before the introduction of livestock in the latter half of the nineteenth century, luxuriant native grasses (mainly western wheat grass) covered extensive areas. Overgrazing and extension of wheat farming into unsuitable areas reduced thousands of square miles to a semi-desert.

The entire region is not grass covered, of course. The isolated upland enclaves are mostly forested, primarily with Rocky Mountain conifers, plus aspen and willow. The largest forest area is in the Black Hills, but tree cover is also dominant in the Raton Mesa area, the so-called Black Forest between Colorado Springs and Denver, portions of the Nebraska Sand Hills, most of the Montana mountain outliers, and almost every hill in the southern Prairie Provinces. A scrubby juniper woodland also covers much of the central Texas hill country and some cap rock escarpment faces in that same state.

Along the northern fringe of the region is a heterogeneous mixture of grasses and trees that serves as a transition zone between the prairies to the south and the boreal forest to the north. It is known as a "parkland" area (Fig. 13-7) with an

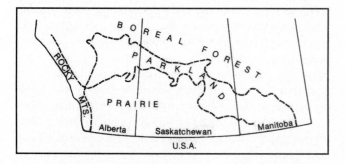

FIGURE 13-7 An area of "parkland" vegetation functions as a transition zone between the boreal forest to the north and the prairie to the south.

erratic variation of dominant species—willows, aspen, conifers, and various grasses.

The major stream valleys of the region are usually marked by a narrow band of riparian timber, nearly all of which consists of cottonwood, willows, poplars, and similar deciduous species.

During the past century much of central and southern Texas had a massive invasion by a deep-rooted, scrubby tree called mesquite (*Prosopis juliflora*); it is native to the area but has greatly expanded its range, presumably as the result of overgrazing, short-term climatic fluctuations, and cessation of recurrent grassland fires.[4] This invasion significantly reduced the grazing forage and encouraged ranchers to undertake stringent control campaigns, involving poisons, burning, and especially uprooting with heavy equipment (Fig. 13-8).

Also within the last century junipers (*Juniperus* spp.) similarly expanded their range over more than 25 million acres of what had been mostly grassland in central and western Texas. There are nine species of these hardy, scrubby, fragrant conifers, which are often inaccurately referred to as "cedars," in the southern plains. Although useful for fence posts and as a source of oil to add an aroma to household detergents, junipers, like mesquite, are generally considered pastoral pests.

Wildlife

The Great Plains and Prairies Region was the principal habitat of the American bison, with an estimated 50 million of these magnificent beasts occupying the region at the coming of the white man. Once white penetration of the region got underway in earnest, almost all the vast herds were exterminated in less than a decade.

Other hoofed animals—pronghorn antelope, deer, elk, and mountain sheep—were also common. They suffered a lesser fate than the bison, mostly being pushed into the mountains to the west as settlement advanced.

Although this is a subhumid region, furbearers were numerous along the streams. Beaver, muskrat, mink, and otter attracted the trappers and fur traders, who were, with the exception of a few explorers, the first whites to penetrate the region.

A tremendous number of small, shallow marshes and ponds dot the glaciated terrain of the Dakotas and Prairie Provinces. These poorly drained areas provide an excellent

[4] David R. Harris, "Recent Plant Invasions in the Arid and Semi-Arid Southwest of the United States," *Annals,* Association of American Geographers, 56 (September 1966), 408–422.

FIGURE 13-8   This scene, near Abilene in west Texas, shows two floral "invaders" in the southern part of the Great Plains. Both mesquite (the taller shrub) and broomweed (the shorter one) are natives to the region, but in recent decades they have spread widely, presumably as a response to overgrazing (TLM photo).

muskrat habitat and are used as summer breeding ground for myriad waterfowl. It is estimated that about half of all the ducks in North America breed in these ponds.

Several exotic species have been introduced to this region, generally as additional prey for hunters. Most important by far is the ring-necked pheasant (*Phasianus colchicus*), which has become well established in every state and province from Colorado and Kansas northward. Because of the money spent by nonresident hunters, pheasant hunting has become so important to the economy of South Dakota that it is one of the prime economic and political factors in this state.

SEQUENT OCCUPANCE OF THE GREAT PLAINS AND PRAIRIES

The human saga of the region, with its varied stages of occupance and settlement and its diversified attempts at satisfactory and profitable land use, is a dramatic and interesting one. Only a few of the highlights are recounted here, with emphasis on sequential occupance.

The Plains Indians occupy a special place in North American history because of their relationship to the Wild West era and their midcontinent position athwart the axis of the westward flow of empire. At the time of European contact

Native Americans of the Great Plains consisted of about two dozen major tribes, most of which were scattered in small semisedentary settlements over a particular territory. Their livelihood was based partly on hunting, especially buffalo, and partly on farming, particularly corn; their chief avocation was combat with other tribes; and one of their major problems was lack of transportation over the vastness of the plains.

By the middle of the eighteenth century essentially all the Plains Indians had obtained horses, and most had become expert in their use. They became much more mobile, much more effective as hunters, and much more deadly as warriors. Some tribes, such as the Dakota (Sioux) and Blackfeet in the north and the Comanche and Apache in the south, became very powerful and for many years exerted a strong influence over parts of the region. Their dominance was eventually challenged and overthrown, however, in part by eastern tribes that were displaced to the plains by whites, in part by the virtual extermination of bison, the principal food supply, but mainly by the overwhelming superiority of white soldiers and settlers.

The last stronghold of Native Americans in the Great Plains was the Indian Territory established between Texas and Kansas (Fig. 13-9). Displaced tribes from the Southeast were settled there in the early 1800s, and midwestern and Plains tribes were relegated there after the Civil War. Eventually Indian Territory became Oklahoma, and all its reservations were dissolved.

A CLOSER LOOK   Biological Dynamism in Texas

Amid yellow green spangles of oak pollen, a male golden-cheeked warbler tilts his head upward to rasp an unmistakable "laysee-daysee" song. His cheek glows like a diminutive moon halved by a black line through his eye; the yellow disc is bordered by black around the head and throat, hence his name. Here, in the so-called Hill Country of Central Texas west of Austin, this member of the New World warbler family, affectionately termed "butterflies of the bird world," is defining his territory. Golden-cheeked warblers nest in "cedar brakes," consisting of mature ash junipers and several species of oak. There are two important things to note about this attractive 6-inch-long spring migrant. First, the golden-cheeked warbler breeds exclusively in extensive tracts of oak and juniper habitat in the incised limestone that constitutes the Edwards Plateau of east-central Texas; and, second, the golden-cheeked warbler is federally listed as an endangered species.

This bird's presence exemplifies the biological diversity of Texas, which has more bird species (almost 600 have been reported, of which about 330 have nested) than most nations along a similar line of latitude. It is the most bird-rich state in the Union. This is not merely a Texas brag, it reflects the state's peculiar location in the American southwest, where temperate and tropical flora and fauna interdigitate and mix.

Texas is literally a biological crossroads. In Texas, overwintering snow geese and sandhill cranes from Alaska and northeast Russia associate with herons, whistling ducks, and similar waterbirds, which fly in from tropical America in early spring. In a day's drive one can catch sight of hummingbirds, woodpeckers, and pigeons common in western states such as California, and also spy bluebirds, cardinals, and wrens typical of states east of the Mississippi

River. The sheer size of Texas [267,339 square miles (692,400, km^2)] helps explain this biological complexity. Its biophysical character defines a different kind of region than normally promoted by Texas boosterism or decribed by cultural geographers. It is a summer home for warblers such as the golden-cheek (which winters in montane uplands of Guatemala, Nicaragua, and Honduras), and a winter home for arctic-nesting geese and cranes. In this sense, Texas is not so much a place unto itself as an integral part of a wider living system, the precise boundaries of which are elastic and elusive.

The golden-cheeked warbler arrives from Central America in early March. It is a harbinger of other trans-Gulf bird migrants whose ranks swell into the millions throughout April and the first half of May. At that time, coastal sites such as High Island (north of Galveston), Rockport, Corpus Christi, and the lower Rio Grande Valley, attract thousands of wildlife enthusiasts, who come to see brightly feathered migrants. Birds drop into oak mottes near the coast after a nocturnal passage across the open sea. Here, birds rest and feed before steering northward.

In storms, bird fallouts along the coast may be spectacular. Tired migrants literally drop into any spot that affords security; many are even cast ashore, drowned within sight of land. Bird migration has it costs. Natural hazards, such as strong rain and headwinds, and brightly lit buildings, overhead wires, high towers, lighthouses, and oil rigs take a toll, especially when dense clouds and precipitation obscure astronomical features by which the warblers orientate. Then, confused night migrants are attracted to powerful lights and crash into towers and buildings, or they appear mesmerized by gas flares sparkling over a darkened sea and fly in

so-called saturn rings around them. Losses in both spring and fall migration must run into millions of birds.

Increasingly, Texans are recognizing and appreciating this biological dynamism. The recently established "Texas Birding Trail" with pull-offs under trees and around other habitats is providing a conservation incentive to local communities. Residents welcome extra cash that birdwatchers spend. Wildlife festivals associated with bird migration and the viewing of wildlife spectacles attract a larger and larger public. More than 3000 people attended the Hummer/Bird Festival in Rockport in September 1995, its seventh year. Currently, the city of Austin, the capital of Texas, boasts the largest known population of bats of any metropolis in the world. City officials have set aside an area from which the public may observe evening flights of Brazilian free-tailed bats from under a downtown bridge over Town Lake. In August, when estimates put numbers in excess of 100 million statewide, these bats appear as plumes of smoke as they quit this municipal roost for a night of foraging for insects.

Public interest in golden-cheeked warblers, hummingbirds, bats and other nongame wildlife signals changing perspectives on what gives character to a place. In this view, interdependence, ecosystem complexity, and biodiversity are paramount. This trend, however, conflicts with an earlier, and still entrenched, vision of individual rights over property and the disposition of natural resources. The tensions inherent in the two modes of defining and approaching the biophysical heritage of Texas will require negotiation and reassessment into the twenty-first century.

Professor Robin W. Doughty
University of Texas
Austin

The first significant movement of white settlers into the region came from the south, from Mexico. Very early in the eighteenth century Spanish settlers moved north of the Rio Grande, following missionaries who had come to the Tejas

Indians in 1690. Their major bastion in this region was San Antonio, founded in 1718 (the same year that New Orleans was founded by the French). For several decades Spanish and later Mexican settlers trickled into what is now southern

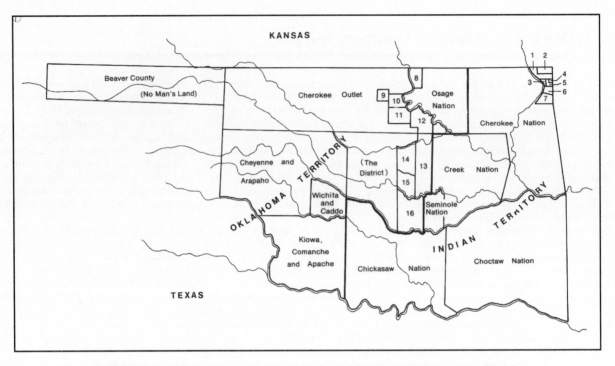

FIGURE 13-9   Tribal divisions of Indian Territory and Oklahoma Territory, generalized for the late 1800s: (1) Peoria, (2) Quapaw, (3) Ottawa, (4) Modoc, (5) Shawnee, (6) Wyandot, (7) Seneca, (8) Kansas, (9) Tonkawa, (10) Ponca, (11) Oto and Missouri, (12) Pawnee, (13) Sauk and Fox, (14) Iowa, (15) Kickapoo, and (16) Potawatomi and Shawnee (from *A Guide to the Indian Tribes of Oklahoma* by Muriel H. Wright; copyright 1951 by the University of Oklahoma Press).

Texas. They were soon joined by Anglos, who were attracted by Mexico's initially generous land-grant policies. Most of the early Anglo settlement of Texas, however, was to the east of the Great Plains Region.

The first white settlers in the central and northern plains were not slow to follow the explorers and fur traders of the early nineteenth century, but most of the major early parties were moving across the region to Oregon, Utah, and California. Meanwhile, prairies of the eastern part of the region were being occupied by farmers who had moved out of the forested Midwest, the major thrust being into Kansas. The westward flow of settlement into the region was soon in full swing, to be interrupted only partially by the Civil War.

After the war the cowboy era came to the Great Plains. The extensive diamond-shaped area of south Texas, between San Antonio and Brownsville, was the home of literally millions of wild Longhorn cattle and thousands of wild mustangs (Fig. 13-10). To the penniless returning Confederate soldiers who were hard working enough, here was the raw material of a livestock enterprise that lacked only one significant factor, a market. This problem was soon solved by trail driving the cattle northward to railheads in Missouri and Kansas. Trail driving started in 1866 and lasted for barely two decades, long enough to establish an enduring legend. During this period much of the northern plains area was also stocked by Longhorns that were overlanded to Wyoming and Montana from Texas.

The open-range trail drives were soon replaced by established ranches, which were made possible by the use of barbed wire for fencing and windmills to augment water supply. A vast cattle kingdom of large ranches soon spread the length and breadth of the region, with sheep introduced into some areas.

Apart from a scattering of Métis settlers in the Red River valley around and south of the present location of Winnipeg in the first half of the nineteenth century, there was virtually no European settlement in the Canadian prairies until the 1870s. For the next 50 years prairie settlement expanded erratically, often based on ethnic (as Ukrainian) or religious (as Mennonite or Mormon) groups.

Railways provided an important catalyst to settlement. It was hoped that they would accelerate all phases of development of the region, but as it turned out, their main function was to provide access to markets for the products of the

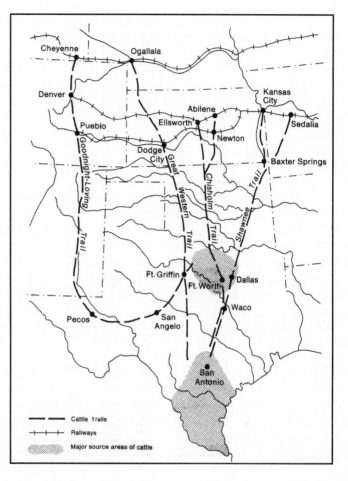

FIGURE 13-10   Major cattle trails and railways of the southern plains during the trail-driving era, from the 1860s to the 1880s.

plains. The railways received huge land grants and sold much of the land to settlers.[5] But only in the Canadian portion of

[5] The Union Pacific, for example, received 20 square miles (52 km^2) of land for every mile of track laid.

the region did the railway, the Canadian Pacific, actually colonize; the others merely sold land (Fig. 13-11).

The flood tide of settlement in the region was most prominent in the last three decades of the nineteenth century. During the 1870s Nebraska's population more than tripled to

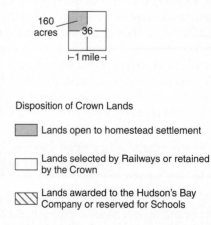

FIGURE 13-11   The land survey system used in most of the Canadian prairies was based on townships and modeled after the U.S. system. It was first introduced in 1870. Depicted here is a hypothetical township (after L. D. McCann (ed.), *Heartland and Hinterland: A Geography of Canada.* Scarborough, ONT: Prentice-Hall of Canada, 2d ed., 1987, p. 298).

nearly half a million; Kansas reached the million mark early in the 1880s. The Great Plains were "tamed," the age of the farmer began, and the frontier moved westward to other regions.

There was only a slight increase in cattle on the plains after 1890. The encroachment of the wheat farmer curtailed the amount of land available for cattle ranches, and overgrazing on the drier western parts still further reduced the area.

After 1910 several new influences arose in the region. The development of the tractor, combine, and other power machinery made feasible the planting and harvesting of a much larger acreage of land. Numerous drought-resistant crops, especially wheat and the grain sorghums, were planted on the plains.

The skyrocketing demand for wheat during World War I ushered in a two-decade period of planting on land that should never have been cultivated. In the spring of 1934 winds began to blow the soil. Great clouds of dust swept eastward from this land largely devoid of anchoring vegetation. Thus the nation paid a high price for having grown wheat on grazing land. And yet it is what might be expected from a people who had inherited the idea that in America land is practically unlimited and soil is inexhaustible.

The Dust Bowl at its greatest extent covered 16 million acres (6,400,000 ha). During December to May, the blow season, fine fertile soil particles were whisked hundreds of miles away, forming "black blizzards." The heavier particles remained as drifts and hummocks. Sand dunes attained heights of 20 feet (6 m). The atmosphere was choked with dust; in some areas people had to put cloth over their faces when going out of doors. The vegetation in the fields was coated and rendered inedible for cattle, and whole groups of counties became almost unlivable.

Although modified, Dust Bowl–like conditions have returned occasionally since the 1930s. The soils of the region were generally treated much more carefully in ensuing decades. Contour plowing, strip cropping, stubble mulching, and a host of other conservation farming techniques were introduced, particularly by the Soil Conservation Service. Most importantly, however, much marginal land that never should have been plowed was returned to grass. Problems of soil depletion still occur in the plains and prairies, but the greatest uncertainties today focus on the use and overuse of water.

CONTEMPORARY POPULATION OF THE GREAT PLAINS AND PRAIRIES

A current map of population distribution would show a fairly open and regular pattern, generally decreasing from east to west. In detail, the irrigated valleys stand out as distinct strings of denser occupance. The topographically unfavorable areas, such as the Sand Hills and various badlands and mountain outliers, are mostly barren of population.

There is considerable ethnic homogeneity to the population of the region. Most people are of European origin and the vast majority are Anglo-Saxon. Asians are almost nonexistent in the populace, and blacks are a smaller minority than in any other major region of the United States. In the Dakotas, for example, less than 0.5 percent of the population is African American.

The principal ethnic "minority" consists of Hispanics, who are prominent in portions of Texas, New Mexico, and Colorado. The major concentration is in southern Texas, where more than half the citizenry of San Antonio (second largest city in the region) is of Hispanic extraction.

There is a varied European ethnic mix in the Canadian portion of the region. The only concentration of French Canadians in western Canada is found in the St. Boniface suburb of Winnipeg, but Ukrainians and other Eastern Europeans are prominent in many parts of the Prairie Provinces.

Native Americans constitute a significant proportion of the population in Oklahoma and South Dakota. The 200,000 citizens of Indian extraction in the former represent some three dozen tribes and make up 6 percent of the state's population. South Dakota has about 50,000 Native Americans, constituting 7 percent of its population total. There are large Indian reservations in South Dakota and Montana, and the Blood Reserve in Alberta is the largest in Canada.

The most significant characteristic of the population of the Great Plains is probably its urbanity. Despite the prominence of a rural way of life in most of the region, the population is nearly 70 percent urban.

CROP FARMING

Although crop growing does not occupy as much acreage as livestock raising, it is the most conspicuous use of the land in the region. The close-knit precision of the rows of irrigated cotton in the Texas High Plains, the gargantuan linearity of strip-cropped wheat in central Montana, and the immense cultivated circles serviced by center pivot sprinklers in western Kansas are representative of the grandiose geometry of Great Plains farming that is obvious even to the traveler jetting 40,000 feet (12,000 m) above the region.

Despite marginal precipitation, the natural advantages of the region for growing certain kinds of crops are not to be denied. Flat land and fertile soil combined with some sort of water source make this a region of prodigious production for grains, oilseeds, and some irrigated crops. Wheat is the keystone on which fortunes are made and governments are elected, but active efforts toward diversification have also made other crops significant.

The region as a whole is characterized by a mixture of intensive and extensive farming. Irrigated vegetables, sugar beets, and cotton typify the former; wheat, other small grains, and oilseeds, the latter. Storage, transportation, and processing facilities are conspicuous in every farming area—cotton gins here, sugar mills there, and grain elevators looming on the flat horizon in every town.

Wheat Farming

Wheat is widely grown throughout the United States and Canada, but the two most important areas are the winter wheat belt in Kansas, Nebraska, Colorado, Oklahoma, and Texas and the spring wheat belt in the Dakotas, Montana, western Minnesota, and the Prairie Provinces (Fig. 13-12).

The two wheat-growing areas are not contiguous. Between them (in southern South Dakota and northern Nebraska) is a belt where little wheat is grown. Much of this is Sand Hill country, a disordered arrangement of grass-covered slopeland that is unsuited to cultivation.

The Winter Wheat Area The seasonal rhythm of winter wheat cultivation begins with planting in late summer or fall. The seeds germinate and growth begins. By the time winter sets in, the green wheat seedlings have raised their shoots several inches above the ground. Winter is a period of dormancy, but the shoots can begin growth with the first warm days of spring. The crop is ready for harvesting by late May or early June in the southern part of the region, and even in Montana it can normally be harvested before August.

Winter wheat is increasingly cultivated in the northern part of the region because, where it can survive winter, it generally gives larger and more valuable yields than spring

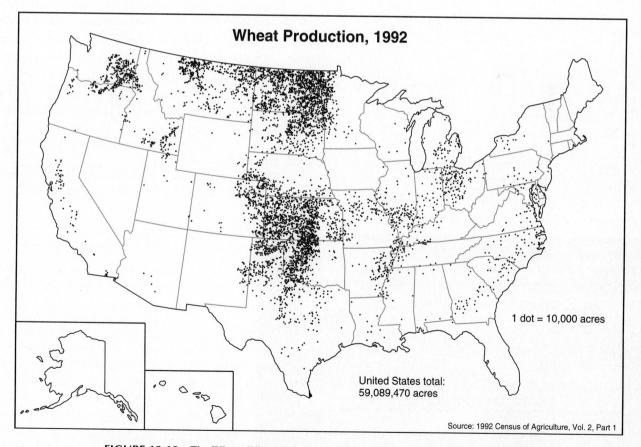

FIGURE 13-12 The Winter Wheat Belt is centered on Kansas, and the Spring Wheat Belt focuses on North Dakota.

wheat. Improved varieties of seed are more tolerant of cold weather and allow a northward shift. Today hardly any spring wheat is grown in eastern Wyoming. In South Dakota spring wheat is still the leader, but winter varieties are rapidly gaining in favor. In the plains area of Montana more winter than spring wheat is actually grown, and even in southwestern Alberta there is increasing cultivation of winter wheat.

The Spring Wheat Area   Only in North Dakota, western Minnesota, and most of the Prairie Provinces does spring wheat still hold undisputed sway. Spring wheat is planted as soon as the ground thaws and dries in late spring or early summer. It grows during the long days of summer, and the harvest takes place in August, September, or occasionally October. Unlike winter wheat, which is cut and threshed in a single combined operation, spring wheat normally is harvested in two steps. The stalks are first cut and raked into long windrows, where they are left for a few days to dry out. Spring wheat usually contains so much moisture that it would mildew in storage if not dried first. After drying, it is safe to collect and thresh the wheat.

Natural Rainfall vs. Dry Farming   Most wheat in the region is grown under natural rainfall conditions. One of the great advantages of wheat growing is that it is capable of producing a plentiful harvest on a minimum of rainfall. In some parts of the region satisfactory yields are obtained where the annual rainfall is only 11 inches (28 cm).

To maximize the effectiveness of the scanty precipitation, it is characteristic of wheat farmers, particularly those in the western part of the region, to use special "dry-farming" techniques. The simplest and most widespread of these techniques is strip-cropping, in which strips of wheat are alternated with strips of fallow land (Fig. 13-13). Care is taken to see that nothing grows on the fallow strips; the year's rainfall is thus "stored" in the soil of those strips so that it can be used the next year when the fallow strips are cropped and the cropped strips left fallow. One of the most distinctive landscape patterns in the Great Plains is the expansive crop-and-fallow stripping of dry-farmed wheat areas. The strips are usually oriented north–south, which is at right angles to the prevailing westerly winds, so as to minimize damage from wind erosion.

Harvesting   Migrant workers have been involved in wheat harvesting for many decades. They operate as custom-combining crews, working northward throughout the long summer and returning to their homes and families, usually in Texas, for the winter. After World War II there was a decided decline in custom combining, for wheat farmers had accumulated sufficient wealth and land to make it feasible for most of them to own their own combines. Since the mid-1950s, however, federal wheat acreage controls have caused a swing back to custom combining. Some 16,000 people work as combine crews, following the northward harvest trail. More than one-third of the total wheat crop is combined by migratory crews; the proportion is twice as great in much of the Winter Wheat Belt (Fig. 13-14).

Marketing Uncertainties   Most kinds of agriculture are uncertain enterprises at best, and nowhere is this principle demonstrated better than in the leading farm activity in this region—wheat growing. Weather conditions, particularly rainfall, are one important area of uncertainty for wheat farmers, but in the last few years marketing difficulties have taken center stage.

The early 1990s were times of good weather—and therefore bumper crops—for much of the region, particularly

FIGURE 13-13   The varied geometry of agricultural field patterns in northeastern Colorado near Fort Collins (TLM photo).

FIGURE 13-14 Until recently one could get a good idea of the importance of the smaller towns in the region by counting the number of grain elevators it had. Cayley, Alberta (shown here), was a five-elevator town. In the last few years, many of the smaller elevators have been consolidated into fewer, larger ones (TLM photo).

the Canadian prairies. However, marketing difficulties negated possibilities of prosperity and brought severe economic distress instead. Canada normally exports 80 percent of its wheat crop (the analogous figure in the United States is 50 percent) and is therefore heavily dependent on the international marketplace for its sales. Despite the bumper crops of the early 1990s, the international wheat price was so low that most Canadian wheat farmers actually lost money or experienced bankruptcy. The low price was blamed primarily on heavy government subsidies provided to wheat farmers in other countries, primarily in the European Economic Union.

The province of Saskatchewan exemplifies the situation. Wheat prices in the early 1990s were less than half what they had been in the early 1980s. This bad economic news affected not only the 63,000 Saskatchewan wheat farm families but the entire rural economy. A University of Saskatchewan study showed that more than 70 percent of the province's 600 towns, villages, and hamlets had lost the basic commercial functions necessary to sustain the communities (Fig. 13-15)

Other Extensive Farming Crops

In addition to wheat, several other grains and some oilseeds are important crops in the extensive farming system of the region. All are grown under irrigation in some instances but the great bulk of their production is under natural rainfall conditions.

Grain Sorghums The principal area of grain sorghum production of the continent is the winter wheat area on the High Plains of Texas, Kansas, Nebraska, and Oklahoma (Fig. 13-16). These drought-resistant crops, introduced from semiarid parts of Africa, have grown in importance in the last quarter century. There is now a greater acreage in

sorghums than in wheat in both Nebraska and west Texas, and it is a major crop in Kansas.

The principal use of grain sorghums (Fig. 13-17) is for stock feed. Some are considered to be 90 percent as good as corn for feeding and fattening. The more important types of sorghums grown on the High Plains include the milos, the kafirs, feterita, darso, and hegari. They provide a good substitute for wheat, corn, and cotton in areas of questionable water availability.

Small Grains The northern plains constitute North America's major producing areas of barley (Fig. 13-18) and rye and rank a close second to the Heartland as a producer of oats. The greatest concentration of production of all three of these feed grains is on the Lake Agassiz Plain of North Dakota, Minnesota, and Manitoba.

Oilseeds The northern plains have long been the stronghold of flax production in North America; 95 percent of the U.S. crop and 99 percent of the Canadian crop are grown here. Flax is grown almost exclusively for its seed, from which linseed oil is extracted. Sunflower seed is also cultivated extensively in the Lake Agassiz Plain.

The most dynamic new development in oilseed production was the development of canola in the mid-1970s by Canadian agronomists and its rapid adoption as a major crop in the Prairie Provinces. Rapeseed had been an important oilseed crop in the prairies since World War II, but concerns about its high level of erucic acid led to the development of canola (Fig. 13-19) as a derivative that compares favorably with other oilseeds in protein quality. It is used as a major ingredient in the production of cooking oil and margarine and is a staple of protein meal for livestock feed. Already it is the

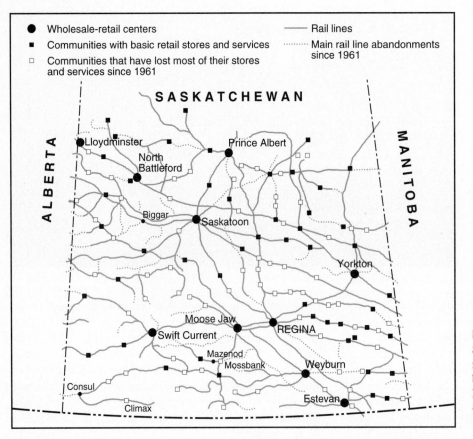

● Wholesale-retail centers
■ Communities with basic retail stores and services
□ Communities that have lost most of their stores and services since 1961

—— Rail lines
········· Main rail line abandonments since 1961

SASKATCHEWAN

ALBERTA

MANITOBA

Lloydminster
North Battleford
Prince Albert
Biggar
Saskatoon
Yorkton
Moose Jaw
Swift Current
REGINA
Mazenod
Mossbank
Weyburn
Consul
Climax
Estevan

FIGURE 13-15 Small towns in Saskatchewan that slipped from commercial self-sufficiency to non-viability between 1961 and 1991 (after Steven Fick, "Harvests of Ruin," *Canadian Geographic*, 112 (January–February 1992), 38).

leading vegetable oil used in Canada, and export markets are developing rapidly. Thus far Canada is the world's only producer, with output particularly concentrated in the prairies and parkland zone of Saskatchewan and Alberta.

Irrigation Agriculture

Irrigation farming is widespread in the region and has been developed in a variety of ways. Most spectacular have been the large-scale government projects developed by the federal Bureau of Reclamation in the United States and by various national and provincial authorities in Canada. The largest single scheme in the Great Plains is the Colorado–Big Thompson Project in northeastern Colorado, which depends primarily on transmountain diversion of water from the western slope of the Rocky Mountains to the valley of the South Platte River, where more than 600,000 (240,000 ha) acres are irrigated. Also particularly notable are the projects associated with the six huge dams of the Pick–Sloan Plan on the upper Missouri River.

A great deal of irrigation in the region, however, is not related to such giant schemes but involves smaller local projects or individual farms that have their own water sources,

often from wells. Many different types of irrigation are practiced. In sprinkler irrigation, by far the most popular in recent years, the water is sprayed on the land from lightweight aluminum pipes that can be shifted from place to place manually or that move automatically in response to motors or piston drive. Of the various sprinkler irrigation techniques, the center pivot is the most prominent recent development.

Center-pivot irrigation uses large self-propelled sprinkling machines that water circular patches with great precision and efficiency. Their rapid proliferation in the last three decades produced a spectacular change in the otherwise rectangular geometry of field patterns in the region as well as fundamental changes in cropping patterns, water use, energy use, and land ownership. Center-pivot systems have been most widely adopted in Nebraska, Kansas, eastern Colorado, and western Texas, but they are used considerably throughout the region. The most conspicuous cropping change associated with center pivots is the expansion of irrigated corn production in the region, especially in Nebraska. There are serious implications, however, to the heavy water requirements of center-pivot systems, no matter how efficiently water is used.

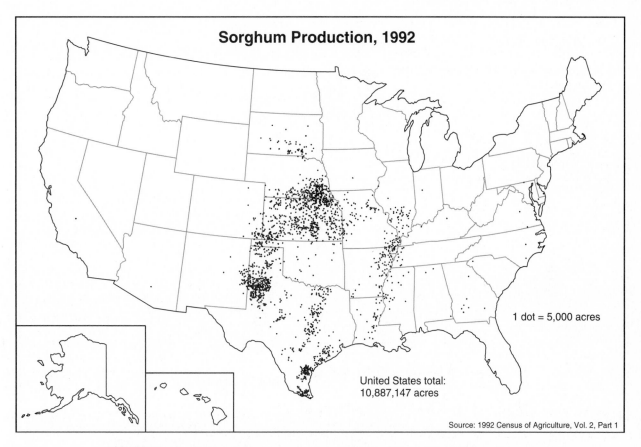

FIGURE 13-16   Grain sorghums are grown mostly in the Winter Wheat Belt, with a secondary concentration along the central Gulf coast of Texas.

FIGURE 13-17   Grain sorghums on the High Plains in west Texas, near Brownfield (TLM photo).

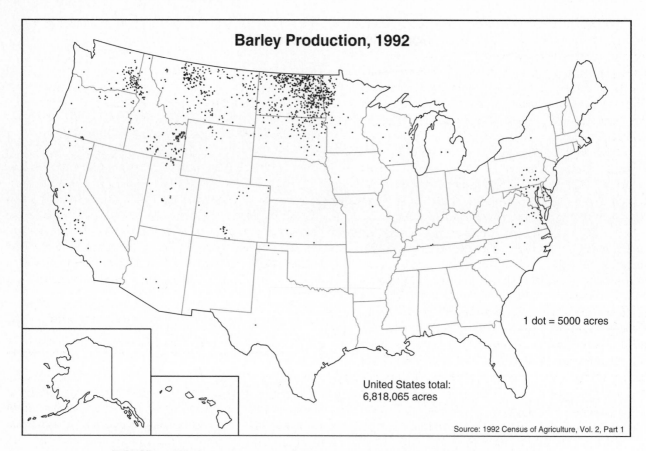

Barley Production, 1992

1 dot = 5000 acres

United States total:
6,818,065 acres

Source: 1992 Census of Agriculture, Vol. 2, Part 1

FIGURE 13-18   Barley is grown mostly in the northern plains, especially in the valley of the Red River of the North.

FIGURE 13-19   Wheat (on the left) and canola growing side by side north of Edmonton, Alberta. Canola is bright yellow in color (TLM photo).

The Major Irrigated Crops   The length of the growing season within this region varies from more than 280 frost-free days in the Winter Garden and Laredo districts of south Texas to less than 120 in northern Montana and southern Canada. The most widely grown crop is *alfalfa,* which is the basic hay crop throughout the West. Alfalfa occupies the largest acreage of any irrigated crop from southern Colorado northward.

The Winter Garden and Laredo areas of southern Texas produce early *Bermuda onions, spinach,* and other *winter truck crops.* On the several irrigated areas of the Pecos Valley in southern New Mexico, cotton and alfalfa are the dominant crops; *peanuts* and *grain sorghums* are also important.

Probably the outstanding irrigated area in the region is the Texas High Plains section in the vicinity of Lubbock, where *cotton* is the principal source of farm income (Fig. 13-20). Irrigation in this district depends on groundwater wells whose water-yielding aquifers are rapidly being depleted. Nevertheless, prodigious yields of cotton, sorghums, wheat, and corn encourage farmers to continue, and even expand, irrigation despite the cost and the specter of depletion.

The Arkansas Valley in eastern Colorado and western Kansas is noted for sugar beets, feed grains, alfalfa, and *cantaloupes.*

In northeastern Colorado, associated with the Colorado–Big Thompson Project and the South Platte River, is the second outstanding irrigated area in the Great Plains. The chief specialty crop is *sugar beets,* but a great variety of other crops are also grown, and Weld County (the Greeley area north of Denver) is a national leader in total value of agricultural output. Vegetables, dry beans, corn, alfalfa, and feed grains also occupy large acreages.

The rapid expansion of cattle feedlots and dairying in northeastern Colorado in the last few years has stimulated production of feed grains. Much *corn* is grown under irrigation and most of it is cut green for silage. Corn is also a major crop all across southern Nebraska, particularly in association with center pivot systems (Fig. 13-21).

Irrigation is also widespread in the northern plains and prairies, but in lesser concentration than in northeastern Colorado or the Texas High Plains. Most river valleys have irrigated sections along their flood plains; more than 1 million acres are irrigated in Alberta, for example, almost all within the drainage area of the South Saskatchewan River. Throughout this northern portion hay crops occupy the bulk of the irrigated acreage, with a considerable share also devoted to grains (particularly wheat), sugar beets, potatoes, and oilseeds.

LIVESTOCK RAISING

The Great Plains and Prairies Region has long been famous for its range livestock industry. Some of the world's largest, best-run, and most productive ranches are located here. Both beef cattle and sheep are widespread; in some cases, both species are raised on the same ranch. By and large, however, the better lands are used by cattle; sheep tend to be restricted mostly to rougher, drier, or otherwise less suitable country. Cattle are distributed relatively uniformly over the region,

FIGURE 13-20 In the High Plains of west Texas cotton often is both irrigated and strip-cropped in order to save moisture. This farmer, near Stamford, also has Permian Basin petroleum under his land (TLM photo).

A CLOSER LOOK   The "Mining" of the Ogallala

Ogallala is a Sioux Indian word that means "to scatter one's own." It was notable historically as a designation for one of the major branches of the Teton Sioux Indian nation, and a town (in western Nebraska) has since been graced with the name. Today, however, the word is most widely recognized as the proper name of a geological formation that underlies a large part of the central Great Plains. The Ogallala is a calcareous and arenaceous sedimentary formation of late Tertiary age that is relatively near the surface under some 225,000 square miles (585,000 km²) of the High Plains stretching from southern South Dakota to west Texas.

The distinctiveness of the Ogallala lies in the fact that it is an enormous *aquifer*—that is, a porous bed that absorbs water and holds it because it is underlain by impervious strata. The Ogallala Aquifer functions as a giant underground reservoir that ranges in thickness from a few inches in parts of Texas to more than 1000 feet (300 m) under the Nebraska Sand Hills; "it is like a 500-mile-(800-km)-long swimming pool with the wading end in west Texas and the deep end in northern Nebraska."[a]

Water has been accumulating in the aquifer for some 30,000 years. By the

[a] James Aucoin and Anne Pierce, "A Curious Piece of Real Estate," *Audubon,* 85 (September 1983), 88.

middle of the twentieth century, it was estimated that it held a total amount roughly equivalent to that of one of the larger Great Lakes. The rate of accumulation is very gradual; it is recharged only by rainfall and snowmelt that trickles down from above, and the climate of the High Plains provides relatively little of both (Fig. 13-A).

If input is tediously slow, outgo is distressingly rapid. Irrigation from this underground source began in the early 1930s. Before the end of that decade it was noticed that the level of the water table was already dropping. After World War II the development of high-capacity pumps, sophisticated sprinklers, and other technological advances encouraged the rapid expansion of irrigation based on Ogallala water. Irrigated acreage expanded more than fourfold in a quarter century.

The results of this accelerated usage were spectacular. Above ground occurred a rapid spread of high-yield farming into areas never before cultivated (especially in Nebraska) and a phenomenal increase in irrigated crops (particularly corn, cotton, sorghums, grain, hay, and vegetables) in all nine Ogallala states. Beneath the surface, however, the water table sank ever deeper, and the rate of extraction could be likened to a mining enterprise because a finite resource was being removed with no hope of replenishment.

Farmers who had obtained water

from 50-foot (15-m) wells now must bore to 150 or 250 feet (45 or 75 m); and as the price of energy skyrockets, the sheer cost of pumping increases operating expenses enormously. Perhaps 100,000 wells tap the Ogallala; many are already played out and almost all the rest must be deepened annually.

Anguish over this continually deteriorating situation is widespread. Some farmers are shifting to crops that require less water (for example, less corn and more sorghums). Others are adopting water-conserving and energy-conserving measures that range from a simple decision to irrigate less frequently to the installation of sophisticated technology that uses water in the most efficient fashion. Many agriculturalists have faced or will soon face the prospect of abandoning irrigation entirely. Millions of dollars have already been spent in a plethora of studies to assess the details and ramifications of the problem and to recommend mitigating measures.

Many farmers, however, have taken another approach. As the cost of water and energy rises, they reason that it is time to concentrate on high-value crops before it is too late. In Lubbock County (Texas), for example, grain sorghum acreage decreased by 80 percent during the 1970s, whereas cotton acreage expanded by 50 percent. Moreover, federal tax laws lessen the need for water-saving techniques by allowing irrigators to deduct the yearly cost of ex-

---

although densities decrease northward and westward (Fig. 13-22). Sheep are much more irregularly scattered, with the principal concentration being in the central Texas hill country and Edwards Plateau; lesser concentrations occur in northeastern Colorado (primarily on irrigated pastures), eastern Wyoming, and adjacent parts of South Dakota and Montana, and the western margin of the prairie in Alberta.

Summer grazing is chiefly on natural grasses, although there is increasing replacement with more nutritive artificial pastures. The most noted natural grasslands of the region are probably in the so-called bluestem belt: the Flint Hills of eastern Kansas and Osage Hills of northeastern Oklahoma. In winter, however, artificial feeding—mostly with hay—is necessary over much of the region because of the long period of snow cover.

Cattle Feedlots

The most dynamic development in the livestock industry of the plains and prairies has been the rapid proliferation of cattle feedlots. In the past nearly all range cattle were shipped out of the region, usually to the Corn Belt, for fattening. More recently the feedlot capacity of the Great Plains has expanded phenomenally and thus beef production has been vertically integrated in the region, from raising on the range through fattening in feedlots to slaughtering in local packing houses.

The Chicago–St. Louis axis of the "beef belt" has now shifted considerably westward and can be considered as being oriented along a Sioux Falls–Amarillo line. Feedlot operations are widespread in Nebraska and Kansas but are

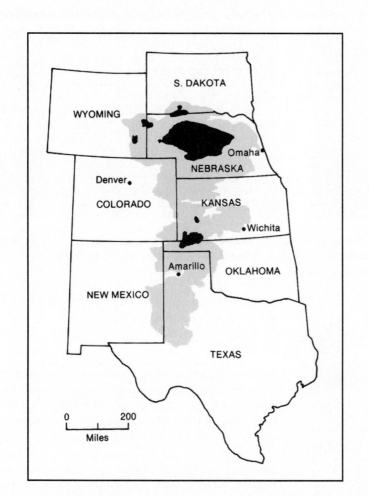

FIGURE 13-A   The Ogallala aquifer. Darker areas indicate greater thickness of the water-bearing strata.

hausted Ogallala water from their gross income, in the form of a "depletion allowance" similar to that given mining companies. Water conservation is further complicated by the obvious fact that groundwater is no respecter of property boundaries. A farmer who is very conservative in his water use must face the reality that less careful neighbors are pumping from the same aquifer and that their profligacy may seriously diminish the water available to him or to her.

In the final analysis, there is no way to escape the inevitable. Ogallala water is a finite resource, and sooner or later it will disappear throughout the subregion no matter which conservation techniques are used. Doomsday is a decade away in some areas, perhaps half a century in others. Irrigation is still on an upward trend in areas where the aquifer is deepest and less used, as in much of Nebraska. But for most of the High Plains subregion Ogallala water is decreasingly available and increasingly costly. Ambitious schemes have been proposed to replace the diminishing groundwater supply by importing surface water from somewhere—the Missouri and Arkansas rivers, for example. Such proposals are multi billion dollar ones however, and clearly their development would be a future project, if ever.

TLM

particularly prominent in west Texas and northeastern Colorado; furthermore, many new feedlots are highly mechanized and have very large capacities. In the Texas Panhandle, for example, more than a million cattle can be accommodated at one time on large feedlots; feedlot capacity in the three principal counties of northeastern Colorado is more than half a million.

Angora Goats

The hill country of central Texas yields more than 85 percent of the nation's mohair. Goat ranching is similar to sheep ranching except that goats, being browsing animals, can subsist on pastures not good enough for sheep. Most of the pasture land used by Angora goats is in the brush country where scrub oak and other small trees and shrubs supply browse (Fig. 13-23).

MINERAL INDUSTRIES

Some of the most flamboyant history in the region revolved around the discovery (1874) and production of gold in the Black Hills. The deep underground Homestake Mine was the only functioning gold mine in the United States until the upsurge of world gold prices in the 1970s made it economical to open (or reopen) several other mines in the West. Even so, the Homestake Mine yielded about one-tenth of all the gold

FIGURE 13-21 The sophisticated machinery of center-pivot irrigation is widespread in the Great Plains and Prairies Region. This scene is near Loveland in northeastern Colorado (TLM photo).

FIGURE 13-22 Hungry Herefords awaiting a delivery of hay on a cold February morning near Denver (TLM photo).

ever produced in the nation and still has the largest output of any single mine.

Most of the region is underlain by sedimentary rocks, in which nonmetallic minerals sometimes occur in significant

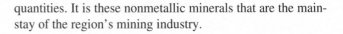

FIGURE 13-23 Angora goats in the central Texas hill country near Fredericksburg (TLM photo).

quantities. It is these nonmetallic minerals that are the mainstay of the region's mining industry.

Petroleum

Because of its extensive size and because most of it is underlain with sedimentary rocks that may contain oil, the Great Plains and Prairies Region has many producing oil fields. Among the 14 states and provinces with portions included in this region, only Minnesota lacks significant production.

Nearly half of all Texas petroleum production—amounting to one-sixth of the national output—is from west Texas fields, particularly the Permian Basin. Major yields are also obtained in parts of Oklahoma and Kansas; in the increasingly prolific fields of Wyoming, particularly the Elk Basin; in scattered Montana localities; in the Williston Basin of North Dakota; and in bountiful production areas of Alberta and Saskatchewan.

Natural Gas

The Prairie Provinces are also prominent producers of natural gas, particularly Alberta, whose output is nearly nine-tenths of the Canadian total. Natural gas is also a major product from Wyoming, Kansas, Oklahoma, and west Texas; the Panhandle field of the last three states is the largest producer of natural gas in the world. In the late 1970s an outstanding new field was developed in the Laredo area of southern Texas, with indications that it might eventually equal the Panhandle field in output.

Helium

The major helium-producing area in the world is located within the Panhandle gas fields. Originally helium was used mostly as a lifting gas for dirigibles, but now is has a multitude of atomic, spacecraft, medical, and industrial uses (particularly for helium-shielded arc welding). Of the 12 plants in the United States producing helium, 11 are in the Great Plains Region. The long-range helium outlook is for increasing demand and decreasing supply.

Coal

It was long known that there were enormous deposits of relatively low grade bituminous and lower-grade lignite coal in the northern Great Plains, but they were largely unexploited until the energy crisis of the 1970s. Although some mining has been carried on for decades in such places as Montana's Judith Basin, and lignite has been extracted in southeastern Saskatchewan for almost a century, it is only in the last two decades that coal mining has abruptly begun to change the landscape, economy, and lifestyle in parts of the Plains (Fig. 13-24).

Most of the coal deposits are extensive and near the surface and so are strip-mined. The deleterious environmental consequences of strip mining are well known, but the relevant states have enacted strict regulations concerning restoration and revegetation of plundered lands, and in some locations cattle are already grazing on rolling grassy swales where there was a yawning open-pit coal mine only 3 years previously.

The abrupt and ambitious plans for mining in Wyoming, Montana, and the Dakotas have occasioned unprecedented economic and population growth and enormous growing pains. Land values have skyrocketed. Sleepy villages have become bustling boom towns (Gillette, Wyoming, for example, experienced a fivefold population increase in 8 years) and "instant" towns sprouted, with the help of acres of mobile homes, on the barren steppe (for example, the population in Colstrip, Montana, grew from 0 to 3000 in 2 years).

Regional coal reserves are stupendous. Montana and Wyoming have the largest known reserves of any states, and Wyoming is already the third-ranking (after Kentucky and West Virginia) coal producer in the nation. Generally Great Plains coal is of low heating value, but it also has a low sulfur content, which makes it highly desirable, especially for electricity generation.

The major area of action is the Powder River Basin of north-central Wyoming and south-central Montana. The only incorporated town in Wyoming's Campbell County, Gillette, has become the classic example of the modern boom town. A dozen enormous mines operate in the vicinity. Every day 40 huge unit coal trains roll into or out of town. Residential subdivisions appear miraculously on the windswept sagebrush flats, but a large proportion of the populace still reside in mobile homes. Debate is endless about the maintenance of a "traditional" ranching way of life versus the easy wealth of selling out to the coal companies. Perhaps nowhere else in North America is there a more clear-cut collision between yesterday and tomorrow.

Potash

The world's two largest suppliers of mineral potash are located at opposite ends of the Great Plains and Prairies Region. Near Carlsbad, New Mexico, on both sides of the New Mexico–Texas boundary, are extensive deposits of polyhalite and sylvite, which have been a major source of potash for many years.

Considerable technological assistance from the New Mexico producers has helped establish a prominent potash mining industry in southern Saskatchewan in the last quarter century. With 10 producing mines and perhaps two-thirds of the world's known potash reserves, Saskatchewan could easily supply total world needs for the next 1000 years. It is a high-cost producing area, however, because of waterlogging problems in the overlying and surrounding rock. The government of Saskatchewan is heavily involved in the enterprise, although more than half the operation is in private hands.

FIGURE 13-24   An open-pit coal mine in southern Montana (courtesy Montana Power Company).

THE EBB AND FLOW OF URBANIZATION IN THE GREAT PLAINS AND PRAIRIES

The Great Plains and Prairies is a region that exemplifies many contemporary trends of North American geography, not the least of which is the pattern of urbanization. Here, better than in any other region, is demonstrated the decay of the small town in juxtaposition with the rapid growth of larger urban centers.

Withering Towns

Proportionally, no other region has as many small towns that have registered population declines during the last four census periods. This decline is partly the result of the changing economy of the Great Plains and Prairies, but in large measure it reflects the specialized origin of many towns and the trend toward larger farm size and smaller farm population.

Originally, numerous towns were established along the advancing tentacles of westering railway lines. The situation is most prominent in the Prairie Provinces, where settlement did not significantly precede the railways and the urban pattern was preconceived and superimposed on the land, but it is evident to some extent throughout the region because a large share of the towns grew up along the east–west railway routes. With regard to the prairies, the scheme has been described as follows:

> The building of the railways accompanied or preceded settlement. Settlement in advance of the railway merely anticipated the railway already projected or under construction, and such towns as were built were usually established in anticipation of the line passing through them. When thwarted, these settlements were frequently moved across the prairie to sites adjoining the line. . . . The distance between the towns was determined by the economic distance for hauling grain at a time when local transportation was horsedrawn. Whenever possible, the railway companies brought production areas to within 10 miles of their lines and, by placing elevators and sidings 7 to 10 miles apart, created a maximum hauling distance of 12 to 15 miles, the upper limit depending on the location and direction of the roads. These transshipment centres became the distributing points for supplies, e.g., agricultural implements, coal, lumber, and general merchandise. . . . The result of this method of settlement is that the towns are arranged along the railway like beads on a string. They appear, heralded by elevators, as regularly as clockwork. . . . Frequently, the names as well as the sites of the towns were chosen by the railway companies, since

there were few existing place names to recognize. This task, too, was executed with characteristic dispatch. One solution was simply to arrange the names in alphabetical order down the line as, for example, on the Grand Trunk railway, which runs from Atwater, Bangor, Cana, to Xena, Young, and Zelma.[6]

With the decreasing importance of railway transportation, most such towns no longer have any functional significance and are left to decay. Other towns that grew up along roads and highways were subsequently bypassed by new interstate or interprovincial highway construction and they, too, have become anachronistic and stagnant.

Burgeoning Cities

More auspiciously located market towns, situated farther apart and with a more diversified economic function, prospered at the expense of the smaller places; thus railway division points, separated from one another by 100 miles (160 km) or so, were able to maintain their vigor by expanding their functional hinterland. Even division points that were sited more or less arbitrarily by the railway companies have been growth nodes—for example, Moose Jaw, Swift Current, Medicine Hat, and Calgary along the Canadian Pacific.

The larger urban centers of the region are few and far between (see Table 13-1 for a listing of the region's largest urban places). Their prosperity in each case represents an extensive trading territory or a local area of high productivity. In almost every instance, their recent rate of population growth has been higher than the national average.

Denver is by far the largest city in the region; it is the commercial and distributing center for much of the plains as well as an extensive portion of the mountain West (Fig. 13-25). In addition to being the state capital of Colorado, it is a major regional center for federal offices and has a rapidly increasing industrial component. Although situated on the plains, it is the tourist gateway to the mountain recreation areas of the Southern Rockies.

The other large urban centers of the southern plains have significant functional differences. *San Antonio*, regarded as the Mexican American cultural capital in the United States,[7] is the oldest city in the region. It is the principal commercial city of southern Texas and is surrounded by an unusually large number of military bases (Fig. 13-26).

---

[6] Ronald Rees, "The Small Towns of Saskatchewan," *Landscape*, (Fall, 1969), 30.

[7] Daniel D. Arreola, "The Mexican American Cultural Capital," *Geographical Review*, 77 (January 1987), 17.

TABLE 13-1

Largest Urban Places of the Great Plains and Prairies
Region, 1995

Name	Population of Principal City	Population of Metropolitan Area
Abilene, TX	110,000	123,000
Amarillo, TX	168,000	202,000
Billings, MT	79,000	118,000
Bismarck, ND	48,000	87,000
Boulder, CO	83,000	256,000
Calgary, AL	692,000	833,000
Casper, WY	43,000	65,000
Cheyenne, WY	55,000	76,000
Colorado Springs, CO	318,000	440,000
Denver, CO	508,000	1,825,000
Edmonton, AL	602,000	886,000
Enid, OK	47,000	59,000
Fargo, ND	73,000	165,000
Fort Collins, CO	80,000	202,000
Grand Forks, ND	52,000	106,000
Great Falls, MT	60,000	72,000
Greeley, CO	59,000	140,000
Laredo, TX	142,000	150,000
Lawton, OK	85,000	123,000
Lethbridge, AL	62,000	
Lubbock, TX	194,000	231,000
Midland, TX	96,000	
Norman, OK	83,000	
Odessa, TX	97,000	237,000
Oklahoma City, OK	469,000	1,007,000
Pueblo, CO	98,000	125,000
Rapid City, SD	57,000	89,000
Red Deer, AL	58,000	
Regina, SK	182,000	199,000
San Angelo, TX	88,000	101,000
San Antonio, TX	1,006,000	1,410,000
Saskatoon, SK	192,000	221,000
Tulsa, OK	403,000	744,000
Wichita, KS	324,000	517,000
Winnipeg, MN	611,000	678,000

FIGURE 13-25 Despite having seemingly endless land over which to sprawl, the cities of the region are marked by a notable vertical dimension. This is Denver (TLM photo).

Oklahoma City and *Tulsa* are both prominent in the petroleum industry; the former's larger size is at least partly related to its role as a governmental and financial center (Fig. 13-27). *Wichita* has a prosperous agricultural hinterland, but its special claim to fame is an aircraft manufacturing center.

The northern plains in the United States have no large cities. Apparently the overlapping hinterlands of Denver, Minneapolis, and Seattle have militated against metropolitan growth in this subregion.

The metropolitan pattern of the Prairie Provinces is still evolving (Fig. 13-28). *Winnipeg* achieved its early dominance on the basis of its splendid gateway location at a river crossing at the apex of an extensive, fan-shaped potential hinterland in the prairies. But Winnipeg never became the primate city of the Prairie Provinces, indicating that a peripheral location can perhaps be a handicap despite its gateway opportunity. After World War II, *Edmonton* and *Calgary* rapidly became the foci of the western prairies as well as the twin centers of Canada's oil and gas industry, and both are now larger than Winnipeg. Curiously, the centrally placed cities of *Regina* and *Saskatoon* have always been the smallest of the prairie metropolitan areas.

A TRANSIT LAND

The Great Plains and Prairies Region has served primarily as a transit land. Most of the freight and passenger traffic passes through the region en route to and from the Pacific Coast; thus the transportation lines of the region show a predominance of east–west railways, highways, and airways. The on-

A CLOSER LOOK   Arena, North Dakota

"Arena, North Dakota, Population 0" is how the welcoming sign at the city limit should read. But there is no sign. A thirsty traveler heading down the two-lane blacktop highway might see Arena's school or its church steeple from the hill north of town and detour onto the gravel road leading to Main Street, expecting to find a cafe or a filling station. The buildings are all there: stores, houses, grain elevator, churches, and a brick school; but Arena is totally deserted. Its last few inhabitants have gone off to live somewhere else and left the town to fade back into the little piece of North Dakota prairie that was first proclaimed "Arena" in a flurry of excitement little more than 75 years ago. The place has become a ghost town, although not of the type usually associated with the American West. Arena had no gaudy dance halls or shoot-'em-up saloons and it never played host to gambling cowboys or roughneck gold miners. Its history was far less exciting than that.

Arena was a trade-center town, one of thousands created in the Middle West and Great Plains during the era of railroad building in the late nineteenth and early twentieth centuries. The student of geography might well ask, "Why were so many towns created?" and "Why were they so small?" The two questions are actually one, since the answers to both come from understanding the role of railroads in shaping the settlement pattern of the interior plains region of North America.

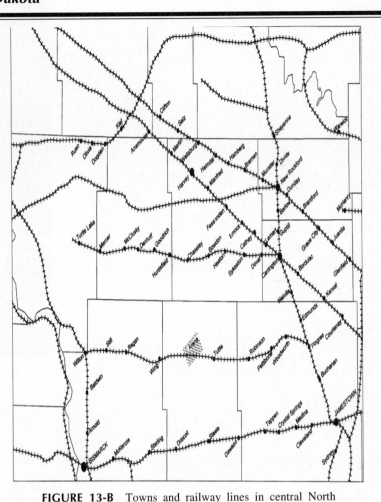

FIGURE 13-B Towns and railway lines in central North Dakota.

The railroad's influence was dominant because it was the new technology of the period when the settlement frontier reached the nation's midsection. In-

---

ly important north–south traffic flows along the western edge, at the foot of the Rocky Mountains, where there are some significant population clusters. The cities of this western margin, such as Roswell, Pueblo, Colorado Springs, Denver, Cheyenne, Billings, Great Falls, Calgary, and Edmonton, serve both the Great Plains and the Rocky Mountains, although all are located in the former region.

Despite the rapid growth of the larger urban areas of the region, the historic pattern of transportation—through rather than to—has been maintained. The major interstate and interprovincial highways continue to have an east-west orientation, and flow patterns are predominantly latitudinal.

LIMITED TOURISM

The Great Plains and Prairies Region, with its extensive area of level to gently rolling lands and its continental climate of hot summers and very cold winters, offers few attractions for the tourist. Thus it is not important as a resort region despite the fact that each summer it is crossed by throngs of tourists seeking a vacation in the mountains to the west. Three areas within the region, however, are of significance: the Black Hills, the Carlsbad Caverns, and the hill country of southwestern Texas.

Of these areas, the most scenic and most important is

vestors in cities like Chicago and Minneapolis saw the possibility of creating a weblike arrangement of railroad lines that would funnel the millions of bushels of wheat and corn produced on the prairies and plains to their cities' mills or export docks. It was to be an agricultural factory of unprecedented scale. Hundreds of thousands of farmers were eager for the income that cash-grain sales would bring them; and the rising industrial cities of the Middle West saw the profits that would come from their role as millers, bankers, and wholesalers to the system. The railroad offered cheap transportation, linking producers to the market.

What was lacking and needed to be supplied immediately was a series of collection points along the railroad where farmers would bring their grain. The horse-team-and-wagon technology of the times dictated that these sites could not be too far apart because there was a limit to the amount of time farmers were willing to spend hauling crops they had grown. Farmers participating in the cash-grain economy also needed stores where they could purchase goods they did not produce, banks where they could arrange the necessary loans that carried them from one harvest to the next and many other services that even the rudimentary way of life on a prairie grain farm required.

The result was the creation of hundreds of small trade-center towns, spaced at 7- to 10-mile (11- to 16-km) intervals along the railroad track, that were designed to attract and keep the trading allegiance of farmers in their vicinity (Fig. 13-B). Sometimes these towns were platted on land the railroad already owned, but most such towns were created by independent businessmen who had a working relationship with the railroad company. They purchased a tract of land where the railroad designated a town would be built and then lured to the new town as many would-be merchants as their advertising wiles could manage. Many new towns were staked out, the lots were sold at auction, and the first business buildings were erected in a matter of weeks. The towns were thus born almost full grown; for many, the first few months were the best times there would ever be.

Arena, North Dakota, like every other town in its vicinity, was created in this manner. And, like nearly all of its neighbors, Arena's population peak came early in its life, before economic and social changes began to render its existence pointless. Automobiles and trucks made it possible for farmers to bypass their local town's small array of businesses in favor of larger, more distant trade centers that had more to offer. The farm population declined steadily, even as the output of local agriculture increased, because fewer farmers were able to use technology to work more acres and produce more per acre than ever before. Arena became a place where the old folks lived—retired farmers and a few older merchants clinging to what remained of businesses that young folks once might have purchased and operated. By 1989, the last of Arena's inhabitants had passed away or moved away.

Many of the movers went only a short distance, to the state capital of Bismarck, a city that has boomed because of energy developments in the Missouri River basin that now make this region an important source of fossil fuels (sub-bituminous coal, lignite, oil) and a center for electricity generation utilizing new technology to transmit power long distances. The fate of the many small, trade-center towns like Arena is thus not the fate of North Dakota or the Great Plains region. But there will be more ghost towns in the years to come. Fewer farmers means fewer retired farmers, and thus fewer retirees to seek a place to live in town. Even the grain elevators in towns like Arena are closing because they are too small to service the 100-ton (0.9-t) capacity grain cars the railroads now use. All these reasons, plus too few children to require a school and too few parishioners to keep a church alive, add up to no reason for anyone to live there anymore. The system that once required Arena and its many look-alike neighbors no longer needs such places. They have become museum pieces of a way of life that probably never will return.

Professor John C. Hudson
Northwestern University
Evanston, Illinois

the Black Hills. With forest-clad mountains, the attractive Sylvan Lake area, Wind Cave National Park, Custer State Park and its abundant wildlife, Mount Rushmore Memorial, and recently legalized casino gambling at Deadwood, the Black Hills area annually attracts hundreds of thousands of tourists (Fig. 13-29).

In southeastern New Mexico, in an area where the surface of the land is harsh, lies the world-famous Carlsbad Caverns National Park with its extensive subterranean caves. As one of the most popular tourist attractions of North America, it is visited by great numbers of people each year. The area, however, does not encourage the tourist to stay long; in most cases, it is visited for a short time by persons on their way to other places.

Several cities on the western margin of the Great Plains have become important summer tourist centers because they are gateways to the mountains beyond. Colorado Springs, Denver, Cheyenne, and Calgary are particularly notable in this respect.

Hunting and fishing activities are limited in this region, with three important exceptions: the central Texas hill country contains one of the largest and most accessible deer herds in North America, the introduced pheasant population of the central and northern plains is a leading hunter's quarry from

FIGURE 13-26   San Antonio was probably the first city in the nation to transform its local stream into a delightful downtown parkland, with the initial work done in the 1930s. The *Paseo del Rio* (River Walk) continues as a focal area of interest for tourists and residents alike (TLM photo).

Colorado northward, and Wyoming's enormous antelope population is a classic example of how a combination of good management and good luck can restore a wildlife species to primeval levels of abundance.

THE OUTLOOK

Despite varying degrees of economic diversification, the prosperity of the Great Plains and Prairies Region has always

FIGURE 13-27   A view of the central business district in Oklahoma City (TLM photo).

been significantly tied to agricultural conditions, and particularly to the balance between grain output and the world market. Output will always fluctuate on the basis of growing-season weather conditions, but it depends even more on the

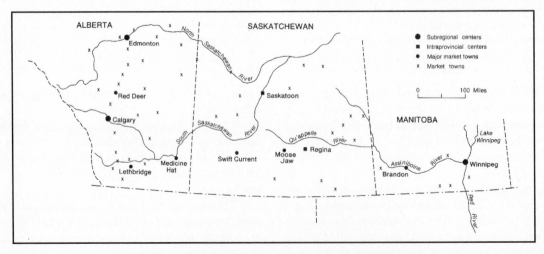

FIGURE 13-28   The urban hierarchy of the Canadian Prairies.

FIGURE 13-29   One of the very few major tourist destinations in the region is Mount Rushmore, with its presidential heads carved in granite, in the Black Hills of South Dakota (TLM photo).

vagaries of the international marketplace and the agricultural policies of the two national governments.

The 1980s were times of bin-busting records for most grain and oilseed crops in the region. This did not make for prosperity, however, as similar conditions prevailed over much of the world, and export markets shriveled. Wheat, the region's leading crop, was hit hardest. The region has enormous capability for crop growing, but future agricultural prosperity is anything but assured.

The area under irrigation will diminish, slowly at first, in the southern part of the region, where water from the Ogallala is so important. From Nebraska northward, however, an expansion of irrigation farming can be expected, for water resources are not yet being used to full capacity. The spread of irrigation will be particularly notable (at least on a relative basis) in the Canadian prairies, where much surface water is available. The established irrigated districts in southern Alberta and Saskatchewan will be the scene of most of the expansion. Manitoba has relatively little irrigation, and much of

it depends on groundwater, which is already being withdrawn at a worrisome rate.

Farms will continue to become larger as small operators sell their land, and large farmers consolidate their holdings. The complexity of farm ownership will probably increase, with more part-owners, absentee owners, and nonagricultural investors—except in the Prairie Provinces, where foreign investment is severely curtailed and there are even some restrictions on corporate farmland purchases.

The frantic pace of energy-related developments in the region has subsided abruptly, and even such previously booming locales as Wyoming and Alberta have experienced out-migration. Coal continues to be in demand, but the outlook for oil and gas is less sanguine.

Alberta can be expected to enjoy many positive residual effects of the recent oil boom due to the farsighted establishment of a provincial "Heritage Fund." A large share of the petroleum income has been channeled into this fund (which now amounts to more than $5 billion), whose purpose is to maintain economic growth and improve the quality of life of the citizenry. Saskatchewan has established a similar Heritage Fund on an obviously more modest basis.

Most of the larger cities in the region are likely to continue rapid population and economic growth. This is particularly true of the largest city—Denver—which has become the national leader in diminishing its atmospheric pollution. Denver's new airport, which is the world's largest in area, experienced horrific problems both in terms of economics and engineering during its beleagured construction in the early 1990s. Already, however, it is the eighth busiest airport in the world, and the Federal Aviation Administration predicts that eventually it will move into second place (behind Chicago's O'Hare Field).

Prosperity can be expected to wax and wane, but the region will maintain its essential character. The landscape will be dominated by a mixture of extensive and intensive agriculture, with notable centers of mineral activity; the population will never be very dense except in a few urban areas; and it will always be an important transit land, lying as it does across all transcontinental routeways.

A CLOSER LOOK   Six Flags Over Norman

A hundred and seven miles north of the Texas line, Interstate 35 crosses the Canadian River. Conceivably—and if you're really looking for it—you'll spot the river; 50 feet wide and knee- or ankle-deep, it flows across a wide floodplain, sandy and heavily obscured by poplars and eastern red cedar. What you'll certainly see, however, is a sequence of precisely spotted billboards, each rising on a massive steel pole so that the signs, elevated against the sky, dominate every motorist's vision for one commercially valuable moment. Here's one for McDonald's; here's one for Thunderbird Lodge; here comes one for Grandy's. A few years ago there was a controversy about allowing signs as big as these, but Norman merchants lined up before the city council and talked about the pressure of competition.

The exit up ahead—that one, Exit 108—is Lindsey Street. It's the one used by thousands of students every day as they go a mile and a half east to the University of Oklahoma campus. At the exit you swing by a Ramada Inn. You have to slow here for a signal; it's the intersection of Lindsey and 24th. The corner by your right elbow has the Ramada; the other three corners all have gas stations. One's new, a Conoco with six islands and a convenience store for smokes and drinks; the other two are old, but one does a business with U-Haul, and the other rents for Ryder.

We're on a grid now: Lindsey follows a section line east, and the corner of 24th in fact is a section corner. All that's history, of course—whatever happened here once with homesteaders and farmers. Lindsey now is a strip.

Just east of the Conoco there's the Del Rancho Drive-In. Corn dogs are a buck, and chicken-fried steak, with Texas toast, is $4.95. Over on the north side is the West Oaks Office Park, which houses builders and lawyers and energy companies. The place has vertical-slit windows and looks like it's made of plywood sprayed with gray gunite. Back over on the south side, there's a three-bay Midas muffler, and next to it is

a chiropractor's office. He operates out of a building that looks like a comfortable ranch-style house. There's a coin-operated carwash, the kind with high-pressure water wands. There's Taco Bell, then a strip mall called West Lindsey Center. Its 10 shops are separated from the street by a parking lot, so we can pull in for the Toy Box, which sells adult gifts, or even for Planned Parenthood, which is conveniently located just a couple of doors down. Getting back and forth across Lindsey is no problem: It has three lanes, with the center one for left turns.

Over on the north side of the street is Washington Square, an apartment complex from the sixties—you can tell, because the steps were built without risers, just treads in a steel frame. For a long time there wasn't much of a Mount Vernon look, but some carpenters last year took a load of plywood and built some gables on the front.

We're coming by a defunct savings and loan: There's no sign, but you can tell it was a bank from the big burglar alarm and maybe from the still well-trimmed grass and shrubs. Now there's another car wash and here's Taco Bell and KFC. They're next to Heritage Plaza—another strip mall, this time with a Subway, a self-service laundry, and a travel agency. We've come half a mile, and so there's another light and three more gas stations. The fourth corner is occupied by the Bethel Baptist Church—new and massive, but with slit windows pointed to look like gothic arches. The church also has a big lighted plastic sign, with timely homilies.

Over on the south side now is the Hollywood Center, the biggest strip mall on Lindsey. It's L-shaped and set back on several acres of roughly seamed asphalt. The anchor here is a supermarket, but there's also a big Sally Ann thrift shop and a franchise restaurant turned into a fitness salon. There's Gallery Nouveau, too; it sells antiques.

We're coming up now on the National Guard. The low brick building sits on the north side across from Holly-

wood Center, and its neighbors on the west are two bright and authentic drive-ins, so you tend to ignore the Guard building. Besides, it's hard to tell what goes on there. A sign in front says something about the 3rd Battalion, 378th Regiment, 95th Division, and there's a sign out back that says "Restricted Area—Warning." That's about it, though on Saturdays the big parking lot is full.

We're still heading east. Here's Mazzio's, and here's an educational publisher who moved into an old building-supply store. Here's a boarded-up place that used to sell tires. It's next to Lindsey Plaza, with a yogurt shop and a drugstore, a barber and a Mexican food store.

Now there's a house on the right. It's big, a ranch house not more than 30 years old, but it's set way back and screened by sycamores and pines. Most people hardly even see it. Across, on the north, is Legend's restaurant; you can tell it's high class because of the berms and monkey grass, the pines and oaks. But then there's the Golden Emperor—Chinese cuisine with Vietnamese and French. Next to it there's a blizzard of Whattaburger, McDonald's, and Dunkin' Donuts. There's one last strip mall; it has no name but is dominated by a Blockbuster. Then there's another light—another section corner. There are two more gas stations on the west, but then, across the street, there's a change in zoning—and half a mile of houses before you hit the OU campus.

Exhausting, you say? This 1 mile? Maybe, but streets like this are part of the daily lives of most Americans. Can so many million people be wrong? Why, just on the one bit of Lindsay near the National Guard there are six American flags. They fly bravely in that famous Oklahoma wind, and they're right at home. You can tell. Nobody minds.

Professor Bret Wallach
University of Oklahoma
Norman

SELECTED BIBLIOGRAPHY

ARREOLA, DANIEL D., The Mexican American Cultural Capital," *Geographical Review,* 77 (January 1987), 17–34.

———, "Plaza Towns of South Texas," *Geographical Review,* 82 (January 1992), 56–73.

BALTENSPERGER, BRADLEY H., *Nebraska.* Boulder, CO: Westview Press, 1984.

———, "A County That Has Gone Downhill," *Geographical Review,* 81 (October 1991), 433–442.

BEATY, CHESTER B., *The Landscapes of Southern Alberta: A Regional Geomorphology.* Lethbridge, Alta.: University of Lethbridge, Department of Geography, 1975.

BORCHERT, JOHN R., "Climate of the Central North American Grassland," *Annals,* Association of American Geographers, 40 (1950), 1–39.

———, "The Dust Bowl in the 1970s," *Annals,* Association of American Geographers, 61 (1971), 1–22.

BRADO, EDWARD B., "Mennonites Enrich the Life of Manitoba," *Canadian Geographic,* 99 (August–September 1979), 48–51.

———, "Red Deer River, Another Alberta Treasure," *Canadian Geographic,* 98 (June–July 1979), 22–27.

BROWN, ROBERT HAROLD, *Wyoming: A Geography.* Boulder, CO: Westview Press, 1980.

CARLYLE, WILLIAM J., "Farm Population in the Canadian Parkland," *Geographical Review,* 79 (January 1989), 13–35.

———, "The Management of Environmental Problems of the Manitoba Escarpment," *Canadian Geographer,* 24 (Fall 1980), 238–247.

COURTENAY, ROGER, "New Patterns in Alberta's Parkland, "*Landscape,* 26 (1982), 41–47.

EISLER, DALE, "Harvests of Ruin," *Canadian Geographic,* 112 (January–February 1992), 32–43.

EVERITT, JOHN, "Social Space and Group Life-Styles in Rural Manitoba," *The Canadian Geographer,* 24 (Fall 1980), 237–254.

FARNEY, DENNIS, "The Last of the Tallgrass Prairie," *Defenders,* 50 (1975), 308–316.

FIDLER, V., "Cypress Hills: Plateau of the Prairie," *Canadian Geographical Journal,* 87 (September 1973), 28–35.

GRIFFITHS, MELAND, AND LYNNELL RUBRIGHT, *Colorado: A Geography.* Boulder, CO: Westview Press, 1983.

HARRIS, DAVID R., "Recent Plant Invasions in the Arid and Semi-Arid Southwest of the United States," *Annals,* Association of American Geographers, 56 (1966), 408–422.

HEWES, LESLIE, *The Suitcase Farming Frontier: A Study in the Historical Geography of the Central Great Plains.* Lincoln: University of Nebraska Press, 1973.

HUDSON, JOHN C., *Crossing the Heartland: Chicago to Denver.* New Brunswick, NJ: Rutgers University Press, 1992.

JANKUNIS, FRANK J., "Perception, Innovation, and Adaptation: The Palliser Triangle of Western Canada," *Yearbook,* Association of Pacific Coast Geographers, 39 (1977), 63–76.

KARPAN, ROBIN, AND ARLENE KARPAN, Should the Souris Be Dammed?," *Canadian Geographic,* 108 (February–March 1988), 12–20.

KATZ, YOSSI, AND JOHN C. LEHR, "Jewish and Mormon Agricultural Settlement in Western Canada: A Comparative Analysis," *Canadian Geographer,* 35 (Summer 1991), 128–142.

LAWSON, MERLIN P., KENNETH F. DEWEY, AND RALPH E. NEILD, *Climatic Atlas of Nebraska.* Lincoln: University of Nebraska Press, 1977.

LEHR, J. C., "The Sequence of Mormon Settlement in Southern Alberta," *Albertan Geographer,* 10 (1974), 20–29.

———, "Ukranian Presence on the Prairies," *Canadian Ceographic,* 97 (October–November 1978), 28–33.

LONSDALE, RICHARD E., ED., *Economic Atlas of Nebraska.* Lincoln: University of Nebraska Press, 1977.

LYNCH, WAYNE, "Prairie Grasslands Preserved in Latest Park," *Canadian Geographic,* (February–March 1982), 10–19.

MACDONALD, JAKE, "The Red River Valley: Manitoba's Storied Waterway is Rich in History and Biodiversity," *Canadian Geographic,* 114 (January–February 1994), 42–53.

MARIL, ROBERT LEE, *Poorest of Americans: The Mexican Americans of the Lower Rio Grande Valley of Texas.* Notre Dame, IN: University of Notre Dame Press, 1989.

MATHER, E. COTTON, "The American Great Plains," *Annals,* Association of American Geographers, 62 (1972), 237–257.

MCKNIGHT, TOM L., "Centre Pivot Irrigation: The Canadian Experience," *Canadian Geographer,* 23 (Winter 1979), 360–367.

OPIE, JOHN, *Ogallala: Water for a Dry Land.* Lincoln: University of Nebraska Press, 1993.

REES, RONALD, *New and Naked Land: Making the Prairies Home.* Saskatoon: Western Producer Prairie Books, 1988.

RICHARDS, J. H., AND K. I. FUNG, EDS., *Atlas of Saskatechewan.* Saskatoon: University of Saskatchewan, 1969.

ROSENVALL, L. A., "The Transfer of Mormon Culture to Alberta," *The American Review of Canadian Studies,* 12 (Summer 1982), 51–63.

SHERMAN, W. C., *Prairie Mosaic: An Ethnic Atlas of Rural North Dakota.* Fargo: Institute for Regional Studies, 1983.

SHORTRIDGE, JAMES R., *Kaw Valley Landscape.* Lawrence: University Press of Kansas, 1988.

SMITH, P. J., ED., *The Prairie Provinces.* Toronto: University of Toronto Press, 1972.

———, AND DENIS B. JOHNSON, *The Edmonton–Calgary Corridor.* Edmonton: University of Alberta, 1978.

STARK, MALCOLM, "Soil Erosion Out of Control in Southern Alberta," *Canadian Geographic,* 107 (June–July 1987), 16–25.

"The Buffalo Commons Debate," *Focus,* 43 (Winter 1993), 16–27.

TIESSEN, H., "Mining Prairie Coal and Healing the Land," *Canadian Geographical Journal,* 90 (January 1975), 29–37.

WATTS, F. B., "The Natural Vegetation of the Southern Great Plains of Canada," *Geographical Bulletin,* 14 (1960), 24–43.

WEIR, THOMAS R., ED., *Atlas of Winnipeg.* Toronto: University of Toronto Press, 1978.

———, ED., *Manitoba Atlas.* Winnipeg: Department of Natural Resources, 1984.

———, *Maps of the Prairie Provinces.* Toronto: Oxford University Press, 1971.

WILKINS, CHARLES, "Amazing Flax: Such a Versatile Crop on the Prairies," *Canadian Geographic,* 108 (October–November 1988), 38–45.

———, "Winnipeg: Tough, Self-Reliant, a Truly Canadian City," *Canadian Geographic,* 104 (December 1984–January 1985), 8–19.

WOODCOCK, DON, "How Big Farms Are Changing the Prairies," *Canadian Geographic,* 103 (April–May 1983), 8–17.

WOTJIW, L., "The Climatology of Hailstorms in Central Alberta," *Albertan Geographer,* 13 (1977), 15–30.

The Rocky Mountains

In the western interior of North America is a great cordillera that constitutes the Rocky Mountain Region. It encompasses one of the more conspicuous highlands of the world, rising between the flatness of the interior plains and the irregular topography of the intermontane West. It is a region of steep slopes, rugged terrain, and spectacular scenery; of extensive forests, abundant wildlife, and deep snows; and of sparse settlements, decaying ghost towns, and busy ski trails. Mine output has long been notable, forest products are significant in some areas, and there are scattered pockets of productive agriculture. The chief functions of the region, however, are as a place for outdoor recreation and as a source of water for most rivers of the West.

EXTENT OF THE REGION

The Rocky Mountain Region, as recognized in this book, includes the lengthy extent of the Rockies from northern New Mexico to the Yukon Territory as well as the various mountains, valleys, and plateaus of interior British Columbia. This latter section, although much of it is not mountainous, is included within the Rocky Mountain Region because its patterns of vegetation, occupance, land use, and economic activities are more akin to those of the Rockies than to those

of either the Intermontane Region to the south or the North Pacific Coast Region to the west.

As thus delimited, the eastern boundary of the Rocky Mountain Region is marked by the break between the Great Plains and the Rocky Mountains (Fig. 14-1). From central New Mexico to northern British Columbia the mountains rise abruptly from the flatlands, except in central Wyoming where the Wyoming Basin merges almost imperceptibly with the Great Plains. The western boundary, which marks the transition from the Rocky Mountain Region to the Intermontane Region, is fairly distinct in the southern and middle sections. In much of British Columbia, however, the demarcation between the interior mountains and plateaus to the east and the Coast Mountains to the west consists of an indefinite transition zone, particularly in north-central British Columbia where the Omineca and Skeena mountains provide a "bridge" across the interior plateaus to connect the Coast Mountains with the Rockies. The northern boundary of the region is also transitional except where the broad lowland of the middle Liard River valley makes an abrupt interruption in the topographic pattern.

It should be noted that the name "Rocky Mountains" has a different usage in Canada, normally being applied only to a specific range along the Alberta–British Columbia border.

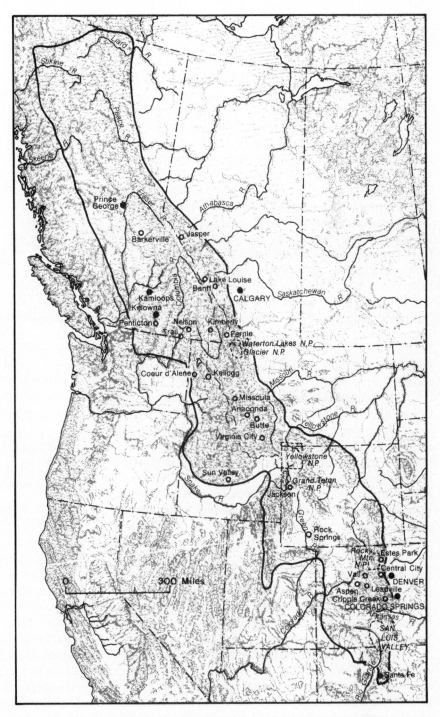

FIGURE 14-1 The Rocky Mountain Region (base map copyright A. K. Lobeck; reprinted by permission of Hammond, Inc.).

ORIGIN OF THE ROCKY MOUNTAINS

During the Cretaceous period most of the area of the Rocky Mountain Region and the Great Plains was covered by a shallow sea that extended from the Gulf of Mexico to the Arctic Ocean. At the close of that period, the Rocky Mountain area was uplifted and the waters drained off. Sediments with a thickness of perhaps 20,000 feet (6000 m) were involved in this first great uplift.

A long period of erosion accompanied and followed this early uplift, during which time much material was removed from the summits and deposited in the basins. In the later Tertiary period the Rocky Mountains were subjected to another period of uplift, accompanied by considerable volcanic activity and followed by still another period of leveling. The region's master streams, flowing over sediments that had buried the mountain roots, established courses that they continued to hold after they cut into older rocks, forming major gorges and canyons through many ranges. In the more recent uplifts many ranges have been raised so high that erosion has stripped away much of the sedimentary cover, exposing the ancient Precambrian rocks (usually granitic) of the mountain core. The sedimentaries that originally extended across the axis of the uplifts are now mostly found as uptilted edges on the flanks of the ranges or as downfolded or downfaulted basins within the mountain masses.

The geologic history varies from range to range, but overall it is a region of crustal weakness and young mountains; consequently, the relief is great and slopes are steep. In many areas, however, there are fairly extensive tracts of land at high elevation—10,000 to 11,000 feet (3000 to 3300 m) in much of the Colorado Rockies, for example—which apparently represent old erosion surfaces that have been uplifted without significant deformation and appear as gently rolling upland summits or accordant ridge crests.

Essentially all the high country was severely reshaped by glaciation during the Pleistocene epoch. In more southerly latitudes abrupt U-shaped valleys, broad cirques, and horn peaks are the most prominent results of mountain glaciation; farther north the glacial features are more complex, for the mountain glaciers were larger, and some continental glaciation also impinged on the ranges.

It goes without saying that the entire region is not mountainous. There are many valleys and some extensive areas of relative flatness. But mountains are everywhere on the horizon, and alpine country almost universally dominates the landscape. The most extensive nonmountainous area is the so-called Wyoming Basin, which represents an extension of Great Plains topography into the Rocky Mountains in southern and central Wyoming. This "basin" is topographically quite heterogeneous, varying from alluviated plains to badlands to steep hills.

Scattered about in the Southern Rockies are a large number of relatively flat-floored basins, generally called *parks,* that are mostly not timbered and present a distinct change in landscape from the surrounding mountainous terrain. The largest is the San Luis Valley. Other notable basins include South Park, Middle Park, and North Park, and there are numerous smaller parks.

North of Wyoming the circular basin is much less common; more distinctive landscape features in this subregion are long linear valleys of structural origin. The most conspicuous is the Rocky Mountain Trench, which extends for 1000 miles from the vicinity of Montana's Flathead Lake to the Liard Plain in northeastern British Columbia.

MAJOR GEOMORPHIC SUBDIVISIONS

In terms of geomorphic association and geographical proximity, the Rocky Mountain cordillera can be subdivided into five principal sections (Fig. 14-2).

1. Southern Rockies, mostly in the state of Colorado
2. Middle Rockies, primarily in Utah and Wyoming
3. Northern Rockies, in Montana and Idaho
4. Columbia Mountains, in southeastern British Columbia
5. Canadian Rockies along the Alberta–British Columbia boundary

The Southern Rockies

The Southern Rockies include a series of linear ranges that extend from north-central New Mexico northward into southern Wyoming as well as a number of mountain masses in central and western Colorado that are less orderly in arrangement. The frontal ranges rise abruptly from the western edge of the Great Plains with just a narrow foothill zone that consists chiefly of uptilted sedimentary strata, often in the form of hogback ridges. In southern Wyoming these mountains are called the Laramie or Snowy Ranges, in most of Colorado they are simply named the Front Range, and in southern Colorado and northern New Mexico they are known as the Sangre de Cristo Mountains.

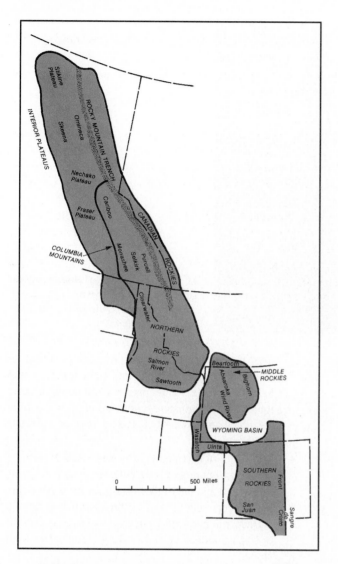

FIGURE 14-2 Principal geomorphic units of the Rocky Mountain Region.

The origin of these frontal ranges is complex, involving huge batholithic intrusions, erosion of the sedimentary cover, further uplift, perhaps peneplanation, and extensive alpine glaciation. The general pattern of the present topography is a high-altitude subdued upland, with peaks up to 14,000 feet (4200 m) standing above the rolling upland surface and deep canyons trenching it (Fig. 14-3).

Just west of the frontal ranges is an interrupted line of the four large-sized basins mentioned, extending from Wyoming into New Mexico. The origin of these extensive "parks" is varied, but all four are flat-floored and broad.

The ranges of central and western Colorado are generally similar in appearance and height to the frontal ranges. Their patterns are much more amorphous, however, and there are significant variations in both structure and origin (Fig. 14-4). The most complicated highland mass is the San Juan Mountains in southwestern Colorado.

The Southern Rockies reach the greatest altitudes of the entire Rocky Mountain cordillera, although greater local relief may be found in some parts of the Middle Rockies. There are no low passes through the Southern Rockies. These ranges functioned much more significantly as a barrier to transportation than did most of the ranges farther north in the region.

The Middle Rockies

The Middle Rockies include the mountains of western Wyoming, northern Utah, and adjacent parts of Idaho and Montana. Four major separate ranges are involved:

1. The Uinta Range in Utah is the only significant range in the entire cordillera that extends east–west rather than north–south. It is a broad and massive range and is less rugged than the others.
2. Utah's Wasatch Range is a massive linear feature with considerable variation in the height of its crest

FIGURE 14-3 The high country of the Southern Rockies is largely a subdued upland with peaks rising above it and canyons carved into it. This is a scene in Rocky Mountain National Park with the Never Summer Range in the distance (TLM photo).

FIGURE 14-4 High altitude, steep slopes, and rugged rockiness characterize the Maroon Bells in central Colorado (TLM photo).

line. It rises fairly gently on the eastern side, and several streams have cut relatively low passes through it. Its precipitous western face, the Wasatch Front, however, is remarkable for its combination of height, steepness, and continuity.

3. The Wind River Mountains in western Wyoming encompass some of the most rugged granitic wilderness that can be found. It is a little-known area characterized by steep slopes and high relief. It was heavily glaciated in the past and still contains more than 60 living glaciers, constituting a greater total ice area than all other U.S. Rocky Mountain glaciers combined.[1]

4. The Bighorn Mountains of north-central Wyoming are a massive range rising between the Great Plains on the east and the Bighorn Basin on the west. There are a steep eastern slope, a broad subdued upland surface, ramparts of higher glaciated country, another broad subdued upland, and a less rugged western slope.

In addition to these major units, a series of smaller ranges extends from northern Utah into northern Wyoming, arranged with a conspicuous north–south linearity. These relatively minor high-lands culminate northward in the majestic Grand Tetons of west-central Wyoming, whose block-faulted and glaciated eastern scarp presents an even more spectacular face than the Wasatch Front (Fig. 14-5).

Also a part of the Middle Rockies is the broad, jumbled mountain mass that makes up the Absaroka Range in Wyoming and the Beartooth Mountains in Montana. They are fairly rugged and steep-sided but with no discernible pattern and no notable crest line.

FIGURE 14-5 The spectacular eastern fault scarp of the Grand Tetons rises abruptly to glaciated peaks above the flat floor of Jackson Hole (TLM photo).

[1] Stephen F. Arno, "Glaciers in the American West," *Natural History,* 78 (February 1969), 88.

The lava plateau of Yellowstone Park and the north-trending valley of the upper Yellowstone River provide a relatively complete break between the Middle and Northern Rockies, although this breach had no transportational significance in the history of the West.

The Northern Rockies

The Northern Rockies occupy essentially all of western Montana and central and northern Idaho. The eastern half consists of a number of discrete linear ranges separated by broad, flat-bottomed, structural valleys. It is almost ridge-and-valley topography on a grand scale. These are often referred to as the Broad Valley Rockies because the spacious intermontane valleys occupy so much (sometimes half) of the total area. This "broad valley" pattern also extends into parts of central Idaho. Some of the mountains are exceedingly rugged, but altitudes are lower than in the Middle and Southern Rockies (Fig. 14-6).

The western portion of the Northern Rockies is a massive jumble of mountains that is almost patternless. Nearly all the land is in slope, with many deep, narrow river valleys. Generalized nomenclature refers to the southern part as the Salmon River Mountains and the Sawtooth Mountains; most of the northern section is considered part of the Clearwater Mountains. The Sawtooths are the highest and most rugged, but no part is easily crossed by transportation routes. The largest of Idaho's numerous lakes—Coeur d'Alene, Pend d'Oreille, and Priest—are found in the extreme north, where continental glaciation occurred.

FIGURE 14-6 The Northern Rockies are lower in elevation than those ranges farther south, but their spectacular scenery is unparalleled. This is Mount Jackson and St. Mary Lake in Montana's Glacier National Park (TLM photo).

The Columbia Mountains

The Columbia Mountains occupy an area in southeastern British Columbia that is quite broad in the south but tapers northward until it pinches out at about the latitude of Edmonton and Prince George (54° north latitude). There are four major ranges: the north–south trending Purcell, Selkirk, and

FIGURE 14-7 Looking north along the Rocky Mountain Trench near Windermere, British Columbia (TLM photo).

A CLOSER LOOK   Montana's Broad Valley Rockies: A Spiritual Geography

Montana is a state with a dual physiographic personality—Great Plains Montana in the east and Rocky Mountain Montana in the western third. Mountains not only dominate the physical scene in this section of the Big Sky State, they help define a Montana mood and are central to the psyche of residents. Mere association with mountains is viewed positively in Montana. The names Bridgers, Tobacco Roots, Big Belts, Missions, Crazy Mountains, and others are second nature to residents of the region. They would no more call one of them the Rocky Mountains than a resident of Chicago would call Lake Michigan the Great Lakes.

Contrary to depictions on eighteenth- and nineteenth-century maps, the Montana Rockies are not a single mountain chain but a surprisingly diverse collection of more than two dozen distinct highlands separated by intervening lowlands. Intermontane areas are nowhere more generous than in the Broad Valley Rockies section, which includes a majority of the Montana Rockies. Here, the basic topography is open mountains, a distinctive and globally rare setting. The topography consists of detached and islandlike mountain ranges, commonly rising to 9000 feet (2700 m) and more in elevation, separated by broad, smooth-floored valleys generally 3500 (1100 m) to 5000 (1500 m) feet above sea level and up to 50 miles (80 km) across. These generous intermontane areas provide abundant living space and allow Montana the distinction of being the only Rocky Mountain state with a majority of its residents living within, not just next to or near, the Rockies. Montana's mountains and their divides provide the spatial framework for both population clustering and distinct regional identities—people view themselves as residents of the Bitterroot, Big Hole, Gallatin, Madison, Flathead, or any of a number of other valleys.

Most ranges rimming western Montana valleys are not of the subdued Ozark or Appalachian variety. These are the real thing, and they look the way mountains are supposed to look—grand, spectacular, overpowering, and majestic. Such prominent landscape features breed familiarity and help instill a sense of place. Although mountains have not always been held in high esteem by American society, our late-twentieth-century culture tells us this is scenic substance of the first order. Both residents and visitors revel in the natural high triggered by mountain vistas.

All towering and striking mountains do not trigger an equal aesthetic rush. On a clear day, metropolitan Los Angeles residents see their own 10,000-foot (3000-m) peaks rising as a not-too-distant backdrop. In the Angeles National Forest north of Los Angeles, polluted air and water, dearth of wildlife and abundance of litter, crowds and trampled landscapes and worn trails stand in stark contrast to the more pristine and wild conditions of western Montana's forested mountains. Much of this rugged Montana back country is de facto or designated wilderness. For some modern Americans wilderness has become a kind of sacred land, unpolluted by contaminating civilization, and a wilderness visit is seen as akin to a religious pilgrimage. Ready access to such spiritual quests, as well as to more traditional outdoor recreation, might help explain why Rocky Mountain counties with federally designated wilderness or adjacent to such counties, have some of the fastest population growth rates.

Like all of Montana, the Broad Valley Rockies area has the added elixir of the Old West as a very real element in its spiritual geography. In Montana the recency and residual aura of the frontier/cowboy complex is omnipresent. Original log ranch and farm buildings, abandoned mining operations, rodeos, outsized belt buckles, cowboy boots, Stockman's Bars, and ubiquitous "Howdy's" are just some of the daily reminders of place. And that Old West, the mythic West of promise, optimism, frontier opportunity and elbow room, individualism, and informality is especially alive and well in the Broad Valley Rockies. Refugees from Los Angeles and escapees from Chicago and other urban combat zones often are quickly acculturated by this potent Montana Old West–cowboy culture. There is some-

---

Monashee Ranges in the south and the knot of the Cariboo Mountains in the north. Unlike the true Canadian Rockies, these mountains consist largely of crystalline rocks. The narrow north–south trenches among the ranges are chiefly occupied by long, beautiful bodies of water, such as Kootenay Lake and Lake Okanagan.

The Rocky Mountain Trench

In many ways, the most remarkable topographic feature of the entire cordillera is the Rocky Mountain Trench (Fig. 14-7). It is a depression that extends in a direct line from Montana almost to the Yukon Territory with regular boxlike sides. Its exact origin is uncertain and variable, although parts are clearly related to faulting. The trench bottom is flat to rolling and even hilly in spots, with four low drainage divides between major north- and south-flowing streams. Eleven rivers, including the upper reaches of both the Columbia and the Fraser, occupy some portion of the trench. New dams now impound two immense reservoirs in the trench: Williston Lake behind W. A. C. Bennett Dam on the Peace River, and McNaughton Lake behind Mica Dam on the Columbia River.

thing exciting about living a part of America's almost century-old, West-centered national creation myth. The Broad Valley Rockies will, for many, remain the epitome of that magically seductive and mythic notion.

Although Montana's distinctive and strong sense of place is a statewide phenomenon, the power of the place seems nowhere more potent than in the Broad Valley Rockies. Far-fetched as it may sound, perhaps residents and visitors who feel a rush, a natural high and empowerment with mere proximity to the Broad Valley Rockies can, in part, thank humankind's family tree. For millions of years survival of members of our genus, *Homo*, depended on closeness to, and correct reading of, nature. It is not unreasonable to assume that the evolutionary environment impacted the creature and the hard wiring of its brain. Let's speculate on how this all might relate to the contemporary spiritual geography of western Montana.

In our presumed evolutionary hearth area of East and southern Africa, unobstructed vistas and a grassland setting were the norm for millions of years. Does this help explain our species' penchant for grasslands and savanna-like park settings? Although farming and domestic animals have greatly changed the vegetation on the floors of Montana's Broad Valleys, most still are classed as grasslands. Perhaps this might help explain a certain level of comfort associated with residency in the region.

But there are many grassland areas with which people do not associate a strong feeling of attachment. Other factors must be involved. British geographer Jay Appleton has noted other additional human environmental preferences, including prospect and refuge. In premodern times an ability to scan large areas and see great distances (prospect) had obvious survival benefits. Similarly, a refuge, or readily accessible hiding place, would be another, perhaps a critical, site attribute. In Montana's Broad Valley Rockies even the low flanks of islandlike ranges provide ideal vantage points for viewing the adjacent grassland, and nearby mountainside forests furnish requisite concealment. Interestingly, the high-premium building sites in the region are panoramic view lots on the flanks of the mountains nestled and sequestered in the lowest limits of the forest. Furthermore, this zone of contact between the better-watered forest above and the drier grassland below, an ecotone, is environmentally rich in terms of variety and number of plant and animal species. Is it possible that our hunting/gathering progenitors programmed us to also favor such ecologically varied sites?

Montana's open mountains landscape also scores highly for what has been called "legibility." According to research psychologists Stephen and Rachel Kaplan, humans have a genetic-based preference for open spaces that are easily comprehended and invite reconnaissance but are sufficiently heterogeneous to afford landmarks. The Broad Valley Rockies with open valleys and always-within-site mountain ranges would seem to combine to offer great legibility. Is it possible that people have a primordial programming to feel reassured and at home in places with this prized compliment of landscape features?

The strong sense of place and spiritual geography of Montana's Broad Valley Rockies probably is best explained by a synergism of cultural conditioning and primordial placeness. Clearly it is a high-energy environment that stimulates the senses, the imagination, and creativity. Perhaps the most tangible evidence of the empowering potential of the place is its rich literary tradition. If you add up the literary activity of neighboring Idaho, Wyoming, North and South Dakota, it is questionable that the total would be comparable to that emanating from Montana's Broad Valley Rockies. The more recent Hollywood invasion by a small army of nationally recognized actors, writers, directors, and other highly creative people, many of whom buy large ranches on the flanks of mountains within the Broad Valley Rockies, also underscores the region's spiritual geography.

Professor John A. Alwin
Central Washington University
Ellensburg

The Canadian Rockies

Along the Alberta–British Columbia border is a lengthy northern continuation of the cordillera known as the Canadian Rockies. Although not as high as the Middle and Southern Rockies, there is great local relief, and the narrow ranges are more rugged and steep-sided. There was alpine glaciation during the Pleistocene, with extensive present-day mountain glaciers and even some upland icefields—the largest is the Columbia Icefield—from which lengthy glaciers extend down-valley.

The mountains consist primarily of conspicuously layered sedimentary rocks that were uplifted and, in some cases, extensively folded and faulted. Ridges and valleys are more or less uniform in altitude and tend to parallel one another for long stretches in a general northwest–southeast trend. These ranges provide some of the most spectacular scenery of the continent.

Interior Plateaus and Mountains of British Columbia

The interior section of British Columbia consists of an extensive area of diversified but generally subdued relief. The

FIGURE 14-8 Central British Columbia consists mostly of forested plateaus. This scene is in Wells Gray Provincial Park (TLM photo).

Fraser Plateau in the south and the Nechako Plateau in the center are characterized by moderately dissected hills, occasional mountain protuberances, and a few large entrenched river valleys (Fig. 14-8). Most notable of these entrenchments are in portions of the Fraser and Thompson Valleys and in the Okanagan Trench with its series of beautiful lakes.[2]

North of the Nechako Plateau is a relatively complicated section of mountains and valleys—the Skeena Mountains to the west and the Omineca Mountains to the east—that flattens out into a series of dissected tablelands, the Stikine Plateau, in the far north. The Stikine country is underlain by lava and surmounted by several volcanic peaks.

VERTICAL ZONATION:
THE TOPOGRAPHIC IMPERATIVE

In a landscape dominated by slopes and considerable relief, the nature of other environmental elements is significantly dictated by the topography; thus climate and vegetation are

[2] This lengthy linear valley extends southward across the international border into Washington State. It has the same name in both countries but different spellings. In Canada the word is spelled "Okanagan"; in the United States "Okanogan."

markedly affected by altitude and less so by exposure and latitude. Within the Rocky Mountain Region are found some of the classic examples of vertical zonation in vegetation patterns, with major variations occurring in short horizontal distances because of significant vertical differences.

Climate

The paradox of the Rocky Mountain climate is that summers tend to be semiarid and winters relatively humid. Summer is a time of much sunshine even though there may be a characteristically brief thundershower each afternoon. Windiness, generally low humidity, and sunshine quickly evaporate the rainfall, partially negating the effectiveness of the precipitation. Summer is therefore a time of relative dryness; after the main runoff of the melting snowpack has subsided, dust is much more characteristic than mud on the mountain slopes except in the Canadian portion of the region, where the summers are rainier.

The inland location of this region means that the prevailing westerly air masses have had to pass over other mountains before reaching the Rockies and that part of their available moisture was dropped there. The western slopes of the high country may receive a considerable amount of summer rain, for the forced ascent squeezes out more moisture from the westerlies. But the lowland valleys, parks, and trenches of the region may experience almost desertlike conditions. So the center of Colorado's San Luis Valley receives only 6 inches (15 cm) of moisture annually and Challis, on Idaho's Salmon River, records 7 inches (18 cm); Ashcroft, in the Thompson River Canyon, with 7 inches (18 cm), is probably the driest nonarctic locality in Canada.

The low temperatures of winter make dry weather seem less dry because the precipitation that does fall is normally in the form of powdery snow; once it is on the ground, it may not melt until spring, even in the basins. Storm tracks are also shifted southward in winter, with the result that stormy conditions are more numerous and frontal precipitation is more commonplace. Usually snow lies deep and long over most of the region, particularly at higher altitudes.

The general pattern, then, is that the lower elevations are quite dry, but with increasing altitude there is increasing precipitation up to some critical level [generally between 9000 and 11,000 feet (2700 and 3300m)] above which there is once again a decreasing trend. West-facing slopes normally receive more rain and snow than comparable levels on east-facing slopes because the prevailing winds are from the west and make their forced ascent on that side. Also, south-facing slopes receive more direct sunlight than north-facing ones, which makes for more rapid evaporation on the former and reduces the effectiveness of the rainfall or snowmelt received.

Natural Vegetation

The region is basically a forested one with coniferous species predominant, but there are many areas in which trees are absent. All the valleys and basins in the Southern and Middle Rockies and some in more northerly localities are virtually treeless except for riparian hardwoods along stream courses. A sagebrush association is widespread throughout the Wyoming Basin and in many lower parks and valleys as far north as southern British Columbia. The higher valleys and basins are more likely to be grass covered.

At the other altitudinal extreme, on the mountaintops, trees are also absent as a rule. These higher elevations have the low-growing but complex plant associations of the alpine tundra. Thus many mountain ranges of the region have a double tree line: one at lower elevation that marks the zone below which trees will not grow because of aridity and one at higher

FIGURE 14-9 There is no question about the direction of the prevailing wind in this area. The irresistible effect of wind shear on tree growth near the treeline is clearly shown in this Colorado scene. Branches are able to sprout and survive on this subalpine fir only on the direct leeward side of the tree trunk (TLM photo).

elevation above which trees cannot survive because of low temperatures, high wind, and short growing season (Fig. 14-9).

The relationship of altitude to latitude is shown quite clearly by the variation in elevation of the upper tree line in the Rocky Mountains. At the southern margin of the region, at 36° north latitude in New Mexico, the upper timberline occurs at 11,500 to 12,000 feet (3450 to 3600 m) above sea level. This elevation progressively decreases northward: about 9500 feet (2850 m) at 45° in Yellowstone Park; about 6000 feet (1800 m) at 49° at the international border; and about 2500 feet (750 m) at 60° at the northern extremity of the region near the Yukon border. Exposure, drainage, and soil characteristics influence the details of this pattern, but the general principle is clear cut.

The varying elevation of the upper tree line represents only one facet of the broader design of vertical zonation in vegetation patterns. As a result of abrupt altitudinal changes in short horizontal distances, various plant associations tend to occur in relatively narrow bands or zones on the slopes of the ranges (Fig. 14-10). The principle of vertical zonation is ever present in the Rockies; only the details vary.

THE OPENING OF THE REGION TO SETTLEMENT

The Rocky Mountains Region has always been characterized by sparse human population. It was no more attractive as a permanent habitation for aboriginal peoples than it is for contemporary Americans and Canadians; the climate was too rigorous and the resources too few.

There was a scattering of semipermanent Indian settlements in the Rockies—particularly in what is now Idaho and British Columbia—before the arrival of the first Europeans, but such settlements always were small and marginal. During summers there were frequent incursions into the Rockies by Plains Indians from the east and by Great Basin and Plateau Indians from the west, but almost always on a temporary basis. Thus, prior to the coming of Europeans various Indian tribes had considerable knowledge of the region, primarily in terms of summer hunting grounds and routes of passage. As a whole, however, the region was marginal to their way of life.

Except in the extreme southern part of the region that was penetrated by the Spaniards at the end of the sixteenth century, the first white men to see the Rocky Mountains were French fur traders who sporadically worked their way up the Great Plains rivers during the eighteenth century. Beginning in the late 1700s, there were several significant exploring expeditions in various parts of the region; some were government sponsored and others were backed as commercial ventures.

FIGURE 14-10 The vertical zonation of vegetation is clearly recognizable on Sierra Blanca in the Sangre de Cristo Mountains of Colorado (TLM photo).

Alexander Mackenzie was the first explorer of note to cross the Rockies; he did so in 1793 by penetrating the Peace River Valley in the far north of the region. The Lewis and Clark Expedition of 1803–1804 made its way up the Missouri River and crossed the Northern Rockies on the way to the mouth of the Columbia. While Lewis and Clark were still encamped on the Pacific shore, Simon Fraser, who was affiliat-

FIGURE 14-11 The presence of beaver attracted the first Europeans to the Rocky Mountain Region. These industrious rodents are still widespread, as evidenced by their lodges and dams (TLM photo).

ed with the North West Company, as was Mackenzie, established the first trading post west of the Rockies, at Fort McLeod on a tributary of the Peace River.

Fur trapping and trading became a way of life throughout the region (Fig. 14-11), with large periodic trading rendezvous at such places as Jackson's Hole at the eastern base of the Grand Tetons in Wyoming and Pierre's Hole in eastern Idaho. This colorful period, however, did not last long because the value of beaver fur declined in the 1840s, following changes in the style of men's hats. From that time until the discovery of gold in the late 1850s the Rocky Mountains served only as a barrier to the westward movement, and pioneers pushed through the lowest mountain passes as rapidly as possible on their way to the Oregon Territory or California.

The 1849 gold rush to California encouraged many people to cross the Rockies. While doing so, some prospected for gold in the stream gravels and found traces of the precious metal. Although most seekers of gold went on to California, many returned within a few years to prospect further in the numerous mountain gulches.

It was not until 1859, however, that gold in paying quantities was found in the Rocky Mountains. At almost the same time a gold rush began to Central City, Colorado, and another to the western flanks of the Cariboo Mountains in British Columbia. By the early 1860s the Cariboo gold fields had the largest concentration of people in western Canada; and Barkerville, center of the find, was the largest western town north of San Francisco. By the early 1870s Central City had grown to become the largest urban center in the Rocky Mountain Region.

A CLOSER LOOK   *Fire in Rocky Mountain Forests*

On a warm, windy July morning, in a small town in pine-covered foothills, a spark from a burning trash pile is lofted into a nearby meadow. With the aid of dry gusts, flames quickly race upslope into a dense thicket of pine, and with an explosive crackle, the forest bursts into flame. Within 3 days, 7000 (2800 ha) acres of pine forest are burned to the ground, along with 32 houses that were nestled among the trees. Across the valley, a simultaneous fire burns in a national park, destroying recreational facilities and archaeological ruins.

This scenario is becoming an increasingly common one in the conifer forests of the Rocky Mountain Region. Throughout the summer of 1988, the nation's attention was focused on the catastrophic fires that burned over a million acres of forest in the Yellowstone National Park area. Winds of up to 100 miles an hour (160 km/h) and drought conditions created raging fire. Only the cool weather of autumn allowed firefighters to finally douse the flames.

Why have forest fires become more destructive after a century of fire suppression and Smokey the Bear ethic? In fact, fires have occurred naturally for many centuries in the flammable pine forests of the West. Many conifers contain resins and esters that burn readily. Summers in the region are generally dry, but thunderstorms bring lightning that can ignite trees. Fires were more frequent in the warmer southern mountains, burning every 5 to 15 years in some places, and less frequent, every 200 to 400 years, in the cooler northern mountains. These natural fires that burned in the landscape created a mosaic pattern of burned, recovering, and unburned patches. When indigenous people arrived in the region, they hunted in mountains rich with game, and increased fire frequency by starting fires to drive the animals in the hunt.

The history of fires in past centuries is known because trees themselves record them. Many large, old conifer trees have thick bark that enables them to survive small fires. But sometimes a wound will be burned into the tree trunk. As the tree continues to grow, bark will begin to heal around the wound, creating a visible scar in the wood. This can happen many times. Historical reconstruction of these scars in the annual growth rings of trees gives us a record of past fires that goes back hundreds of years.

When European settlers came to the Rocky Mountains, their early activities often increased the number of forest fires. The new railroads threw sparks alongside the tracks, and mining for gold and silver brought people into the forests in large numbers for the first time. Soon, the new Westerners began to see forest fires as dangerous and wasteful, and by the late 1800s, efforts were made to extinguish them. By this time there were increasing numbers of cattle and sheep grazing the grasses of mountain meadows that helped spread the fires. By the turn of the century, fires were successfully suppressed in most of the Rocky Mountain region.

The mountain forests changed rapidly with the elimination of fire. Seedlings that would have been killed by small flames now survived to maturity. Forests had once been open and parklike; it was said that early explorers could ride at a gallop on horseback through the trees. After fire suppression, many of these forests grew into dense thickets. There was also a dramatic change in the species of trees present. Fire-sensitive trees, those easily killed by flames, now grew in abundance. Logs and woody debris that had been formerly turned to ash in fires, accumulated on the ground. In many places where there previously had been light fires that burned only underbrush, the stage was set by the dense stands of trees and woody fuel on the ground for catastrophic, destructive fires.

The altered forest landscape also had implications for wildlife. Forests that occasionally burned were a mix of young- and old-growth stands. Forests recovering from fire are often rich in grasses, berries, and shrubs—essential food sources for bear, deer and the rodents that feed coyote, mountain lion, hawks, and eagles. Natural patterns of fire played an important role in preserving the diversity of wildlife.

For a century, the care and management of these mountain forests has meant the elimination of fire, and the principal responsibility has fallen to the Forest Service. The familiar, friendly face of Smokey, with his Forest Service hat, has sternly reminded us for decades that only we can prevent forest fires. Now it appears that fires are an important ecological process in conifer forests, and the management of forests, parks, and wilderness area in the Rocky Mountains is undergoing an important shift. Restoration of presettlement forests, conservation of biodiversity, and even the safety of suburban homes in rustic forest settings, may depend on the necessary skills for the reintroduction of forest fire.

Several research tools can help managers reintegrate fire into the forests. The history of fires in a locale can be reconstructed from fire scars, so that the natural interval of years between fires can be known. Computer models of forest fire can also help predict the influence of fire in order to create the most effective management plan. Already, forest managers are using prescribed fires, those set under special circumstances, to reduce tree density and reduce woody debris. Prescribed fires are usually set during cool and moist conditions to avoid a severe burn. Selective thinning of trees by local communities for firewood or small mills can also help prevent destructive fires. Most important is the recognition that fire is an essential ecological force in western conifer forests and in the development of a long-term management plan to restore forest fires into the regional landscape.

Professor Melissa Savage
University of California
Los Angeles

FIGURE 14-12 Remnants of hundreds of ghost towns still exist in the region. This is Zincton in the Columbia Mountains of British Columbia (TLM photo).

Within a short time more people had settled in the mountain country than in all its previous history. Practically every part of the region was prospected and many valuable mineral deposits were found, especially in the Southern Rockies. Boom towns sprang up in remote valleys and gulches of the high country, which, in turn, led to the development of a series of narrow-gauge railroads, built at great expense per mile, for hauling out gold ore. As the higher-grade ores became exhausted, production declined in these camps; in time most became ghost towns (Fig. 14-12).

Lumbering and logging, grazing activities, irrigation agriculture, and the tourist trade later brought additional population to the mountains, but none was so significant in the early peopling of the region and in bringing its advantages to the attention of the rest of the country as gold mining.

THE MINING INDUSTRY

Wherever rich mineralized zones were found, mining camps developed. Colorado was especially important, with its Central City, Ouray, Cripple Creek, Victor, Leadville, Aspen, Georgetown, and Silver Plume. Wyoming and New Mexico were relatively insignificant, but farther north there were major discoveries around Virginia City in southwestern Montana, in the vicinity of Butte and Anaconda, in the Coeur d'Alene area of northern Idaho, and in the Kootenay and Cariboo districts of British Columbia. Although gold was the mineral chiefly sought, valuable deposits of silver, lead, zinc, copper, tungsten, and molybdenum were also found.

The region is highly mineralized. However, in common with many mining areas, there is much ebb and flow as mines are opened and closed with some frequency, usually in response to the fickle forces of the international market. A few of the major mining districts of the region are discussed here.

The Leadville District   One of the oldest and most important mining areas of the Rocky Mountain Region is located at Leadville, near the headwaters of the Arkansas River at an elevation of about 10,000 feet (3000 m). Placer gold was discovered in this remote valley in 1860, and within four months more than 10,000 people were in the camp. Silver–lead mining began a dozen years later, and zinc was discovered in 1885. Soon after the turn of the century the value of zinc output exceeded that of any of the other three metals.

The Leadville area continues as an erratically important mining district, producing all its former metals as well as molybdenum and tungsten. The molybdenum comes from North America's highest major mining complex [11,000 feet (3300 m)] at Climax, which is 13 miles (21 km) northeast of Leadville and is the world's largest producer of this valuable ferroalloy ore.

The Butte–Anaconda District   In western Montana is located one of the most famous of all North America's mining districts but one that has now ceased operations in a "temporary suspension" that may well become permanent. With mines at Butte and a gigantic smelter 23 miles (37 km) away at Anaconda, this area was one of the great copper producers of the world and also yielded significant amounts of silver, gold, lead, and zinc. Butte's "richest hill on earth" is honeycombed with tunnels that once had as many as 20,000 workers digging ore from 150 mines, although in recent years most production was concentrated in a single large open-pit operation.

For more than a century the Anaconda Minerals Company dominated the economy, and often the politics, of the state, besides owning most daily newspapers in Montana. In 1983, however, it was decided that the losses were too great, and the operation was shut down. Many mining enterprises in North America have suffered severe financial stress in recent years, but the Butte–Anaconda district represents perhaps the most spectacular example of the basic economic equation of rising mining costs and soft market prices for the ore. To compound the unpleasantness of the situation, serious water pollution from both the mining and smelting operations have created a major mess for Superfund cleanup.

The Coeur d'Alene District

This mining area, one of the richest of the Rocky Mountains, lies in northern Idaho. The two dozen or so underground mines of the district have yielded more than $2 billion worth of various ores in their century of operation. The history of mining in northern Idaho is a story of technological triumphs. The structural geology of the district is complicated and the mineralogy of the ores is complex. Techniques were devised to meet every challenge, however, in terms of mining, concentrating, and smelting.

Silver and lead are the principal ores mined, with byproduct output of zinc, antimony, copper, and gold. As of the mid-1990s only two mines were operating in the district. The principal urban place and smelter site is Kellogg.

The Kootenay District

In the southeastern corner of British Columbia lies one of Canada's foremost mining districts. The great Sullivan mine at Kimberly has been dominant, but a dozen other mines are active. The most important minerals have always been lead and zinc, with silver as the usual associated mineral. Other byproducts are tin, gold, bismuth, antimony, sulfur, and cadmium.

Concentrates from the Kootenay district, as well as from other parts of western Canada and some foreign sources, are treated in an enormous refinery at Trail, which is one of the world's greatest nonferrous metallurgical works (Fig. 14-13).

Recent Developments

Mining is inherently an unstable economic activity. As technologies change and demands fluctuate, there are almost immediate responses from the mineral industries. What is a valueless sulfide may become a valuable ore. Today's boom can become tomorrow's bust. The rise in value of one mineral may generate a frenzy of exploration and development, whereas a declining market for another produces a ghost town.

These characteristics are prominent in recent developments in the Rocky Mountain Region. Many established mining centers have retrenched their operations significantly and some, like Butte, have actually closed down. On the other hand, many new mineral extraction activities have begun, and the chance that more valuable deposits will be discovered is considerable. Major emphasis has been on energy minerals (coal, petroleum, natural gas), but not exclusively.

Southwestern Wyoming is a major focus of development. Open-pit coal and iron mines are now in production,

FIGURE 14-13 One of the largest metal refining complexes on the continent is found adjacent to the Columbia River at Trail, British Columbia (TLM photo).

and the continent's largest supply of trona (soda ash) is being mined on a large scale. Oil and gas extraction was begun in the 1970s, but collapsed in the 1980s.

The Canadian portion of the region has seen significant new development. Coal mining has expanded remarkably at several places in the Alberta foothills and at scattered localities in British Columbia, with the old coal-mining area of Crow's Nest Pass–Fernie (situated on the Alberta–British Columbia border) leading the way. Copper, however, is British Columbia's most valuable mineral product, with several new copper mines opening in the central part of the province. Significantly increased output of molybdenum has also occurred in British Columbia.

FORESTRY

Although there has long been small-scale local cutting of timber for mine props, firewood, cabin construction, and other miscellaneous purposes, large-scale commercial forestry has been restricted to certain sections of the Rocky Mountain Region, particularly the northern half. For example, less than 2 percent of total U.S. timber cut is from the Southern and Middle Rockies whereas nearly 10 percent comes from the Northern Rockies. Large sawmills and pulp mills operate at several localities in western Montana and northern Idaho.

The traditional logging area of western Canada has been coastal British Columbia, but lately increasing timber supplies have come from the interior of the province. Sawmilling is a major activity at such places as Fernie, Cranbrook, Nelson, Golden, and McBride. Pulping, although growing, is much more limited, but there are large mills at Prince George, Kamloops, and Castlegar. New pulping operations have also been started at several locations in the Alberta (eastern) foothills of the Rockies, in some cases processing aspen rather than conifers.

AGRICULTURE AND STOCK RAISING

In any consideration of Rocky Mountain agriculture and ranching, it must be remembered that the region consists chiefly of forested or bare rocky slopes. The occasional level areas are used for irrigation farming, dry farming, or ranching and thus attain an importance out of proportion to their size.

The earliest agricultural use of Rocky Mountain valleys involved livestock grazing and subsistence farm plots. Generally agricultural and pastoral settlement was late to occur because of inaccessibility, harsh winters, a short growing season, and Native American opposition. The southern San Luis Valley became two large Spanish land grants before the mid-nineteenth century, but no significant settlement arose there until after the Mexican War. Elsewhere in the region, farming and ranching in the valleys and basins were even later developments.

Today there are livestock enterprises in most valleys of the region. Both beef cattle and sheep ranches (Fig. 14-14) are scattered throughout the Rockies, but the former are much more widespread. The typical ranching operation involves *transhumance:* during the summer the animals are taken (either by walking or by truck) to high-country pastures while valley ranch lands are used for growing hay; in early autumn the stock is returned to the ranch, where the hay is used as basic or supplemental feed. Most summer grazing is done under purchased permit on national forest lands; a long-established policy by the two principal land management agencies of the West—the Forest Service and the Bureau of Land Management—affirms the principle of grazing by permit on federal land, thereby maintaining continuity and coherence in these ranching operations.

Although a certain amount of crop growing is done under natural rainfall conditions in the valleys of the region, most farming involves irrigation. Most of this irrigation is on a small scale and simply involves diversion of water from the valley-bottom stream into adjacent fields. The basic crop throughout the region is hay (Fig. 14-15). Irrigated valleys yield valuable supplies of alfalfa and other hay crops for winter feeding of livestock. Some specialty crops are also grown; pinto beans and chili peppers in New Mexico, miscellaneous vegetables in Colorado and Utah, sugar beets and grains in Idaho and Montana, fruits around Flathead Lake and in some Kootenay valleys, and wheat in the Kootenay River valley and around Cranbrook (Fig. 14-16).

The most valuable specialty crop is a relative newcomer—ginseng. This is a parsniplike root that is used as an herbal medicine (pills and teas). First introduced in the early 1980s, it has become one of the most lucrative crops in British Columbia. Ginseng plants are grown with irrigation under protective shade tarps. It is a labor-intensive crop but very valuable ("the most expensive legal crop in the world"). It occupies only limited acreage but is now well established in the Fraser Valley around Lillooet, the Thompson Valley near Kamloops, and the Okanagan Valley around Vernon and Kelowna.

The most notable irrigated valley in the Rocky Mountain Region, and one of Canada's most distinctive speciality

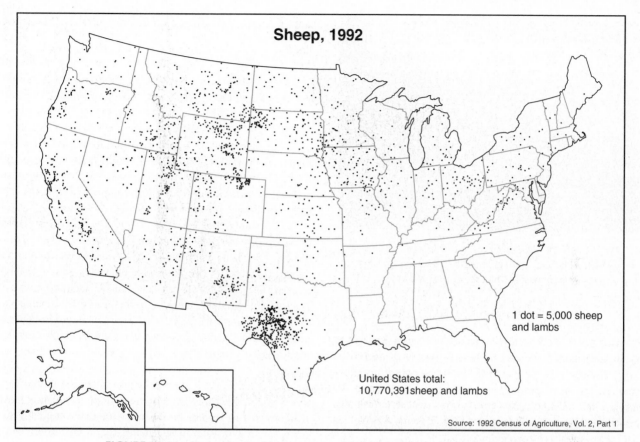

Sheep, 1992

1 dot = 5,000 sheep and lambs

United States total:
10,770,391 sheep and lambs

Source: 1992 Census of Agriculture, Vol. 2, Part 1

FIGURE 14-14 Sheep are raised in many parts of the country, but particularly in the West.

crop areas, occupies the long narrow trench of the Okanagan Valley (Fig. 14-17). Extending north for 125 miles (200 km) from the international border, the valley is only 3 to 6 miles (5 to 10 km) wide except where it broadens a bit into tributary valleys in the north. Large lakes occupy most of the valley floor, and farming is limited to adjacent terraces. Irrigation is necessary in the desertlike conditions [annual rainfall, 9 inches (23 cm) and temperatures up to 110°F (43°C)] of the southern part of the valley, but general farming can be carried on under natural rainfall in the north [precipitation, 17 inches (43 cm)].

Various field crops, feed crops, and vegetables are grown, but fruit growing is the distinctive activity in the valley. The Osoyoos section in the south produces the earliest fruits in Canada. Apples are the major crop (occupying about two-thirds of the valley's orchard area and yielding about one-third of Canada's total output), but the valley is particu-

larly noted as one of only two areas in the nation that has a sizeable production of soft fruits—peaches, plums, pears, cherries, and apricots. Lately there has been a significant increase in the acreage devoted to vineyards.

Because of its relatively mild climate and abundant sunshine, the Okanagan is an attractive place to live. Consequently, increasing population pressure exacerbates problems of congestion and pollution and raises land prices in an ever-higher spiral that makes it difficult for small farmers to continue farming.

The other conspicuous farming area in the region is the San Luis Valley. Although it has a short growing season because of its high elevation [above 7000 feet (2100 m)] and its rainfall totals are low, much of the earlier ranch land has been converted to irrigated farming. An underground water resource has been easily tapped by artesian wells. Most of the valley's farmers have been attracted to center-pivot technolo-

FIGURE 14-15   Wherever there are valleys with relatively flat land; hay is usually grown in the summer. This scene is near Gunnison, Colorado (TLM photo).

gy, with the result that the San Luis Valley now has one of the greatest concentrations of center-pivot irrigation to be found anywhere in the world. Crop options are limited because of the short growing season and harsh winters. Although hay is the most widely grown crop, potatoes constitute the principal source of farm income. Vegetables and barley are also grown in quantity.

WATER "DEVELOPMENT"

The Rocky Mountain Region is sometimes spoken of as the "mother of rivers" because so many of the major streams of western North America have their headwaters on the snowy slopes of the Rockies. This water-collecting function is generally considered one of the two leading economic assets (tourism is the other) of the region. From the Rocky Mountains the Rio Grande and Pecos flow to the south; the Arkansas, Platte, Yellowstone, Missouri, South Saskatchewan, North Saskatchewan, Athabasca, Peace, and Liard flow to the east; and the Stikine, Skeena, Fraser, Columbia, Snake, Green, and Colorado flow to the west (Fig. 14-18).

The capriciousness of river flow has long been recognized, and the alternation between flood and low-water stages has been deplored. Dam building is the chief tool to smooth these imbalances of flow and make the waters more "usable" for various purposes. In the past, most "development" of Rocky Mountain rivers was deferred to downstream locations, particularly in the Great Plains and Intermontane regions. But later, dam building came to the high country, modified, in some cases, by wholesale water-diversion schemes.

The principle of transmountain diversion is that "unused" water, generally from western slope streams, is taken from an area of surplus by means of an undermountain tunnel to an area of water deficit, normally on the eastern slope

FIGURE 14-16   Dairying is an important activity in the Broad Valley Rockies of western Montana. This is the Mission Valley near Ronan (TLM photo).

FIGURE 14-17   The Okanagan Valley is western Canada's premier fruit-growing district. Shown here are Lake Osoyoos, the town of Osoyoos, and orchards (darker areas) covering most of the valley bottom lands (TLM photo).

of the Rockies or in the western edge of the Great Plains. Potential western-slope users are compensated for this loss by storage dams built there to catch and hold flood-stage flow for later release when the river is low.

Until the 1960s there had been almost no dam building in the Canadian portion of the region except for five small hydroelectric dams along the lower Kootenay River to supply power to the huge smelter at Trail. In 1964, however, the United States and Canada ratified the Columbia River Treaty, which provided for the construction of major dams on the upper Columbia River and its tributaries to control floods and permit increased hydroelectric power generation in both countries. The dams were paid for by the United States, as an advance against power generated in the future in Canada but sold in the United States.

An even more grandiose scheme has been constructed on the upper reaches of the Peace River near Finlay Forks, British Columbia. The W. A. C. Bennett Dam backs up a reservoir for 70 miles (110 km) on the Peace and another 170

miles (270 km) on two major tributaries, the Finlay and the Parsnip Rivers. Much of the power is transmitted 600 miles (960 km) to metropolitan Vancouver, the largest population concentration in western Canada.

The long, narrow, structural valleys of the Columbia Mountains and the Rocky Mountain Trench are thus increasingly being filled with reservoirs, resulting in undeniable economic benefits and equally undeniable environmental degradation.

THE TOURIST INDUSTRY

The other major role of the Rocky Mountain Region is in providing an attractive setting for outdoor recreation. The tourist industry is undoubtedly the most dynamic segment of the regional economy.

Summer Tourism

With its high rugged mountains, spectacular scenery, extensive forests, varied wildlife, and cool summer temperatures, the Rocky Mountain Region is a very popular summer vacationland. The location of the region between the Great Plains on the east and the intermontane and Pacific coastal areas on the west places it directly across lines of travel. Nearby flatlanders flock to the mountains for surcease from the summer heat of Dallas, Kansas City, or even Denver. And people come from greater distances to sample the scenic delights of Banff (more than 3 million tourists per year) or Yellowstone (200,000 visitors in a midsummer week, a number equal to more than one-third of Wyoming's resident population).

FIGURE 14-18   Rivers flow out of the Rocky Mountains in all directions. This is the Little North Fork of the Clearwater River in central Idaho (TLM photo).

Throughout the Rockies there are places of interest for visitors whose interests and activities are varied, but spectacular scenery is the principal attraction. Many of the outstanding scenic areas have been reserved as national parks, which generally function as the key attractions of the region. Seven of the most popular tourist areas in the region, in terms of numbers of visitors, are described next.

New Mexico Mountains The southernmost portion of the Rocky Mountains, located in northern New Mexico, is not particularly rugged or spectacular and consists primarily of pleasant forested slopes. Its summers, however, are considerably cooler than those of the parched plains of the Southwest. It is an area with a rich historical heritage that is manifested in the pervasive Native American and Hispanic character of the cultural landscape. As a result, the narrow twisting streets of Santa Fe (the principal focal point of the area) are jammed with visitors' vehicles during the summer (Fig. 14-19). The Taos area is a center for dude ranch and youth-camp activities, and Red River has developed into a year-round tourist resort.

Pikes Peak Area Colorado Springs is a city of the Great Plains, but its site is at the eastern base of one of the most famous mountains in America. Pikes Peak, with a summit elevation of 14,110 feet (4233 m), is far from being the highest mountain in Colorado, but its spectacular rise from the plains makes it an outstanding feature. The surrounding area contains some of the most striking scenery (Garden of the Gods, Seven Falls, Cheyenne Mountain, Cave of the Winds, Rampart Range) and some of the most blatantly commercial (Manitou Springs) tourist attractions in the region. Few tourists visit the Southern Rockies without at least a brief stop in the Pikes Peak area, as overcrowding of even the unusually wide streets of Colorado Springs gives eloquent evidence.

Denver's Front Range Hinterland The Front Range of the Southern Rockies rises a dozen miles (16 km) west of Denver and the immediate vicinity provides a recreational

FIGURE 14-19 Santa Fe is the oldest (1609) European settlement in western United States. Shown here is the oldest house in the city, a prominent tourist attraction (TLM photo).

area for the residents of the city as well as visitors from more distance places. Denver maintains an elaborate and extensive group of "mountain parks," which are actually part of the municipal park system. There are thousands of summer cabins for rent; many streams to fish; deer, elk, mountain sheep, and bear to hunt; dozens of old mining towns to explore; and countless souvenir shops in which to spend money.

Rocky Mountain National Park   Following many years of agitation by the people of Colorado for the establishment of a national park in the northern part of the state to preserve the scenic beauty of that section of the Continental Divide, a rugged area of 400 square miles (1036 km^2) was reserved by Congress in 1915 as Rocky Mountain National Park (Fig. 14-20). It includes some of the highest and most picturesque peaks, glacial valleys, and canyons of the region, as well as extensive forested tracts.

Spectacular Trail Ridge Road traverses the park, connecting the tourist towns of Grand Lake and Estes Park, and reaches an elevation of 12,185 feet (3656 m). Automobile touring is the principal activity in the park, but hiking, climbing, and trail riding are also popular.

Yellowstone–Grand Teton–Jackson Hole   In the northwestern corner of Wyoming is an extensive forested plateau of which a 3500-square-mile (9065-km^2) area has been designated as Yellowstone National Park. It was established in 1872 as the first national park in the world. It is lacking in spectacular mountains but contains a huge high-altitude [elevation 7700 feet (2300m)] lake, magnificent canyons and waterfalls, and the most impressive hydrothermal displays—geysers, hot springs, fumaroles, hot-water terraces—in the world (Fig. 14-21). A few miles to the south is Grand Teton National park, a smaller and more recently re-

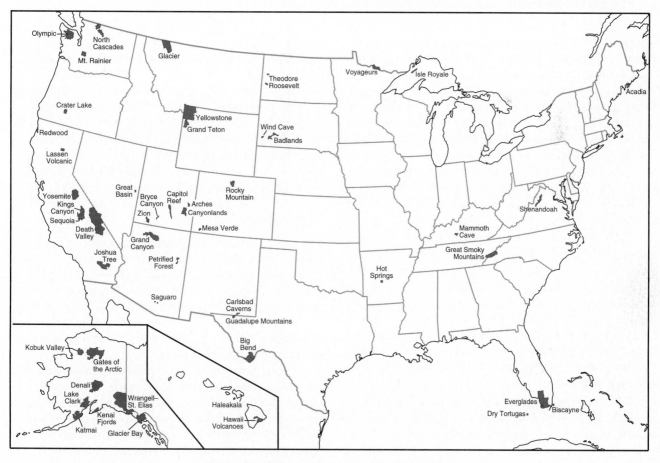

FIGURE 14-20   National parks of the United States.

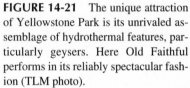

FIGURE 14-21 The unique attraction of Yellowstone Park is its unrivaled assemblage of hydrothermal features, particularly geysers. Here Old Faithful performs in its reliably spectacular fashion (TLM photo).

served area that encompasses the rugged grandeur of the Grand Teton Mountains, a heavily glaciated fault block that rises abruptly from the flat floor of Jackson Hole. Jackson Hole was an early furtrappers' rendezvous that is now the winter home of the largest elk herd on the subcontinent.

The Grand Tetons are particularly attractive to hikers and climbers; Yellowstone is a motorists' park. The ubiquitous bison and a great variety of other species of wildlife add to the interest of the area. In spite of its relatively remote location and the brevity of the tourist season, the Yellowstone–Grand Teton country annually attracts more than 4 million visitors.

Waterton–Glacier International Peace Park Glacier National Park in Montana and Waterton Lakes National Park in Alberta are contiguous and have similar scenery. The mountains are typical of the Canadian Rockies, with essentially horizontal sediments uplifted and massively carved by glacial action. The area is a paradise for hikers, climbers, horseback riders, and wildlife enthusiasts, and contains one of the most spectacular automobile roads on the subcontinent, the Going-to-the-Sun Highway that traverses Glacier Park from east to west.

The Canadian Rockies The most magnificent mountains in the region are found west of Calgary and Edmonton on the Alberta–British Columbia boundary. The national park system of Canada was started in 1885, when a small area in the vicinity of the mineral hot springs at Banff was reserved as public property (Fig. 14-22). The famous resorts of Banff (Fig. 14-23), Lake Louise, and Jasper were de-

veloped by the two transcontinental railways, and the Canadian Pacific and Canadian National are still major operators in the area. Today there are four national parks and six provincial parks with a contiguous area totaling nearly 11,000 square miles (28,500 km^2), probably the largest expanse of frequently visited nature recreational areas in the world (Fig. 14-24). Heavily glaciated mountains, abundant and varied wildlife, spectacular waterfalls, colorful lakes, deep canyons, the largest ice field in the Rockies, and luxurious resort hotels characterize the area.

Winter Sports

The region possesses superb natural attributes for skiing and snowboarding. High elevation ensures a long period of snow cover; many areas can provide skiing from Thanksgiving to mid-May. Because winter storms have lost much of their moisture by the time they reach the Rockies, the snow is often of the fine, powdery variety preferred by skiers. And there is an abundance of different degrees of slopeland to accommodate all classes of skiers.

In the past, the region was relatively remote from large population centers, and few skiers came to the Rockies. To some extent this is still true, but many regional ski areas, especially in the Southern Rockies, have experienced a rapid increase in the number of users. Nearby populations have grown, and skiers travel much greater distances to ski than they did in the past. About 20 percent of the citizens of Colorado are skiers, and some two-thirds of the users of ski areas in the state are local people (Fig. 14-25). In the other

FIGURE 14-22   National parks of Canada.

large winter sports areas of the region about half the users are nonresidents.

Fishing and Hunting

No other generally accessible region in North America provides such a variety of faunal resources to tempt the hunter and fisherman. The fishing season normally lasts from May to September, with various species of trout as the principal quarry. Only diligent artificial stocking can maintain the resource in the more accessible lakes and streams, where overfishing is rampant. The hunting season lasts from August to December in various parts of the region. The list of legal game is extensive, running from cottontail rabbit to grizzly bear and from pronghorn antelope to mountain goat. In an av-

FIGURE 14-23   The tourist town of Banff is situated in a valley surrounded by spectacular mountains (TLM photo).

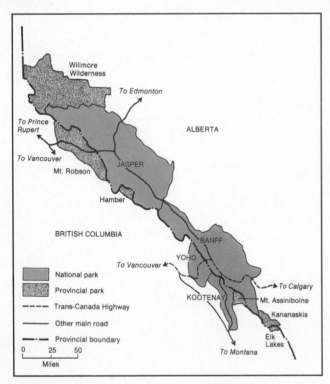

FIGURE 14-24   There is a splendid collection of national and provincial parks in close jutaposition along the trend of the Canadian Rockies in Alberta and British Columbia.

erage year more than a third of a million big-game animals and two dozen hunters are shot in this region.

Problems

The flocking of visitors to these high-country scenic areas is a mixed blessing. By its very nature, a pleasurable outdoor experience can be ruined by overcrowding of people and overdevelopment of facilities to cater to the crowds. The national parks and other prime scenic attractions of the Rockies have become centers of controversy between the advocates of wilderness preservation on one hand and developers on the other. Important, precedent-setting decisions are now being made. Should a complete summer–winter resort town site be constructed at Lake Louise? Should the roadway system of Yellowstone Park be converted to one-way traffic? Do we want national parks or national parking lots?

TRANSPORTATION

The Rocky Mountains have always functioned as a conspicuous barrier to east–west travel. The principal early trails and later routeways across the Rockies either passed around the southern end in New Mexico or crossed through the Wyoming Basin between the Southern and Middle Rockies. The first "transcontinental" railroad, the Union Pacific, used

FIGURE 14-25   There are many developed winter sports areas in the Rocky Mountain Region, with the greatest concentration in north-central Colorado. This is Breckenridge (TLM photo).

the Wyoming Basin route. It continues as the busiest east–west rail line, with as many as 75 freight trains per day in operation. The paralleling highway, Interstate 80, vies with Interstate 40 (which passes through New Mexico around the southern end of the Rockies) as the most heavily used east–west roadway. Only the Denver and Rio Grande Railway built a line across the high-altitude Southern Rockies, and it did not become an all-weather route until the 6-mile (10-km)-long Moffat Tunnel was constructed west of Denver in 1927.

North of the Wyoming Basin there are six railway routes across the Rockies. Three cross the Northern Rockies in Montana and Idaho and three are Canadian lines. The Canadian Pacific built the pioneer route westward from Calgary through Banff; its descent on the western side of the Canadian Rockies is through the famous spiral tunnels down into the Rocky Mountain Trench. This route continues westerly through the precipitous Selkirk Mountains by a long tunnel and then continues toward Vancouver along river valleys and through canyons. A second Canadian Pacific line was subsequently constructed over the more southerly Crow's Nest Pass route. The government-owned Canadian National Railway crosses the Rockies by means of the northerly but low-level Yellowhead Pass in Jasper National Park.

Canada's most ambitious road-building program was the Trans-Canada Highway, which extends from the Atlantic Ocean to the Pacific. The most expensive and difficult section of the highway to construct was that crossing the Rockies, especially in the Selkirk Mountains where the Rogers Pass segment is blocked several times a year by avalanches despite the protection offered by lengthy concrete snowsheds (Fig. 14-26).

SETTLEMENT NODES

This is virtually a cityless region. No urban agglomeration in the Rocky Mountains has as many as 80,000 people. The relatively modest existing population nodes are chiefly associated with major lumbering, pulping, and mining-smelting activities or with agricultural valleys. The greatest concentration of population in the entire region is in the Okanagan Valley, where a dense farming population is clustered around the three urban centers of Kelowna, Penticton, and Vernon in an area that is also popular with summer vacationers. Three hundred miles (480 km) north of the Okanagan is one of the region's fastest growing urban places, Prince George, which is a notable forest-processing center as well as being the commercial hub of British Columbia's northern interior (Table 14-1).

During the busy tourist season the population of some resort towns swells to many times the normal size. Estes Park in Colorado, Jackson in Wyoming, and Banff in Alberta are prime examples.

FIGURE 14-26 The most difficult stretch of the Trans-Canada Highway to build was that traversing the Selkirk Mountains through Rogers Pass. This monument in Rogers Pass commemorates the construction feat (TLM photo).

TABLE 14-1

Largest Urban Places of the Rocky Mountain Region, 1995

Name	Population of Principal City	Population of Metropolitan Area
Bozeman, MT	27,000	
Butte, MT	29,000	
Coeur d'Alene, ID	25,000	
Kamloops, BC	68,000	
Kelowna, BC	72,000	
Missoula, MT	39,000	
Penticton, BC	35,000	
Prince George, BC	71,000	
Rock Springs, WY	22,000	
Santa Fe, NM	64,000	131,000

THE OUTLOOK

Permanent settlements in the region are based mostly on mining, forestry, limited agriculture, and tourism. Farming and ranching are developed almost to capacity and cannot be expected to change to any great extent. Logging activities have also probably reached their limit except in British Columbia, which has considerable capability for expansion, provided that demand is sufficient. The British Columbia lumber industry depends considerably on the home-building market in the United States, whereas its pulp and paper industry is particularly affected by varying competition from Scandinavia.

Mining will undoubtedly fluctuate in different areas, in a boom-and-bust syndrome that has great historical precedent in the region. The abrupt hydrocarbon expansion of the 1970s was followed by an equally sudden collapse in the 1980s.

Contrasting prosperity in mining highlights the instability of the regional economy and is counterpointed by gyrating population fluctuations. Regional population growth was three times the national average (in the United States) in the 1970s, and well below the national average in the 1980s. Wyoming, for example, was the fastest-growing state (proportionally) in the 1970s (expanding by 40 percent), owing to in-migration, and virtually stagnated in the 1980s, owing to out-migration.

The dynamic future of the Rocky Mountains appears to be intimately associated with that seasonal vagabond, the tourist. Natural attractions are almost limitless; recreational developments on government lands have been accelerated by the National Park Service, the Forest Service, and other federal, state, and provincial agencies; and improvements in transportation facilities and accommodations are being made haphazardly but continually. Despite the high cost of gasoline, tourism continues to be a growth industry.

An offshoot of tourism is the growth of "second home" developments, which are occurring with increasing frequency in recreational or scenic areas throughout the continent and can be expected to increase in the future. There are many foci of such developments in the Rockies; notable examples include the area around Taos (New Mexico), many prominent ski resorts (especially Vail, Aspen, and Steamboat Springs),

Estes Park (Colorado), Jackson (Wyoming), the Flathead Lake area and the Bitterroot Valley of Montana, the Banff area of Alberta, and Golden (British Columbia).

A major philosophical controversy with significant social, political, and economic overtones has arisen in the region: How is it possible to reconcile the pressures of rapidly expanding development while maintaining an environment that visitor and resident alike can enjoy? The problem surfaces most conspicuously in connection with open-pit mining, mineral boom towns, large-scale recreational developments, and national parks in general. The Rocky Mountains constitute a region of remarkable aesthetic appeal, but the balance of effective use without destructive abuse seems increasingly difficult to attain.

SELECTED BIBLIOGRAPHY

ALWIN, JOHN A., *Western Montana: A Portrait of the Land and Its People.* Helena: Montana Magazine, 1983.

ARNO, STEPHEN F., "Glaciers in the American West," *Natural History,* 78 (1969), 84–89.

CHENG, JACQUELINE R., "Tourism: How Much Is Too Much? Lessons for Canmore from Banff," *The Canadian Geographer,* 24 (Spring 1980), 72–80.

CROWLEY, JOHN M., "Ranching in the Mountain Parks of Colorado," *Geographical Review,* 65 (1975), 445–460.

———, "The Rocky Mountain Region: Problems of Delimitation and Nomenclature," *Yearbook,* Association of Pacific Coast Geographers, 50 (1988), 59–68.

FARLEY, A. L., *Atlas of British Columbia: People, Environment, and Resource Use.* Vancouver: University of British Columbia Press, 1979.

GEORGE, RUSSELL, "More Energy for British Columbia in Peace River Coal," *Canadian Geographic,* 100 (February–March 1980), 26–33.

GRIFFITHS, MEL, AND LYNNELL RUBRIGHT, *Colorado.* Boulder, CO: Westview Press, 1983.

HARRINGTON, LYN, "The Columbia Icefield, *Canadian Geographical Journal,* 80 (June 1970), 202–205.

IVES, JACK D., ED., *Geoecology of the Colorado Front Range.* Boulder, CO: Westview Press, 1980.

———, ET AL., "Natural Hazards in Mountain Colorado," *Annals,* Association of American Geographers, 66 (1976), 129–144.

KRUEGER, RALPH R., AND MAGUIRE, N. G., "Protecting Specialty Cropland from Urban Development: The Case of the Okanagan Valley, B.C.," *Geoforum,* 16 (1985), 287–300.

MATHER, COTTON, P. P. KARAN, AND GEORGE F. THOMPSON, *Beyond the Great Divide: Denver to the Grand Canyon.* New Brunswick, NJ: Rutgers University Press, 1992.

ROE, JOANN, "Our Beautiful Okanagan," *Canadian Geographic,* 102 (December 1982–January 1983), 25–26.

TODHUNTER, RODGER, Banff and the Canadian National Park Idea," *Landscape,* 25 (1981), 33–39.

TRENHAILE, A. S., "Cirque Elevation in the Canadian Cordillera," *Annals,* Association of American Geographers, 65 (1975), 517–529.

———, "Cirque Morphometry in the Canadian Cordillera," *Annals,* Association of American Geographers, 66 (1976), 451–462.

VEBLEN, THOMAS T., AND DIANE C. LORENZ, *The Colorado Front Range: A Century of Ecological Change.* Salt Lake City: University of Utah Press, 1991.

WALLACH, BRET, "Sheep Ranching in the Dry Corner of Wyoming" *Geographical Review,* 171 (January 1981), 51–63.

WYCKOFF, WILLIAM, AND LARRY M. DILSAVER, EDS., *The Mountainous West: Explorations in Historical Geography.* Lincoln: University of Nebraska Press, 1995.

The Intermontane West

15

The western interior of the United States makes up the Intermontane West Region. As the term *intermontane* implies, it primarily encompasses the vast expanses of arid and semiarid country between the Rocky Mountain cordillera on the east and the major Pacific ranges (Sierra Nevada and Cascade) on the west (Fig. 15-1).

The boundary of the region is relatively clear-cut in most sections, for there are prominent geomorphic units that are usually associated with obvious land-use changes. In only two areas is the regional boundary indistinct.

1. In northwestern Colorado there is an eastwest transition from Rocky Mountains to Intermontane Region that is broad and indistinct.

2. In southern California there is a very clear environmental boundary between the desert portion of the Intermontane Region and the various Transverse and Peninsular mountain ranges of the California Region. But the spillover of urban population from the Los Angeles basin into the Palm Springs area of the Colorado Desert and the Antelope Valley section of the Mojave Desert is so pronounced and the resultant functional connection of these two areas with the Los Angeles metropolis is so strong that it

seems clear to this observer that these two areas should be considered outliers of the southern California conurbation and thus part of the California Region. The boundary of the Intermontane Region is therefore drawn east of these two areas.

The vast extent of the Intermontane Region and its topographic diversity have led some regionalists to consider it as several regions rather than one. For the scale of generalization appropriate to this book, however, it is felt that a single regional designation is warranted. The three prominent subregions identified in Figure 15-2 include the Columbia Plateau in the north; the Colorado Plateau, occupying parts of four states, in the eastern portion of the region; and the Basin-and-Range section, the largest subregion, extending in a crescent from southern Oregon to western Texas.

ASSESSMENT OF THE REGION

In broad generalization, the Intermontane West can be thought of as a sparsely populated region whose vast extent, relatively isolated inland location, rugged but varied terrain, and paucity of freshwater make it best suited to serve the na-

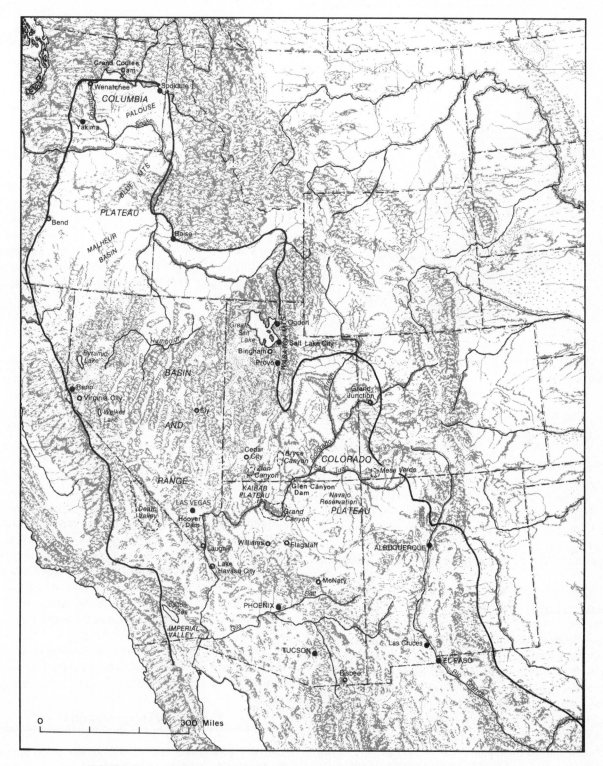

FIGURE 15-1   The Intermontane West (base map copyright A. K. Lobeck, reprinted by permission of Hammond, Inc.).

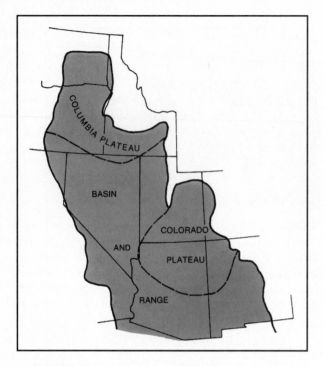

FIGURE 15-2   Major topographic subdivisions of the Intermontane Region.

tion as a limited source of primary resources and as a recreational ground.

Where people have attempted economic endeavor in this region, the bulldozer has been their instrument, the cloud of dust their symbol, and a drastically altered and "tamed" landscape the result. Where humans are found in any numbers, there are the ordered fields of an irrigated farming development, the giant amphitheater of an open-pit copper mine, the massive wall of a major dam, the extensive runways of a military airfield, or the bright lights of a gambling town.

The region's economy is based partly on primary production, especially from irrigated agriculture, pastoralism, and mining, and partly on such tertiary activities as tourism and government expenditures. Where water is available for irrigation and land is sufficiently level, intensive farming prevails as an oasis type of development. Sheep and beef cattle are grazed widely throughout the region, usually on extensive ranches, and the region, in general, is well endowed with economic mineral resources. The greater part of the feeding and slaughtering of livestock, refining of ores, and marketing of both occurs outside the region.

It is significant that most of the land in the Intermontane West is still part of the public domain, and government (primarily federal government) expenditures for manage-

ment, development, construction, and exploitation constitute a major contribution to the local economy. The vast extent of federal ownership also has become an increasingly volatile sore point for some of the people of the region, and fires of the "sagebrush rebellion" burn stronger here than elsewhere in the nation. Efforts to transfer land ownership from federal to state, local, or private control are continuing, and some of the most acrimonious confrontations between "conservation" and "development" occur here.

Inaccessibility was long a hallmark of the region. The terrain is characterized by deep gorges, abrupt cliffs, and steep mountainsides. The climate is noted for temperature extremes. And almost everywhere there is a scarcity of water. Civilization was slow to reach many areas. The last fights with Native Americans occurred just over a century ago. Some post offices were still being served by packhorses as recently as 60 years ago. The last sizable portion of the country to be provided with all-weather roads was northeastern Arizona–southeastern Utah in the early 1960s.

The historical movement of population in the region has been from both east and west, generally *across* rather than *into*. The region served as a barrier to westward expansion and only incidentally as a goal for settlement except in the Great Salt Lake basin. In recent years, however, this pattern has been modified. The interior West still serves in part as a transit region, but more and more it is the chosen destination of automobile nomads. Tourists, hunters, skiers, and other recreationists visit the region for a few days or weeks of vacation, and increasingly people are settling on a long-term or permanent basis, particularly in the sunny southern section.

The explosive rate of population growth, combined with rapid economic "development" of various kinds, ushered in an era of unprecedented conflict in the region. The equation is simple: Attractive landscape plus fragile environment plus large-scale development schemes equal unending controversy.

If there is an eternal verity for the Intermontane Region, it is scarcity of water. The fact that evaporation exceeds precipitation throughout the region is critical to all forms of life. John Wesley Powell, the most notable explorer of the inland West, probably said it best: "All the great values of this territory have ultimately to be measured to you in acre-feet."

TOPOGRAPHIC VARIETY

There is a great deal of topographic variety in the Intermontane Region. Each of the three subregions has its distinctive geomorphic personality, which can be easily recognized and described.

The Columbia Plateau

The Columbia Plateau lies between the Cascade Mountains on the west and the Rocky Mountains on the east and north and grades almost imperceptibly into the Basin-and-Range section to the south. Although called a plateau, which popularly suggests a rather uniform surface, the area has quite varied relief features of mountains, plateaus, tilted fault blocks, hills, plains, and ridges. In general, this intermontane area is covered with basalt lava flows that originally were poured out over a nearly horizontal landscape and interbedded with a considerable quantity of silts that were deposited in extensive lakes. After the outpouring of the sheets of lava and the deposition of the lake beds, the surface of much of the region was strongly warped and faulted, so that the present surface of the lava varies from a few hundred feet above sea level to nearly 10,000 feet (3000 m) in elevation (Fig. 15-3).

In central Washington, steep-sided, flat-floored, streamless canyons, cut the plateau into a maze known as the *channeled scablands*. These abrupt gorges were eroded by raging floodwaters released by the breaking of the huge ice dam that formed an extensive Pleistocene lake (Lake Missoula) in northwestern Montana. This massive discharge is estimated to have been 10 times the combined present flow of all the world's rivers. Moreover, it is believed that the ice dam formed at least 40 times, with giant discharges after each breakage.

In eastern Washington is the rolling Palouse hill country, deeply mantled with loess. Northern Oregon has an irregular pattern of faulted and folded mountains, generally referred to as the Blue and Wallowa Mountains. Southeastern Oregon and southern Idaho have variable terrain, ranging from the lava-covered flatness of the Snake River Plain in southeastern Idaho to the irregular basins and hills of the Malheur Basin in south-central Oregon to the spectacularly deep canyons of the lower Snake drainage.

The Colorado Plateau

This large area consists of several strongly differentiated parts but has sufficient unity to justify separation from adjacent subregions. It stretches outward from the Colorado River and its tributaries in Colorado, Arizona, New Mexico, and Utah. The greater part consists of a series of flattish summit areas slightly warped or undulating as a result of earlier crustal movements and interrupted by erosion scarps in the eastern portions and fault scarps in the western parts. Physiographically the area is distinguished by the following features.

1. All the subregion except the bottoms of canyons and the highest peaks has an elevation of 4000 to 8000 feet (1200 to 2400 m). Some high plateau surfaces

FIGURE 15-3   A lava landscape in southern Idaho. This is a portion of the lengthy gorge of the Snake River (TLM photo).

FIGURE 15-4 Canyon de Chelly (near Chinle in northeastern Arizona) is a classic example of mesa-and-scarp terrain (TLM photo).

reach 11,000 feet (3300 m), and a few mountain ranges have still higher peaks.

2. Hundreds of remarkable canyons (Fig. 15-4) thread southeastern Utah, northern Arizona, and the Four Corners country in general. They make this subregion the most dissected and difficult to traverse part of the country.

3. Numerous arroyos, which cut some parts of the subregion into mazes of steep-sided chasms, are dry during most of the year but filled from wall to wall during the rare rains.

4. Mesas, flat-topped islands of resistant rock, rise abruptly from the surrounding land.

The basic topographic pattern might be described as mesa and scarp—that is, flat summits bordered by near-vertical cliffs. Some summit areas, such as the Kaibab Plateau in northern Arizona and Mesa Verde in southwestern Colorado, are remarkably extensive. The scarps, too, sometimes extend to great lengths; the Book Cliffs of Colorado–Utah, for example, are more than 100 miles (160 km) long.

Throughout the Colorado Plateau subregion the land is brilliantly colored, particularly in the exposed sedimentary surfaces of the scarp cliffs. The Painted Desert of northern Arizona, which is badlands terrain, is especially noted for its rainbow hues, but throughout the mesa-and-scarp country the landscape is marked by colorful rocks and sand.

The Basin-and-Range Section

To the northwest, west, southwest, south, and southeast of the Colorado Plateau, from southern Oregon to western Texas, is a vast expanse of desert and semidesert country that has notable physiographic similarity. Throughout this extensive area the terrain is dominated by isolated mountain ranges that descend abruptly into gentle alluvial piedmont slopes and flat-floored basins (Fig. 15-5).

The mountain ranges are characteristically rough, broken, rocky, steep sided, and deep canyoned. They tend to be narrow in comparison with their length, distinctly separated from one another, and often arranged in parallel patterns. Although their origins are somewhat diverse, most consist of tilted and block-faulted masses of previously folded and peneplained rocks. The canyons and gullies that drain them are waterless most of the time, harboring intermittent streams only after a rain.

Near the base of the mountains there is normally an abrupt flattening out of the slopes. As the streams reach the foot of the mountains, their gradient is sharply decreased so that they can no longer carry the heavy load of silt, sand, pebbles, and boulders that they have brought down from the

FIGURE 15-5   The basin-and-range subregion consists mostly of alternating mountains and valleys in parallel arrangement. This is the Snake Range in east-central Nevada. The high point shown here is Wheeler Peak, at 13,083 feet (4360 m), the second highest point in the state (TLM photo).

highlands, and considerable deposition takes place. (Although the streams flow only intermittently, they are subject to violent floods, and the amount of erosion that they can accomplish is tremendous.) This piedmont deposition generally occurs in fan- or cone-shaped patterns (called alluvial fans) that become increasingly complex and overlapping (piedmont alluvial plains) as the cycle of erosion progresses.

The fans become increasingly flatter at lower elevations and eventually merge with the silt-choked basin floors. The basins themselves frequently are without exterior drainage. Shallow lakes, mostly intermittent, may fill the lowest portion of the basins. They are saline because they have no outlet and because the streams that feed them, like all streams, carry minute amounts of various salts. As the lake waters evaporate, the salts become more concentrated; the complete disappearance of the water leaves an alkali flat or salt pan.

There are several large and relatively permanent salt lakes in the subregion, particularly Lakes Walker and Pyramid in Nevada and Great Salt Lake in Utah (Fig. 15-6). The

last is a shrunken remnant of prehistoric Lake Bonneville, a great body of fresh water that was as large as present Lake Huron. Although Lake Bonneville and other Pleistocene lakes in the region began to shrink and disappear thousands of years ago, old beach lines still remain strikingly clear on the sides of the surrounding mountains. The highest Bonneville shoreline lies about 1000 feet (300 m) above Great Salt Lake. The present lake expands and contracts according to the variation in precipitation in the mountains its water source, and according to the rate at which irrigation water is drawn off. Because the lake is shallow [average depth, 14 feet (4 m)], its area fluctuates remarkably; the known areal extremes are 2400 square miles (6200 km^2) in 1873 and 1000 square miles (2600 km^2) in 1963. In the 1990s the lake was in an expanding phase, creating havoc for adjacent settlements and transportation routes.

There are only a few permanent streams in the region and generally they can be classified as "exotic" because the bulk of their water supply comes from adjacent regions. Most conspicuous are the Colorado River and the Rio Grande. The

FIGURE 15-6　Look readers: no hands! The renowned buoyancy of Great Salt Lake is its principal attraction for swimmers. Less attractive are the salt flies that abound, and the salt itches that result. In the background the slopes of the Oquirrh Mountains are partly blotted out by the fumes from the nonferrous metal smelter at Garfield (TLM photo).

former and its left-bank tributary, the Gila, provide a significant amount of water for irrigation and domestic use. The Salton Basin in southeastern California was partially flooded in 1906 when attempted irrigation permitted the Colorado River to get out of control. The river was reestablished in its original channel the following year, but the Salton Sea still exists as a "permanent" reminder of the incident.

AN ARID, XEROPHYTIC ENVIRONMENT

The greater part of the region is climatically a desert or semi-desert and the vegetation shows a variety of xerophytic (drought-resisting) characteristics.

Climate

Moisture is the most critical element of the climate. On the basis of precipitation–evaporation ratios, there are four moisture realms: (1) the subhumid, (2) the semiarid, (3) the moderately arid, and (4) the extremely arid.

The *subhumid* portion of the region occurs only in limited highland areas, primarily Washington and Oregon. More precipitation on the upland slopes and less evaporation because of lower summer temperatures result in a climate that shows little evidence of precipitation deficiency. Winters are long and cold; summers are short and cool. Precipitation is concentrated in summer or is evenly distributed.

The *semiarid* climate is typical of most of the Columbia Plateau. Precipitation ranges from 10 to 20 inches (25 to 50 cm) per year and falls chiefly in late autumn, winter, and spring.

The *moderately arid* climate, characteristic of most of the Great Basin, has periodic rainfalls that are fairly regular, although limited, and during which vegetation bursts into life and the water table is replenished. The precipitation at Elko, Nevada, a typical station, is 9 inches (23 cm). The frostless season varies from 100 to 180 days.

In the *extremely arid* climate the rainfall is episodic, coming largely in summer at irregular intervals and usually as cloudbursts. The Mojave–Gila Desert exemplifies this type. Its annual precipitation is less than 5 inches (12 cm), too little even for grazing. Almost the entire annual rainfall may come in a single downpour lasting but a few moments. So much water falls so quickly that little can penetrate the soil deeply.

The diurnal range of temperature throughout the region is high. The days are generally hot to very hot in summer, but radiational cooling in the dry atmosphere decreases the temperature rapidly at night except at low elevations in the southern part of the region, which has the highest summer nighttime temperatures to be found on the continent. In winter the nights are usually quite chilly, following daytime temperatures that may be relatively mild or even warm.

Natural Vegetation

In such a large area and in one varying so greatly in landforms marked differences in the natural vegetation occur. On the whole, however, low-growing shrubs and grasses predominate.

Forests　Ponderosa pine and Douglas fir forests are mostly confined to higher elevations where rainfall is relatively heavy (Fig. 15-7). Where precipitation is somewhat less, forest is replaced by woodland, a more open growth of lower trees, particularly piñon and juniper.

Grasslands　More extensive than might be supposed, grasslands characterize the uplands of southeastern Arizona, New Mexico, and the Columbia Basin. Short grass occupies large areas in the high plateaus of New Mexico and Arizona, as does bunchgrass in the Columbia Plateau. The noxious cheatgrass is almost everywhere.

Desert Shrub　Xerophytic plants dominate the deserts. *Sagebrush,* the principal element in the vegetation complex of the northern part of the region, grows in pure stands where soils are relatively free from alkaline salts. It is especially abundant on the bench lands that skirt mountains and on the alluvial fans at the mouths of canyons (Fig. 15-8).

FIGURE 15-7 On the higher mountains in the Intermontane Region are found forest and woodlands. This ponderosa pine scene is in the Sheep Mountains north of Las Vegas, Nevada (TLM photo).

FIGURE 15-8 Even the most stressful environments often contain distinctive and conspicuous plants. Rising above this sagebrush slope in the parched Mohave Desert of southeastern California are blooming Joshua trees (TLM photo).

Shadscale, a low, gray spiny plant with a shallow root system, grows on the most alkaline soils but never in dense stands. Much bare ground lies between the plants. It is especially prominent in Utah and Nevada. *Greasewood,* bright green in color and occupying the same general region as sagebrush and shadscale, grows from 1 to 5 feet (0.3 to 1.5 m) in height and is tolerant of alkali. *Creosote bush,* dominating the southern Great Basin as sagebrush does the northern, draws moisture from deep under the surface. Creosote bush is a large plant, attaining a height of 10 to 15 feet (3 to 4.5 m).

Within the last century, a number of woody plant species have greatly expanded their range in the arid and semiarid Southwest, mostly at the expense of grassland communities. As in the southern Great Plains, *mesquite* has occupied the greatest area of new territory, particularly in the Rio Grande and Tularosa valleys of New Mexico and the Colorado, Gila, Santa Cruz, and San Pedro Valleys of Arizona. There has also been considerable expansion of the acreage of native *juniper* and introduced *tamarisk,* the latter having extensively colonized islands and sandbars along most southwestern rivers, especially the Colorado and its tributaries.

Various types of *cacti* are widespread in the more arid portions of the regions, especially Arizona. The giant saguaro, symbol of the desert, is most prominent, although many smaller species of cactus are more numerous.

Exotics   The sparser plant associations (grasses and shrubs) of the region are highly susceptible to invasion and replacement by alien annual species, usually grasses, that have been introduced, chiefly from Asia. Severe livestock-grazing pressure in much of the region produced a "biological near-vacuum" that in many areas has become dominated by such grasses as downy brome (*Bromus tectorum*) and such weeds as Russian thistle and tumble mustard. Restoring these degraded range lands has proved difficult, even if the grazing livestock were removed.

Fauna

In spite of considerable barrenness and scarcity of water, the Intermontane Region has a surprisingly varied fauna. It was never an important habitat for bison, but the plains-dwelling American antelope, or pronghorn, is still found in considerable numbers in every state. In the mountains and rough hills other ungulates are notable, including deer, elk, desert mountain sheep, feral burros (particularly in California and Arizona), feral horses (especially in Nevada, Utah, and Oregon), and javelinas (in Arizona) (Fig. 15-9).

Furbearers are common only in forested portions of the northern half of the region. Most large predators (cougar,

FIGURE 15-9   Larger animals are generally scarce in the desert, but no part of the region is devoid of wildlife. These feral horses are crossing a road north of Goldfield, Nevada (TLM photo).

coyote, bobcat, and fox) are scarce as a a result of a systematic poisoning and trapping. The relatively few rivers and lakes provide important nesting and resting areas for migratory waterfowl.

SETTLEMENT OF THE REGION

The pre-European inhabitants of the Intermontane Region were extraordinarily varied. Most aboriginal tribes in the northern part of the region eked out a precarious existence as seminomadic hunters. Yet in the arid Southwest some of the highest stages of Native American civilization developed, mostly in the form of sedentary villages based on self-contained irrigated farming. These settled tribes, the Pueblos, Hopis, Zuñis, and Acomas, were islands of stability in an extensive sea of nomadic hunting and raiding tribes, most notably Apaches and Utes.

Arrival of the Spanish

The Spanish, the first European arrivals, were brought to the area by tales of great wealth. They pushed up the Rio Grande four centuries ago and established several settlements throughout what is now New Mexico. The Spanish explored widely and ruled most of the Southwest for more than 200 years. The major early Spanish settlements were in the Socorro–Albuquerque–Santa Fe–Taos area in the Upper Rio Grande Valley, with another important concentration in the

El Paso oasis. Many decades later and at a much lower level of intensity they occupied that part of southern Arizona called *Pimería Alta,* mostly in the Santa Cruz Valley as far north as Tucson.

The Spaniards left an indelible influence on both the history of the Southwest and on American civilization. Their livestock formed the basis of the later American cattle and sheep industry, and their horses gave mobility to the Native Americans, the importance of which can hardly be overestimated. Small Spanish settlements and trading posts, such as Albuquerque, housed most of the Caucasian population of the Southwest until the middle of the nineteenth century.

Explorers and Trappers

British and American explorers began to filter into the region in the early nineteenth century. Lewis and Clark entered the Pacific Northwest in 1804–1805. The Astorians were active in 1811–1813. Smith penetrated the Great Basin in 1826; and Wyeth, the Pacific Northwest in 1832–1833. Bonneville, in 1832 and 1836, traded furs and casually explored the area drained by the Bear River. Fremont, in 1845–1846, entered the Salt Lake Basin by way of the Bear River, becoming the first white man to examine it systematically. These are but a few of the many who explored the region.

Trapping, a powerful incentive to exploration, was the main object of many of the men who explored the West in the early nineteenth century. The trappers were a special breed—self-reliant, solitary, largely freebooters—who strove to outwit their rivals, to supplant them in the goodwill of the Native Americans, and to mislead them in regard to routes. They lasted until fashion suddenly switched from beaver to silk for men's hats. The trappers were then through; nevertheless, they played a major role in the region's history.

The Farmer Invasion

The outstanding example of farmer invasion was the Mormon migration to the Salt Lake Basin in 1847. The Mormons had trekked from New York into Ohio and Illinois, and then Missouri to escape persecution and find a sanctuary where they might maintain their religious integrity. To do so, they felt impelled to establish themselves on the border of the real American Desert. The agricultural fame of the Deseret colony was soon known far and wide.[1] Utah is the only state in the Union that was systematically colonized. The leader, Brigham Young, sent scouts into every part of the surround-

ing area to seek lands suitable for farming. Throughout the latter half of the nineteenth century Mormon pioneers participated in a major colonizing effort that established settlements, usually based on irrigated agriculture, in valleys and oases throughout the Intermontane West.

The California Gold Rush

Following the explorers, trappers, and farmers came the gold seekers of 1849. So large was the movement that it led to the establishment of trading posts and stations where the migrants rested and refreshed themselves. The Salt Lake Oasis especially became a stop for the weary and exhausted.

Other precious-metal discoveries had a significant influence on early settlement. Outstanding were the silver lodes of western Nevada, dating from 1859.

The Graziers

Much of this region was marginally favorable for the grazier. For some years after the Spaniards came, cattle raising was almost the only range industry, although Navajo Indians and Mexican colonists herded some sheep. Northward in Utah and Idaho, as well as in the Oregon country, cattle raising held sway in nonfarming areas. In fact, the Columbia grasslands were major cattle-surplus areas for many years and shared the stocking of the Northern Great Plains ranges with Texas.

In Utah the self-sufficing Mormons raised sheep for homespun, and as early as the 1850s nearly every farmer possessed a few head.

In the 1870s and early 1880s bands of Spanish and French Merino sheep were driven into the Southwest from California, furnishing a fine short-staple wool in sharp contrast to the coarse long wool of Navajo sheep. Transhumance (seasonal movement of livestock) was practiced: In Arizona the cool northern mountains were used from May until August, then the flocks were moved to the lower desert ranges. Late spring found them once more in the mountain pastures.

Spread of Settlement in the Southwest

The movement of people into the Southwest was erratic and variable and extended over a long period of time. Least noticed but fundamentally very important was the gradual influx of Hispanics.

> The gradual contiguous spread of Hispano colonists during the nineteenth century is a little-known event of major importance. Overshadowed in the public mind and regional history by Indian wars, cattle kingdoms, and mining rushes, this spontaneous unspectacular folk

---

[1] *Deseret* is a word from the Book of Mormon, meaning honeybee and symbolizing the hard work necessary for the success of their desert settlements. The Mormons organized the State of Deseret, but it was not accepted by Congress, which later formulated the Territory of Utah.

A CLOSER LOOK   Greater New Mexico's Hispano Island

One of the amazing facts about the geography of the United States is that as recently as 1900 an area the size of Tennessee was everywhere a minimum 90 percent ethnic and averaged an incredibly high 97 percent ethnic. This area was the inner half of greater New Mexico's Spanish American or Hispano "island." To be sure, a few communities within this area formed pockets of lesser percentages. Santa Fe in 1900 was only 72 percent Hispano, for example. But with these pockets removed, what is so amazing is that nowhere in America has so large an area been so purely ethnic so recently. How this island rose up, as it were, and what happened to erode much of it away is the story of five peoples: Hispanos, Pueblo Indians, nomad Native Americans, Anglos, and Mexican Americans.

Spaniards arrived in New Mexico in 1598. They quite literally moved right in with the Pueblos Indians as they created enclaves of Franciscan missionaries and their soldier guards within many Pueblo villages. The only Spanish community, also the capital, was Santa Fe. Predictably, Pueblo resentment against the Spanish mounted, and in a successful revolt in 1680 the Pueblos forced nearly all 2500 Spaniards to retreat to the El Paso area. But in 1693 the Spaniards returned to reoccupy Santa Fe and to rebuild their missions, and during the next century they established many new communities, like Albuquerque. Gradually, during the 1700s, the one-time Pueblo realm was transformed into a Spanish province within which the Pueblos' villages were now the enclaves.

Until about 1790 all this Spanish–Pueblo activity was confined to the upper Rio Grande Valley by the warlike Apache, Comanche, and Navajo Indians. Successful Spanish military campaigns against these nomad peoples brought relatively peaceful times, however, and Hispanos, a stock-raising people who always sought additional grazing land for their sheep, during the 1800s quite dramatically expanded their territory. They pushed north into the Rocky Mountains of Colorado, east onto the high plains of Texas, west over the Colorado Plateau to Arizona, and south down the Rio Grande Valley. By 1900 the Hispano island reached almost its greatest areal extent, stretching into parts of five states and covering 85,000 square miles (220,000 km^2), or an area the size of Utah (Fig. 15-A).

In 1900, living on the island were 140,700 Hispanos, 8500 Pueblos, 5300 once-nomadic Native Americans, and 66,100 Anglos. Anglos began to arrive in 1821 when traders journeyed to New Mexico over the Santa Fe Trail. In 1846 Anglo soldiers invaded New Mexico, and by 1848 the island was under American control. Under this new regime Anglo merchants, ranchers, farmers, lumbermen, and miners arrived. Their destination was not so much the inner half of the island, where the Hispanos were so numerous, but the island's outer half, where resources could be exploited. The arrival of Anglos reduced the Hispano overall percentage to 64 percent by 1900.

Thus in 1900 the Hispano island was composed of an outer half that was largely Anglo but everywhere a minimum 10 percent Hispano and an inner half that was remarkably Hispano. Containing 42,300 (109,600 km^2) square miles or an area the size of Tennessee, the island's inner half, except for a few pockets, was everywhere at least 90 percent Hispano. Indeed, 29 census precincts located in this inner half were 100 percent Hispano. And so it was that in 1900 the island's inner half stood tall and flat like a lofty plateau, so uniformly high were its Hispano percentages.

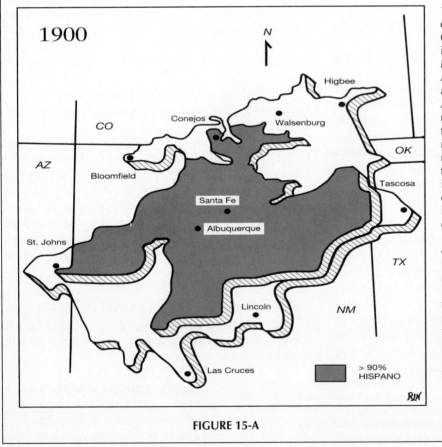

FIGURE 15-A

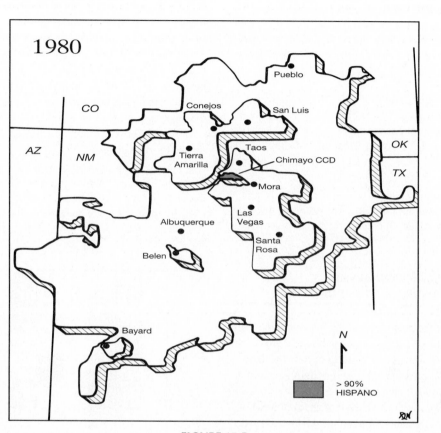

1980

FIGURE 15-B

Weighed down as it were by Anglos, the outer half was a lower plain, and the island's outer edge was like a steep cliff, so few were the Hispanos who had left the island (Fig. 15-A).

After 1900 the number of Anglos arriving at the island greatly accelerated. Substantial numbers of Mexicans, who before 1900 were almost everywhere a minimal part of the island's population, were now being pulled north from Mexico to economic opportunities in places like Albuquerque, Pueblo, and Denver. Although the ranks of the Hispano continued to grow, Hispanos were increasingly engulfed, especially by Anglos. In 1980, in a special computer-generated population count made by the Bureau of the Census, it was determined that 1,260,000 Anglos and relatively small number of Native Americans represent-

ed 69 percent of the island's population, and that 201,000 Mexican Americans accounted for 11 percent. Hispanos, who now numbered 365,000, represented only 20 percent of their own island's population. However, in the island's core, Hispanos were still very much in the majority.

Thus, the Hispano island today has three "tiers" (Fig. 15-B). The highest is the small Chimayó Census County Division (CCD), the one area where Hispano percentages still exceed 90. Below it are parts of 11 counties where Hispano percentages exceed 50 percent and Hispanos continue to have political and economic clout. And below this tier is a vast plain where Hispanos are everywhere a minority yet also a minimum 10 percent of each CCD. In the twentieth century, then, it was as

though the island came under a torrential shower that found mainly Anglos falling on and all around the island. The lofty plateau of 1900 was eroded to but one small butte. And as Anglo jobs pulled Hispanos off their island, the steep outer cliffs became severly eroded banks. But like a monolith of differential erosion, the core of the island—the homeland of America's only surviving Spanish colonial subculture—continues to stand tall and proud. Such is the island today: less majestic, but a part of the United States that continues to stand apart.

Professor Richard L. Nostrand
University of Oklahoma
Norman

movement impressed an indelible cultural stamp on the life and landscape of a broad portion of the Southwest. It began in a small way in Spanish times, gathered general momentum during the Mexican period, and continued for another generation, interrupted but never really stemmed until it ran head on into other settler movements seeking the same grass, water, and soil.[2]

In the latter part of the nineteenth century the influx of Anglos from Kansas, Colorado, California, and especially Texas was the major force in the region, quickly dominating both the economy and the political pattern. Mining camps and pastoral enterprises were particularly prominent, but the coming of the two major east–west railroad corridors—a northerly one through Albuquerque and Flagstaff, and a southerly route through El Paso and Tucson—signaled the beginning of a more diversified economy and the growth of urban nodes.

LAND OWNERSHIP IN THE INTERMONTANE REGION

A striking feature of the geography of the Intermontane West is the large amount of land that is in the public domain. In the 11 conterminous western states more than half the land is owned by the federal government.[3] And in the Intermontane Region the proportion is much higher. To illustrate, in Utah, Arizona, and Nevada, the three states that are almost wholly within the region, more than 75 percent of the land is owned by either federal or state government.

The basic reason that so little land is in private ownership is that nonirrigated agriculture is impractical over most of the region; thus there was little opportunity for dense rural settlement and little practical demand for freehold ownership. Although successful homesteading occurred in many localities, the homestead laws (which made land available either without cost or inexpensively to legitimate rural settlers) were designed to apply to more humid regions and did not function well in the arid West.

The two principal categories of public land are national forests, which include the great majority of all forest land in the region, and Taylor grazing lands, which were withdrawn from homesteading and are reserved for seasonal grazing use. Also notable in the region are Native American and military reservations.

[2] D. W. Meinig, *Southwest: Three Peoples in Geographical Change, 1600–1970* (New York: Oxford University Press, 1971), p. 30.

[3] East of the Mississippi River, only Florida and New Hampshire have as much as 10 percent of their land in federal ownership.

A major problem associated with the large amount of government-owned land is the great complexity of the ownership pattern. Much of the land that is not federally owned occurs in a bewildering checkerboard arrangement (derived from granting scattered but designated sections of land to states, railroad companies, and other institutions either as reward or to generate revenue or as a stimulus to rural settlement) within a broad matrix of public domain (that is, federal) land. This fragmentation often precludes any rational development of use (Fig. 15-10).

The largest category of scattered, nonfederal land is owned by state governments. When western states were originally formed, the federal government made extensive land grants to them (generally four disconnected sections of land out of each township—4 square miles out of each 36) as trust land to support public education. Some 42 million acres (16.8 million ha) of this school trust land (in all states except Neva-

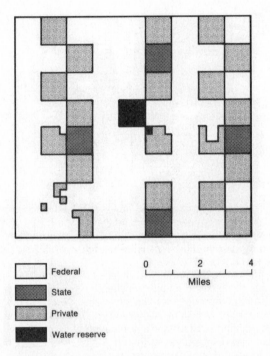

Federal

State

Private

Water reserve

0   2   4
Miles

FIGURE 15-10   In much of the Intermontane West, the land ownership pattern resembles a gigantic, imperfect checkerboard. This sample is from San Bernardino County, California, near the Nevada border. The federal land mostly is administered by the Bureau of Land Management. Private land includes that owned by railroad companies, other corporations, and individuals (data source: Amboy Quadrangle of Bureau of Land Management map series, 1978).

da, which sold all its acreage) remain scattered across the West, particularly in the Intermontane Region.

A concerted legal effort is now underway in Utah to rationalize this hodgepodge ownership pattern by a complicated series of land exchanges between federal and state governments that would consolidate the holdings into large, coherent blocks. Other western states are watching these negotiations with great interest, for the results might set a precedent that would significantly alter both land ownership and land-use patterns across the West.

THE CONTEMPORARY POPULATION: VARIED AND RAPIDLY INCREASING

What population there is in the Intermontane West congregates mostly in "islands" where (1) precipitation is adequate, (2) water is available for irrigation farming, (3) ore deposits permit commercial mining, (4) transportation routes converge, (5) some special recreational attraction exists, or (6) there is Sunbelt retirement. Cities are few and scattered, but rapidly growing. The principal nonmetropolitan population concentrations are in the middle Salt River valley and parts of the lower Colorado River valley in Arizona, the Imperial Valley of southernmost California, along the western Wasatch Piedmont in Utah, at various places along the Snake River in Idaho, and along the middle Columbia Valley in central Washington.

This is a region of considerable population movement—migration into, out of, within, and across. Many of the inhabitants are both restless and mobile. Numerous population clusters are characterized by an extraordinary number of mobile homes, travel vans, campers, and recreational vehicles—which emphasizes their impermanence. Even in the metropolitan areas there is frequent movement from one home to another. For the region as a whole, some 60 percent of the population change domicile at least once every 5 years.

The net rate of population growth has been very rapid in the southern part of the region but somewhat slower in the north. More than 2 million inhabitants were added to the regional total during the 1980s. Nevada and Arizona, the two states wholly within the region, grew by 37 percent and 36 percent, respectively, during that decade.

There are three significant and readily identifiable "minority" elements in the contemporary population of the Intermontane Region. All three represent subcultures that are more prominent here than in any other region in the United States. There is a Mormon culture realm centered in Utah, a Hispanic American borderland along the southern margin of

the region, and Native American lands covering vast areas, particularly in Arizona and New Mexico.

Mormon Culture Realm

As the earliest white settlers in the central part of the Intermontane Region, Mormons (members of the Church of Jesus Christ of Latter-day Saints) have dominated the human geography of the Deseret area for a century and a half. Their cohesive and readily distinguishable culture is manifested in various social patterns, in economic organization and development, and in certain aspects of settlement.[4] Today most Mormons, like most other North Americans, are urbanites; nevertheless, distinctive cultural characteristics set them apart and set their realm apart as a cultural subregion.

The Mormon culture realm (Fig. 15-11) was defined in Meinig's classic 1965 study and has since been refined and updated in the *Atlas of Utah.*[5] The *core* is that intensively occupied and organized section of the Wasatch oasis that focuses on Salt Lake City and Ogden; it is the nodal center of Mormonism. The *domain,* which encompasses most of Utah and southeastern Idaho, includes the area where Mormonism is dominant but with less intensity and complexity of development. The *sphere* is defined as an area in which Mormons live as important nucleated groups enclaved within Gentile (non-Mormon) country. Despite the continuing expansion of Mormonism, its Utah focus is shown by the fact that more than three-fourths of the total county population in 20 of the state's 29 counties are members of the Mormon church, whereas in only 5 other counties in the nation (all in southern Idaho) does such a proportional membership exist.

Hispanic American Borderland

Along the southern margin of the Intermontane Region, from the Imperial Valley in California to the Pecos River in Texas, there are concentrations of varying intensity of Hispanic peo-

---

[4] The traditional Mormon town was a small, nucleated settlement with large lots, extraordinarily wide streets, a network of irrigation canals alongside the streets, relic agricultural features, unpainted barns, and houses of Greek Revival style constructed of bricks. Such settlements were totally unique in western North America but actually represented a re-creation of the "New England nucleated village and the persistence of nineteenth century structures and . . . patterns in the twentieth century" [Richard H. Jackson, "Religion and Landscape in the Mormon Cultural Region" (unpublished manuscript, 1977)]. Small towns in which these characteristics persist today are relatively rare and occur in remoter parts of the Mormon culture realm.

[5] D. W. Meinig, "The Mormon Culture Region: Strategies and Patterns in the Geography of the American West, 1847–1964," *Annals,* Association of American Geographers, 55 (June 1965), 191–220; Deon C. Greer et al., *Atlas of Utah* (Provo, UT: Brigham Young University Press, 1981), pp. 140–143.

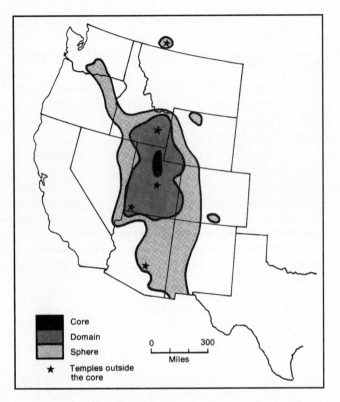

FIGURE 15-11 The Mormon culture realm (after Meinig and *Atlas of Utah*).

ple. Their presence is numbered in the millions. Their proportional size is so great that they are in the majority in some towns and counties, including El Paso, the third largest urban center of the Intermontane Region.

The legacy of Hispanic settlement in the Southwest is long and notable. Architecture, settlement patterns, language, and cuisine are but a few of the more prominent elements of this heritage.[6] The continuing high rate of immigration from Mexico and the rapidity of increase among Hispanic Americans ensure that this portion of the Intermontane Region should maintain its Hispanic subculture indefinitely. Thus from a demographic and cultural standpoint Latin America merges northward into the United States. In contrast, however, the virtual absence of "Americans" south of the international boundary suggests that North America ends abruptly at that line. In a real sense the United States–Mexico border is a frontier where First World meets Third World. As J. W. House has noted, "Along its entire length, the U.S.–Mexican

boundary is one of the most remarkable and abrupt culture contact-faces in the world. . . . Nowhere else are there such steep economic and social gradients across an international boundary."[7] The accompanying box ("The U.S.–Mexico Border") highlights the situation.

Land of the Native Americans

More than in any other region outside the Subarctic, the Intermontane West is the land of the Native American. Some 300,000 Native Americans of various tribal affiliations are scattered over the region, although predominantly in Arizona and New Mexico, where some 7 percent of the bistate population is of Native American origin. There are several large reservations in the northern portion of the region: Yakima in Washington, Umatilla in Oregon, Pyramid Lake and Walker River in Nevada, and Uintah and Ouray in Utah (Fig. 15-12). It is around and south of the Four Corners country that Native American lands are most prominent, however. The Navajo Reservation is by far the largest, but there also are extensive reservations for the various Apache tribes, the Hualapais, the Hopis, the Tohono O'odhams (known as Papagos before 1986), and the Utes. In addition, there are many smaller reservations in the Intermontane Region; some are densely populated, particularly the Pueblo reservations in north-central and northwestern New Mexico.

In general, the Native Americans of the Intermontane Region have been economically poor, socially deprived, and politically inactive. On reservations they have usually maintained cohesive tribal identities, although their livelihood is often near or below the poverty level. Those who have left the reservation—as all are free to do—often find that adjusting to life in a harsh Anglo world is difficult.

There are, however, many pleasant exceptions to this generally depressing picture. Many off-reservation Native Americans have adjusted to living in southwestern cities, as the rapidly growing Indian populations of Los Angeles, Phoenix, and Albuquerque attest; furthermore, economic and social conditions on many reservations have been improving rapidly. The Apaches of the Fort Apache, San Carlos, and Mescalero reservations have developed prosperous logging industries and have shrewdly organized outdoor recreational advantages to attract tourists. The 30,000 Pueblo Indians have generally been able to adjust to the pressures of modern civilization because their ancestral lands were legally restored to them, and each Pueblo village is thus surrounded by a protective girdle of farmland that reinforces its insularity.

[6] For additional details on the concept of an Hispanic-American borderland, see Richard L. Nostrand, "The Hispanic-American Borderland: Delimitation of an American Culture Region," *Annals,* Association of American Geographers, 60 (1970), 638-661.

[7] J. W. House, *Frontier on the Rio Grande: A Political Geography of Development and Social Deprivation.* Oxford: Oxford University Press, 1982, p. 37.

A CLOSER LOOK   The U.S.–Mexico Border—A Line, or a Zone?

The Mexico–U.S. border, or *la frontera* (the frontier) as it is called in Mexico, is popularly if somewhat derisively referred to as the "Tortilla Curtain." This moniker not only suggests the role the border plays as a convenient symbolic divide between North America and Latin America but also between the First and Third Worlds. Few international boundaries separate, at least politically, such fundamentally different countries and prevailing cultures.

Born of conflict and increasing U.S. domination of the region during the first half of the nineteenth century, the border was created by the Treaty of Guadalupe Hidalgo, which concluded the war between the two countries in 1848. From the Pacific Ocean to the Gulf of Mexico, the border runs for 2076 miles (3322 km) over granite-studded mountains, deserts, high plateaus, and coastal plain; it passes through land as inhospitable and desolate as any in either country but also bisects some of the fastest-growing, most dynamic urban centers in the Western Hemisphere. All along its path it generates political complexities and controversies as it cuts through 25 counties in four U.S. states (California, Arizona, New Mexico, and Texas) as well as 35 *municipios*—municipalities that administratively are like counties in the United States—in six Mexican states (Baja California, Sonora, Chihuahua, Coahuila, Nuevo Leon, and Tamaulipas). For approximately 1325 miles (2120 km) it follows the course of the Rio Grande, or Rio Bravo del Norte as the river is known in Mexico. Westward from El Paso and Ciudad Juárez it is marked by barbed wire and wire mesh fence, and 258 white marble obelisks. Yet while the location of this invisible boundary is precise, the perception of it is not. In general, it is

viewed from one of two perspectives—as being either a line, or a zone.

Those who support the former concept argue that the border is an abrupt demarcation, a sort of cultural fault, separating two countries with vastly contrasting material and nonmaterial elements of culture. They point to the differences in history, tradition, religious affiliation, values and symbols, language, ethnic composition, patterns of social organization, lifestyles, political and legal systems, architecture and design, cuisine, music, and levels of economic development. To support this position, they suggest one need only cross the border, in either direction, to immediately perceive the distinctions: The look, the smell, the character, the sense of the two places are quite different. These differences in the cultural landscape—the composite of all manufactured features—are not merely cosmetic but rather, it is argued, reflect the deeper cultural-historical currents of the respective societies. That the landscapes of Matamoros, say, are for the most part unlike those of neighboring Brownsville, or that the landscapes of Tijuana would not likely be confused with those of San Diego, are both readily apparent, and culturally meaningful.

Conversely, there are others who see the border as a zone that straddles the international boundary. It functions, so the argument goes, as a kind of linear third country or special domain with its own identity and character. This zone of "overlapping territorality," as it has been labeled, has produced a hybrid culture, one that is part Mexican and part American, similar to yet different from the cultural mainstreams found in the interiors of the two nations. Here, for instance, Spanish and English words are liberally and spontaneously mixed together in everyday conversations, creat-

ing a dialect and a language usage unique to this land between, it might be called. It is suggested that the so-called *fronterizos*, the border people, share a common experience and are tied not only by geographical proximity but also by interdependence, mutual interests, and transborder concerns, including a host of social, economic, and environmental issues.

Perhaps the most persuasive evidence in support of the border-as-zone position is the large and increasing volume of goods, services, people, and ideas that flow to and fro over the boundary from adjacent communities. Agricultural products and manufactured goods move in both directions. The populations on both sides, often linked by family ties and long-standing friendships, intermingle as they cross the border to work, shop, and play. Yearly, there are over 200 million *legal* crossings from Mexico to the United States, and a majority of these, it is estimated, are made by border residents. For them, if not for others, this highly permeable border is indeed, for all practical purposes, a zone of binational interaction and bicultural attractions.

So whether the border is a line or a zone depends finally on what perspective is taken. Seen from the ground (a microlevel view), the empirical evidence, especially the cultural landscape characteristics, tend to support the former; whereas seen from above (a macrolevel view), the complex functional patterns of transborder movement and interdependence become dominant, thus supporting the latter. Places, it should be remembered, are not one-dimensional entities that can be easily classified.

Professor James R. Curtis
California State University
Long Beach

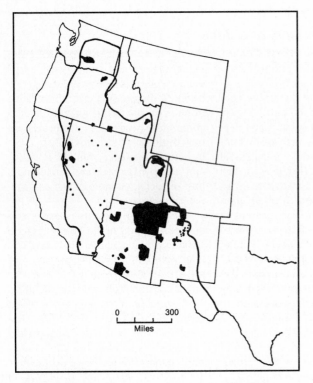

FIGURE 15-12   There is a greater extent of Indian reservations in the Intermontane Region than in the rest of the country combined.

Of all the large tribes of the subcontinent, it is the Navajos[8] who have led the way in adapting to a capitalistic society while still maintaining their cultural and tribal integrity. To do so, they have had wise leaders, but they also have the immeasurable advantage of valuable natural resources to exploit. The massive and abrupt effort to establish Navajo lands as a functional part of modern America, without loss of the traditional Navajo culture, has been rife with problems; the effort continues and the complexities increase.

The Navajo Reservation of Arizona, New Mexico, and Utah is the largest (15 million acres) on the continent, and the Navajo Nation is also the largest (more than 200,000 members). Until recently, pastoralism was the dominant occupation of the Navajos. Their mixed flocks of sheep and goats range widely over the reservation, usually tended by women or children. In the last few years, however, exploitation of

minerals in payable quantities has changed the basis of Navajo life. Coal, petroleum, natural gas, uranium, and helium are being extracted, many jobs have been created, and the Navajo Nation Council now has an annual budget of more than $100 million with which to work. A forest products company has been organized, factories have been attracted to or near the reservation, and tourist facilities have been expanded.

A major and longstanding problem is the acrimonious relationship between the Navajos and their closest Native American neighbors, the Hopis. Some 6500 Hopis occupy a reservation that is virtually in the center of, and completely surrounded by, the Navajo Reservation (Fig. 15-13). Most Hopis live in agricultural villages that are situated on three high mesas. They are much more farming oriented than the Navajos, but the Hopis, too, engage in extensive pastoralism with sheep and goats on and around their mesas. There are many facets to the Navajo–Hopi dispute, but it centers around the use of grazing lands. Each tribe claims land that is used by the other, and much stock that strayed beyond disputed boundaries has been confiscated by both Navajo and Hopi police who patrol the disputed areas.

There are various other internal controversies in Navajo land, but much more visible to outsiders are the problems associated with the exploitation of energy resources, particularly coal. During the 1960s the Tribal Council signed several long-term contracts with mineral and utility companies for the mining and transport of coal and the construction of power-generating facilities to use the coal. Although the Navajos receive large sums of money for the leases and royalties involved, many believe that the value received is inadequate for the resources given up and the environmental deterioration that results. The coal is strip-mined, and thousands of acres of admittedly poor grazing land are lost as a result. The coal is used to generate electricity and produce coal gas in several large plants that spew enormous amounts of conspicuous pollutants into the heretofore pristine air of the Four Corners country. The electricity and gas produced are transported outside the region, mostly to southern California, for sale. Probably the most serious problem, however, is the huge amount of water required for these operations. Ultimately water is the most critical of all resources in the arid reaches of the Navajo Reservation and the long-term commitment of this precious commodity, as specified in the contracts, may cause problems in the future.

THE WATER PROBLEM

The limitations imposed by paucity of water are felt throughout the Intermontane Region. Limited rainfall makes stream water desirable, but rivers are scarce and often located in

---

[8] The Navajo Nation (as it has been known officially since 1969, when the name "Navajo Tribe" was replaced) is considering the changing of its name from "Navajo" to "Dine" (pronounced "Di-nay"). The term *Navajo* has no clear meaning and was bestowed by the Spanish conquistadores. The term *Dine* derives from the group's traditional Athabaskan language and can mean both "man" and "people of the Earth." At the time of this writing, the name change was not yet official.

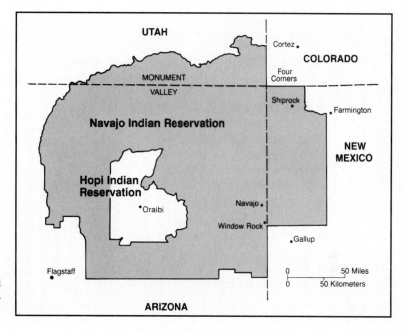

FIGURE 15-13   The Navajo and Hopi reservations of the Four Corners country.

deep gorges, making their water relatively inaccessible. Well water has been obtained in some areas, but the principal hope for increasing the natural water supply of the region has always been river catchment and diversion.

In this region, as in other parts of the West, the federal government has not hesitated to become "enlisted on the side of The People vs. The Desert" by building dams and blocking the rivers to create tiny islands of moisture availability in the sea of aridity.[9] Flood control and irrigation have been the twin purposes of most river development projects in the region, although some dams have more specifically been designed to provide an urban water supply or generate hydroelectricity.

The Dammed Columbia

Of all rivers in North America, the Columbia has an annual flow exceeded only by those of the Mississippi and the St. Lawrence. The Columbia, however, has a steeper gradient and hence the greatest hydroelectric potential on the continent. The Columbia's average runoff is about 10 times that of the Colorado, but only one-third that of the Mississippi. Along its 740-mile (1184-km) course in the United States, the Columbia descends 1290 feet (387 m); it has now been so completely dammed that only 157 feet (47 m) of this "head" is still free flowing.

[9] Walter Prescott Webb, "The American West: Perpetual Mirage," *Harper's Magazine,* 214 (May 1957), 28.

Beginning with Rock Island Dam in 1929–1931, 11 dams have been built along the Columbia in Washington and Oregon. Except for Grand Coulee, the dams are primarily for hydroelectricity generation and navigation on the lower Columbia. Grand Coulee is a dam of superlatives (Fig. 15-14). It houses the largest hydroelectric power plant in the world; impounds the sixth-largest reservoir, F. D. Roosevelt Lake, in the United States; is the fifth-highest dam in the nation; and is designed to irrigate more than 1 million acres (400,000 ha).

Several other dams were built on tributaries of the Columbia, particularly the lower Snake (for hydroelectricity and navigation) and several short streams issuing from the Cascade Mountains (for irrigation).

The Continuing Controversy of the Colorado

Although many rivers in North America carry more water, the Colorado is particularly important because it is the only major river in the driest part of the subcontinent. The river flows for 1400 miles (2240 km), its drainage basin encompassing about one-twelfth of the area of the conterminous states. Seven states and Mexico clamor to use the Colorado's waters.

The first major dam in the Colorado watershed was the Roosevelt Dam on the Salt River near Phoenix, which was begun soon after Congress passed the Reclamation Act of 1902 authorizing the Department of the Interior to establish large-scale irrigation projects (Fig. 15-15). Before long it was realized that basin-wide planning was needed for efficient use of the waters of the basin. The Colorado River Compact

FIGURE 15-14 Rocky Reach, near Wenatchee, is one of the smaller of the Columbia River dams (TLM photo).

was finally hammered out, taking effect in 1929; its main provision apportioned the use of Colorado River water between the upper basin states (Colorado, Wyoming, Utah, and New Mexico) and the lower basin states (Arizona, Nevada, and California) on a 50–50 basis.

Later emendations provided a share for Mexico and subdivided the upper basin total among the four states involved. The lower basin states, however, could not agree on division of their share, and complex litigation finally ended with subdivision of the Lower Basin allotment by the Supreme Court.

Several major problems persisted, not least of which was the fact that the Colorado River was bankrupt; the various agreements called for an annual use of 3 million more acre-feet than the river normally carried. Four dams, starting with the mammoth Hoover Dam, were built along the lower course of the Colorado for various purposes but particularly to stabilize the river's flow and provide maximum usage in California and Arizona (Fig. 15-16).

After years of planning, construction began on the Central Arizona Project in 1973, and the first water was delivered to the Phoenix area in 1985. By the mid-1990s the $2.5 billion project (the largest ever undertaken by the U.S. Bureau of Reclamation) was essentially complete, and the reality was of the situation recognized in southern California. Prior to 1985 the Central Arizona Project's share of

Colorado River water had been diverted to Los Angeles and vicinity. Now that water will increasingly go to Arizona.

AGRICULTURE

Farming is sparse and scattered in this vast, dry region, but it is nevertheless the most prominent activity in most nonurban settled areas. Only a fraction (3 percent in Utah, for example) of the total land area is in farms, and little of this is actually in crops.

Precipitation is so sparse that growing crops under natural rainfall conditions is restricted mostly to the Columbia Plateau subregion. Most important by far is the Palouse country of eastern and central Washington and adjacent parts of northern Oregon (Fig. 15-17). This area has rich prairie soils enhanced by the deep accumulation of loess (windblown silts), which makes it the highest-yielding wheat-growing locale on the continent. Whitman Country, just north of the Snake River in southeastern Washington, is the leading wheat-producing county in the United States, primarily because of very high yields per acre. The bulk of the output is soft white winter wheat, which is commonly used for pastry, crackers, and cookies rather than bread. Normally more than three-quarters of the output is exported, particularly to Japan and India.

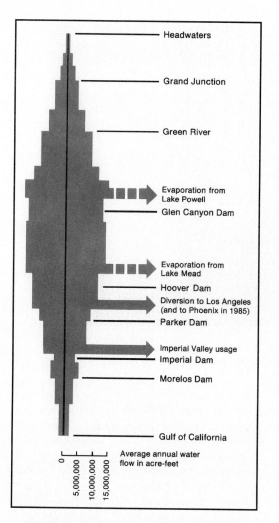

Headwaters

Grand Junction

Green River

Evaporation from
Lake Powell

Glen Canyon Dam

Evaporation from
Lake Mead

Hoover Dam

Diversion to Los Angeles
(and to Phoenix in 1985)

Parker Dam

Imperial Valley usage

Imperial Dam

Morelos Dam

Gulf of California

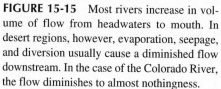

Average annual water
flow in acre-feet

FIGURE 15-15 Most rivers increase in volume of flow from headwaters to mouth. In desert regions, however, evaporation, seepage, and diversion usually cause a diminished flow downstream. In the case of the Colorado River, the flow diminishes to almost nothingness.

out irrigation in central New Mexico, southwestern Colorado, and southeastern Utah where there is sufficient summer rain.

Most crop-growing in the region, however, is dependent on irrigation, which has been expanding ever since it was introduced by the Mormons in Deseret in 1847. Some 10 million acres (4 million ha) of cropland are now under irrigation, of which about one-third are in Idaho and one-sixth in Washington (Fig. 15-18). The principal areas of irrigated farming are summarized next.

The Salton Trough This hot, flat, below-sea-level valley is occupied in part by the saline waters of the Salton Sea and in part by two highly intensive irrigated farming areas—Imperial Valley to the south and Coachella Valley to the north (Fig. 15-19).

The 470,000 irrigated acres (190,000 ha) of Imperial Valley yield about 750,000 acres (300,000 ha) of crops each year as a result of the widespread adoption of double-cropping. A great deal of labor is required on most farms, and the area depends heavily on migratory workers. The valley is watered from the Colorado River via the All-American Canal and produces a remarkable variety of crops, ranging from high-value iceberg lettuce (which dominates the winter market in the United States) to mundane alfalfa. With up to seven cuttings a year, Imperial County produces almost twice as much hay as any other county in the United States. It is also a major producer of sugar beets (the second-ranking U.S. producer), but its most valuable output is beef from cattle that are fattened in the area before marketing.

Coachella Valley also obtains its water from the Colorado River and grows a variety of crops, but the bulk of farm income comes from the four-level agricultural pattern of vegetables (especially carrots), vineyards (mostly table grapes), grapefruit (California's principal area), and dates [the only significant commercial source in the nation (Fig. 15-20)].

The Salt River Valley This was the first major federal irrigation project and one of the most successful economically. The chief cash crop is short-staple cotton, but there is considerable acreage in a great variety of other crops, especially hay, wheat, barley, sorghums, citrus, and safflower (Fig. 15-21, p. 339).

Much irrigation water is also obtained from wells, which have been so depleted by heavy pumping that the water table has dropped alarmingly. No other state has such deep irrigation wells, on the average, as Arizona. Indeed, Ari-

The Palouse area also leads the nation in production of dry peas and lentils and has considerable acreage in barley, clover, and alfalfa. The Blue Mountains district of northeastern Oregon and southeastern Washington is the leading source of green peas in the nation.

There are no other major areas of dry-land farming in the region. Scattered patches of dry-land wheat are found in Idaho, Oregon, and Washington. Dry beans are raised with-

FIGURE 15-16 Many segments of the Colorado River are dammed, thereby creating lengthy reservoirs. This is Lake Havasu, with California on the left and Arizona on the right (TLM photo).

zona has the highest-cost irrigation of any state, but it has outstanding yields to compensate.

The Rio Grande Project The Middle Rio Grande Valley, which is above and below El Paso, constitutes one of the oldest irrigated areas on the continent, having been initiated by pre-Columbian Native Americans. Sporadic private irrigation diversions in the late nineteenth and early twentieth centuries severely diminished the water available below El Paso, which was an important Mexican irrigated district. After considerable international negotiation, the Rio Grande Project was developed under federal auspices. The key structure is Elephant Butte Dam on the Rio Grande in southern New Mexico. Some 180,000 irrigated acres (72,000 ha) are included within the project. The greatest acreage is devoted

to hay and feed grains for cattle fattening. Other prominent farm enterprises involve cotton, poultry, pecans, and grapes.

Colorado's Grand Valley West-central Colorado has several major irrigated areas that use water flowing westward from the Rocky Mountains in such rivers as the Colorado and the Gunnison. Most notable is the Grand Valley project near Grand Junction. Many different field crops, such as corn, small grains, alfalfa, sugar beets, potatoes, and vegetables, are grown. The area's reputation, however, is based on its fruit crop, primarily peaches but also other orchard fruits.

Salt Lake Oasis The valley of the Great Salt Lake was occupied by Mormon pioneers in the 1850s. Within the

FIGURE 15-17 The fertile rolling plain of the Palouse country in western Washington. These are the most productive wheat lands in the world (TLM photo).

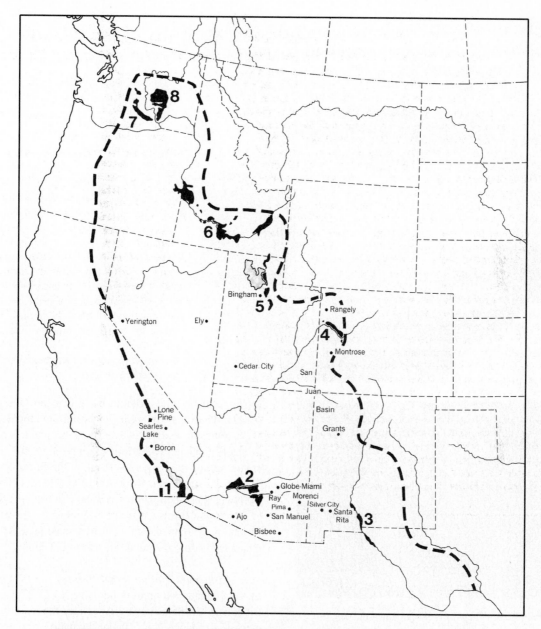

FIGURE 15-18   Major irrigated areas and mining towns in the Intermontane Region: (1) Salton Trough, (2) Salt River Valley, (3) Rio Grande Project, (4) Colorado's Grand Valley, (5) Salt Lake Oasis, (6) Snake River Plain, (7) Columbia Plateau Fruit Valleys, and (8) Columbia Basin Project.

first decade of occupance, the land-use pattern was established, and relatively little expansion of irrigated farming has occurred since then.

The lofty Wasatch Mountains tower above the oasis on the east, their snow-clad slopes providing life-giving water for the dry lands at their base. At the mouth of almost every stream canyon, as it emerges from the Wasatch, is located a city or village girdled by green fields and adorned by orchards and shade trees. The cropping pattern is extremely diverse, although the greatest acreage is devoted to hay and grains (especially wheat). Other notable crops are sugar beets and fruits (particularly apples, peaches, and cherries). Livestock are also abundant and varied, most notably cattle and chickens.

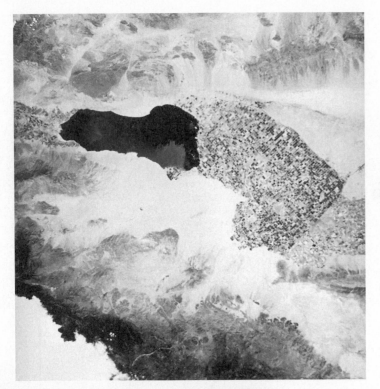

FIGURE 15-19   Probably the most famous space image of North America is this scene in Southern California and adjacent Mexico. The checkerboard pattern of intensive, irrigated agriculture south (Imperial Valley) and north (Coachella Valley) of the Salton Sea contrasts strongly with the surrounding barrenness of desert basins and mountains. The international boundary shows as an east-west line at the south end of the Imperial Valley; its abruptness is caused by a different intensity of land use in Mexico, as well as seasonal differences in planting and harvesting practices (Landsat image).

FIGURE 15-20   The only area of commercial date production in the United States is in the Coachella Valley near Indio, California (TLM photo).

The Snake River Plain   Southern Idaho contains a lengthy sequence of irrigated areas scattered across the sagebrush flats above the abrupt gorge of the Snake River. The basic Idaho crops are hay (by far the most acreage), potatoes (Idaho produces more than one-fourth of the national crop), and sugar beets (Idaho is the second-ranking producer).

The three principal irrigation projects are the Minidoka in southeastern Idaho, and the Boise and Owyhee in south-western Idaho and adjacent southeastern Oregon (Fig. 15-22). The most dynamic irrigation developments in the West in recent years, however, have been extensive, mostly privately financed operations relying on electric- or gas-operated pumps in various locations in the Snake River Plain. The water is pumped from 400 to 600 feet (120 to 180 m) from wells, or a comparable vertical rise from the Snake River. These new farms tend to be enormous in size, corporate in structure, and quite capital intensive, making heavy use of sophisticated equipment and other farm machinery.

Columbia Plateau Fruit Valleys   In the rain shadow of the Cascade Mountains lies a series of Columbia River tributaries whose valleys—especially the Yakima and Wenatchee—produce two-fifths of the national apple crop (Fig. 15-23). There is also notable output of hay and field corn for cattle feeding, hops, potatoes, and enormous recent expansion of vineyards.

The Columbia Basin Project   Grand Coulee Dam was built in the 1930s as the first high dam on the Columbia River. Water from its impounded reservoir is diverted to central Washington, where it is planned that 1.2 million acres (480,000 ha) of semiarid sagebrush country will be irrigated. Only two-thirds of the project was completed by the early

FIGURE 15-21   High-quality cotton is a major product of the irrigated valleys in central Arizona. This cotton gin, with acres of bales and mountains of seeds, is near Phoenix (TLM photo).

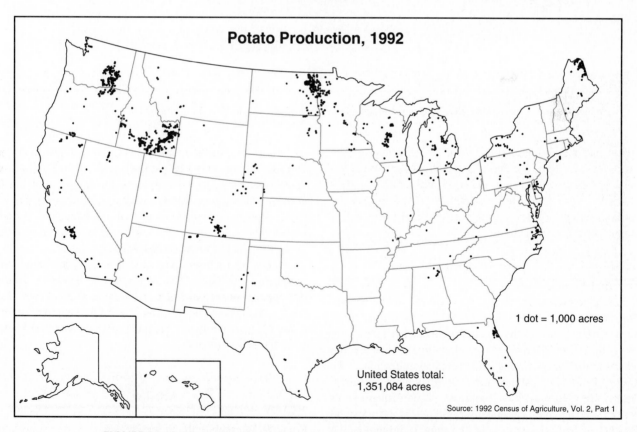

FIGURE 15-22   Two of the five major potato-growing areas of the country are in the northern part of the Intermontane Region.

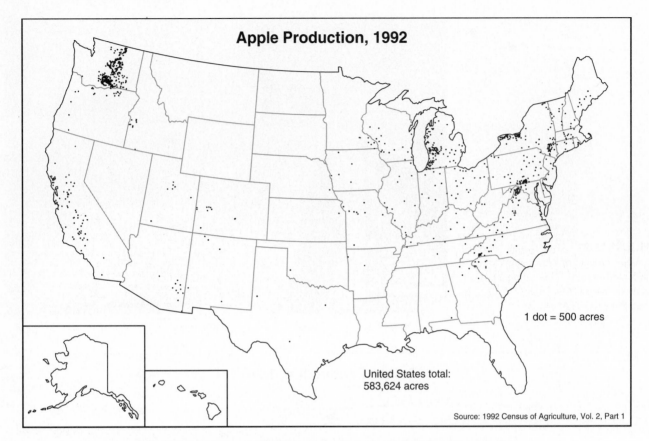

FIGURE 15-23 Central Washington is by far the leader in U.S. apple production.

1990s, and the agricultural pattern is still developing. The most recent major development is the introduction of center-pivot systems to disperse the water, thereby achieving great efficiency of water use and reducing labor costs. This has become a leading center-pivot area, with a notable expansion of output of potatoes and corn.

PASTORALISM

In this region of rough terrain, light rainfall, sparse vegetation, and poor soils most of the land (if it is to be used at all) must serve as range for livestock. Pronounced differences in elevation cause differences in precipitation and vegetation, which, in turn, are reflected in the seasonal utilization of the range. Mountain pastures are strictly summer pastures; deserts are used mostly in winter, when snowfall provides water for sheep and occasionally cattle. Oasis pastures and feedlots are handling more and more animals throughout the year.

The establishment of federal grazing districts by the Taylor Grazing Act of 1934 had a significant effect on the pastoral pattern of the region. This legislation put an end to unrestricted grazing on public lands and helped stabilize the balance between forage resources and numbers of stock. Ranchers may lease portions of a grazing district for seasonal use. It is up to the Bureau of Land Management, the administering agency, to harmonize the carrying capacity of the range with the economic realities of the ranchers.

Ranches for both sheep and cattle are widespread in the region (Fig. 15-24). In recent years there has been a rapid proliferation of feedlots, mostly for cattle. Feedlots have become a big business in the Phoenix area, the Imperial Valley, around Yuma, and in several parts of Utah, southern Idaho, and central Washington.

MINING

From the Wasatch to the Sierra Nevada and from Canada to the Mexican border the region is dotted with communities located solely to tap the mineral resources. These communities enjoy viability as long as the mines produce but decline pre-

FIGURE 15-24   Cattle and cowboys on the sagebrush plains of eastern Oregon (TLM photo).

cipitously and become ghost towns once the ores are worked out or relative price changes make mining unprofitable.

Copper

The most notable mineral resource of the Intermontane Region has long been copper, despite frequent wild gyrations in the international market place, which have caused many mine closures and openings as the price has fluctuated. During the late 1970s and early 1980s there was a drastic decline in demand for copper, and about half the copper mines in the region were shut down. A firmer market in the late 1980s and early 1990s occasioned a decade-long upward trend in production as well as the reopening of several of the closed mines. The United States was the world leader in copper output until about 1980. At present it ranks second (behind Chile).

Arizona has been the leading copper mining state for more than a century. Utah is the second-ranking state, with most production from the largest individual copper mine in the country, at *Bingham,* until it was "temporarily" closed in the 1980s (Fig. 15-25). New Mexico is the third leading state; it experienced rapid production increases in the 1970s from several mines in the *Santa Rita–Silver City* district, but these mines were among the most severely affected by recent closures. Nevada's principal copper mines are at *Ely* and *Yerington.*

Other Metals

There has been a spectacular increase in precious-metal mining in the Intermontane Region ever since prices began to rise significantly in the 1970s, with most of the action in Nevada. Nevada gold output rose from virtually nothing to an annual output of nearly 400,000 troy ounces (120,000 kg) by the mid-1990s, which is more than half of the national total. Nevada also produces more than one-third of the national output of silver.

Other metal ores that are mined in the region include iron in southwestern Utah and tungsten in several locales.

Salts

Several places in the Intermontane deserts yield salts of one kind or another, with major output from southeastern California. Borax minerals are obtained from a voluminous pit at *Boron,* borate and potash compounds are scraped from the surface and extracted from brine wells at *Searles Lake,* various sodium and calcium salts are mined in *Death Valley,* and soda ash has been taken from the dry bed of Owens Lake near *Lone Pine.* Various potash salts are obtained from the *Bonneville Salt Flats* west of Great Salt Lake.

Coal

This region is fairly well endowed with coal. Among the states included, Utah ranks first in reserves, production, and

FIGURE 15-25   The open-pit mine in Bingham Canyon, west of Salt Lake City, has been producing ore for more than three-quarters of a century and is the largest of its kind in the world. It has experienced some shutdown periods in recent years, but at this writing is back in production again (TLM photo).

the importance of coal in a state's total economy. There has been production for several decades in Carbon and Emery counties, an area with extensive reserves.

Coal output has also been expanding in the Four Corners country, with mines in Arizona, New Mexico, and southern Utah. Most of the production is used by thermal electric generating plants or coal gasification plants. These facilities are heavy users of precious water resources and create an abundance of obvious air pollution.

Petroleum

Although the map of oil lands in this region is expanding and the amount of drilling is increasing, the Intermontane Region contributes less than 2 percent of the national output. Principal production comes from the *Rangely* field in northwestern Colorado, several fields in the *Uinta Basin* of northeastern Utah, and northwestern New Mexico.

Oil Shale   Some day oil shale may serve as a great source of petroleum. It is widely scattered over that part of the region in Utah south of the Uinta Mountains, in adjacent west-central Colorado, and in southern Wyoming (partly outside the Intermontane Region). However, high costs and potential production problems have shelved more than a dozen

heralded development projects, and at present only a small amount of oil shale is actually being converted to crude oil. Oil shale certainly will be an important energy source in the future.

Uranium

The history of uranium mining in the United States is one of remarkable ebbs and flows. The frantic boom of the early 1950s and the great decline of the late 1950s were followed in the late 1970s by another major boom. About half the nation's uranium is found in a 100-mile strip of northwestern New Mexico, centering on the town of *Grants*. There is scattered production elsewhere in New Mexico as well as in western Colorado and eastern Utah.

FORESTRY

Forests are generally absent from this region except on the higher mountains and in the north, so logging is not a major activity. Only two areas are notable: central Arizona and the intermontane fringe areas in Oregon, Washington, and Idaho.

The high plateaus and mountains of the Mogollon Rim country and the Coconino Plateau of Arizona are clothed with forests of ponderosa pine, Douglas fir, and other coniferous species. Exploitation is limited mostly to a few large timber-cutting operations and their associated sawmills, although a pulp mill has been brought into operation.

A considerable amount of relatively open forest is found around the margins of the Intermontane Region in the three northern states. The principal species involved is ponderosa pine. Logging here is usually on a small scale, except in a few instances, such as at Bend, Klamath Falls, and Burns, Oregon, which are major pine sawmilling centers.

TOURISM

Tourism has become a major activity in the Intermontane Region in recent years. Its scenic splendors are unmatched on the subcontinent, and there are many other types of attractions for visitor interest. The region also has become a major battleground of environmental concerns. Its scenery, particularly in the Colorado Plateau section, is unique, spectacular, and fragile. The construction of open-pit coal mines and their accompanying thermal-fired plants creates major visual pollution as well as other detrimental environmental impacts. The conflicts are clear-cut and the controversies will undoubtedly proliferate.

Scenery

The Intermontane Region abounds in scenic grandeur. Within it are the Grand Canyon (Fig. 15-26), Monument Valley, Zion Canyon, Cedar Breaks, Bryce Canyon, Death Valley, the Painted Desert, Utah's Canyonlands, the Petrified Forest, Great Salt Lake, gorges of the Snake and Columbia rivers, and the Big Bend canyons of the Rio Grande. Most of these and many other beauty spots have been set aside as protected reserves by federal or state governments. The Colorado Plateau section is particularly magnificent and contains the greatest concentration of national parks (eight) in the nation.

Colorado River Dams and Reservoirs

Although environmental purists may shudder at the thought, the reservoirs created by damming the Colorado River at various locations have become major tourist attractions. Lake Powell behind Glen Canyon Dam, Lake Mead behind Hoover Dam, and Lake Havasu behind Parker Dam are prominent foci for water-oriented recreation—boating, fishing, and skiing—and the dams themselves serve as educational attractions of some note.

Other Tourist Attractions

This region of colorful and spectacular scenery has an almost unlimited number of natural attractions. Many, however, are difficult to reach, and most tourists invest only a limited amount of time in sightseeing off the beaten track. Consequently, various constructed attractions, which almost invariably are reached by excellent roads, draw considerable attention from visitors to the region.

Many ancient *cliff dwellings* and cliff-dwelling ruins are to be found in the Southwest. The most elaborate and best

FIGURE 15-26 The mighty Grand Canyon has been deeply and abruptly incised into the level-surfaced plateaus of northern Arizona. This view is from Mather Point on the South Rim (TLM photo).

FIGURE 15-27   The most extensive and most famous cliff dwellings in the nation are in Mesa Verde National Park in southwestern Colorado. This is Cliff Palace (TLM photo).

interpreted are in Mesa Verde National park in southwestern Colorado (Fig. 15-27). Many others are preserved (and sometimes restored) in national monuments in Arizona and New Mexico. The present-day Native Americans, with their traditional culture and handicrafts, also draw visitors from other regions. The Navajo Reservation and several of the Pueblo villages attract the largest crowds (Fig. 15-28).

There are many *historic towns* in the region, remnants of that romanticized and immortalized period in American

FIGURE 15-28   Many of the Pueblo people of central New Mexico live in villages of traditional architecture. This is Taos Pueblo (TLM photo).

history, "the Old West." Some, such as Bisbee, Tombstone, and Jerome in Arizona or Virginia City in Nevada, have capitalized on their heritage and built up a steady trade in historically minded tourists.

The remarkable history of *Mormonism* in Utah and the continued importance of Salt Lake City as headquarters of the Mormon Church are compelling tourist attractions. Various edifices and monuments in and around Salt Lake City are visited by hundreds of thousands of visitors annually (Fig. 15-29).

Although Nevada has a colorful past, its present is in many ways even more flamboyant. As the only state that has systematically used *gambling* as a major source of revenue, it has developed games of chance in amazing proportions. Almost every town in the state has its cluster of "one-armed bandits" and minicasinos, but the chief centers are Las Vegas (60 casinos) and Reno, although the greatest recent development has been in the tiny town of Laughlin, at the southern tip of the state (and therefore the closest casino location to the major population clusters of Los Angeles, San Diego, and Phoenix). As added attractions these gambling cities also specialize in glamorous entertainment, quick marriages, and simple (although not speedy) divorces (Fig. 15-30). The relentless, headlong rush to attract gambling tourists has begun to focus more on extravagant entertainment and shows that are suitable for the entire family. This thrust apparently is paying off, as ever-bigger and more elaborate casino–hotels continue to appear in Nevada, especially on the Las Vegas "Strip," where the latest hotel (at the time of this writing) has more than 5000 rooms (the largest hotel in the world), 3500 slot machines, and a 33-acre (13-ha) theme park outside the back door, and cost $1 billion to build.

SPECIALIZED SOUTHWESTERN LIVING

The rapid population growth of the southern part of the Intermontane Region in recent years, matched only by that of Florida and southern California, is an obvious tribute to sunshine and health. Many present-day Americans feel that sunny, mild winters and informal outdoor living provide sufficient satisfaction to counteract the problems of moving to a distant locality, even if that locality is characterized by scorching summers. Sufferers from respiratory afflictions also derive some real and some imagined health benefits from the dry air of the Southwest.

In Arizona, southern Nevada, southern New Mexico, and southeastern California particularly, the ordinary summer tourist is a relatively minor element in comparison with the frequent winter visitor, the new resident eagerly anticipating

FIGURE 15-29   The international headquarters of the Church of Jesus Christ of Latter-day Saints towers high above the more famous Mormon Temple in Salt Lake City (TLM photo).

FIGURE 15-30   Las Vegas at night. The flamboyance of gambling casinos and the relatively inexpensive power of nearby Hoover Dam combine to give Fremont Street the brightest lights in the Intermontane Region (TLM photo).

opportunity in a growing community, and the retired couple content to spend their last years in sunny relaxation. It is on these three groups that the social and, to a considerable extent, the economic structure of the Southwest is turning. The significance of these groups is demonstrated nowhere quite as pointedly as in the growth of suburban Phoenix. Scottsdale, on the northeast, is a semiexclusive residential and re-

sort suburb whose luxury hotels and elaborately picturesque shops and restaurants are geared specifically to the winter visitor. Mesa, on the east, is a sprawling desert community scattered with large factories and ambitious subdivisions for the migrant from eastern states. Sun City, on the northwest, is a specifically planned retirement community without facilities for children but with abundant amenities for senior citizens. Litchfield Park, on the west, is a grand design for a totally planned community in which last year's cotton fields will give way to next decade's city of 100,000 people.

SUBURBIA IN THE SUN: THE SOUTHWEST'S RUSH TO URBANISM

The extremely rapid population expansion of the southern part of the Intermontane Region is primarily manifested in the burgeoning of cities and extensive urban sprawl (see Table 15-1 for a listing of the region's largest urban places). The spectacular growth of major southwestern cities is readily apparent (Table 15-2).

TABLE 15-1
Largest Urban Places of the Intermontane West, 1995

Name	Population of Principal City	Population of Metropolitan Area
Albuquerque, NM	437,000	662,000
Boise City, ID	186,000	362,000
El Paso, TX	588,000	683,000
Glendale, AZ	182,000	
Idaho Falls, ID	47,000	
Las Cruces, NM	64,000	165,000
Las Vegas, NV	353,000	1,142,000
Mesa, AZ	324,000	
Ogden, UT	81,000	
Phoenix, AZ	1,099,000	2,716,000
Pocatello, ID	48,000	
Provo, UT	84,000	299,000
Reno, NV	158,000	291,000
Richland, WA	36,000	182,000
Salt Lake City, UT	175,000	1,208,000
Scottsdale, AZ	148,000	
Spokane, WA	202,000	401,000
Tempe, AZ	147,000	
Tucson, AZ	430,000	754,000
Yakima, WA	52,000	208,000

TABLE 15-2
Percentage of Metropolitan Area Population Increase, 1980–1995

Las Vegas	111
Phoenix	58
El Paso	42
Tucson	40
Albuquerque	35
Salt Lake City	31

If the magnitude of recent southwestern urban growth has been remarkable, it is the character and form of this growth that have been even more eye-catching.

Their physical structure is looser than in older cities; their average density is low, they consist mostly of detached single-family houses or garden apartments, they expand rapidly at their edges, and they often enclose a crazy-quilt pattern of unbuilt-upon land. Not even the slum areas, backward as they might be, approach traditional urban densities. Mass transportation is inadequate or nonexistent. . . .

These cities have not only grown to maturity in the time of the automobile, they live by the automobile—and it is for the most part a pleasant and convenient way of life. Traffic jams are rare; parking, if not always well designed, is at least plentiful and inexpensive. Because of the automobile, the strip has often long ago replaced downtown as a center of business. Not just a competitor of the central core, it has become the vital economic area—if not in terms of quantity of money handled, then certainly in terms of daily shopping activity. [10]

Many "new" cities have been condemned by urban planners; they are different and do not resemble the "old" and great cities of the world. They lack the important attributes of high population and building density and a centric orientation.

But downtown in these new cities is not the same downtown that we remember from other places and times. To revitalize or preserve a downtown that contains excellent stores and restaurants, museums and schools as well as banks and offices, a downtown that is served by an adequate or expandable rapid-transit system, and that has an emotional meaning to the people of a city is one thing. *Creating* a downtown in an area having no good stores and few good restaurants, no cultural or education facilities, an area in which even the movie theaters are second-rate with the cinerama-size screens located in the suburbs, where the only unique facilities are more old and cheap office space, bank headquarters, and the bus and railroad stations, a downtown located in a city of a density too low to support mass transit, in a city whose inhabitants' nostalgic memories are of a downtown in a far-off place that they have left—this is a very different matter (Fig. 15-31).

[10] Robert B. Riley, "Urban Myths and the New Cities of the Southwest," *Landscape,* 17 (Autumn 1967), 21.

... This new form of urban structure ... is a result of increasing affluence and mobility, vastly improved communication, greater flexibility of transportation, and the increased importance of amenity in residential, commercial, and site location ...

What is happening [in Southwestern cities] ... is precisely what is happening in the megalopolises of the eastern and western seaboard and the urban regions of the Midwest—with one important difference. In the latter areas the new developments take place over, around, or between strong and still vital industrial urban forms, forms which both dampen and distort the growth of radically new patterns. In the Southwest, where no such strong earlier forms exist, the new forms, as yet neither fully developed or understood, can at least be seen more clearly and studied for what they are or want to become.[11]

One of the most conspicuous features of the "new" southwestern cities is their expansive sprawl. Their booming growth has a spatial expression of erratic, leapfrogging development (for instance, nearly 40 percent of the land area within the Phoenix city limits is open space), their population density is quite low (Phoenix has a population density that is only about one-third that of Los Angeles, long the epitome of low density), and their areal extent is likely to be enormous (Tucson, for example, encompasses twice as many square miles as Boston).

Apart from this "new" form of urban development in the region, two types of specialized communities have become prominent:

1. Paired towns have grown up at a number of locations on opposite sides of the international border with Mexico. There are 17 pairs of these international twins, 11 of which are on the southern margin of the Intermontane Region.[12] Although the individual towns of each pair have a different cultural origin, they constitute a single spatial unit with a symbiotic economic and social relationship (Fig. 15-32). In almost all cases, the Mexican town is more populous than its American counterpart (Ciudad Juarez, for instance, has nearly twice the population of El Paso), but the central shopping district for both twins is in the American community. The symbiotic relationship has been particularly enhanced since 1965 when the Mexican government initiated its *maquiladora* (twin-plant) border indus-

FIGURE 15-31   The central areas of most of the larger Intermontane cities have been revitalized and metamorphosed by ambitious and imaginative building projects in recent years. This splashy scene is in Albuquerque's Civic Center (TLM photo).

try program, which was designed to encourage the establishment of U.S. assembly plants just south of the border, using inexpensive Mexican labor but with essentially all the finished products being shipped to the United States for marketing. More than 500 such plants have been established in Mexican border towns, about half of them in communities adjacent to the Intermontane Region. The largest *maquiladora* development has been in Cuidad Juarez.

2. Specialized recreation-retirement towns are blossoming widely in the Intermontane Region. In every case, their site has some physical attraction—characteristically, the shoreline of some water body, otherwise a mountain location or a mild-winter desert spot. These towns experience relatively heavy tourist traffic, but stability is provided by a growing number of "permanent" residents. There are many examples in the region, such as Ruidoso (New Mexico) and Show Low (Arizona), but the prime exam-

---

[11] Ibid., p. 23.

[12] From west to east, they include Tecate, CA/Tecate, Baja California; Calexico, CA/Mexicali, Baja California; San Luis, AZ/ San Luis Rio Colorado, Sonora; Lukeville, AZ/Sonoita, Sonora; Sasabe, AZ/Sasabe, Sonora; Nogales, AZ/Nogales, Sonora; Naco, AZ/Naco, Sonora; Douglas, AZ/Agua Prieta, Sonora; Columbus, NM/Las Palomas, Chihuahua; El Paso, TX/Ciudad Juarez, Chihuahua; and Presidio, TX/Ojinaga, Chihuahua.

FIGURE 15-32   Looking south across downtown El Paso into smog-covered Ciudad Juarez, Mexico (TLM photo).

ples of a self-sustaining, nongovernmentally developed, recreational retirement new town are Lake Havasu City, Arizona, which grew from nothing to a population of nearly 20,000 in two decades, making it the largest urban place in its county, and upriver Bullhead City (Arizona's fastest-growing community in the late 1980s and early 1990s), opposite Laughlin, Nevada's newest casino center.

THE OUTLOOK

People have accomplished much in this restrictive environment. No one can stand on the steps of the State Capitol Building at Salt Lake City and gaze at the green island that is the Oasis without being impressed. Nevertheless, there is a limit to what human beings can accomplish against a stubborn and relentless nature. Because water, which means life, is scarce and much of the terrain is rugged, the greater part of the region is destined to remain one of the emptiest and least used on the continent.

Agriculture should become more important, but the development of additional large reclamation projects is unlikely, simply because most feasible dam sites in the region have already been developed (except at very controversial locations in the Grand Canyon). Irrigated crop acreages will ex-

pand most in central Washington and southern Idaho, but modest expansions will be widespread, particularly in Oregon and Arizona, often associated with center-pivot technology.

Livestock raising will probably increase. More feeding will be carried on, both at local ranches and at centralized feedlots. Hay and sorghum feeding will continue to dominate, but grains, often brought into the region, will increase in importance. More attention will be paid to breeding, too, with improved Hereford and Angus strains in the north and more emphasis on Santa Gertrudis and Charolais in the south.

Mining is an industry of fluctuating prosperity in the region and will continue to be so. Copper, historically the most important intermontane mineral, often experienced instability due to an erratic market and antagonistic labor relations, a situation that is unlikely to change. Prospecting for, and mining of, energy minerals was the booming activity of the 1970s, but in the 1980s this was dramatically superseded by precious-metal mining.

The North American Free Trade Agreement (NAFTA) was approved by the U.S., Mexican, and Canadian governments in the early 1990s. Its immediate effect on the Intermontane Region was to stimulate economic activity and to clog the highways near the international border with Mexico. El Paso is the most active port of entry on the Mexican border, with more than a million commercial vehicles passing through in 1995. It is anticipated that NAFTA will stimulate business widely in the region, with the most active developments taking place in the border zone.

Tourism in this region, as in most, is bound to expand. Summer is tourist time in the northern three-fourths of the Intermontane West; winter visitors are more important in the southern portion. An abundance of natural allurements, a variety of constructed attractions, and improving transportation routes combine to ensure a steady flow of tourists.

Population increase in the northern half of the region was slow in the 1980s, expressed principally by net out-migration. The southern, or Sunbelt, portion of the region continued to grow apace. These trends probably will persist. Mild winters, few clouds, dry air, informal living patterns, and the mysterious attraction of the desert will continue to exert their magnetic effects on dissatisfied, snow-shovel-weary citizens of the northern states.

Such a migration-fostered population growth will probably become overextended at times, outstripping a sound economic base. Generally, however, it is likely to grow with soundness, for capital will accompany people in the migration. Water may be a long-run limiting factor, but in the short run it is no barrier; urban growth is often at the expense of irrigated agriculture, and the former uses less water than the latter.

The southern Intermontane Region, then, is in functional transition from desert to metropolis. Today one can find smog in Phoenix that would make a Los Angeleno proud, traffic jams in Albuquerque that would do credit to Chicago, and tension-induced psychiatric treatments in Salt Lake City that would be suitable for New York. Indeed, the "new" cities of the interior West are already embarking on imaginative urban renewal projects to revitalize their downtowns. The developments are most striking in Tucson and Albuquerque, but almost every city of note from Spokane to El Paso has made at least a start on a mall–park–fountain complex in association with sparkling modern high-rise buildings in the heart of the central business district.

Rapid growth, however, is accompanied by growing pains. Cities are not immune to urban problems merely because they are new and different. Moreover, it is in rural areas that the major conflicts and controversies are likely to occur. Development versus preservation marks a battle line throughout North America, but perhaps in no other region are the opposing interests and values so clear-cut and the opportunities for compromise so limited.

SELECTED BIBLIOGRAPHY

COOKE, RONALD U., AND RICHARD W. REEVES, *Arroyos and Environmental Change in the American South-West.* London: Oxford University Press, 1976.

FRADKIN, PHILIP L., *A River No More: The Colorado River and the West.* Berkeley: University of California Press, 1996.

FRANCAVIGLIA, RICHARD V., *The Mormon Landscape.* New York: AMS Press, 1979.

GILBERT, BIL, "Is This a Holy Place?" *Sports Illustrated,* 58 (May 1983), 76–90.

GOODMAN, JAMES M., *The Navajo Atlas.* Norman: University of Oklahoma Press, 1982.

GREEN, CHRISTINE, AND WILLIAM SELLERS, *Arizona Climate.* Tucson: University of Arizona Press, 1964.

GREER, DEON C., ET AL., *Atlas of Utah.* Provo, UT: Brigham Young University Press, 1981.

HARNER, JOHN P., "Continuity Amidst Change: Undocumented Mexican Migration to Arizona," *Professional Geographer*, 47 (November 1995), 399–410.

HART, JOHN, *Storm over Mono: The Mono Lake Battle and the California Water Future.* Berkeley: University of California Press, 1996.

HECHT, MELVIN E., "Climate and Culture, Landscape and Lifestyle in the Sun Belt of Southern Arizona," *Journal of Popular Culture,* 11 (Spring 1978) 928–947.

———, AND RICHARD W. REEVES, *The Arizona Atlas.* Tucson: University of Arizona, Office of Arid Lands Studies, 1981.

HERZOG, LAWRENCE A., *Where North Meets South: Cities, Space, and Politics on the U.S.–Mexico Border.* Austin: University of Texas Press, 1990

HUNDLEY, NORRIS, JR., *Dividing the Waters.* Berkeley: University of California Press, 1966.

———, *Water and the West: The Colorado River Compact and the Politics of Water in the American West.* Berkeley: University of California Press, 1975.

JAEGER, EDMUND C., *The California Deserts.* Stanford, CA: Stanford University Press, 1965.

JETT, STEPHEN C., "The Navajo-Homestead: Situation and Site," *Yearbook,* Association of Pacific Coast Geographers, 42 (1980), 101–117.

KERSTEN, EARL W., "Nevada Then and Now: Forging an Economy," *Yearbook,* Association of Pacific Coast Geographers, 47 (1986), 7–26.

McGREGOR, A. C., *Counting Sheep: From Open Range to Agribusiness on the Columbia Plateau*. Seattle: University of Washington Press, 1983.

MEINIG, D. W., *The Great Columbia Plain—An Historical Geography, 1805–1910*. Seattle: University of Washington Press, 1968.

———, "The Mormon Culture Region: Strategies and Patterns in the Geography of the American West, 1847–1964," *Annals,* Association of American Geographers, 55 (1965), 191–220.

———, *Southwest: Three Peoples in Geographical Change, 1600–1970*. New York: Oxford University Press, 1971.

NOSTRAND, RICHARD L., "The Hispanic-American Borderland: Delimitation of an American Culture Region," *Annals,* Association of American Geographers, 60 (1970), 638–661.

———, *The Hispano Homeland*. Norman: University of Oklahoma Press, 1992.

———, "The Hispano Homeland in 1900," *Annals,* Association of American Geographers, 70 (September 1980), 382–396.

———, "Spanish Roots in the Borderlands," *The Geographical Magazine,* 52 (December 1979), 202–210.

PEASE, ROBERT W., "Modoc County, A Geographic Time Continuum on the California Volcanic Tableland," *University of California Publications in Geography,* 17 (1965), 1–304.

RUHE, ROBERT V., "Landscape Morphology and Alluvial Deposits in Southern New Mexico," *Annals,* Association of American Geographers, 54 (1964), 147–159.

SAUDER, ROBERT A., "Owens Valley's Abandoned Landscape," *California Geographer*, 32 (1992), 61–76.

SPEAR, STEVEN G., "The Climate of Death Valley," *California Geographer*, 32 (1992), 39–50.

SYMANSKI, RICHARD, *Wild Horses and Sacred Cows*. Flagstaff, AZ: Northland Press, 1985.

VALE, THOMAS R., "Forest Changes in the Warner Mountains, California," *Annals,* Association of American Geographers, 67 (1977), 28–47.

———, AND GERALDINE R. VALE, *Western Images, Western Landscapes: Travels Along U.S. 89*. Tucson: University of Arizona Press, 1989.

The California Region

16

The California Region, one of the smallest major regions of the continent, is probably the most diverse. This diversity is manifested in almost every aspect of the region's physical and human geography, from landforms to land use and from soil patterns to social patterns.

The region's location in the southwestern corner of the conterminous states has been a major long-range determinant of its pattern and degree of development (Fig. 16-1). The location is remote from the area of primary European penetration and settlement of North America and from the heartland of the nation that emerged after colonial times. Adjacency to Mexico has been significant from early days, even though the region is relatively distant from the heartland of that country, too. The region is well positioned for contact across the Pacific, but the Pacific has been, until recently, the wrong ocean for significant commercial intercourse. California has thus been denied ready access to the core regions of the United States and Mexico, and even to the Pacific Northwest, by pronounced environmental barriers (mountains and deserts). This factor has had an important effect on the population and economic development of the region; it is remarkable that it has not had an even greater effect.

The California Region encompasses most settled parts of the state, excluding only the northern mountains (north Coast Ranges, Klamath Mountains, southern Cascades), the

northeastern plateaus and ranges, and the southeastern deserts except the Antelope Valley and Palm Springs areas, which are functionally integrated with the southern California conurbation and are therefore considered part of this region. The region, then, is essentially a California region and includes the intrinsic California. More than 97 percent of the inhabitants of the most populous state, numbering about 32 million persons in the early 1990s, reside within this region.

THE CALIFORNIA IMAGE: BENIGN CLIMATE AND LANDSCAPE DIVERSITY

The image of the region is synonymous with the image of the California lifestyle. It is the never-never land of contemporary American mythology. It is focused on Hollywood, Disneyland, and the Golden Gate; flavored with equal parts of glamour and smog; and populated by a mixture of sun-bronzed beach lovers and eccentric night people.

That the actuality is less exciting than the image is not too important. The fact that most Californians live in the same sort of suburban tract homes as other North Americans, watch the same television shows, vote for the same political candidates, and complain about the same taxes is convenient to overlook, for the California lifestyle has been sugarcoated,

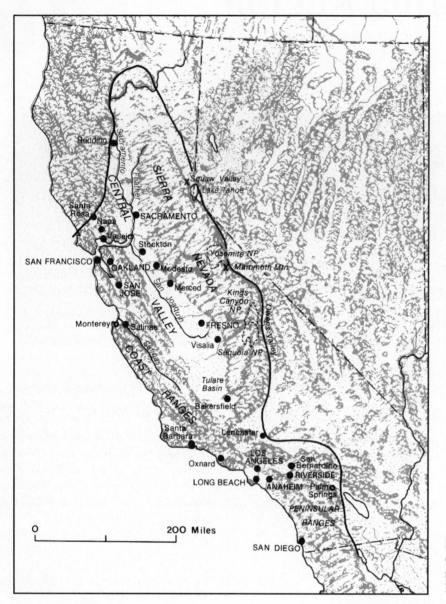

FIGURE 16-1 The California Region (base map copyright A. K. Lobeck, reprinted by permission of Hammond, Inc.).

packaged, and marketed to the world as a thing apart, a destiny with a difference.

The image does, of course, have some substance. It is due in part to the relatively late development of the region's urban economy; in part to the boom-and-bust psychology and flamboyant nature of some of the staple industries—for example, gold mining, oil drilling, real estate promotion, the motion picture industry, television and radio empires, aircraft and spacecraft production; in part to the unusual natural endowments of this southwestern corner of the country; and in part to its residents, people who are drawn from every corner of the globe and who come seeking an elusive opportunity

that they failed to find in their homeland and that they expect to discover in California.

Of utmost importance, both physically and psychologically, is the regional climate. This is the only portion of the continent with a dry-summer subtropical climate, generally called "mediterranean." Its basic characteristics are simple and appealing: abundant sunshine, mild winter temperatures, absolutely dry summers, and relatively dry winters. No other type of climate is so conducive to outdoor living.

And the diverse characteristics of the California outdoors multiply the opportunities and enhance the appeal of the region. High mountains are adjacent to sandy beaches and dramatic sea cliffs; dense forests rise above precipitous canyons that open into fertile valleys. Nearby to the east is the compelling vastness of the desert, and to the south is the charm of a different culture in a foreign land. And yet there is much more to the regional character than a flamboyant image, a benign climate, and a diverse landscape.

The California Region is outstanding in agriculture, significant in petroleum, unexcelled in aerospace and electronics, important in design, trendsetting in education, innovative in urban development, and increasingly significant in decision making. It is also the world champion in air pollution, the national leader in earthquakes and landslides, preeminent in both traffic movement and traffic jams, and the destination of a population inflow that is unparalleled in the history of the continent.

THE ENVIRONMENTAL SETTING

The diverse terrain of the California Region engenders much variety in most other environmental components.

Structure and Topography

The region occupies an area of great crustal instability, due primarily to its location at the interface of two major tectonic plates (North American and Pacific). Thus there has been, and continues to be, much diastrophic movement in the region, with a widespread occurrence of active faults. These faults are many and varied, but have two prominent characteristics: (1) The principal ones mostly have a NNW–SSE trend; and (2) many of them are extraordinarily lengthy. As a result, fault lines and fault scarps are prominent in the landscape (Fig. 16-2), and the general trend of most topographic lineaments (including the coastline) is NNW–SSE. Moreover, much of the topography has a "youthful" appearance, with a prevalence of steep slopes and high relief.

In gross pattern there are three broad topographic complexes within the region—coastal mountains and valleys, Central Valley, and Sierra Nevada.

The western portion of the region contains a series of mountain ranges with interspersed valleys. The Coast Ranges (roughly from Santa Maria northward) are strikingly linear in arrangement and are separated by longitudinal valleys of similar trend. The topography is structurally controlled, with prominent fault lines—of which the San Andreas is the most conspicuous and famous—tilted fault blocks, and some folding of the predominantly sedimentary strata. The nearly even crests of the ranges average 2000 to 4000 feet (600 to 1200 m) above sea level and are notable obstacles to the westerly sea breezes (Fig. 16-3).

FIGURE 16-2   An aerial view of one of the conspicuous faults in the California landscape. This is the San Andreas Fault as it crosses the Carrizo Plain northwest of Los Angeles (TLM photo).

FIGURE 16-3   Along most of the seafront of central California the Coast Ranges drop precipitously to the ocean. This is in the Big Sur section (TLM photo).

South of the Coast Ranges are higher and more rugged mountains (the Transverse and Peninsular ranges) that are less linear and parallel in pattern.

The Central Valley consists of a broad down warped trough averaging about 50 miles (80 km) in width and more than 400 miles (640 km) in length. The trough is filled with great quantities of sand, silt, and gravel [to a depth of more than 2000 feet (600 m) in places] washed down from the surrounding mountains. The Sacramento River flows south through the northern half of the Central Valley, and the San Joaquin flows north through the southern half. They converge near San Francisco Bay and empty into the Pacific Ocean through the only large break in the Coast Ranges. Their delta originally consisted of a maze of distributaries, sloughs, and low islands. It has now been diked and canalized and is one of the state's leading truck-farming and horticultural areas.

The Sierra Nevada is an immense mountain block 60 to 90 miles (100 to 150 km) wide and 400 miles (640 km) long, situated just east of the Central Valley. It was formed by a gigantic uplift that tilted the block westward. The eastern front is marked by a bold escarpment that rises 5000 to 10,000 feet (1500 to 3000 m) above the alluvial-filled basins of the Intermontane Region to the east; this escarpment marks one of the most definite geographical boundaries on the continent. The western slope, although more gentle, is deeply incised with river canyons and was greatly eroded by glaciers, forming such magnificent canyons as those of Yosemite (Fig. 16-4) and Tuolumne.

The summits of the Sierra Nevada suffered severe glacial erosion and consist of a series of interlocking cirques (Fig. 16-5). Complex faulting, mountain glaciation, and stream erosion account for most details of the mountain terrain. Some block-faulted valleys contain lakes, the most noted of which is Lake Tahoe; smaller but often spectacularly sited lakes occupy some glaciated valleys.

The long western slope of the range contains many magnificent glaciated valleys in which the glacial debris was completely washed away by the subsequent meltwater and

FIGURE 16-5   A representative view of sawtooth peaks in the glaciated high country of the Sierra Nevada. This is the Great Western Tiers in Sequoia National Park (TLM photo).

deposited in the Central Valley. On the abrupt eastern slope, however, drier conditions resulted in less ice and less melting, with the result that glacial till accumulated in great moraines and now often encloses lovely alpine lakes. The high country of the Sierra Nevada still contains some 70 glaciers, all small.

Climate

The distinctive characteristics of the climate of the California Region are mild temperatures and a mediterranean precipitation regime.

Summer is dominated by a subsiding, diverging air flow that results in calm, sunny, rainless days and mild rain-

FIGURE 16-4   The long western slope of the Sierra Nevada is cut by numerous spectacular canyons that were shaped by Pleistocene glaciers, the most striking of which is Yosemite Valley (TLM photo).

less nights. Temperature inversions are frequent, and inland locations can experience much hot weather.

Winter is the rainy season; this is due to recurrent cyclonic disturbances, but lowland areas receive only limited amounts of precipitation. The annual total increases northward from 10 inches (25 cm) in San Diego to 20 inches (50 cm) in San Francisco. West-facing mountain slopes receive much more rain and, particularly in the Sierra Nevada, considerable snow (Fig. 16-6).

Natural Vegetation

There is much localized variation in the vegetation pattern, but basic generalizations can be made. The original plant association of most of the immediate coastal zone was dominated by low-growing shrubs of coastal sage. Chaparral (a close growth of various tall broad-leaved evergreen resinous shrubs) was the characteristic vegetation of most of the valleys and lower mountain slopes of southern and central California, although there were also extensive areas of grassland dotted with oak trees. The Central Valley itself was mostly a natural grassland. The middle and upper slopes of the mountain ranges were (and still are) mostly forested, largely with a variety of conifers. Ponderosa pine, lodgepole pine, and various firs are the principal species. The world-famous giant sequoias occur in about 75 scattered groves at middle elevations [4500 to 7500 feet (1350 to 2250 m)] along the western side of the Sierra Nevada (Fig. 16-7).

Natural Hazards: Shake and Bake in California

Unstable Earth   The region is in a very unstable crustal zone, seamed in profusion by faults. Thus, earthquakes are experienced, sporadically and unpredictably, but frequently. Some tremors have had spectacular and tragic results (for example, the Loma Prieta/San Francisco quake of 1989 killed 65 people and caused $8 billion worth of property damage, and the Northridge quake of 1994 killed 57 people and caused $7.2 billion worth of damage).

Other types of earth slippages, on a much smaller scale, occur with greater frequency. The steep hills and unstable slopes of the subregion often afford spectacular views but precarious building sites. Every year there are dozens of small slumps and slides, and a few unfortunate residents "move down to a new neighborhood" whether they want to or not.

Flood and Fire   The relatively small amount of precipitation falls almost entirely in the winter. As water flows

FIGURE 16-6   The Sierra Nevada normally is deeply inundated with snow every winter. Here a snowblower clears the road in Giant Forest of Sequoia National Park (TLM photo).

FIGURE 16-7   Giant sequoias are found in several dozen discrete groves at middle elevations on the western slope of the Sierra Nevada. This is Crescent Meadow in Sequoia National Park (TLM photo).

down the steep hill slopes into urban areas, which are so extensive that they are inadequately supplied with storm drains, destructive floods and mud slides often result. During the rainless summers, on the other hand, forest and brush fires are ever imminent. The tangled chaparral, chamisal, and woodland vegetation of the abrupt hills and mountains that abut and intermingle with the urbanized zones is readily susceptible to burning. Only carefully enforced fire precautions, a network of firebreaks and fire roads, and efficient suppression crews are able to hold down the damage. And where burning strips away vegetation in the summer, flooding becomes even more likely in the following winter.

SETTLEMENT OF THE REGION

The region's physical and cultural diversity is further reflected in its sequential occupance pattern. The prominent waves of Native American, Spanish, Mexican, and Anglo settlers have been augmented more recently by an influx of African Americans and Asians.

California Native Americans

Natural resources were relatively abundant in most parts of the region in pre-European times, particularly along the coast. The aboriginal inhabitants essentially were hunters and gatherers, and the basic foods for most tribes were either seafoods or plants that gave seeds and nuts that could be ground into flour and then boiled as gruel or baked into bread.

Tribal organization was loosely knit and generally involved small units. Often just a dozen or so families, perhaps related in some sort of patriarchal lineage, would form a wandering band. There were few large tribes or strong chiefs, and internecine warfare was the exception rather than the rule. This made it possible for California to support a relatively high density of Native American population compared with most other parts of the country.

The lack of strong organizational unity also made it easy for the invading Europeans to dominate. Most Native Americans in the region tamely gave way to white occupance and left few marks on the landscape to signal their passing. Yet all the earliest recorded history of California revolves around them, and the entire pre-Anglo period was shaped by relations with the Native Americans.

The Spanish Mission Period

The European "discoverer" of the region was Rodriguez Cabrillo, on a maritime mission from Mexico in 1542. Following this expedition came a long era of occasional exploration, but more than two centuries passed before there was any attempt at colonization of Alta California by the Spaniards. California seemed to offer little attraction to the adventurous, gold-seeking conquistadors. Renewed interest was stimulated not by the possibilities of the region itself but by the rapid advance of Russian domination of the Pacific Coast south from Alaska. Although Spain considered the California area economically worthless, it wanted a buffer state to prevent possible Russian encroachment on the more valuable colony of Mexico.

The first of Spain's famous missions in California was founded in 1769 at San Diego, and within 3 years an irregular string of missions, both in protected coastal valleys and directly on the coast, had been established as far north as Monterey. Eventually 21 continuing missions were established, the last in 1823 (Fig. 16-8). The purpose of the mission system was to hold the land for Spain and to Christianize the Native Americans. These steps were accomplished by settling the nearby Native Americans at each mission to carry on a completely self-sufficient, sedentary crop-growing and livestock-herding existence. Each of the missions garnered an attached population of from 500 to 1500 Native Americans.

The basic results of the system were that some 85,000 Native Americans were "civilized," at least that many and probably more were exterminated by the inadvertent introduction of exotic diseases, various European plants and animals were introduced to the region, and a number of settlement nuclei and transportation routes were established.

Two other types of settlements, *presidios* and *pueblos,* were established during the mission period. Four presidios were set up as army posts for protection against Native Americans and pirates; each developed as a major settlement nucleus (San Diego, Monterey, San Francisco, and Santa Barbara). Pueblos, three in number, were planned farming villages; Los Angeles and San Jose survived as future major cities.

Generally the missions reached the height of their prosperity in the 1820s in spite of problems with fire, flood, earthquake, and unbelievers. The mission–presidio–pueblo system, however, did not succeed in attracting many settlers to California. Land grants were few, small, and allotted chiefly to retired soldiers.

The Mexican Period

Mexico, including California, achieved independence from Spain in 1822, and the missions were gradually secularized. The result was a rapid alienation of the land into private ownership, neglect of property and buildings, and lapsing of the mission Indians into degraded poverty or demoralized frontier life. The Mexican government embarked on a large-scale scheme of land grants to individuals; within only a few years most of the land from Marin and Sonoma counties (just north

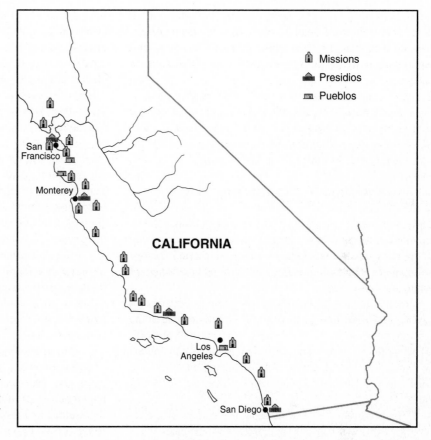

FIGURE 16-8 Early Spanish settlements were along or near the coast. (Contemporary place names are included for reference.)

of San Francisco Bay) southward along the coast was held in *ranchos.*

Ranchos quickly replaced missions as the focus of life in California. They were primarily self-sufficient cattle empires, with beef as the chief food and hides and tallow as the principal exports. Towns were few, small, and far apart. Los Angeles was the largest; Monterey was the capital and the chief seaport. Most of the inland portion of the region remained unsettled except for a few ranchos in the Central Valley.

The Early Anglo-American Period

California Territory was annexed to the United States at the end of the Mexican War in 1848. By a remarkable coincidence, gold was discovered in the Sierra Nevada foothills only 2 months before the peace treaty was signed. The impact of gold discovery was sensational. The total population of California in 1845 was about 5000, of whom less than 8 percent were Anglo-Americans. Half a decade later the population had reached nearly 100,000, of whom 90 percent were Anglos.

It was San Francisco that prospered most as a result of gold. Its population of 800 almost vanished in the first rush to the gold fields. Its outstanding location soon gave it ascendancy, however, and in a short time it was the largest city in western North America. The population increased to 35,000 in 2 years and was more than 50,000 in 1860 despite having been leveled by fires five times during that period. The population of California was 380,000 by then.

Los Angeles at that time had less than 5000 people; the influence of the gold strike had been much less significant in the "cow counties" of southern California. Most of the land was devoted to cattle raising and, except for the pueblo of Los Angeles and settlements around a few presidios and missions, there were no towns at all.

The Beginnings of the Commercial-Industrial Era

The first California land boom started slowly but grew abruptly in the 1880s. The boom spread from Los Angeles to San Diego and Santa Barbara and was felt over much of the region. The population of Los Angeles County tripled to 100,000 in that decade. Separate colonies, based on irrigation farming of specialty crops, were founded in many places. There were orange colonies at Anaheim, Pasadena, Riverside, and Redlands; grape colonies in the Fresno area; and other varied colonies in different parts of the San Joaquin Valley.

The availability of "transcontinental" railway connections provided access to distant markets and proved a major solution to marketing problems. The first transcontinental line, from Omaha to San Francisco, was completed in 1869, and a railway was extended from San Francisco to Los Angeles in 1876.

Irrigated agriculture spread and intensified rapidly in the Los Angeles lowland, in the numerous coastal valleys, and in the Central Valley. Agricultural progress was accelerated by the development of refrigerated railway cars, canning and freezing of produce, improved farm machinery, pumps and pipes for irrigation and drainage, and the beginnings of agribusiness enterprises.

Industrial growth began with petroleum discoveries toward the end of the nineteenth century. The emergence of the motion picture business was a major catalyst, both physical and promotional, to economic and population growth and urban expansion. Manufacturing was stimulated during World War I and experienced a phenomenal growth during World War II.

The present occupance and land-use pattern of the region, however, was basically in existence by the early years of the twentieth century. Changes since that time have mostly involved the extension of irrigated acreage in lowland areas and the never-ending sprawl of the cities.

POPULATION: SENSATIONAL GROWTH SLOWING DOWN

The keynote of the geography of the California Region during most of this century was the rapid growth of its population. The rate of natural increase (excess of births over deaths) was only average, but the trend of net in-migration (excess of in-migrants over out-migrants) has been nothing short of sensational.

Although there have been fluctuations, for most of the past half-century California's population has been increasing at an annual average rate approximately twice as great as that for the nation as a whole. In the 1980s the California population experienced a net gain of about 800,000 people, half attributed to natural increase and half due to net in-migration. There were about 600,000 births, offset by only 200,000 deaths, for a net gain of 400,000. In terms of migration there was a net gain of 300,000 foreign in-migrants and 100,000 domestic in-migrants.

However, in the 1990s California's lagging economy tarnished the region's once-golden lure, with the result that domestic net migration became negative, and the population growth rate slowed considerably. In every year of the 1990s (to the date of this writing) there was a net loss of people to other states; that is, more people moving from California to other states than vice versa. More than 1.5 million Californians moved away in the first half of the decade.

The total regional population continues to grow, however, because of natural increase and in-migration of foreigners. California continues to be the destination of choice for most foreign migrants. More than one-third of all foreign-born persons in the United States live in California. Whereas about 9 percent of the total U.S. population is foreign-born, the proportion in California is 24 percent.

The combination of these three factors—domestic net out-migration, foreign in-migration, and natural increase—produces a slow rate of population growth for the region as a whole. In the mid-1990s California population increase amounted to 0.6 percent per year, in comparison with the national growth rate of 0.9 percent per year.

The ethnicity of the regional population mix is also extraordinarily varied. Of the 50 major ancestry groups tabulated in the 1990 census, California contained the largest population of 21 of them. People of non-European origin are represented in unusually large proportions; more than a third of the national population of each of the following ancestry groups reside in California: Armenian, Chinese, Filipino, Iranian, Japanese, Mexican, and Vietnamese. As a corollary, it might be noted that in California only about three-quarters of the residents speak only English in their homes, whereas for the nation as a whole, more than 90 percent of all people speak only English at home.

One other prominent characteristic of the regional population is its urbanness (Table 16-1). More than 94 percent of all Californians live in urban areas. This figure is exceeded proportionately only by two small states in the Megalopolis Region and compares with a national average of 74 percent urban.

THE RURAL SCENE

Although most of the population of California is urban and most employment is in urban-oriented activities, three prominent nonurban activities are significant contributors to the regional economy. Each is discussed in turn.

Agriculture

Although many other states harvest greater acreages of crops than California, throughout the past half-century the total val-

TABLE 16-1

Largest Urban Places of the California Region, 1995

Name	Population of Principal City	Population of Metropolitan Area
Anaheim	288,000	2,571,000
Bakersfield	208,000	618,000
Berkeley	101,000	
Chula Vista	157,000	
Chico	41,000	195,000
Concord	118,000	
El Monte	110,000	
Escondido	124,000	
Fremont	186,000	
Fresno	403,000	847,000
Fullerton	117,000	
Garden Grove	152,000	
Glendale	179,000	
Hayward	119,000	
Huntington Beach	189,000	
Inglewood	113,000	
Irvine	134,000	
Lancaster	89,000	
Long Beach	454,000	
Los Angeles	3,498,000	9,166,000
Merced	53,000	194,000
Modesto	184,000	412,000
Moreno Valley	152,000	
Oakland	375,000	2,195,000
Oceanside	158,000	
Ontario	148,000	
Orange	118,000	
Oxnard	150,000	712,000
Pasadena	135,000	
Pomona	152,000	
Rancho Cucamonga	126,000	
Redding	57,000	164,000
Riverside	255,000	2,947,000
Sacramento	404,000	1,451,000
Salinas	124,000	381,000
San Bernardino	184,000	
San Diego	2,006,000	2,652,000
San Francisco	732,000	1,654,000
San Jose	830,000	1,570,000
Santa Ana	285,000	
Santa Barbara	81,000	383,000
Santa Clarita	128,000	
Santa Cruz	49,000	237,000
Santa Rosa	123,000	424,000
Simi Valley	109,000	
Stockton	235,000	525,000
Sunnyvale	120,000	
Thousand Oaks	114,000	
Torrance	140,000	
Vallejo	123,000	487,000
Visalia	72,000	348,000
Yuba City	24,000	133,000

ue of farm products sold has been higher in California than in any other state (Fig. 16-9). Fertile soil, abundant sunshine, supplies of irrigation water, a long growing season, and careful farm management result in high yields of high-value products. California leads the nation in the production of 58 different crops and livestock products, including the total national production of 10 crops, more than 90 percent of the output of another 10 crops, and more than half the yield of another 16 farm products.

By almost any measure, California is the leading agricultural state. It produces more than 40 percent of the national total of fresh fruits and vegetables, for instance, which is more than the combined total of the next three ranking states. With 2 percent of the nation's farms, it earns nearly 10 percent of gross national farm receipts. Eight of the 10 most productive agricultural counties in the nation are in the California Region.

Irrigation is the backbone of the region's agriculture, and California is the leading irrigated state. Irrigation projects and techniques are many and varied. Complete integration and totally effective utilization of water supply are difficult to achieve but have been more closely approached in this region than in just about any other irrigated area in the world. The most ambitious projects—Central Valley Project and California Water Plan—both transfer water from the surplus north to the deficient south. The principle of operation is that surplus Sacramento River water is stored and then diverted by canal to the San Joaquin Valley to the south. Many smaller valleys have integrated water projects, and there is widespread pumping from wells.

Agricultural areas Three-fourths of the farm output of the California Region come from the *Central Valley*. The *Sacramento Valley* (which consists of the northern third of the Central Valley) receives more rainfall and therefore requires less intensive irrigation; its major products are rice, almonds, tomatoes, sugar beets, wheat, hay, cattle, milk, and prunes. The heavily irrigated *San Joaquin Valley* is an agricultural cornucopia that has major output of dozens of farm products, particularly cattle, milk, grapes, cotton, hay, oranges, and almonds (Fig. 16-10). The *Salinas Valley,* south of Monterey, has the richest farmland in the nation and specializes in sensitive and expensive vegetables such as lettuce, artichokes, broccoli, cauliflower, and brussels sprouts. The *Oxnard Plain,* between Los Angeles and Santa Barbara, concentrates on vegetables and citrus fruits. A dozen other coastal valleys and basins are noted for high-quantity output of high-value crops—ranging from nursery products on the San Diego terraces to the famed vineyards of the Napa Valley just north of San Francisco—largely grown under irrigation (Fig. 16-11).

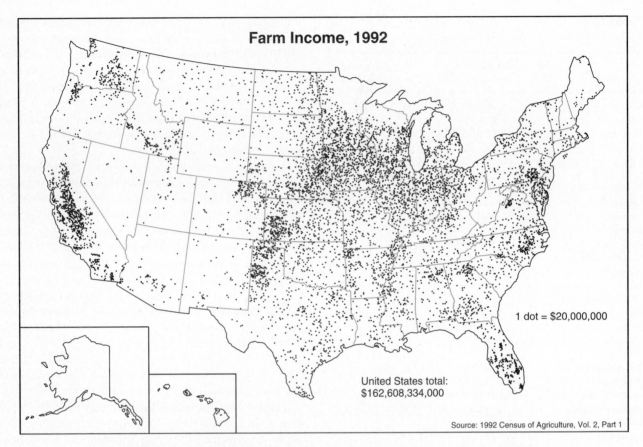

FIGURE 16-9   Gross farm income is higher in the California Region than in any other. The Central Valley concentration is particularly obvious.

FIGURE 16-10   Harvesting potatoes in the southern portion of the San Joaquin Valley near Bakersfield (TLM photo).

FIGURE 16-11   A newly planted vineyard in the Napa Valley near St. Helena (TLM photo).

(a)

(b)

FIGURE 16-12 The Harris Ranch feedlot from the air (a) and at ground level (b). This feedlot in the western part of the San Joaquin Valley has the capacity to feed 100,000 head of cattle at one time. The animals usually are kept in the feedlot for 4 months, during which time they gain 300 to 400 pounds (140 to 180 kg) each (TLM photos).

Agricultural products *Milk* is the most valuable agricultural commodity produced in the California Region, and California ranks second only to Wisconsin as a dairy state. Production is widespread but comes principally from the San Bernardino–Riverside area and the San Joaquin Valley. *Cattle* constitute the second most valuable farm product and come most notably from the San Joaquin Valley (Fig. 16-12). The region's third-ranking farm product and leading crop is *grapes,* of which California produces nearly 90 percent of the national total, primarily from the San Joaquin Valley (Fig. 16-13). California is also the nation's leader in *nursery products,* the region's fourth-ranking farm product, with most production from Los Angeles, Orange, and San Diego counties. *Cotton,* primarily from the San Joaquin Valley, ranks fifth, but is on a declining trend because of a cutback in government price supports. *Alfalfa* is the sixth-ranking farm product, and California is the leading producing state. Other major farm products from the California Region are *flowers, lettuce, almonds, processing tomatoes, strawberries, oranges,* and *chickens* (Fig. 16-14).

One other high-value California crop is missing from the list of leading farm products—*marijuana.* As an illicit narcotic, its cultivation is illegal, but available information suggests that marijuana cultivation in California may be worth $5 billion annually, more than the state's top five legal crops combined.[1]

Petroleum and Natural Gas

Only three other states exceed California in output of oil and natural gas. Most production is in the southern California coastal area, but there are also two dozen small oil fields in the southern part of the San Joaquin Valley. Four of the 10 leading oil fields in the United States, in terms of cumulative

[1] Blake Gumprecht, "Marijuana Cultivation in California: A Geographic Perspective," unpublished manuscript, 1995.

A CLOSER LOOK Rearranging the Waters: Complex and Controversial

The enormous and continuing growth of California population, especially southern California's, has placed enormous demands on the natural resources of this and adjacent regions. Most lowlands of the California Region have a semiarid-to-arid climate and would be absolutely devoid of surface waters if precipitation was not collected in adjoining mountains. Despite this inherent paucity of water, these lowlands continue to support some of the largest cities and most intensive irrigated agriculture to be found on the continent.

Water has long been cited as the principal limiting factor for the development of California, but actually, the California Region has experienced fewer water shortages than most humid regions of the continent. This fact is even more anomalous when it is realized that California's natural water supply is awkwardly located, as far as its principal users are concerned. The northern third of the state, much of which is outside the California Region, receives 70 percent of the average annual runoff, but the southern two-thirds of the state has almost 80 percent of the water need on the basis of population and land-use patterns (Fig. 16-A).

Stated simply, the problem of a serious water shortage has been avoided because of careful planning and ruthless implementation of large-scale projects to shift water from places where there is a surplus to places where there is an inadequate natural supply. Such efforts were Herculean; construction expenditures amounted to more than $12 billion in major projects alone. The result was the establishment of the most ex-

tensive and sophisticated water storage and delivery systems in the world. A listing of the larger projects exemplifies the magnitude of these systems (Fig. 16-B):

1. Los Angeles Aqueduct—diversion from Owens Valley and Mono Valley to Los Angeles, 300 miles (480 km)
2. Hetch Hetchy Aqueduct—diversion from upper Tuolumne drainage to San Francisco area, 140 miles (225 km)
3. Mokelumne Aqueduct—diversion from upper Mokelumne drainage to East Bay area, 100 miles (160 km)
4. Colorado River Aqueduct—diversion from Colorado River to Los Angeles and San Diego areas, 250 miles (400 km)
5. All-American and Coachella canals—diversion from Colorado

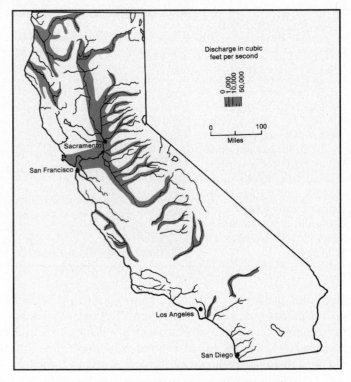

FIGURE 16-A   The average flow of California rivers.

production, are in southern California. The Wilmington Field is second only to the fabulous East Texas Field in output, the Huntington Beach Field ranks sixth, Long Beach is seventh, and Ventura is tenth.

There are large known and suspected reserves of petroleum in the southern California offshore waters, but exploitation was limited until 1965 when the city of Long Beach sold the first major leases to oil companies. Several hundred new wells have been drilled since then, most of them whipstocked from four islands that were built in Long Beach harbor. The

wells have been totally camouflaged by a covering of pastel shielding, and the artificial islands have been made as attractive as possible by palm trees plantings, waterfalls, and night lighting (Fig. 16-15).

Offshore drilling is more expensive in southern California than in the Gulf of Mexico because the producing horizons are deeper. Furthermore, several costly and controversial leaks—most notably in the Santa Barbara Channel—occurred and resulted in protracted litigation and stringent regulation. Nevertheless, the reserves of this off-

River to Imperial and Coachella Valleys, 175 miles (280 km)

6. Central Valley Project—diversion from Sacramento Valley to San Joaquin Valley

7. California Water Project—diversion from Sacramento Valley to San Joaquin Valley and various coastal locations in central and southern California.

Yet even these efforts are not enough. Despite immense costs and remarkable efficiency in intrastate water diversion, the burgeoning cities and thirsty farmlands of the region have an almost insatiable demand for water.

Feasible sources for further storage diversion are limited, and strident opposition based on both environmental and economic objections is becoming increasingly powerful. Moreover, the drought years of the mid-1970s exposed the Achilles heel of the water diversion syndrome. Below-normal precipitation over an extended period of time undermines the whole concept; if there is inadequate water to store, diversion is at best inefficient and at worst impractical.

The water future of the California Region is problematical and the long-predicted "serious" water shortage may not be far in the future. Grandiose ideas have been conceived to import water from such distant locales as the Columbia River, but the difficulties seem insurmountable. It is hoped that desalination of seawater will eventually provide for the water needs of the region, but this process is still too costly for the immediate future. So water conservation and reuse increasingly appear to be the most practical measures for the near future.

Under "normal" weather conditions lack of water should not inhibit sustenance and growth, although the profligate consumption of the past will become increasingly intolerable. The specter of future drought years is an implacable reality, however, and the ingenuity of planners and politicians may be tested to the utmost.

FIGURE 16-B   Major water diversion projects in California.

TLM

shore area are among the largest in North America. By the early 1990s some 2000 wells were producing in the Santa Barbara Channel, whipstocked from 24 drilling platforms and seven artificial drilling islands. This area is second only to Alaska's Prudhoe Bay in recent production expansion.

Commercial Fishing

Major fishing grounds extend from southern California waters southward to Peru and westward to Hawaii. Los Angeles is the only prominent fishing port in the region, and its significance is on the decline, as lower-cost Asian fishing and fish-processing activities provide harsh competition for West Coast operators.

Tuna is easily the most important catch of southern California fishermen, although anchovies (for fish meal, oil, and bait), mackerel (especially for pet food), and bonito are also significant. In overall landings, California ranks fifth among the states in quantity and eighth in value of its fishing industry.

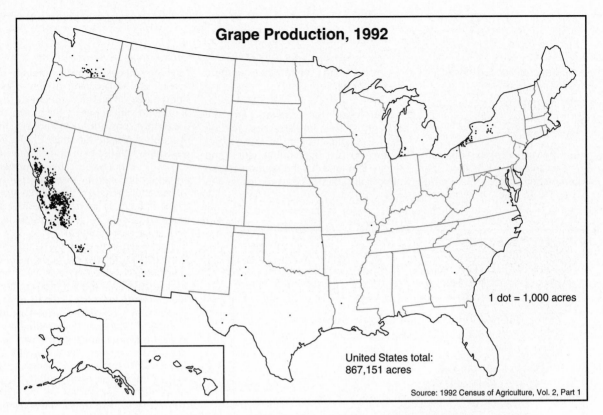

FIGURE 16-13   California, especially the Central Valley, produces most of the nation's grapes.

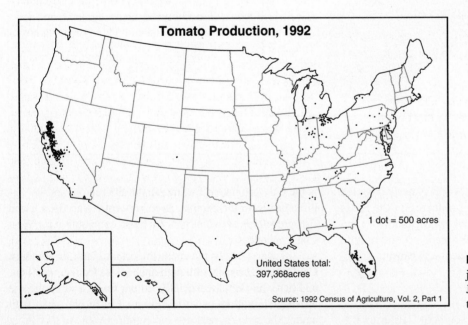

FIGURE 16-14   The delta area and adjacent parts of the Sacramento and San Joaquin Valleys grow the bulk of the nation's tomatoes.

FIGURE 16-15   A cluster of offshore oil wells near Long Beach, California. For aesthetic purposes, the wells are disguised to look like colorful buildings (TLM photo).

URBANISM

The California Region is second only to Megalopolis in its degree of urbanism. Some 94 percent of its inhabitants are urban dwellers, and most of them live in large metropolitan areas.

The Southern California Metropolis

The development of a southern California "megalopolis" is now an accomplished fact (Table 16-1). The urban assemblage is an extensive one: The east–west axis from Palm Springs to Santa Barbara is 170 miles (270 km); the north–south extent from Lancaster, in the Antelope Valley, to the Mexican border is approximately the same distance; and the southeast–northwest coastal frontage from San Diego to Santa Barbara covers more than 200 miles (320 km). Urbanization is by no means complete within these dimensions. Massive hills and even a few mountain ranges are encompassed; urban areas are restricted mostly to the lowlands and valleys except where favored hill slopes are being carved and terraced for residential construction.

Characteristics of the Urban System   Urban southern California consists of a loosely knit complex of people, commerce, and industry—all fused in a single system by a highly developed freeway network, a common technology, a common economic interest, and by numerous other shared values. Los Angeles is the major center, and its name serves as a toponymic umbrella for most of the urban system; however, Los Angeles cannot properly be classified as the focus of the metropolis, for this sprawling urban complex really has no focus. Its development has been polynuclear, and with each passing year the other nuclei become proportionately stronger and more self-contained.

The Los Angeles–Long Beach node is the largest and most prominent; the other major nuclei, in descending rank-order of population, are San Diego, Anaheim–Santa Ana, Riverside–San Bernardino, Oxnard–Ventura, and Santa Barbara–Santa Maria–Lompoc. All these nodes are interrelated within the subregional economy and within the general sphere of Los Angeles influence, but they are also essentially separate entities that dominate their own cluster of lesser cities. Separating the major nodes is a mixture of nonurban land, attenuated and irregular string-street commercial development, and low-density residential sprawl.

The complexities of the polynuclear pattern are varied. Important commercial centers, for example, are not limited to central business district locations but are widely dispersed over the metropolis. A visual indication of this dispersion is provided by the relatively large number of localities—more than a dozen—where there are concentrations of high-rise commercial buildings. Although southern California skyscrapers do not rise as high as those in many eastern cities, their clustering is much more thoroughly disseminated, which has a markedly different effect on traffic patterns and subregional economic relationships.

The Economic Base The southern California economic structure is highly diversified, as with most metropolises, but it has certain distinctive elements. The change from an industrial to a post-industrial economy has been striking. "Defense" industries, most notably *aerospace*, dominated the economy for several decades, and are still important, although they experienced a 50 percent shrinkage in the 1980s and early 1990s. The new economy of southern California is based on electronics, high-tech research and development, and a rapidly expanding service base—particularly in finance, health care, professional services, and information technologies.

There also has been an abrupt expansion of low-tech manufacturing that depends on immigrant labor (at least some of it illegal) to produce clothing, furniture, and other products in hundreds of small factories. "In effect, Los Angeles is emerging as a garment manufacturing center, not because New York has moved westward, but because Hong Kong and Singapore have moved eastward . . . ."[2]

[2] Dan Walters, *The New California: Facing the 21st Century* (Sacramento: California Journal Press, 2d ed., 1992), p. 24.

FIGURE 16-16 Los Angeles has become a melange of diverse ethnicity. At the time this photo was taken the Kosher Burrito shop was owned by a Japanese (TLM photo).

The most distinctive major economic activity in the subregion, however, is the *motion picture and media entertainment industry.* Hollywood was the birthplace and longtime capital of movie making and is still a focal point, although there has been decentralization of the major studios and network production facilities, both within the metropolitan area and far afield. Still, more than one-third of America's total employment in film production and distribution is in southern California and more than 90 percent of the world's recorded entertainment is produced within 5 miles (8 km) of the corner of Hollywood and Vine.

The list could go on and on. There are now more people employed in health care in southern California than in any other single industry, including aerospace. Los Angeles has even become the dominant financial center in the West; its 200,000 workers in financial institutions are half again as many as in San Francisco, the traditional banking center.

The Population Mix The spectacular long-run population growth of southern California has slowed a bit in the most recent years but is still very impressive. The 1990 population of southern California (15,500,000) was larger than that of 47 of the 50 states.

Within the last three decades there has been a remarkable upsurge of in-migration of ethnic minorities (Fig. 16-16). Anglos continue to move out to the suburbs or beyond, whereas there is an inflow of Hispanic and Asians. The result is an increasingly obvious two-tier society: a mostly white, affluent overclass and a mostly minority, relatively poor underclass.

The ethnic trends are clear—Anglos diminishing, Hispanics and Asians increasing rapidly, African Americans staying at about the same level. Los Angeles itself is a "minority" city, with no major ethnic group amounting to as much as half the total. The ethnic composition of Los Angeles County in the mid-1990s was 40 percent Hispanic, 35 percent Anglo, 12 percent Asian Pacific, and 11 percent African American.

The Changing Urban Pattern For many years the form and pattern of urban areas of southern California, in general, and Los Angeles, in particular, have been acclaimed as unique, their development diverging from that of other North American metropolises. The most conspicuous abnormalities have included a lack of focus on the central business district, emphasis on detached single-family housing, low population density, and overwhelming dependence on the automobile for local transportation (Fig. 16-17). More recently, however, the distinctiveness of southern California urbanism has been significantly muted. This reflects, in part, the fact that the older cities of North America have been exhibiting some of these same characteristics—declining CBD focus, increasing urban sprawl, greater dependence on automobiles—in their recent development.

More pertinent to southern California, however, is the fact that opposite tendencies have become apparent within the subregion. Population density is increasing, multifamily housing is expanding rapidly, and CBDs are developing "new" central functions. In short, a centripetal trend is setting in. In and around the principal urban cores of southern California multifamily housing units, often high-rise, are now being constructed in greater numbers than are single-family detached units. Rising land costs and an expanding population are (perhaps inevitably) producing a housing and business profile, and possibly an overall urban profile, that is increasingly similar to that of cities in the East and Midwest.

These centripetal tendencies are particularly prominent in Los Angeles but can also be seen in San Diego, Santa Monica, Long Beach, Santa Barbara, and even Pasadena. The functional complexity of the CBD increases. There is no intensification of the commercial function, but there are an increasing focalization and centralization of corporate headquarters, financial services, government offices, and cultural facilities. Furthermore, population and building densities are increasing in residential areas, owing to high-rise apartments, smaller subdivisions, and other factors.

The southern California metropolis, long a harbinger of things to come in urbanism, has thus lost much of its distinctiveness. In varying ways the urban area has evolved into a reasonably "average" metropolis. This is not to say that uniformity has set in but merely that some tendencies have been changed and even reversed.

There are still many unique and perhaps futuristic characteristics to the urban pattern and lifestyle of southern California. Where else does one find such a conspicuous example of cellular urban development on a polynuclear framework, spatially integrated by a highly developed freeway system, and seemingly capable of indefinite expansion? Where else can one find a meticulously detailed plastic Matterhorn rising 146 feet (44 m) above the floor of Disneyland, only to be topped by the $21 million, 230-foot-high (69-m) scoreboard of a nearby baseball stadium, which is, in turn, overshadowed by the 250-foot (75 m) fluorescent cross of a neighboring drive-in church?

Smog

Air pollution is a persistent fact of life in most cities, but its most infamous occurrence is in the southern California me-

FIGURE 16-17   The world's first four-level freeway interchange was "The Stack" in Los Angeles. It is still the focal point of the city's freeway system (TLM photo).

tropolis. Los Angeles is the undisputed champion, with more than twice as many "smog alert" days as any other city in the country, but Riverside, San Bernardino, and San Diego are also among the leaders.

The frequent high level of smogginess in Los Angeles is due primarily to the unique physical characteristics of the metropolitan site. The Los Angeles lowland is sandwiched between desert-backed mountains and a cool ocean. At this latitude there is persistent subsidence of air from upper levels, which acts as a stability lid over the lowlands, inhibiting updrafts and keeping vertical air motions at a minimum (Fig. 16-18). Significant horizontal air motion is also limited by the combination of hot desert, high mountains, and cool waters, with the result that Los Angeles has a much lower average wind speed than any other major metropolitan area in North America. Local winds—land and sea breezes, mountain and valley breezes—help to break the stability from time to time, but the most prominent feature of the air over the metropolis is its relative lack of movement. This stagnant condition enables air pollution to build up with annoying frequency.

Many varieties of pollutants are scattered into the stagnant air from factory smokestacks, electricity generation, human lungs, and especially automobile exhausts. It is

A CLOSER LOOK   Southern California's Chinese Ethnoburb

Ethnoburbs, or ethnic suburbs, are suburban clusters of ethnic residential areas and business districts in large metropolitan regions. They are multiracial communities in which one ethnic minority group is dominant but does not necessarily constitute a majority of the population. Ethnoburbs are created through deliberate efforts of a dominant ethnic group, played out within changing demographic, socioeconomic, and geopolitical contexts operating at multiple spatial scales (global, national, and local). They function as a settlement type that replicates some features of an enclave and some features of a suburb lacking a specific ethnic identity. Ethnoburbs coexist alongside traditional ethnic ghettos and enclaves in inner cities, forming an alternative type of ethnic settlement in contemporary urban America. Ethnoburbs are characterized by a unique spatial form and internal socioeconomic structure and involve racial and intraethnic class divisions and conflicts.

Southern California's San Gabriel Valley, centered around the City of Monterey Park in Los Angeles County, is a prime example of a contemporary Chinese ethnoburb. The San Gabriel Valley ethnoburb emerged within a

framework of global and national and place-specific conditions. Changing geopolitical and global economic contexts, and shifting national immigration policies, were key to this ethnoburb's establishment and growth. Local demographic dynamics and economic restructuring, however, also stimulated ethnoburb formation as well as determined its particular location within the region. These place-specific conditions, and the heterogeneity of ethnoburban Chinese population in terms of nativity and class, have led to subtle racialization of subgroups within the ethnoburb and a variety of expressions of ethnicity observable in the local urban landscape.

Starting in the 1960s, many upwardly mobile Chinese in the Los Angeles area moved from Chinatown to the suburbs to secure better housing and neighborhood environments. New immigrants with higher educational attainment and professional jobs also settled directly in the suburbs without ever having experienced life in an inner-city ethnic enclave. Some Chinese dispersed across the suburban landscape and became socioeconomically assimilated into mainstream society. However, the Chinese ethnoburb, complete with economic activities and social life, also be-

gan to emerge in the San Gabriel Valley suburbs. By 1990 there were more than 158,000 Chinese in the San Gabriel Valley, making it the largest suburban Chinese concentration in the nation. In many parts of the San Gabriel Valley, Chinese residents, most of whom are Mandarin-speaking immigrants from Taiwan, mainland China, and Hong Kong, account for more than 25 percent of total population.

The community of Monterey Park formed the center of this suburban Chinese zone, which eventually extended into neighboring Alhambra, Arcadia, Rosemead, San Gabriel, San Marino, and Hacienda Heights, Rowland Heights, and Walnut (Fig. 16-C). Located 7.5 miles (12 km) east of downtown Los Angeles, Monterey Park was first incorporated as a municipality 80 years ago and remained a typical white American suburban bedroom community for about a half-century (Fig. 16-D). But after the 1965 Immigration and Nationalization Act for the first time gave China the same annual immigration quota as that of European nations, larger numbers of Chinese began to immigrate to the United States on the basis of both professional employment and family reunification provisions of the act. The in-

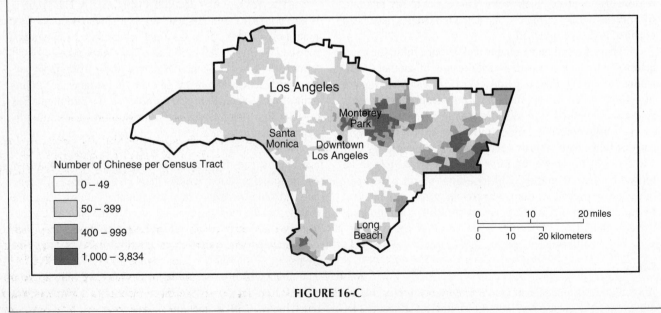

Number of Chinese per Census Tract

- 0 – 49
- 50 – 399
- 400 – 999
- 1,000 – 3,834

FIGURE 16-C

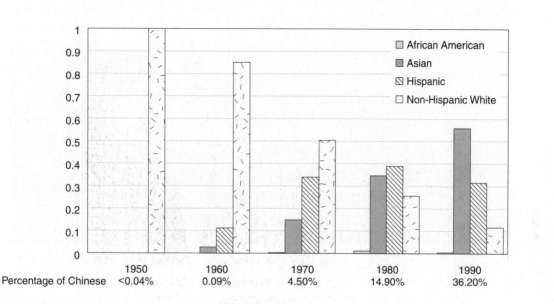

	1950	1960	1970	1980	1990
Percentage of Chinese	<0.04%	0.09%	4.50%	14.90%	36.20%

FIGURE 16-D

flux of Chinese immigrants after 1965 was also promoted by geopolitical changes in the international arena, including the ouster of the Republic of China from the United Nations; President Nixon's visit to mainland China in 1972; and the establishment of diplomatic relations between the U.S. and People's Republic of China in 1980; and the 1984 agreement between Britain and China concerning the return of Hong Kong to China.

The increasing number of Chinese immigrants coming to Monterey Park in the 1970s led Chinese real estate investors to deliberately plan the transformation of the city into a Chinese concentration. Chinese investors purchased properties to set up businesses or to convert them to multi-family dwellings for the new arrivals from overseas. In many cases they offered sellers above-market prices and convinced the owners to sell. Long-time local residents were aware of, and had some concerns about, this situation. Aware of the potential for conflict, a group of key Chinese business people gathered with local leaders at the Monterey Park Chamber of Commerce to inform them of their intentions to slowly transform

the city through their investments of capital and real estate developments into a hub for the Chinese community. They assured local residents that Chinese interests were not going to "take over" the city, but that they simply sought to provide decent housing and neighborhoods for the many Chinese flowing into the metropolitan region and to promote their assimilation into the mainstream. Using their economic resources, powers of persuasion, and sometimes their reserves of personal trust established with the Anglo community, Chinese leaders were able to quell local fears and insure an orderly, peaceful demographic transition.

Monterey Park is perhaps now better known to many Americans as "the nation's first suburban Chinatown," with the nickname "Mandarin Park." But the sheer geographical scale of San Gabriel Valley's Chinese ethnoburb far surpasses the earlier Chinatown model, making a "suburban Chinatown" label inappropriate and misleading. Moreover, the ethnoburb differs dramatically from traditional Chinatowns in terms of demographic characteristics, socioeconomic profile, and business structure, being far more heterogeneous and boasting a

more diversified economic base. These sharp contrasts between Chinatowns and the ethnoburb are clearly evident in the built environment. Local residential and business landscapes—of building forms and styles, street scenes and signs—make today's San Gabriel Valley Chinese ethnoburb a multiethnic, multicultural, multilingual area with a strong ethnic Chinese signature, but one that retains a distinctly suburban character.

More concerted research is needed on the nature of new suburban ethnic settlements and their development contexts and trajectories. The United States is a multiethnic society. With escalating racial and class conflict in American cities and new concerns over immigration issues in California and the nation as a whole, it is imperative that we understand the attitudes, behaviors, and cultures of ethnic groups; evaluate their contributions and problems; and search for the meanings of new forms of ethnic settlement, such as San Gabriel Valley's Chinese ethnoburb, to American society.

Ms. Wei Li and Professor Jennifer Wolch
University of Southern California
Los Angeles

FIGURE 16-18   A typical late-summer inversion lid over the Los Angeles lowland. Air pollutants are trapped below the lid (marked by the shallow cloud layer), with clear skies above (TLM photo).

estimated that some 90 percent of all pollutants emanate from moving sources, primarily automobiles and trucks.

The Bay Area Metropolis

The remarkable urban metropolis that grew up around San Francisco Bay had its antecedents in the Gold Rush period but reached its present form and significance in association with the rapid population expansion of the last few decades. The bayside location has been an outstanding economic advantage from earliest days. As the best large natural harbor on the West Coast, it functioned as the nation's funnel to the Pacific, serving an extensive western hinterland but especially oriented toward the Mother Lode Country and then the Central Valley.

San Francisco was the largest city in the western United States from the time of the Gold Rush until World War I and served as the dominant focus of the Bay Area metropolis until well after the Golden Gate and Oakland Bay bridges were completed in the 1930s; indeed, San Francisco is still "The City" to most people in northern California (Fig. 16-19).

Since World War II, however, the metropolitan focus has become much more diffused. The metropolis has become noncentric; the three generalized foci of development are San Francisco, the East Bay district, and San Jose, around each of which further diffused growth has taken place. The movement of workers and commuters is not predominantly from the periphery to a "core" but rather from one outlying area to another.

San Francisco itself is a place of charm and variety and usually ranks high on lists of favorite cities for people all across the country. The beauty of its setting (sloping streets readily providing extensive bay or ocean views) combines with the unusual nature of its weather patterns (brisk breezes and the alternation of brilliant sunshine and moving fog), the rich diversity of its culture, and the cosmopolitan lifestyle to make it one of the few cities in North America with a valid claim to urban uniqueness.

Its economic base is also unusual, primarily in the relatively limited role of manufacturing (San Francisco is one of the least industrialized cities—proportional to its size—in the nation) and the heavy dependence on government employment (San Francisco is second only to Washington in number of federal employees). Further distinction is provided by a style of residential development that is different from that of other western cities. Most homes are packed next to one another; the tall narrow white stucco row houses are set close to the street, with a tiny back yard, minuscule front yard, and no side yard (Fig. 16-20). Population density is significantly higher in San Francisco than in any other city west of Chicago.

The East Bay area consists of a number of separate but adjacent communities, with Oakland as the largest political entity. This is a very heterogenous urban complex with several important commercial cores, many affluent residential hillsides, prominent African American ghettos, major port facilities (the port of Richmond greatly outranks the port of San

FIGURE 16-19   The imposing skyline of San Francisco (TLM photo).

FIGURE 16-20   Densely packed housing in a typical San Francisco neighborhood (TLM photo).

Francisco in total tonnage because of the volume of petroleum shipments, and Oakland, one of the largest container ports in the world (Fig. 16-21), surpassed San Francisco in general cargo handling more than two decades ago), notable counterculture complexes, and an endless horizon of middle-class subdivisions.

San Jose is the focal point of South Bay urban sprawl (Fig. 16-22). This Santa Clara County area has experienced a phenomenally rapid transition from an agricultural and food-processing economy to one based on durable goods manufacture and related urban services. The so-called Silicon Valley of San Jose and vicinity is the principal U.S. center for the

production of computers and other high-technology equipment. San Jose has a greater share of its work force employed in factories than any other major city in the region and is one of the 10 leading industrial centers in the nation. The rapid outward spread of urban sprawl has triggered much controversy with regard to both the philosophy and practicality of maintaining large green spaces around mushrooming metropolises.

Concern about urban sprawl is only one of many catalysts that led to innovative changes in social attitude and lifestyle. As southern California has been a trendsetter in recent years, so the Bay Area may be destined to make a major

FIGURE 16-21 A portion of the expansive container terminal in Oakland's inner harbor (TLM photo).

impact on American society in the near future. In the words of one observer,

> Here seems to be the most concentrated awareness of many national problems and concern for solutions or alternatives. From here has come the main impetus of the new environmental consciousness. . . . Here, certainly, is the major hearth of the "counter-culture" which has mounted a comprehensive critique of American society and markedly influenced national patterns of fashion, behavior, and attitudes. . . . Although na-

tional in scope, the impact of such movements is regionally varied, and their prominence and power in [the Bay Area] serves to set that . . . diverse metropolitan area apart from other Western regions, reinforcing earlier cultural distinctions.[3]

Central Valley Urbanism

Urban places are dotted with relative uniformity over the established agricultural sections of the Central Valley. The eastern portion and the central "trough" have, therefore, a hierarchical scattering of small and large market towns. The drier western portion, where agriculture was insignificant until recently, has a much more open urban network, and the towns are mostly associated with a crossroad location or petroleum production.

Nine urbanized areas, from Redding in the north to Bakersfield in the south, exceed the statewide population growth rates. Although their functions are broadening beyond "farm town" image, they mostly are immature as metropolitan areas.

The largest city of the Central Valley, *Sacramento*, is the commercial center for the northern third of the subregion. Its commercial function is almost overshadowed by the administrative importance of being the capital of the most populous state. There is also a significant amount of manu-

FIGURE 16-22 Urban sprawl is facilitated by freeways such as I-17 in San Jose (TLM photo).

[3] D. W. Meinig, "American Wests: Preface to a Geographical Interpretation," *Annals,* Association of American Geographers, 62 (1972), 182.

FIGURE 16-23 Fresno is a typical medium-sized fast-growing sprawling California city. In this residential scene there are no fewer than 105 backyard swimming pools (TLM photo).

facturing in Sacramento, particularly aerospace production. Sacramento experienced explosive population and economic growth in the 1980s and 1990s, along with concomitant growing pains.

In the San Joaquin Valley are three prominent urban centers, spaced equally apart. *Fresno,* in the center, is the largest (Fig. 16-23), but *Stockton* and *Bakersfield* are of the same magnitude and rapidly growing.

CALIFORNIA AND THE PACIFIC RIM

Long disparaged as the wrong sea for trade, the Pacific Ocean recently has emerged as a dynamic ocean of commerce. The movement of merchandise and people among the countries of the Pacific Rim is the most vigorous of long-distance international flows. The U.S.–Japan link is the principal transpacific route, but connections with China, South Korea, Taiwan, Singapore, and Hong Kong are not far behind, and there are energetic interchanges with Australia, New Zealand, the Philippines, and the countries of Southeast Asia as well.

Some 30 ships depart from West Coast U.S. ports every day on transpacific voyages, and an equal number arrive. The twin ports of Long Beach and Los Angeles dominate the United States end of Pacific Rim trade. Indeed, in 1994 their combined trade exceeded that of New York for the first time. In terms of general cargo, most of which is now handled in containers, the Long Beach–Los Angeles duo is exceeded only by the container superports of Hong Kong and Singapore. Oakland also benefits mightily from Pacific Rim trade, ranking in the top 20 among world container ports.

TOURISM

The California Region, with its mild and sunny climate, is an American "Riviera." It offers tourists mountains, beaches, citrus groves, the Golden Gate, Disneyland (Fig. 16-24), and Hollywood. Generally speaking, the natural attractions of the region (mountains, beaches, islands) are the haunts of local residents on vacation; the constructed attractions draw the out-of-state visitors. Disneyland, for example, has become a

FIGURE 16-24 Disneyland in southern California is the first and most famous of the world's theme amusement parks (TLM photo).

tourist goal that ranks with the Grand Canyon, Yellowstone Park, and Niagara Falls in popularity. Movie and television studios are always swarming with visitors, and such specialized commercial spots as Knott's Berry Farm, the San Diego Zoo, and Sea World are nationally famous.[4]

The great extent of the Sierra Nevada, its height, deep canyons, waterfalls, fish-laden streams and lakes, forests, historical interests, good roads, and heavy winter snows at well-developed ski resorts, and the tremendous nearby urban population make the mountains a great attraction for tourists and vacationists. In the Sierra Nevada are located three of the nation's famous national parks: Sequoia, Kings Canyon, and Yosemite. All are noted for their magnificent groves of giant sequoia trees, spectacular glaciated valleys, waterfalls, and wildlife. Another particularly scenic area is Lake Tahoe, which nestles in a large pocket between the double crest of the Sierra Nevada.

Another major tourist attraction is San Francisco, a city of great beauty and charm. Its hill-and-water site, its cosmopolitan air, and its many points of scenic and historic interest have given it an international reputation as a place to visit.

The central coast is stirringly beautiful when not fogbound and is a favorite holiday area, particularly for Californians. Principal interest focuses on the Monterey Peninsula, where spectacular scenery, marine fauna (sea lions, sea otters, water birds), and a reputation for glamour (Monterey,

---

[4] Disneyland, Knott's Berry Farm, and Universal Studios normally receive more visitors than any other commercial attractions in North America except Florida's Walt Disney World.

Carmel, Big Sur, Pebble Beach) exist in close conjunction (Fig. 16-25).

THE OUTLOOK

As the 21st century approaches, the California Region is emerging as a society of very complex ethnicity, more distinct socioeconomic classes, and high technological sophistication. Its once-powerful industrial economy has partly given way to a hybrid postindustrial economy that is broadly but erratically based. An expanding overclass of non-Hispanic whites and Asians and an exploding underclass of Hispanics and African Americans are giving rise to a distinct two-tier society and economy.

The Anglo component of the population is virtually stagnant in numbers and is aging. The African American component is neither shrinking nor growing from a proportional standpoint but is increasing only at the rate of the total population. It appears that perhaps three-fourths of the near-term population growth and nearly all the long-term growth will be among Hispanics and Asians.

Year after year California ranks as the leading agricultural state, and many counties are in the vanguard in national standing. More and more, the agriculture of this region will be devoted to the production of specialized fruit and truck crops and to dairy products.

Farmers are innovative and imaginative in their choice of crops, and one can anticipate that the list of current favorites (almonds, pistachios, kiwis, pecans, table grapes) will vary considerably from year to year. Displacement of

FIGURE 16-25 Attractive beaches are the hallmark of the California coastline. This is the municipal beach at Carmel (TLM photo).

farm land by urban sprawl seems to be a permanent feature of the land-use pattern of the region. Pressure to subdivide farm acreage undoubtedly will intensify and expand, particularly in southern California and in the Bay Area. The Central Valley will continue to be the principal recipient of displaced agriculture, although increased irrigation and intensified farming will also be experienced in various coastal valleys. A spread of sophisticated, water-conserving irrigation techniques can be anticipated. California already contains nearly half of North America's drip irrigation acreage, primarily in tree crops, and more will undoubtedly be added. Center-pivot technology, so popular in most other irrigation states, has been little adopted in California, but its even more sophisticated offshoot (linear move) literally was invented for San Joaquin Valley conditions and is spreading in the region.

The expansive sprawl of urbanization will continue its steady march in the region. Its most conspicuous impact should be in the coastal zone between Los Angeles and San Diego, where the ambitious planned development of the huge Irvine Ranch and many lesser individual schemes will eventually result in an unbroken conurbation except for the Camp Pendleton Marine base in San Diego County. Other major growth areas will be in the hills and valleys between the San Fernando Valley and Ventura, in the desert margin of the Antelope Valley, in the Santa Clara lowland, and in the Concord–Walnut Creek area east of the Berkeley Hills.

The decline in the region's leading manufacturing activity, aerospace, has ended, but the industry still depends heavily on government expenditures and thus has an uncertain future. Aerospace manufacturers are increasingly diversifying into consumer and commercial electronics production as a hedge against this problem. Still, California's dominance in high-technology and aerospace industries is not as great as it once was and it is likely to continue to diminish from a relative standpoint. The region still has the basic advantages of skilled workers and excellent academic institutions, but it has the increasing disadvantages of high costs for land, labor, housing, and taxes. Moreover, numerous environmental regulations and anti-growth restrictions have been imposed in most parts of California, with the result that an increasing stream of manufacturers is moving out of the state.

From the broad economic point of view the region has been an important trend-setter for the nation, and its shift to a sequence of slower growth thus has implications for the broader national economy as well as for its own future. California has declined in economic importance relative to the rest of the nation since the mid-1960s, but it still generates about one-ninth of total national income and would be among the world's 10 largest countries in personal income if it were a nation in itself.

Continued population growth and accompanying urban expansion are predictable, at least in the near future. Eventually there must be a limit to frantic urban expansion. Perhaps the potential water problems can be solved by desalination of seawater; but smog, transportation difficulties, urban crowding, energy problems, and the sheer mass of humanity may combine to destroy the "California way of life" and, with it, the principal reasons for continued long-term growth.

SELECTED BIBLIOGRAPHY

ARON, ROBERT H., "The Changing Location of California Almond Production," *California Geographer,* 28 (1988), 69–94.

BAILEY, HARRY P., *The Climate of Southern California.* Berkeley: University of California Press, 1966.

BUCZACKI, STEFAN T., "The Land That Is the North American Salad Bowl," *Geographical Magazine,* 53 (February 1981), 297–299.

BULMAN, TERESA L., "Development of the Grape Monoculture of Napa County," *Yearbook,* Association of Pacific Coast Geographers, 53 (1991), 61–86.

CANTOR, LEONARD M., "The California State Water Project: A Reassessment," *Journal of Geography,* 79 (April–May 1980), 133–140.

COOK, DOUGLAS D., "The Fight to Conserve California's Coast," *Geographical Magazine,* 54 (November 1982), 623–629.

DAVIS, MIKE, *City of Quartz: Excavating the Future of Los Angeles.* New York: Verso, 1990.

DILSAVER, LARY M., "Taking Care of the Big Trees," *Focus,* 37 (Winter 1987), 1–8.

GODFREY, BRIAN J., *Neighborhoods in Transition: The Making of San Francisco's Ethnic and Nonconformist Communities.* Berkeley: University of California Press, 1988.

HALLINAN, TIM S., "River City—Right Here in California," *Yearbook,* Association of Pacific Coast Geographers, 51 (1989), 49–64.

HORNBECK, DAVID, *California Patterns: A Geographical and Historical Atlas.* Palo Alto, CA: Mayfield Publishing Company, 1983.

KIRSCH, SCOTT A., "California's Redistributive Role in Interstate Migration, 1935–1990," *California Geographer,* 33 (1993), 59–78.

KLING, ROB, SPENCER OLIN, AND MARK POSTER, EDS., *Postsuburban California: The Transformation of Orange County since World War II*. Berkeley: University of California Press, 1991.

LLOYD, WILLIAM J., "Changing Suburban Retail Patterns in Metropolitan Los Angeles," *Professional Geographer*, 43 (August 1991), 335–344.

LUKINBEAL, CHRISTOPHER L., AND CRISTINA B. KENNEDY, "Suburban Landscapes of the East Bay," *California Geographer*, 32 (1992), 77–94.

MCKNIGHT, TOM L, "Center Pivot Irrigation in California," *Geographical Review,* 73 (January 1983), 1–14.

MILLER, CRANE, AND RICHARD HYSLOP, *California: The Geography of Diversity*. Palo Alto, CA: Mayfield Publishing Co., 1983.

MITCHELL, MARTIN D., "Land and Water Policies in the Sacramento–San Joaquin Delta," *Geographical Review*, 84 (October 1994), 411–423.

MOUNT, JEFFREY L., *California Rivers and Streams: The Conflict Between Fluvial Process and Land Use*. Berkeley: University of California Press, 1996.

PARSONS, JAMES J., "A Geographer Looks at the San Joaquin Valley," *Geographical Review,* 76 (October 1986), 371–389.

PETERS, GARY L., DAVID W. LANTIS, ARTHUR E. KARINEN, AND RODNEY STEINER, *California*. Dubuque, IA: Kendall/Hunt Publishing Co., 1995.

PRESTON, WILLIAM, "The Tulare Lake Basin: An Aboriginal Cornucopia," *California Geographer*, 30 (1990), 1–24.

PRYDE, PHILIP R., "Thirty Million Californians Can't be Wrong: Reflections on Reaching a Dubious Milestone," *Yearbook*, Association of Pacific Coast Geographers, 54 (1992), 7–22.

TURNER, EUGENE, AND JAMES P, ALLEN, *An Atlas of Population Patterns in Metropolitan Los Angeles and Orange Counties: 1990*. Northridge: California State University, Department of Geography, 1991.

VALE, THOMAS R., "Vegetation Change and Park Purposes in the High Elevations of Yosemite National Park, California," *Annals, Association of American Geographers*, 77 (March 1987), 1–18.

———, AND GERALDINE R. VALE, *Time and the Tuolumne Landscape: Continuity and Change in the Yosemite High Country*. Salt Lake City: University of Utah Press, 1994.

WALTER, DAN, *The New California: Facing the 21st Century*. Sacramento: California Journal Press, 2d ed., 1992.

The Hawaiian Islands

17

The Hawaiian Islands form the smallest major region of North America. Situated some 2100 miles (3360 km) southwest of California in the Pacific Ocean, Hawaii is tied to the mainland by political affiliation and commercial dependence, which are strong enough to prevail over the volcanic base, tropical climate, Polynesian history, and Oriental population that in the past have tended to set the islands apart from North America.

The state of Hawaii encompasses a 1600-mile (2560-km) string of islands, islets, and reefs (except for the Midway Islands, which are administered by the U.S. Navy), that extends westward across the Pacific from the island of Hawaii (at 155° west longitude) to Kure (at 178° west longitude). For practical purposes, the term *Hawaiian Islands* is normally restricted to a group of two dozen islands that stretch 400 miles (640 km) from Hawaii to Niihau (Fig. 17-1). The total land area of the Hawaiian Islands is approximately 6500 square miles (16,835 km^2)—about the same area as New Jersey—nearly all of which is made up of eight major islands, listed here in order of size: Hawaii (4021 square miles or 10,415 km^2), Maui, Oahu, Kauai, Molokai, Lanai, Niihau, and Kahoolawe (45 square miles, or 115 km^2). The area of Hawaii is nearly twice the combined area of the other seven islands.

THE PHYSICAL SETTING

Origin and Structure of the Islands

The Hawaiian islands represent isolated tops of a submarine mountain range that were built up so much by volcanic action that they protrude above sea level, where their surface was modified by further vulcanism and subaerial erosion. The volcanoes are of the quiescent shield type that develop by emission of lava rather than by explosive ejection of rock fragments. Essentially the larger islands are basaltic domes in various stages of dissection. The two highest peaks of the islands, Mauna Kea (elevation 13,784 feet, or 4135 m) and Mauna Loa (elevation 13,679 feet, or 4104 m), are sometimes described as the highest in the world from base to summit, for their bases are set some 19,000 feet (5400 m) below sea level on the floor of the Pacific. Mauna Loa is the world's largest active volcano. It is an extraordinarily massive mountain; the volume of Mauna Loa above its seafloor base is estimated to be 125 times greater than that of Washington's Mount Rainier (Fig. 17-2).

Vulcanism continues to the present; there are two major active volcanoes on "the Big Island" (Hawaii) and several smaller areas of geothermal activity. Periodic lava flows are

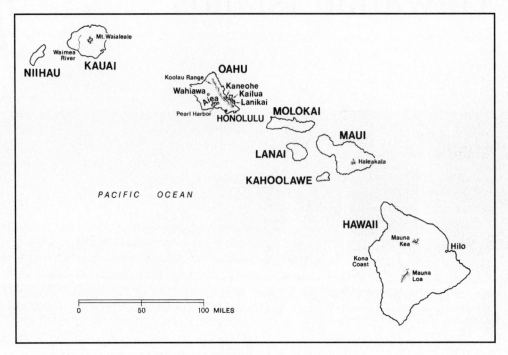

FIGURE 17-1   The Hawaiian Islands Region.

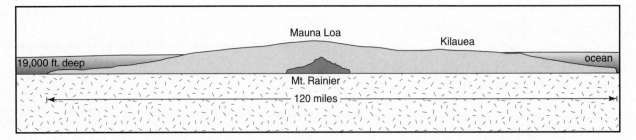

FIGURE 17-2   Schematic profiles of Mauna Loa and Mount Rainier. The massive bulk of Mauna Loa is striking. (Robert I. Tilling, Christina Heliker, and Thos. L. Wright, *Eruptions of Hawaiian Volcanoes: Past, Present, and Future.* Washington: United States Geological Survey, n.d., 42.)

actually expanding the area of the Big Island, forming new peninsulas on the southeast coast. A single eruption of Kilauea in 1960 added 500 acres (200 ha) of new land to the island (Fig. 17-3). Lava ejections from the northeast rift zone on Mauna Loa's flank pose a continuing potential danger for Hilo, the island's largest city. Some 24 flows have entered the present area of Hilo in the last 2000 years, and it is predicted that an average of one flow per century can be expected to penetrate Hilo in the future. Indeed, Kilauea's continuing eruptions (it is called the world's "most active volcano") expelled lava that destroyed more than 100 homes in 1990.

Coralline limestone has been uplifted in a few places to form flattish coastal plains of modest size. In addition, submerged fringing coral reefs partially but not completely surround most of the islands. Coral sand has frequently been washed up to form beaches in bays that are sheltered between rocky lava headlands.

Surface Features

The islands are all dominated by slopeland, and most are distinctly mountainous. Sheer cliffs, called *pali,* and rugged,

FIGURE 17-3   This is the lava-rimmed southeast coast of the Big Island. Steam rises in the distance where hot lava from Kilauea is pouring into the sea (TLM photo).

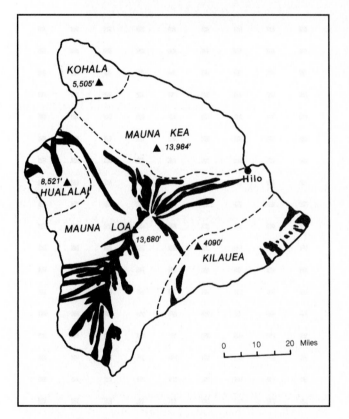

FIGURE 17-4   Shield volcanoes of the Big Island, showing the extent of major lava flows within historic time.

steep-sided canyons provide the most abrupt changes in elevation. Flat land is scarce, even around the coastal fringes.

The Big Island was formed by the overlapping union of five gently sloping volcanic cones (Fig. 17-4), of which Mauna Loa and Mauna Kea are the highest and most massive

(Fig. 17-5). Only Kilauea and Mauna Loa are currently in an active phase (Fig. 17-6). There are a few short rivers on the northern and eastern sides of the island, some of which drop into the sea as waterfalls over sheer cliffs.

Maui is composed of two volcanic complexes separated by a narrow lowland isthmus. The extinct volcano Haleakala (elevation 10,032 feet, or 3000 m) towers over the eastern part of the island. Its gigantic eroded summit depression, which resembles a giant caldera but is not, contains several dormant cinder cones and other volcanic forms. The West Maui Mountains are much lower but are rugged.

Oahu consists of two mountain ranges separated by a rolling plain. The Waianae Mountains parallel the west coast and the more extensive Koolau Range parallels the east coast. The Nuuanu Pali of the latter range is one of the most spectacular terrain features in North America, its sheer cliffs descending from cloud-shrouded peaks to a fertile coastal fringe (Fig. 17-7). The lowland between the mountains is abruptly dissected by deeply incised steep-sided gorges in several places, and at its southern end is the embayment of Pearl Harbor, one of the largest and finest harbors in the North Pacific. Honolulu's two conspicuous natural landmarks, Diamond Head and Punchbowl, are the stumps of extinct volcanoes.

Kauai is completely dominated by Mount Waialeale, which rises to 5170 feet (1550 m) in the center of the island. Heavy rainfall results in numerous short rivers that plunge coastward, often cutting deep canyons, of which Waimea Canyon is the most spectacular. The coastal fringes have little flat land, with the steep Na Pali of the northwest coast prohibiting the building of a complete circumferential roadway.

Kahoolawe, a barren hilly island with a maximum elevation of nearly 1500 feet (450 m), is uninhabited and used only as a military firing-range target. Lanai is also hilly. The long, narrow island of Molokai consists of a rugged mountain mass on the east and a broad sandy plateau on the west. Niihau consists of a moderately high tableland in the center, with low plains at either end of the island.

Climate

In its basic characteristics, the climate of the Hawaiian Islands is controlled by three factors:

1. Its tropical fringe location in a vast ocean accounts for generally mild, equable temperatures and an abundance of available moisture.

2. The northeast trade winds blow almost continually across the islands during the summer but are less pronounced, although persistent, during the winter.

FIGURE 17-5 Mauna Loa (on the left) and Mauna Kea rise to great heights on the Big Island, but their slopes are long and relatively gentle (TLM photo).

FIGURE 17-6 The crater of the world's most active volcano, Kilauea (TLM photo).

FIGURE 17-7 The stark and spectacular cliffs of Nuuanu Pali on the windward side of Oahu (TLM photo).

3. The terrain's height and orientation are the major determinants of temperature and rainfall variations.

The dominance of the trade winds throughout the region means that usually there are some clouds in the sky, the humidity is relatively high on the average, temperatures are mild to warm, there is some wind movement that hastens evaporation, and showers of brief duration are to be expected with some frequency.

The most striking aspect of the climate is the variation in rainfall from place to place. In general, a windward (essentially northeast in this region) location receives considerably more precipitation than a leeward one. As moisture-laden air rises over a topographic barrier, it expands and cools and becomes incapable of retaining all the moisture it contains; rainfall results. As the same air descends the lee side of a mountain, it contracts and becomes warmer and can hold more moisture; thus rainfall is unlikely. Specifically, it can be seen that the higher mountains receive most rain on the northeast flanks at moderate elevations of 2000 to 4000 feet (600 to 1200 m), whereas the lower mountains receive most rain along or near the crest line.[1] This difference is the result of the movement of the onshore trades that blow *over* the lower mountains but *around* the higher ones.

This windward-leeward relationship results in extraordinary rainfall variations within short horizontal distances. A weather station on the northeast slope of Mount Waialeale on

[1] David I. Blumenstock, "Climate of Hawaii," *Climates of the States,* U.S. Weather Bureau Publication (Washington, DC: Government Printing Office, 1961), p. 8.

Kauai (said to be the rainiest spot on earth) records an average of 476 inches (1209 cm) of rain annually, but 15 miles (24 km) away the average is only 20 inches (52 cm) . Variations of the same order of magnitude also occur on Oahu, Molokai, Maui, and Hawaii. Only the three small and relatively low lying islands do not have similar situations. Within the urbanized area of Honolulu it is possible to choose a building site with a 93-inch (241-cm) average rainfall or one with a 25-inch (65-cm) average only 5 miles (8 km) away.

Temperatures in the lowlands are uniformly mild. Honolulu's January average of 72°F (23°C) is close to its July average of 78°F (25°C), and the highest temperature ever recorded in the city is only 88°F (31°C), in comparison with an absolute minimum of 57°F (13°C). Only in locales of great altitude do temperatures drop markedly below the mild range; for example, the Mauna peaks of the Big Island are sometimes snowcapped.

Tropical hurricanes are experienced in Hawaii about twice a decade, on overage. They originate off the west coast of Mexico or Central America and normally approach Hawaii from the east or southeast. Extensive damage was caused most recently by Hurricane Iwa in 1982 and Hurricane Iniki in 1992.

Soils

Heavy rain and steep slopes have been the principal determinants of soil, development. In general, the slopelands have only a thin cover of soil, and the flattish lands have deep soil development. Most of the mature soils are lateritic in nature, having been leached by percolating water. The average soil in agricultural areas is red in color, moderately fertile, and relatively permeable, so that irrigation is often necessary even in places of considerable rainfall.

Biota

The Hawaiian chain is the most remote group of high volcanic islands in the world—remote from any continent or any other islands of appreciable size. This isolation has engendered a native flora and fauna that not only are limited in variety but also exhibit many unusual characteristics. To cite but two prominent examples among many:

1. More so than in any other area in the world, many Hawaiian plant groups have developed arborescence, an evolution of growth form from small nonwoody plants into large shrubs, even trees.

2. There is an exceptionally high proportion of flightless insects.

The fragility of island ecosystems is well known and Hawaii is a classic example of this principle. More than 95 percent of the biotic species originally endemic to Hawaii were found nowhere else in the world; yet within a relatively short time the vast majority were exterminated, significantly endangered, or thoroughly displaced. The statistics are overwhelming: "24 of the 69 species of birds known from Hawaii are extinct, and another 26 are either rare or endangered. Both native species of mammals—the hoary bat and the monk seal—are endangered, and half the native land mollusks are extinct. A partial list of endangered insects contains over 250 entries; the list of endangered plants includes 227 species."[2] This situation is chiefly the result of human carelessness or indifference and the introduction of exotic plants and animals that often proliferated at the expense of native biota.

For example, when Captain James Cook arrived in 1778, there were perhaps 1500 species of seed-bearing plants on the islands. Since then, more than 4000 other varieties have been introduced, and many have become naturalized.

As a result, Hawaii accounts for more than 70 percent of all plant and animal extinctions that have taken place in the United States, and it harbors more than one-fourth of the nation's endangered birds and plants.

Mild temperatures and abundant precipitation provide conditions for lush vegetation. The better-watered areas are noted for thick growth of tropical trees and shrubs (Fig. 17-8). Most areas of thick forest, however, have been denuded by commercial logging or overgrazing. In areas of inter-

FIGURE 17-8   A view of rain jungle in the Pali area near Honolulu (TLM photo).

[2] Warren King, "Hawaii: Haven for Endangered Species?" *National Parks and Conservation Magazine,* 45 (October, 1971), p. 9.

mediate rainfall the flora often reflects more arid conditions because the highly permeable volcanic soil permits water to percolate rapidly to great depths—frequently beyond the reach of plant roots. Xerophytic plants characterize such areas. The introduction of exotic plants has been particularly characteristic of this region, so that now a large proportion of the total vegetative cover represents earlier imports.

Animal life has always been limited on the islands. Native fauna was mostly restricted to insects, lizards, and birds. The most conspicuous wildlife today consists of feral livestock (livestock that reverted to the wild). Tens of thousands of feral sheep, goats, and pigs roam the islands, and there are considerable numbers of feral cattle, dogs, and cats. These animals destroy many plants and other animals and are a major nuisance in some areas, particularly on the Big Island. But they provide an important recreational resource for sport hunting. Other exotic species, such as mouflon, axis deer, and mongoose, are also present in some profusion.

POPULATION

Early Inhabitants

Little is known of the pre-Polynesian inhabitants of the Hawaiian Islands except that they are reported to have been short in stature and peaceful by nature. Presumably they were either destroyed or assimilated by waves of Polynesian settlers, the first of whom were thought to have arrived from the western Pacific between A.D. 750 and 1000. After several hundred years of isolation, another great Polynesian immigration occurred in the fourteenth and fifteenth centuries, followed by a second lengthy period of insular seclusion.

These people have been known through recent history as Hawaiians and are characterized by bronze skin, large dark eyes, heavy features, and dark brown or black hair. Although their social and community life was intricately complicated by restrictions and regulations, making a living was relatively simple. They had domesticated pigs and chickens and a variety of cultivated food plants. Fruits and vegetables were common, but dietary staples were fish and poi (the cooked and pounded root of the taro plant).

European Penetration

The islands were officially discovered by Captain James Cook of England in 1778; however, it is thought that a Spanish captain named Gaetano had been there more than two centuries previously, and still other seafarers may have touched the islands before Cook. It is clear, nevertheless, that Cook's visit opened up the "Sandwich Islands," as he called

them, to the world. Before long the islands became important bartering, trading, and refreshment stops for merchant vessels of England, France, Spain, Russia, and the United States and for whalers and pearlers of many nationalities.

British influence was strong for many years, but few colonists were attracted. Missionaries from New England arrived during the 1820s, and these dedicated people became very influential by the 1840s. Moderate but increasing numbers of U.S. settlers migrated to the islands during the nineteenth century.

Other Immigrants

Asians came and were brought to Hawaii in considerable numbers during the last half of the nineteenth century, usually in response to a need for cheap labor. The first Chinese were brought to work on the sugar plantations in 1852. The Japanese began to arrive in 1868, first as fishermen blown off course and later as plantation workers. In spite of restrictions, Japanese immigration greatly exceeded that from any other country. The first Filipino sugar workers came in 1906, and many more were brought in during succeeding years. Other immigrants came from Korea, Samoa, other Pacific islands, and Portugal's Atlantic islands (Azores and Cape Verde).

The Contemporary Melting Pot

The present population of the islands is more complex and varied than that of any other region in North America except perhaps California. All ethnic groups have intermarried, particularly in recent decades.

Hawaiian society is becoming increasingly open and has an unusual degree of social and economic mobility. Although there are visible cracks in the region's widely acclaimed racial harmony, Hawaii still stands as North America's most successful melting pot. Caucasians (called *haole),* mostly from mainland North America, were significantly outnumbered by Japanese in the past, but in the last few years a large *haole* influx has made Caucasians the most numerous element in the region's population, more than one-fourth of the total. Japanese are almost as numerous; Hawaiians and part-Hawaiians constitute about 20 percent; Filipinos, sparked by rapid immigration in recent years, about 15 percent; and Chinese, about 6 percent. Nearly half the population is of Asian ancestry; in no other state is that figure as much as 10 percent.

CENTURIES OF POLITICAL CHANGE

Throughout this region's early history the islands were politically fragmented. Various kings and chiefs ruled different is-

A CLOSER LOOK   *Renaissance of the Hawaiian Language*

A renaissance of the Hawaiian language began in the late 1960s and developed, so that today the native language of the Hawaiian Islands is again being written and pronounced accurately. This movement to restore the language is linked to the drive for self-determination, land ownership, and cultural preservation by the Hawaiian people. Efforts are being made to compensate for the harm done to the language by more than a century of disparagement and distortion.

Disenfranchised from their land and marginalized from the economic development of the islands, the Hawaiians are moving to recapture their heritage and a portion of the wealth being generated by haoles and the Asians who are longtime residents of the islands. Language purity has become one of the flags of renaissance waving over the islands.

King Kamehameha I united the islands, and his name is not to be shortened. It is not politically correct today to call the Kamehameha Highway the "Kam Highway" or Kamehameha Schools "Kam Schools." The Hawaiians are concerned with all aspects of ancient culture and correct spelling and pronunciation of the language is a prime example of this cultural renaissance, which includes such diverse indicators as signage, the purity of the hula, and the issue of sovereignty.

Restoration of the language involves correcting faulty spellings that have come into common usage. Hawaiians are properly using words with the macron (a bar over a vowel indicating greater than normal duration and stress) and the ʻokina (the glottal stop represented by a reversed apostrophe). Improperly placed, these signs completely change the meaning of the word. For many years, non-native speakers did not recognize that the glottal stop was, in fact, a consonant. For example, o, ō, ʻo, and ʻō all have different meanings, as do the following words: pau, paʻu, paʻū, and pāʻū. Linguists have known this fact for years, however it took Hawaii until the 1990s to correctly position the macron and the ʻokina in such basic

items as street signs, magazines, and government publications.

The state government has decided to spell all new signs correctly and to replace the older signs as money is made available. Computer software programs have been designed for the Hawaiian language and are readily available throughout the islands. Whether this will change the rest of the world's pronunciation of Oʻahu and Hawaiʻi remains to be seen.

Even within the islands, compliance with rules for the proper use of the Hawaiian language is not guaranteed. Hotels and tourist attractions are major violators. Recently, a major restaurant and hotel complex opened an eatery named Hoku or "star"; however, hoku, pronounced in the manner indicated by this spelling does not mean star. The word for star is hōkū, with both the "ō" and "ū" receiving equal stress in spoken Hawaiian. Whether an embarrassing mistake or a deliberate misuse of the language, the name was in either case an insult to the Hawaiians and their culture.

Tourists have long had problems with pronouncing local names without proper diacritical markings. Laughter

rises from the locals when a visitor tries to locate Likelike (they pronounce it Like Like) Highway. It is pronounced like ʻlike in Hawaiian. Proper and complete markings will eventually make the Hawaiian language "learnable" by visitors and will help them to correctly pronounce it. Hawaiians argue that cultural respect comes from proper usage of local languages by visitors and native speakers. Therefore, the movement to purify the language is by necessity forcing its way into schools, government publications, street signs, university research, and publishing.

Only a small percentage of the islands' total population speak Hawaiian, a fact that emphasizes the significance of language renaissance to the preservation of the Hawaiian culture. Widespread use of diacritical markings and "proper" pronunciation, however, may be long in coming, as most visitors are uninformed, and many residents are uninterested.

Mr. Charles W. Berry
Kamehameha Schools
Oʻahu, Hawaiʻi

FIGURE 17-A (*TLM photo*).

lands and parts of islands with a sporadic pattern of warfare and change. The uniting of the region under one ruler was accomplished by Kamehameha I, but it required 28 years of war, diplomacy, and treachery. In 1782, he began a bloody civil war on the Big Island that lasted for 9 years. Later he conquered Maui and Molokai and overwhelmed Kalanikupule's army to seize control of Oahu. The other islands came under his rule by 1810.

Six kings and a queen successively carried on the monarchy after Kamehameha's death in 1819. The first constitution was promulgated during the reign of Kamehameha III and the kingdom became a constitutional monarchy. Most influential Hawaiians became increasingly interested in some sort of liaison with Britain, but a location near the United States and increased trade with California (sugar, rice, and coffee) in the 1850s and 1860s, as well as the establishment of regular mail service with San Francisco, foreshadowed the manifest destiny of annexation by the United States. The monarchy declined after 1875, and Queen Liliuokalani was deposed in 1891.

An American immigrant served as president of the interim republic until annexation was completed in 1900. For Hawaii, the basic motive for annexation was to increase trade, especially in sugar; and for the United States, to secure a major Pacific naval base. In spite of much agitation for statehood, the islands remained a territory for nearly six decades. In 1959 Hawaii was admitted as the 50th state.

THE HAWAIIAN ECONOMY: SPECIALIZED, LIVELY, ERRATIC

Economic opportunities were always limited in the region because of its insular position and lack of natural resources. Although bauxite and titanium have been discovered, no mineral deposits have ever provided a significant income. Commercial fishing has been only partially successful. Logging of sandalwood was once very important and formed the basis of a thriving trade with the Orient, but the stands of sandalwood have long since been depleted.

Although only one-tenth of the land is arable, the soil is generally productive, there is no danger of frost, and natural rainfall and abundant groundwater provide sufficient moisture. The first foreign cash crop was tobacco, which flourished during the first half of the nineteenth century and then faded out. During the Gold Rush period in California, foodstuffs of various kinds were exported to the West Coast. This trade was significant to the Hawaiian economy for only about a decade and then virtually ceased. The provisioning of whalers and other ships bolstered the production of local

crops from the 1820s until the 1870s, but it also declined. Eventually it became evident that the best hope for agriculture was to specialize in growing subtropical crops for the mainland market (Fig. 17-9).

The growing of specialized plantation crops carried the brunt of the regional economy for a long time. Nevertheless, the single largest component of the economy for many decades was federal government expenditures, primarily for the construction and maintenance of military facilities. Today the relative significance of agriculture has declined and the importance of tourism has soared. The latter, however, is an erratic producer of income, for it reacts quickly to the ebbs and flows of the national economy. In recent years, however, fluctuations have been less pronounced because of the dramatic impact of increased tourism from Japan.

Cane and "Pines"

Sugarcane has been the leading crop of Hawaii for more than a century, and pineapples have ranked as a strong second crop for almost as long. Indeed, the Hawaiian image has long been almost as much "sugar and pineapple" as it has been "grass skirts and Waikiki." At present, however, the steady decline of both staple crops clearly signals their relegation to a position of only supplementary importance, fading perhaps to even less than that.

Other Crops

For many years there was talk of the need to diversify the region's agriculture, for almost all foodstuffs must be imported from a distant source.[3] Nevertheless, diversification was slow to occur, and most efforts were aimed at growing other tropical specialties for export rather than producing food for local consumption.

The third-ranking crop (and some sources place it second only to sugar) on the islands in terms of revenue is probably marijuana, or *pakalolo,* as it is called locally. Officially, however, macadamia nuts are considered to be most valuable after sugar and pineapple; an enormous expansion of this unique Hawaiian crop has occurred in the last few years. By the mid-1990s, macadamia sales were already half as great as those from pineapple. The only other major crop category in Hawaii is floriculture and nursery products, whose total sales are approximately equal to those from pineapple. The bulk of these specialties are cultivated on the Big Island. The well-known Kona coffee of the leeward coast of the Big Island has declined since the 1960s, owing to a variety of problems, most notably the cost and availability of labor.

---

[3] The only staples in which the region is self-sufficient are sugar and milk.

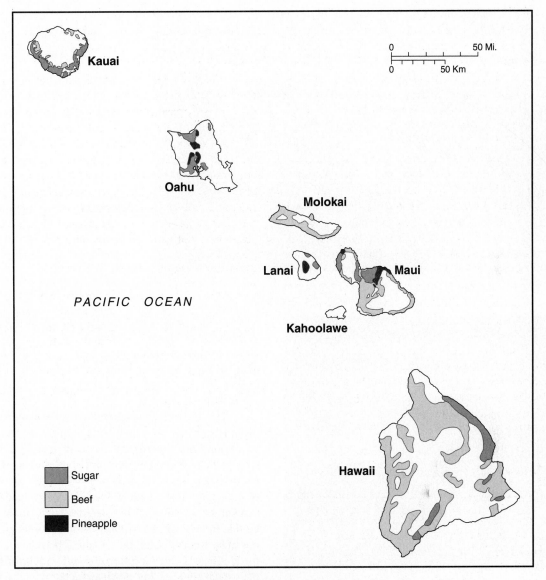

FIGURE 17-9 The land-use pattern of the principal islands. Cattle ranching and the intensive growing of sugarcane and pineapple are striking.

Livestock

Cattle are the principal domestic livestock in the region, and ranches occupy more than three-quarters of the agricultural land (Fig. 17-10). Most ranches are large—one is reputed to be among the five largest in North America—and concentrate on the raising of beef cattle. Grain feeding is uncommon and, although some hay is produced, generally the animals subsist on pasturage. Nearly all the meat is consumed on the islands, but this amount satisfies less than one-half the local demand for beef. Hides are exported. Cattle ranches are most notable on the Big Island but also occupy much of Maui and Molokai, and the entire island of Niihau is owned and operated as a single ranch.

Both dairying and poultry raising are expanding on the four larger islands.

Tourism

The largest source of income for the Hawaiian economy is tourism. Beaches, climate, scenery, ceremonies, hospitality, and transpacific crossroads location are the major assets.

FIGURE 17-10 Cattle ranching is widespread, and a cowboy culture is well established. Here some of the locals are having a Sunday afternoon competition in calf-roping on the Big Island (TLM photo).

FIGURE 17-11 The Hawaiian image in all of its reality. High-rise hotels overlook sun-seekers crowding narrow Waikiki Beach, with Diamond Head in the distance (TLM photo).

These items are exploited by one of the most thorough and best-organized publicity efforts anywhere; the renown of a holiday in Hawaii is worldwide.

Much of the business life of the region is geared to the visitor. Companies that cater to sleeping, eating, entertainment, and transportation services are continually expanding their operations. Hawaii is the major stopping point for transpacific passengers. The great majority of all passenger ships and planes crossing the Pacific call at Honolulu. It is unrivaled as the major terminal city within the entire Pacific Basin, excluding the marginal centers of California, Japan, and Australia. Summer is the busiest season, with June decidedly the peak month. A smaller secondary peak occurs in December and January.

The Waikiki area of Honolulu is the unquestioned center of island tourism (Fig. 17-11). It contains more than half the region's hotel rooms and is a seething hive of restaurants, elegant shops, sparkling beaches, and fashionable sunburns. Most visitors, however, also manage to see some other parts of the region. Interisland air carriers offer frequent and convenient service among the six larger islands, and it is estimated that more than three-fourths of the visitors go to at least one other island in addition to Oahu. The volcanic features of the Big Island, the exceptional scenic beauty of Kauai, and the outstanding beaches of Maui are the principal attractions among the "outer islands" (Fig. 17-12).

The visitor industry, as tourism is referred to officially, experienced a downturn for several years in the late 1980s and early 1990s. This trend, however, was reversed in 1993, and expansion continues at present. The annual number of visitors to the region is approaching 6.5 million, which is more than five times as large as the permanent population. On any given day, more than 125,000 tourists are in the region. Nearly two-thirds of the visitors are "westbound," arrivals from the North American mainland, with the majority originating in California. In the North American market Hawaii is a mature destination in which repeat visits dominate more than three to two over first-time visits. "Eastbound" visitors have increased significantly in recent years,

FIGURE 17-12   A portion of the Napili Bay Resort on the island of Maui (TLM photo).

three-quarters of them from Japan. On the average, west-bound visitors stay twice as long, but eastbound visitors spend 2.5 times as much money per day.

Federal Government Expenditures

Second only to the visitor industry as a generator of wealth in the islands is the federal government. Because of the strategic value of its mid-Pacific location, Hawaii contains some of the nation's largest military bases. It is the headquarters for the U.S. Pacific Command and the administrative center for the Pacific operations of each of the three individual services.

Approximately one of every five members of the region's labor force, including military personnel, is employed by the federal government, mostly in military establishments. The U.S. Armed Forces annually spend more than $2 billion in the islands; however, in every year since 1975 the volume of federal nonmilitary spending has been even greater than that figure, primarily because of the vast increase in direct health, education, and welfare benefit payments.

URBAN PRIMACY: A ONE-CITY REGION

No single city dominates any other region in North America as thoroughly as Honolulu dominates the Hawaiian Islands (Table 17-1). Not only does it contain most of the region's population—some 80 percent of the total—but it also has the great preponderance of all economic, political, and military activities.

TABLE 17-1

Largest Urban Places of the Hawaiian Islands, 1995

Name	Population of Principal City	Population of Metropolitan Area
Hilo	41,000	
Honolulu	385,000	883,000

There has been a long-continued drift of population from the outer islands in general and rural areas in particular to Honolulu. Since 1930 outer-island population has been on a declining trend, which only recently was slightly reversed. Oahu's share of the Hawaiian population total has grown from just over half in 1930 to more than four-fifths in the early 1990s. Honolulu's population continues to boom, whereas other parts of the region show only irregular and sporadic growth.

Despite its exotic location, Honolulu is much like other rapidly growing North American urban areas and exhibits both the best and the worst of urban patterns and trends. There are many delightful residential areas, and various parks and beaches provide almost unparalleled recreational opportunities. Transportation routes, however, are congested, and some Honolulu traffic jams would be impressive in cities twice its size. The cost of living is high, most people cannot afford to own their own homes, unregulated high-rise construction has blighted the most cherished views of Waikiki and Diamond Head, and the city's raw sewage still pours

A CLOSER LOOK   The De-Sweetening of Hawaii

Few crops have been as intimately associated with both the image and the economic well-being of a region as have sugarcane and pineapples with the Hawaiian Islands. Brought to Hawaii by Polynesian migrants, cane was being produced commercially in the 1830s. Cultivation was primarily by smallholders until the 1850s when a swelling influx of contract workers (initially from China; later from Japan, Portugal, the Philippines, and other places) provided the labor force needed for large-scale plantation farming. By the 1860s sugarcane had become the economic mainstay of Hawaii, a position it retained for three-quarters of a century.

Pineapples were also introduced from elsewhere in Polynesia, but there was no significant commercial production until 1890. The industry became notable after Dole's development of improved varieties in 1903 and Ginaca's invention of a mechanical peeling, coring, and slicing machine 10 years later.

The islands have a few critical physical advantages for these crops, most important of which is the semitropical climate. Low temperatures are never a problem, nor is searing heat for any length of time. Both crops are heavy users of water (it takes about 1 ton (0.9 t) of water to grow enough cane to produce 1 pound (0.45 kg) of raw sugar), but the Hawaiian growing areas either receive adequate rainfall or can be supplied by irrigation from adjacent mountains (Fig. 17-B). The deep lateritic soils, although not particularly fertile, are certainly not impoverished and take well to fertilizers.

An insufficient labor supply was an early disadvantage, but a continuing flow of contract workers from various countries soon eliminated this problem. There was no local market to speak of, but the relative closeness of the United States provided a rich market potential, and aggressive advertising did the rest. Thus during the first half of the twentieth century sugarcane and pineapple were

dominating elements in the Hawaiian economy.

During the 1930s there were more than 50,000 workers in the cane fields of the islands, another 5000 in the pineapple plantations, and perhaps 10,000 employees in the sugar mills and pineapple canneries. This total represented nearly 40 percent of all jobs in the Hawaiian Islands. After World War II the number of workers declined steadily, but output continued to rise, as did acreage. The area devoted to pineapples peaked at about 70,000 acres (28,000 ha) in 1948; the greatest sugarcane acreage—250,000 (100,000 ha)—was cultivated in 1969.

Then came the decline. Pineapple experienced the most disastrous reversal (Fig. 17-C). Hawaii had easily been the world's dominant producer for years; in the early 1940s some 80 percent of the world's pineapples were grown on the islands. By the 1980s both pineapple acreage and employment were less than half as much as in peak years. By the

FIGURE 17-B   Harvested cane is trucked to this sugar mill near Waimea on the island of Kauai (TLM photo).

FIGURE 17-C   A pineapple plantation on the island of Oahu (TLM photo).

early 1990s this share had dropped to less than 25 percent. Various factors were involved in the decline, but the most important was the cost of labor. New plantations in the Philippines, Taiwan, Malaysia, and other tropical countries pay wages that are only 5 to 10 percent as much as those paid to pineapple field workers in Hawaii the disparity of cannery workers' wages is even greater. The principal producing companies, chiefly large conglomerates, such as Del Monte and Dole, operate in foreign countries as well as in Hawaii. So it is in their interest to shift production to lower-cost areas.

Two (Kauai and Molokai) of the five islands where pineapple was once prominent are no longer producers, and output has diminished on all three of the others (Lanai, Maui, and Oahu). Only two canneries are still in operation compared with nine a half-century ago. In fact, if not for vigorous marketing of pineapple as a fresh fruit item, the decline would have been even more precipitous. Until the last decade almost all Hawaiian pineapple was sold in cans; now, however, more than three-fourths of the fruit is marketed fresh.

The downturn in sugar is more recent and less abrupt but no less real. In 1974 the U.S. Congress failed to renew the Sugar Act, which for four decades had provided a strong measure of protection for domestic sugar producers against foreign imports. Higher production costs (primarily labor; Hawaiian canefield employees are among the highest-paid farm workers in the world) in Hawaii mean that many foreign producers have a competitive advantage, and thus sugar acreage, production, and employment have all been on a downward trend in this region. In the early 1980s Congress enacted some price supports for domestic sugar that mitigated, but did not stop, the decline.

The region now produces only one-fourth of the total U.S. sugarcane output. The principal cane-growing areas are on the windward coast of the Big Island, in the lowland of Maui, in several places on Oahu, and around the coastal margin of Kauai. Most production comes from a dozen large sugar ranches. The refining, most of which is done in California, and the marketing of the raw sugar are handled by a producers' cooperative, the California and Hawaiian Sugar Corporation.

Although growing and processing sugar and pineapple are still significant in Hawaii, their relative importance has diminished notably. Receipts from sales of these two products combined are far below the income generated from any major nonagricultural activities (government expenditure, tourism, commerce, services, manufacturing, transportation, construction) of the region. And total employment in all sugar- and pineapple-related activities now amounts to only 1 percent of the labor force.

TLM

FIGURE 17-13 Looking eastward across the central business district of Honolulu, with the Koolau Range towering in the distance (TLM photo).

undiminished into the blue Pacific. Urban sprawl is extensive, although constricted by the steep slopes of the Koolau Range (Fig. 17-13). The extent of the urbanized area has rapidly spread eastward toward Koko Head and westward partially to surround Pearl Harbor and has jumped the Koolau Range to encompass the shores of Kaneohe Bay on the east coast (Fig. 17-14).

PROBLEMS AND PROSPECTS

Compared with the U.S. mainland Hawaii's unique environment and location have helped to engender an unusual pattern of economic-social-political relationships that have had significant effects on the geography of the region.

Population Explosion

Overshadowing all else today is the high rate of population growth. Although not overwhelming in relation to other states (Hawaii's population increased by 23 percent during the 1980s, a rate exceeded by 11 other states), associated facts make such growth of great concern.

The burgeoning population is largely focused in Honolulu and it is there that the unpleasantness of overcrowding has an almost smothering effect. In portions of the Ala Moana–Waikiki census tract the population density exceeds 60,000 people per square mile. Although such numbers would not be unusual on Manhattan Island or in Chicago's

Loop, they represent an abrupt shattering of an image of tranquil Hawaii. And the concomitant deterioration of the quality of life shows clearly.

Vehicular congestion is a particular nuisance. The region's population has doubled in the last quarter-century, whereas the number of motor vehicles has tripled. Hawaii has fewer miles of streets than the District of Columbia, for example, but twice as many cars.

Another disturbing aspect of the population increase is the escalating welfare load. Nearly 10 percent of the total population is on welfare, and welfare costs accelerated ninefold from 1985–1995. Sample studies show that only 30 percent of the welfare cases are Hawaii-born; most represent relatively recent in-migrants. The high cost of living in Hawaii is generally unanticipated by potential immigrants, whether foreigners or mainlanders.

Although crowding, congestion, pollution, staggering welfare costs, and related problems are not restricted to Hawaii, they pose a particularly menacing situation in a region that is both insular and isolated. Hawaii has limited resources and a finite land base, but the crux of this whole complex of problems is that this region, unlike any other in North America, must face the problems alone, without feasible interaction with neighboring regions or states. A mainland region can supplement its own resources by bringing products in from elsewhere; water or gas or electricity can be obtained by pipeline or transmission line. This option, of course, is unavailable for Hawaii. Ships and planes can move goods into and people out of the region to alleviate the situation, but for Hawaii even these normal flows are greatly complicated by cost and distance.

Palliatives, such as land-use restrictions and zoning, have been instituted, but they do not get to the root of the problem. Overpopulation can be solved only by some type of limitation on growth. Yet freedom of interstate (and international) migration is a cherished U.S. principle that is clearly upheld by the Constitution.

Hawaii has long been noted for its activist, liberal stance on socioeconomic issues. Today, however, authorities face the need to consider some very repressive and conservative approaches to population limitation. What is the best way to react to the immutable problem of overpopulation in a fragile island environment, exacerbated by the region's unique position with relation to mainland United States and, increasingly, to Asia?

Development Versus Preservation

A prominent and continually escalating controversy in the region revolves around the desirability and propriety of development. Development in the Hawaiian context chiefly

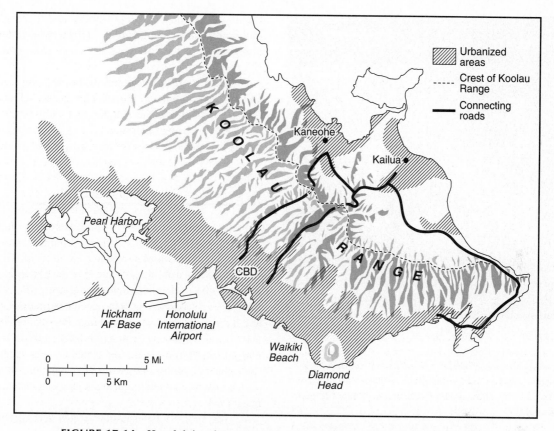

FIGURE 17-14 Honolulu's urban sprawl has encompassed much of the leeward coast of Oahu and jumped across the Koolau Range to occupy a considerable portion of the Kaneohe lowland on the windward side.

involves building resort complexes. Many projects are in the already supercrowded environs of Waikiki, but increasingly the emphasis is on the construction of condominium-hotel resorts, with their associated infrastructure of roads and golf courses, in remoter locations on Oahu and in the outer islands.

The visitor accommodation inventory (hotel rooms and condominiums) in the region increased from 5000 in 1960 to 70,000 in 1990, but the occupancy rate continues to be high even in recession years. Thus the demand for more rooms and facilities appears to be unabated. Can the fragile environment of the island state withstand such pressures (Fig. 17-15)?

Hawaii was the first state to enact a statewide land-use plan (1961),[4] but its principal thrust was to preserve agricultural land. New laws, including a constitutional amendment, in 1978 addressed a different set of land-related concerns—

those associated with rapid development in general and urbanization in particular. Moreover, many areas have local regulations that strictly control zoning, building heights, parking, landscaping, and building design. Still, it would appear that without an agonizing reappraisal of land-use plans and more regulatory safeguards, the charm of uncluttered beaches and valleys, not to mention the perpetuation of local lifestyles, may be dissipated or destroyed by the ambitious schemes of land developers, highway lobbies, and construction unions.

Land Ownership

One of the most unusual aspects of Hawaiian geography involves the system and pattern of land ownership. Approximately 42 percent of the total land area is under government control. It is a much smaller proportion than in many states west of the Mississippi, but, in addition, another 47 percent is held by a mere 70 estates, trusts, and other large owners. Less

[4] The 1961 Land-Use Law put the state government in charge of all land-use administration, regardless of ownership of the land.

FIGURE 17-15   For all its problems, Hawaii continues to be a major destination for visitors from all over the world, particularly from continental United States. Warm water, fine beaches, and a semitropical climate provide the basis for its attractiveness. This is Poipu Beach on the island of Kauai (TLM photo).

than 11 percent of the land is therefore subject to general private ownership. Many plans were advanced, particularly by the state legislature, to enable individuals to obtain small parcels of land. As a result, the number of farms is increasing and the average farm size is decreasing, both in direct opposition to the trends on the mainland. More than half the farms in Hawaii are now less than 10 acres (4 ha) in size.

Homesteads proliferated on Molokai, Hawaii, and Oahu, but generally the land is in large estates. Traditionally these estates do not pass freely to heirs; instead, trusts of various types are set up to administer them. This step results in "freezing" the ownership, and the land is leased in large blocks rather than being sold. Half the private land in the islands is owned by nine major estates. Such a situation is not inherently unsavory, but the long-range effect may prove deleterious to economic growth as well as to social and political conditions.

Transportation

Here, as for other islands far removed from the mainland, the problem of transportation is always notable. Because the region consists of a group of islands, the physical matter of moving people and goods from one place to another within the region can be intricately complicated. Factories in Honolulu must have materials that originate on the other islands, and citizens of the outer islands need goods produced in or shipped through Honolulu.

Most tourists arrive by air, but the shipment of food and other commodities is handled by surface transport, which is slow, expensive, and subject to disruption by labor disputes, weather, and other factors. The problem of transportation is one of the immutable facts of life in the region.

THE OUTLOOK

Perhaps in no other region are the hazards and the potentials so clear-cut. From an economic standpoint there is cause for concern, but there is also reason for optimism. The sugar industry is well established, but it is caught in a cost–price squeeze that portends an increasingly precarious existence. Several cane plantations were closed during the mid-1990s, and indications are that all the remaining plantations in Oahu and Hawaii will cease production by the turn of the century. Only plantations on Maui and Kauai are expected to survive the industry's downsizing, primarily because they have invested heavily in technological advancement and cost-cutting programs.

The future for pineapples is also dark, despite the expansion of the fresh pineapple market. Lanai's last pineapple plantation is now out of business, leaving Oahu, Hawaii, and Maui as the last bastions of pineapple production in the region.

Much discussion has concerned agricultural diversification, but little was done to diminish the vast need for importing basic foodstuffs. Expanded production of subtropical specialties, such as macadamia nuts and papayas, is likely to continue. It is not clear what will become of the land that is going out of cane and pineapples, but there are optimistic portents in the prospect of developing an agroforestry industry. Prospects for wood-chip exports, wood-products manufacturing, and—longer term—tropical hardwoods plantings have improved as desirable prime agricultural lands have been freed up.

Hawaii's cost of living is significantly higher than that of any other state except Alaska, and the state budget is strained in many directions. The unemployment rate has been at a high level for some time even though such occupations as garment workers and coffee harvesters are in short supply. Manufacturing, in general, shows some prospects for growth but primarily in fields that are not basic to the total economy—that is, those that supply goods for the Hawaiian market.

As agriculture declined in recent years the tourist boom took up much of the economic slack in this resource-poor region. The visitor industry dramatically expanded during the 1980s, so that its revenues doubled from 20 percent to more than 40 percent of the gross state product. This was followed by a 5-year downturn, beginning in 1988. By the mid-1990s, however, tourism had become healthy again, and the outlook is for a continuing modest growth, based partly on the completion of a mammoth new convention center, which is now under construction. A prominent lesson of recent years is that Hawaiian tourist visitation is more dependent on inexpensive air fares than on general economic prosperity. If the public perceives that it is relatively cheap to go to Hawaii, then the public will go to Hawaii in considerable numbers.

Although tourism passed military expenditures as a source of regional revenue two decades ago, Hawaii is still to a considerable extent a garrison state. Cutbacks in federal defense spending throughout the United States in the 1990s did not affect Hawaii much. It is likely that the present level of defense spending will be maintained, which means that its rela-tive importance to the regional economy will slowly diminish as the population and economy of Hawaii continue to grow.

From a social standpoint Hawaii has been an American showcase for racial assimilation. Continued intermarriage will probably blur individual ethnic strains into a more widespread Hawaiian blend.

The region's population should grow faster than the national average because of a relatively high birthrate and continued immigration of people attracted by the prospect of island living. Greatly expanded urbanization is likely, particularly on Oahu, where Greater Honolulu, along with its extended suburbs of Lanikai–Kailua–Kaneohe, will spread north and south on both sides of the Koolau Range.

Hawaii's ancient motto is "The life of the land is perpetuated in righteousness." But how is it possible to preserve righteousness toward the land in a tropical island milieu that is being overwhelmed with the incessant pressures of civilization? The delicate balance between an economy dependent on boom-growth tourism and the maintenance of a pleasant environment and attractive lifestyle to lure tourists is an almost imponderable dilemma.

SELECTED BIBLIOGRAPHY

ARMSTRONG, R. WARWICK, ED., *Atlas of Hawaii.* Honolulu: University Press of Hawaii, 1973.

BRYAN, E. H., JR., *The Hawaiian Chain.* Honolulu: Bishop Museum Press, 1954.

CARLQUIST, SHERWIN JOHN, *Hawaii: A Natural History.* Garden City, NY: Natural History Press, 1970.

CUDDIHY, LINDA W., AND CHARLES P. STONE, *Alteration of Native Hawaiian Vegetation: Effects of Humans, Their Activities and Introductions.* Honolulu: University Of Hawaii Press, 1989.

FARRELL, BRYAN H., *Hawaii, the Legend That Sells.* Honolulu: The University Press Of Hawaii, 1982.

FOSBERG, F. R., "The Deflowering of Hawaii," *National Parks and Conservation Magazine,* 49 (October 1975), 4-10.

KING, WARREN, "Hawaii—Haven for Endangered Species?" *National Parks and Conservation Magazine,* 45 (1971), 9–13.

LIN, GONG-YUH, "Spectacular Trends of Hawaiian Rainfall," *Proceedings,* Association of American Geographers, 8 (1976), 12–14.

MACDONALD, GORDON A., AND AGATIN T. ABBOTT, *Volcanoes in the Sea: The Geology of Hawaii.* Honolulu: University of Hawaii Press, 1970.

McDougall, Harry, "Volcanoes of Hawaii, *Canadian Geographical Journal,* 80 (June 1970), 208–217.

MORGAN, JOSEPH R., *Hawaii: A Geography.* Boulder, CO: Westview Press, 1983.

NORDYCKE, ELEANOR C., *The Peopling of Hawaii.* Honolulu: University of Hawaii Press, 1989.

SANDERSON, MARIE, ED., *Prevailing Trade Winds: Weather and Climate in Hawaii.* Honolulu: University of Hawaii Press, 1993.

STONE, CHARLES P., CLIFFORD W. SMITH, AND J. TIMOTHY TUNISON, EDS., *Alien Plant Invasions in Native Ecosystems of Hawaii.* Honolulu: University of Hawaii, 1992.

The North Pacific Coast *18*

Extending latitudinally for more than 2000 miles (3200 km) along the northwestern fringe of the continent is the North Pacific Coast Region. The region encompasses the continental margin from northern California to southwestern Alaska, nowhere penetrating inland more than 200 miles (320 km) from the sea.

The attenuated, coast-hugging shape of the region is due largely to the topographic pattern (Fig. 18-1). The major mountain ranges of far western North America are oriented parallel to the coastal trend, lying directly athwart the prevailing currents of midlatitude atmospheric movements and severely restricting the longitudinal penetration of oceanic influences. The interior (eastern and northern) boundary of the region is thus approximately coincidental with the crest of the principal mountains: the Cascade Range in the conterminous states, the Coast Mountains in British Columbia, the St. Elias Mountains in the Yukon Territory, and the Wrangell and Alaska ranges in Alaska. The southern margin of the region is just north of the San Francisco Bay Area conurbation in California, and the western extremity is in the Alaska Peninsula, where forest is replaced by tundra.

Such coastal proximity ensures that the influence of the sea is pervasive throughout the region, although it is some-what subdued in such sheltered lowlands as Oregon's Willamette Valley and Vancouver Island's eastern coastal plain. Human activities and the physical environment are significantly affected by the maritime influence, which is most conspicuously reflected in climatic characteristics. Winters are unusually mild for the latitude, and summers are anomalous in their coolness.

The most memorable climatic characteristics are associated with moisture relationships. Abundant precipitation, remarkably heavy snowfalls in the mountains, high frequency of precipitation, considerable fogginess, and the widespread and relatively continuous occurrence of overcast cloudiness produce a climatic regime that, although not extreme, is exceedingly drab.

Another prominent characteristic of the North Pacific Coast Region is that its natural resources occur in limited variety but often in great quantity. Partly as a result, the economy of the region is not diversified but is dependent on specialties of production; furthermore, the limited resource base is a continual arena of controversy. The exploitation and development of resources frequently involve major conflicts of interest. How to dam the rivers for hydroelectricity generation without ruining the salmon fishery? How to exploit the timber resources without despoiling the unparalleled

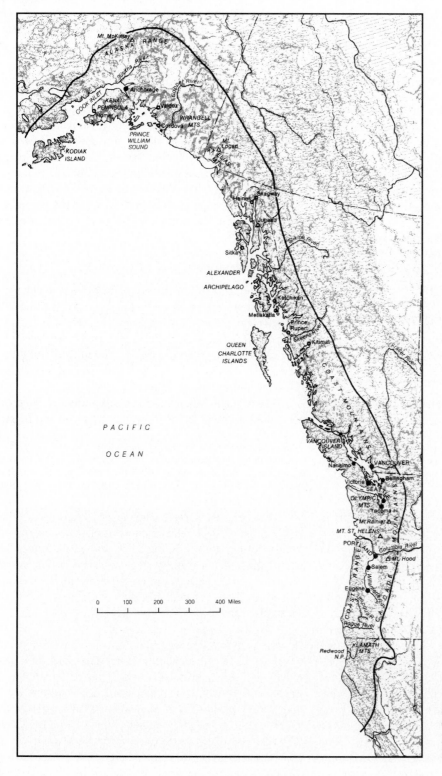

FIGURE 18-1 The North Pacific Coast Region (base map copyright A. K. Lobeck; reprinted by permission of Hammond, Inc.).

scenery? How to develop the national parks for visitor convenience without destroying the wilderness?

This is a region in which the people have had to live with remoteness—which is a joy to some but despair for others. The North Pacific Coast is remote from the heartland of both nations and is separated by significant topographic barriers from all external population centers. Access to the interior is limited to a relatively few routes, and north–south connections are difficult. No railroad runs along any part of this coast and no highway is found along most of the British Columbia and Alaska coastline. This difficulty of access and connectivity has been a significant deterrent to economic growth, resulting in high transportation costs for goods brought into the region and in one of the continent's highest cost-of-living indexes.

Another effect of remoteness has been psychological, a provincialism of attitude that rivals the more celebrated parochialism of Texans, Californians, New Englanders, French Canadians, or Southerners.[1] But this aspect of the regional character is changing as improved communications tend to smooth out regional differences throughout North America.

THE TERRAIN: STEEP AND SPECTACULAR

The entire region is dominated by mountains. They vary in height from the comparatively low coastal ranges of northern

California, Oregon, and Washington, to the higher ranges of the Cascade Mountains with their superb volcanic peaks, and to the great alpine ranges of western Canada and Alaska surmounted by Mount McKinley[2] (elevation 20,300 feet, or 6090 m), the highest peak on the North American continent (Fig. 18-2).

The general topographical pattern consists of three very long landform complexes, northwest–southeast in trend and generally parallel to one another throughout the region. The westernmost zone consists of low mountains that become higher, more rugged, and more severely glaciated toward the north. Just to the east is a longitudinal trough that is prominent from Oregon to the Yukon. The easternmost zone consists of complex mountain masses surmounted by spectacular volcanic peaks in the south and extensive ice fields in the north.

Coastal Ranges

The coastal ranges within the U.S. portion of this region are a series of somewhat distinct mountain areas. California's northern Coast Ranges have the same parallel ridge-and-valley structure as those of the central coast of the state. The topography is strongly controlled by structure, with folding and faulting dominant. Ridge crests are even, if discontinuous; slopes are generally steep.

The Klamath Mountains mass appears as a complicated and disordered complex of slopeland. It is wild and rugged

---

[1] A classic example of this attitude was the headline in the *Vancouver Sun* a few years ago when a severe blizzard had disabled major highways, both railway lines, and all wired communications extending east from Vancouver. The headline read, "Canada Cut Off."

[2] Although the former Mt. McKinley National Park now is officially designated Denali National Park, the name of the mountain itself has not been changed. At the time of this writing, the official name is still Mount McKinley.

FIGURE 18-2 Mount McKinley is the highest peak on the continent and is the focal point of Denali National Park (TLM photo).

country extending inland to connect with the southern part of the Cascade Range. Peaks reach to almost 9000 feet (2700 m), and much of the high country was heavily glaciated in Pleistocene time.

The Coast Range of Oregon and Washington has relatively low relief; crest lines average about 1000 feet (300 m) in elevation, with some peaks another 1000 feet (300 m) above that. North of the Columbia River the range is much more subdued. Unlike the California Coast Ranges, it is crossed by a number of prominent transverse river valleys—notably those of the Columbia, Rogue, and Umpqua Rivers.

The Olympic Mountains are a massive rugged area of high relief and steep slopes. Although their highest peak (Mount Olympus) is less than 8000 feet (2400 m), heavy snowfall has given rise to more than 60 active glaciers in the range (Fig. 18-3). The margins of the Olympic Mountains are abrupt and precipitous, particularly on the east side.

The Strait of Juan de Fuca, which separates the Olympic Peninsula from Vancouver Island, provides a broad passage across the coastal mountain trend. Most of Vancouver Island is composed of a complex mountain range that descends steeply to the sea on the southwestern side, where the coastline is deeply fiorded and embayed. This range, too, has been heavily glaciated and contains a number of active glaciers, although its maximum elevation is less than 7500 feet (2250 m).

The mountains of the Queen Charlotte Islands and the Alexander Archipelago are relatively low and less rugged. Farther north, the coastal mountains unite with the inland ranges in the massive and spectacular knot of the St. Elias Range.

Interior Trough

Between the Cascades to the east and the Olympic and Coast ranges to the west lies the structural trough of the Willamette Valley and Puget Sound. The trough was formed by the sinking of this landmass at the time the Cascades were elevated. In glacial times a large lobe of ice advanced down Puget Sound and was instrumental in shaping the basin that holds that body of water. Today the Willamette Valley is a broad alluvial plain, 15 to 30 miles (24 to 48 km) wide and 125 miles (200 km) long; the Puget Sound lowland is somewhat smaller in area, for a large part of it has been submerged. North of Puget Sound the structural trough persists as the Strait of Georgia and Queen Charlotte Strait, which becomes a continuous waterway (the "Inside Passage") almost to the Yukon Territory.

Inland Ranges

The Cascade Range, extending from Lassen Peak in northern California to southern British Columbia, is divided into a southern and a northern section by the deep gorge of the Columbia River. The relief features of the southern part are predominantly volcanic in origin, with a subdued and almost plateaulike crest on which several conspicuous volcanic cones were superimposed. From south to north, prominent peaks are Mount Lassen, one of only two volcanos in the conterminous states to have erupted in this century; the 14,162-foot (4250-m) Mount Shasta; collapsed Mount Mazama, whose caldera is now occupied by Crater Lake, one of the world's deepest lakes; The Three Sisters; Mount Jefferson; and the 11,225-foot (3368-m) Mount Hood.

FIGURE 18-3   The rugged crest of the Olympic Mountains (TLM photo).

The northern Cascades are granitic rather than volcanic and are much more rocky and rugged. Surmounting the mass of the range are five prominent, old, ice-capped volcanoes: Mount Adams, Mount St. Helens, Mount Rainier, Glacier Peak, and Mount Baker (Fig. 18-4). Heavy snowfall and cool summers combine to produce extensive glaciation (Figs. 18-5 and 18-6). Some 750 glaciers exist in the north Cascades, which is two-thirds of all the active glaciers in the conterminous states (Fig. 18-7). Indeed, more than 40 glaciers occur on Mount Rainier alone.[3]

North of the Fraser Valley, the ranges that correspond to the Cascades are known as the Coast Mountains. They average 100 miles (160 km) in width, are nearly 900 miles (1440 km) in length, and have been severely eroded by mountain ice caps and glaciers. Deep canyons of the Fraser, Skeena, Stikine, and Taku Rivers have cut across the range, forming features similar to the Columbia Gorge. These mountains plunge directly down to the coast without a fringing lowland, where they are incised by a remarkable series of long fiords and inlets. The high country contains innumerable glaciers and some ice fields that are hundreds of square miles in extent (Fig. 18-8).

Inland from the Gulf of Alaska, where Alaska, the Yukon Territory, and British Columbia come together, is the ice-and-rock wilderness of the St. Elias Mountains. This

[3] Glaciers have developed wholly within the forest zone in some parts of the North Cascades, a phenomenon that apparently does not occur elsewhere in the Northern Hemisphere. See Stephen F. Arno, "Glaciers in the American West," *Natural History*, 78 (1969), 86.

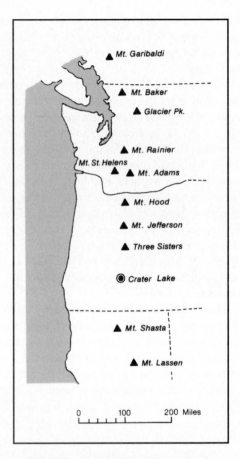

FIGURE 18-4 The major volcanic peaks of the Cascade Range.

FIGURE 18-5 Mount St. Helens (foreground) and Mount Adams in the tranquil days prior to the former's 1980 eruption (TLM photo).

FIGURE 18-6   Mount St. Helens after the eruption. The changed shape and size of Mount St. Helens are notable (TLM photo).

FIGURE 18-7   The high country of the North Cascades includes dozens of jagged peaks. Lake Ann is in the foreground (TLM photo).

West and northwest from the St. Elias massif the mountain trend again bifurcates. Along the coast is the long, remarkably rugged, and heavily glaciated Chugach Range, which eventually gives way to the less extensive but equally rugged Kenai Mountains. The inland ranges include the massive Wrangell Mountains and the Alaska Range, which is the continent's highest, culminating in Mount McKinley. The Alaska Range is crescent shaped, with its western extremity terminating at the base of the Alaska Peninsula. Between the Alaska Range and the Kenai Peninsula is the only extensive lowland in the northern part of the region; about half of it is occupied by the broad bay of Cook Inlet, and the remainder consists of the valleys of the Susitna and Matanuska Rivers.

CLIMATE: MOIST AND MONOTONOUS

The North Pacific Coast Region has a temperate marine climate in which the downwind relationship with the ocean markedly ameliorates temperatures. High summer temperatures are almost unknown, and low winter temperatures occur with frequency only in the lowlands of south-central Alaska and in highlands throughout the region.

Precipitation is characteristically abundant, but this factor is sharply modified by altitude and exposure. In general, there is a fairly even seasonal regime in the north but a decided winter maximum in the south. During the winter gigantic cyclonic storm systems, which migrate eastward across the Pacific Basin, bring simultaneous rains for 1000 miles (1600 km) north and south along this coastal region. Mountainous terrain influences the areal distribution of precipita-

mountain fastness contains Canada's highest peak, Mount Logan (19,850 feet, or 5955 m), and has more than a dozen peaks that are higher than any in the conterminous states or elsewhere in Canada (Fig. 18-9). These are the highest coastal mountains in the world, and they encompass the world's most extensive glacial environment outside the polar regions (Fig. 18-10).

FIGURE 18-8   There are dozens of high mountain icefields in this region. The Juneau Ice-field, pictured here, straddles the Alaska/British Columbia border for nearly 100 miles (160 km) (TLM photo).

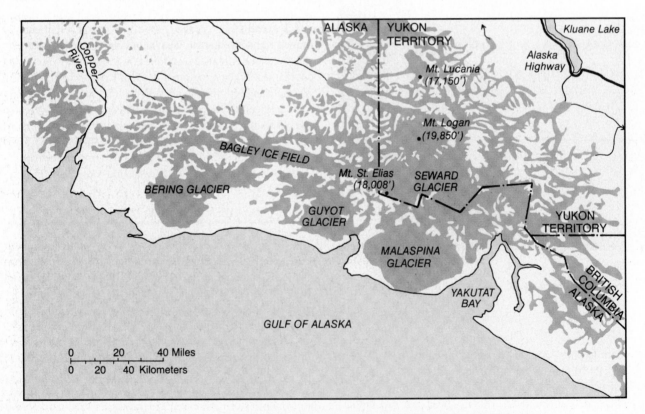

FIGURE 18-9   North America's most spectacular corner, the St. Elias–Wrangell–Malaspina area where Alaska, British Columbia, and the Yukon Territory come together. The shaded portions represent glaciers and icefields.

FIGURE 18-10   Hundreds of glaciers flow down valleys in southern and southeastern Alaska. Here the surface of Casement Glacier (near Haines) appears candy-striped by the longitudinal debris of several lateral moraines (TLM photo).

tion, southwest slopes receiving copious rainfalls and northeast sides receiving scant ones. The southwest flank of the Olympic Mountains, saturated in winter, has an average annual precipitation of 150 inches (380 cm), the maximum for the conterminous United States. In contrast, the northeast side of these mountains, only 75 miles (120 km) away, has an annual rainfall of 16 inches (40 cm), which is too little to support even good pastures without the aid of irrigation. The northwestern corner of Vancouver Island records 250 inches (635 cm) annually and is considered the wettest spot in continental North America. Snow accumulates to great depth, especially on exposed mountain slopes; some localities experience annual snowfalls in excess of 80 feet (24 m).

Modified by the terrain, the east–west precipitation pattern falls into four easily recognizable belts:

1. The coastal strip, with abundant rainfall and little snow
2. The windward side of the coastal ranges, with excessive precipitation
3. The leeward side of the coastal mountains and the interior trough, with only a moderate rainfall and sporadic snows
4. The western slope of the Cascades in the United States and Coast Mountains in Canada, with nearly 100 inches (254 cm) of precipitation—mainly winter snows

The winter season is cloudy, monotonously damp, and protected from chilling continental winds by a double barrier of mountains to the east. Summer—the dry season—is characterized by mild temperatures, light surface winds, coastal fogs (Fig. 18-11), and low clouds. Throughout most of the region, the average number of clear days each year is less than 100; Juneau, for example, records only 45.

FIGURE 18-11   Fogs and low clouds are frequent along the North Pacific coast. This scene is near Mendocino in northern California (TLM photo).

THE WORLD'S MOST MAGNIFICENT FORESTS

Except in some of the interior valleys, the heavy precipitation throughout this region makes it a land of forests. In the northern California Coast Ranges the dominant tree is the redwood and within this area may be found some of the most magnificent forests of the world. Along the coasts of Oregon and Washington and in the Cascade Mountains the Douglas fir is dominant, constituting one of the major lumber trees of the continent (Fig. 18-12). Other notable trees within this area are western hemlock, western red cedar, and Sitka spruce. Douglas fir is the most valuable species, followed by hemlock, in the Canadian section. The Sitka spruce is the leading tree in southeastern Alaska. The forests of this region are almost exclusively coniferous.

A marked contrast in natural vegetation exists between the Willamette Valley of Oregon and the Puget Sound area of Washington. In the former, the barrier nature of the Coast Range tends to produce a light summer rainfall; in the latter, the rainfall is slightly heavier in summer because the gaps in the Coast Range allow more rain-bearing winds to enter from

FIGURE 18-12 Massive trees dominate the forests of this region. This Douglas fir scene is near Chemainus on Vancouver Island (H. Brown/Crown Zellerbach Corporation).

the Pacific. The cooler temperature of the Puget Sound area also causes it to be somewhat more humid. As a result, the native vegetation of the Puget Sound area was a dense stand of giant Douglas fir trees with limited expanses of prairie; the Willamette Valley's original vegetative cover apparently was largely prairie grass, although some scholars believe that the original prairies were artifically generated and maintained by deliberate annual fires set by the Native Americans.[4] This contrast of vegetation types at the time of European contact profoundly influenced the settlement of the two areas.

Much of the region, particularly in the north, is above the tree line; in such areas, alpine tundra rock, and ice constitute the ground cover.

OCCUPANCE OF THE REGION

Aboriginal inhabitants of the North Pacific Coast Region consisted of a great variety of tribes: small tribes of Algonquian speakers in northern California; various Chinook groups in the Columbia River area; Salishan, Nootka, Kwakiutl, Bellacoola, and Tsimshian in British Columbia; and the redoubtable Haida and Tlingit in southeastern Alaska. From the aboriginal standpoint this was a region rich in resources: salmon and other seafoods in great quantity, berries in profusion in summer, and great forests of useful trees. The natives built impressive buildings of evergreen planks, fashioned large dugout canoes from the native cedars, and, particularly in the north, became skilled woodcarvers—with giant totem poles as their most lasting achievements (Fig. 18-13).

The Chinooks and Bellacoolas did considerable trading with the encroaching whites, but for the most part Native Americans and First Nations of the region had minimal effect on European penetration and settlement, with the important exception of the Tlingits in the north. The Tlingits were proud, well organized, and resourceful; for several decades they exerted both political and economic hegemony in the Alaska panhandle area, controlling trade between the more primitive Athabaskans of the interior and the European mercantile interests. But even the Tlingits were soon swept aside in the flurry of prospecting and settlement.

Today Native Americans and First Nations are relatively insignificant in numbers in the southern part of the region. In north coastal British Columbia and southeastern Alaska, however, they constitute much of the work force in fishing, canning, and forestry.

The voyages of Vitus Bering between 1728 and 1742 led to the advance of Russian trappers and fur traders south-

[4] For example, see Carl L. Johannessen et al., "The Vegetation of the Willamette Valley," *Annals*, Association of American Geographers, 61 (June 1971), 302.

FIGURE 18-13 Totem poles played an important role in the culture of some of the Native American tribes in this region. The scene is at Totem Bight near Ketchikan (TLM photo).

ward along the Alaskan coast. In 1774 the Spaniard Juan Perez sailed as far north as latitude 55°. Another important voyage was that of the English Captain James Cook in 1778, who explored the coast between latitudes 43° and 60° north and further complicated the claims to this strip of coast. In 1792 a New England trading vessel reached the mouth of the Columbia River and established the fourth claim to the region. Thus by the end of the eighteenth century Spain, Russia, Great Britain, and the United States had explored and claimed in whole or in part the Pacific Coast of North America from San Francisco Bay to western Alaska.

Permanent Spanish settlements were never established north of San Francisco Bay, but by 1800 Russia was entrenched on Baranof Island in southeastern Alaska and had its seat of colonial government at Sitka. Further settlements were made to the south, but agreements were signed with the United States and Great Britain in 1824 and 1825 limiting the Russians to territory north of the 54°40′ parallel. Spain abandoned its claim to all land north of the 42nd parallel. This left the Oregon country—between the Spanish settlements in California and the Russian settlements in Alaska—to the United States and Great Britain.

Following the Lewis and Clark expedition, the American Fur Company in 1810 established a trading post at Astoria, at the mouth of the Columbia River, but this settlement was seized by agents of the Hudson's Bay Company during the War of 1812. British forts on the lower Columbia dominated the area until 1818, when an agreement was reached for joint occupance by English and American traders.

At first the only Americans who reached this far-off land were a few trappers and traders, but in the early 1830s

New England colonists came overland via the Oregon Trail, and soon a mass migration began. The great trek along the Oregon Trail took place in the early 1840s. These pioneers, determined to establish a Pacific outlet for the United States, had the slogan "Fifty-four forty or fight." Most of them located in the prairie land of the Willamette Valley of Oregon and by 1845 there were 8000 Americans in the Oregon country. In the settlement of the "Oregon Question" in 1846 the United States got the lands south of the 49th parallel (except for Vancouver Island) and Great Britain got the land between there and Russian America. The final status of the San Juan Islands, now a part of the state of Washington, was not decided until 1872.

Victoria was chosen as a settlement site by the Hudson's Bay Company in 1843, but there was little activity in the British part of this region until the great Cariboo gold rush that began in 1858. Victoria prospered as the transshipment point and funnel for the British Columbia gold fields in the same way that San Francisco did for the California mining areas. In the 1850s two other settlements were founded in southwestern British Columbia: Nanaimo on Vancouver Island had the only tidewater coal field on the North American Pacific coast, and New Westminster in the lower Fraser Valley became the mainland commercial center. British Columbia joined the Canadian confederation as a province in 1871, but there was little population or economic growth until the arrival of the transcontinental railway in 1886 and the founding of Vancouver as its western terminus. Within a decade of its origin, Vancouver had surpassed Victoria as the largest city in western Canada (population 25,000).

After furs became depleted in the 1840s, the Russians

began to lose interest in their far-off American possession. Although they had leased or sold some of their posts to the Hudson's Bay Company, they were loath to sell Alaska to Great Britain because of the Crimean War and therefore offered it to the United States. The purchase was made in 1867 for the sum of $7.2 million, or less than 2 cents per acre.

The Puget Sound country still was remote from populous centers of the continent and until the completion of the Northern Pacific Railroad in 1883 its only outlet for bulky commodities of grain and lumber was by ship around Cape Horn. Between 1840 and 1850 a number of small sawmills were erected in the area to export lumber to the Hawaiian Islands and later to supply the mining camps of California. The Canadian Pacific Railway reached its Vancouver terminus in 1886; in 1893 the Great Northern completed its line across the mountains to Puget Sound; and sometime later the Chicago, Milwaukee, St. Paul, and Pacific Railroad built into the region. Meanwhile, the Union Pacific established direct connection with Portland, and the Southern Pacific linked Portland with San Francisco. These rail connections made possible the exploitation of the great forest resources, which became important about the beginning of the present century, and also contributed to the industrial development and urban growth of that part of the region. Not until 1914 was the other Canadian transcontinental railway, originally called the Grand Trunk Pacific and now the Canadian National, completed to its terminus at Prince Rupert, although a subsequent terminal at Vancouver was much more important to the Canadian National.

When gold was discovered in the Klondike in 1897 and at Nome in 1898, a stampede began that closely rivaled the California rush of 1849. It was a long, hard, dangerous trip; the most direct route to the Klondike field was by ship through the Inside Passage from Seattle to Dyea and later Skagway, thence over Chilkoot Pass or White Pass to the headwaters of the Yukon River, and finally by riverboat or raft about 500 miles (800 km) downstream to Dawson. When gold was found in the beach sands at Nome, the trip was made entirely by ship, but in each case Seattle and Vancouver profited by being the nearest ports with railway connections to the rest of the continent.

Settlement significantly expanded and intensified in older centers of the region—that is, the Willamette Valley, Puget Sound lowland, lower Fraser Valley, and the Victoria area. The most spectacular relative growth in recent decades, however, has been in the Cook Inlet area of south coastal Alaska. Anchorage and its immediate hinterland, the Matanuska Valley and the Kenai Peninsula, attracted a great many settlers from various parts of the United States. Nearly half the people of Alaska live within 25 miles (40 km) of Anchorage.

WOOD PRODUCTS INDUSTRIES: BIG TREES, BIG CUT, BIG PROBLEMS

The North Pacific Coast Region, with its temperate marine climate, contains the most magnificent stand of timber in the world. The trees decrease in size from the giant redwoods of northwestern California and the large Douglas firs and western red cedars of Oregon, Washington, and British Columbia to the smaller varieties of spruce, hemlock, and fir along the coast of Alaska. At one time probably 90 percent of the region was covered by these great forests.

The Douglas fir has a greater sawtimber volume than any other tree species on the continent, the size of the individual trees and the density of the stand being exceeded only by sequoias and redwoods. Douglas fir attains its best development in western Oregon, Washington, and British Columbia and constitutes about half of both the sawtimber volume and cut of this region.

The North Pacific Coast Region has long been dominant in lumber production on the continent, although its share has been diminishing over the past few years. Oregon, Washington, and California were the leading lumber-producing states for many decades but are now facing leadership challenges from several southeastern states, most notably Georgia. British Columbia normally produces more than half of Canada's lumber output, although much of the production is from the interior of the province and thus outside the North Pacific Coast Region (Fig. 18-14).

This region is also a major producer of plywood, particleboard, and pulp and paper, normally ranking just behind the Inland South in each category. In recent years there has

FIGURE 18-14   A logger felling a Douglas fir on Vancouver Island (TLM photo).

been a great increase in "production" of logs and wood chips from the North Pacific Coast region; the major destination for both products is Japan.

Logging Operations

Most timber harvesting in this region has long been done by the clearcutting technique, in which every tree in a designated section of an area is removed but none is removed from surrounding sections until later years. A typical acreage may be divided into 70 sections, with one section being clearcut each year and immediately reforested. At the end of the 70th year all sections have been cut and the first section, now supporting a mature, even-aged stand, is ready for harvest again (Fig. 18-15).

Most of the timberland in the region is government owned. In the United States the Forest Service is the principal administering agency; it conducts regular timber sales and tries to cooperate with the forest products companies in coordinating the areas of operation for both efficient utilization and sustained yield production. Similar management is carried out on both provincially and federally owned timberlands in British Columbia.

Most privately owned timberland in the region is either owned outright by major forest products corporations or is leased by them. Many large tracts have been organized into tree farms for perpetual forestry production.

Conservation Controversies

Despite a generally good record of forest conservation policies, the logging interests in the North Pacific Coast Region have long been the subject of attack for despoiling the environment. Much ill will is engendered by the technique of clearcutting because it is perceived differently by different people.

> To the forest products manufacturer, clearcutting is the cheapest and most efficient method to cut down timber and open the way for intensive new growth of certain species. To the professional forester, it is an effective tool of scientific management of a renewable resource. To the wildlifer, it creates new kinds of habitat which attract greater numbers and varieties of wildlife. To the outdoors enthusiast, it jars the serene landscape with stripped, ravaged surfaces that in some areas cause soil erosion and water polution. To the ecologist, the single species of tree-growth . . . that follows clearcutting can be more vulnerable to pests or fire than a diversified, multiaged forest.[5]

The principle of clearcutting is considered by most foresters as a sound silvicultural practice in stands of big, shallow-rooted trees, such as the Douglas fir (Fig. 18-16); nevertheless, it is a practice that lends itself to abuse in many instances (Figs. 18-17 and 18-18).

Clear-cutting is a long-established practice in the North Pacific Coast Region, and recently it has become

[5] "Clearcutting Moves into Congress," *Conservation News,* National Wildlife Federation, 41 (1 May 1976), 10.

FIGURE 18-15   Logs can be airlifted by helicopter from difficult sites, eliminating the necessity to build haul roads (TLM photo).

FIGURE 18-16   Clearcutting normally is done in relatively small patches. This scene is just east of Mount St. Helens in Washington (TLM photo).

FIGURE 18-17   This clearcutting of immense proportions, far from prying eyes in the back country of central Oregon, is a highly questionable approach (TLM photo).

FIGURE 18-18   Clearcutting leaves the landscape with a devastated look, particularly when done over a large area, as in this scene from the central part of Vancouver Island (TLM photo).

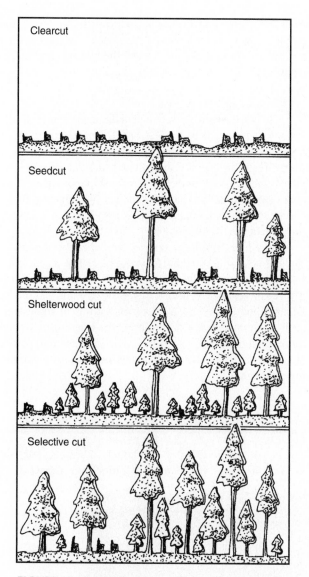

Clearcut

Seedcut

Shelterwood cut

Selective cut

FIGURE 18-19   Clearcutting and some of its alternatives.

widespread in other parts of the continent. It is an "even-aged management system" that has several attractive attributes, but it is coming under increasing attack, particularly by environmentalists, because of various concerns, most noticeably reflected by its scarring effect on the landscape. Accordingly, forest managers in the region, for the first time, are now planning the use of other types of management systems on a large scale. Some alternatives are indicated in Figure 18-19.

1. *Clearcut.* Entire stands of trees are removed at one time.

2. *Seedcut.* Five to 10 trees per acre are left to provide source of seeds.

3. *Shelterwood cut.* Up to 30 trees per acre are left to shelter seedlings.

4. *Selective cut.* Mature trees are harvested individually, leaving behind trees of varying ages. This has been the most common management system outside the North Pacific Coast Region, until recently.

In many parts of the region clearcutting has been the only system in use. Responding to public outcry, however, other approaches are now being taken. In northern California, for example, the Forest Service announced in 1990 that clearcutting would be used on only about half of all timber operations and that more attention would be given to preservation and to the maintenance of "biodiversity" (perpetuation of all plant and animal species naturally found in forests).

Throughout the region timber-cutting controversies abound. The local citizenry generally favors accelerated harvesting because the economy is so dependent on this resource, but the wisdom of prudent long-range planning for sustained yield and scenic preservation is hard to deny. In some cases the controversies involve economics versus environment; in others, short-term gain versus long-term sustainability.

In the continental U.S. portion of the region, the logging of old-growth forests has created the bitterest feelings and has had the greatest impact on the local economy. The federal government has decreed a moratorium on timber cutting in some old-growth forests and has greatly reduced the allowable cut in others because of the apparent danger to the survival of the northern spotted owl, an endangered species.

The U.S. Forest Service finds itself under increasing criticism in the region for three other management practices: allowing "too much" timber to be cut selling timber at prices below market value, and building "too many" roads in the forests for use of timber harvesters. These objections apply particularly to the situation in Tongass National Forest (the nation's largest national forest) in southeastern Alaska.

The twin controversies of logging old-growth forests and using clearcutting practices have also reached a fever pitch in British Columbia. The principal disputation involves the watershed that drains into Clayoquot (pronounced Klakwut) Sound on the west coast of Vancouver Island. Four years of "consultation" among the principal stakeholders (provincial government, forest products industry, local citizenry, First Nations representatives, and conservation organizations) produced the "Clayoquot Compromise" in 1993. None of the stakeholder groups are happy with the covenant, but that is the nature of a true compromise, and it shows that agreements can be reached even in the bitterest and most delicate situation.

Forest Products Industries

Although most large sawmills of the region have always been at coastal locations, in the early days of the industry many small sawmills were located on remoter, landlocked sites. As time passed, the trend was to phase out the small mills and concentrate activities at ever-larger sawmills situated on tidewater or at strategic inland locations (Fig. 18-20).

A prominent pulp and paper industry has also developed in the region, particularly in recent years. Although a considerable quantity of pulp is shipped out of the region, local paper mills are consuming an increasing quantity in the manufacture of kraft paper and newsprint.

Big corporations and huge mills are characteristic of the wood products industries in this region. Vertical and horizontal integration of production facilities is commonplace. It enables many economies of scale and more efficient use of each log.

Integration of logging, transportation, processing, and manufacturing operations is widespread in western Oregon and Washington but is most notable in the Strait of Georgia area of British Columbia, where functional interconnection in terms of both areal space and corporate organizations is tightly knit.

Economic Trends

As a major income generator of the region, the forest products industry is critically important to economic well-being. Yet it has experienced hard times for much of the last two decades. Home construction, particularly in the United States, is the largest market for the lumber producers of both the United States and Canada, and the housing demand has been weak for some years. Expansion of pulping and plywood production has helped to ease the problem, but it has failed to take up all the slack.

This industrywide decline would have been even more serious if not for the increasing role of Japanese purchases of logs, lumber, and other wood products. Japanese capital has been infused in several Canadian and Alaskan forestry enterprises, and Alaskan lumber mills now ship more than 90 percent of their output to Japan. Whether the Japanese market will be so significant in the future is conjectural, but its short-run impact has been favorable.

The forest products industry has experienced many periods of economic fluctuation in the past. The recent industrywide recession of the 1980s and early 1990s was unusually severe and resulted in the closing of many mills and the loss of many jobs. The situation was disastrous in some areas, such as northwestern California. At the time if this writing, however, the recession seems to have bottomed out, particularly in British Columbia, where the forest industry is booming once again.

AGRICULTURE: SPARSE AND SPECIALIZED

The North Pacific Coast is still largely in timber, relatively little of the land being suited to agriculture. Dairying, the

FIGURE 18-20   Large sawmills are scattered through the region. This huge mill is near Enumclaw in western Washington (TLM photo).

dominant farming activity of the region, accounts for a large part of the agricultural land being in pasture. Hay and oats occupy more than half of the land in crops.

Some of the locally important agricultural areas that contain most of the crop land are

1. The Umpqua and Rogue River valleys of southwestern Oregon
2. The Willamette Valley
3. The Cowlitz and Chehalis Valleys and the lowlands around Puget Sound in Washington
4. The Bellingham lowland
5. The lower Fraser Valley of British Columbia
6. Southeastern Vancouver Island.

Of these areas, the Willamette Valley, with more than 2 million acres (800,000 ha) in crops is by far the largest and best developed.

Agriculture in Oregon and Washington

The older settled parts of the Willamette Valley and Puget Sound lowlands present an agricultural picture of a mature cultural landscape that can be found in few places in the West. The Willamette Valley has miles of fruit farms that grow prunes, cherries, berries; hop fields; fields of wheat and oats; excellent pastures; and specialized farms that produce commercial grass seeds or mint (for the oil). The Willamette Valley, occupied by farmers of the third or fourth generation on the same farms, is the old, long-settled prosperous heart of Oregon that grows most of the fruit, berry, vegetable, and grain crops of the North Pacific Coast Region. Dairying is the principal occupation, but diversified horticultural and general farms are also common (Fig. 18-21).

In the Puget Sound lowlands, where considerable land is diked or drained, are the region's best dairy and pasture lands. Market gardening is an important agricultural activity that has increased in proportion to the growth of the large urban centers. Vegetables are grown in this area for the local urban markets, but the surplus is shipped to other parts of the United States.

The lower floodplains and deltas of four medium-sized rivers, often collectively referred to as the Bellingham Plain, occupy most of the eastern fringe of Puget Sound north of Seattle. Dairy farming is the principal rural activity in this area, but there is a major concentration of green pea cultivation; many kinds of vegetables and various berries are also grown here (Fig. 18-22).

Agriculture in British Columbia

Farming in the British Columbia portion of this region is concentrated on the floodplain and delta of the Fraser River and on southeastern Vancouver Island. Agriculture here serves the urban markets of metropolitan Vancouver and Victoria, but the

FIGURE 18-21   Many fields in the Willamette Valley are burned after the late-summer harvest (TLM photo).

FIGURE 18-22   A berry farm in the Nooksack Valley near Bellingham, Washington (TLM photo).

small area of suitable farmland is inadequate to produce sufficient food for the urban population. General mixed farming and dairying are carried on in these areas; there are also many specialty crops, such as fruits, berries, vegetables, and flowering bulbs. The leading agricultural industry of the Lower Fraser Valley is the production of whole milk for the Vancouver market. Pasture occupies the largest proportion of farm land.

Other important agricultural activities include horticulture, especially berries; poultry and beef cattle raising; and hops. Major farm problems in the valley include small and uneconomic farm size, too little summer rain (the annual average is 70 inches, or 178 cm, but dry summer periods often cause crops to wilt) occasional floods, and inadequate seasonal labor supply.

The southeastern lowlands of Vancouver Island contain about 50,000 acres (20,000 ha) of cultivated land. Temperatures are similar to those of the Fraser delta, but rainfall is considerably lower. In addition to dairying, poultry raising, and the cultivation of fruits and vegetables, this area has specialized in growing spring flowers for eastern Canadian markets. Many farms are marginal, and Vancouver Island is not even self-sufficient in dairy products.

The Matanuska Valley of Alaska

The Matanuska Valley, a fairly extensive and well-drained area of reasonably fertile silt-loam soils, lies at the head of Cook Inlet inland from Anchorage. Nearly two-thirds of Alaska's cropland acreage is in the valley, but it comprises only a few dozen farms, many of them marginal (Fig. 18-23). Production costs are high and crop options are limited, with the result that Matanuska farming, never very important, continues to decline in significance.

THE UPS AND DOWNS OF COMMERCIAL FISHING

The North Pacific Coast is one of the major fishing regions of the world. The bulk of the catch consists of only a few varieties of fish, but they are taken in tremendous quantities,

FIGURE 18-23   A Matanuska Valley farm scene near Palmer, Alaska (TLM photo).

and the contribution of the industry to the total economy of the region is a major one.

Salmon has long been the commercially dominant fish in North Pacific waters, but the last few years have seen a large and continuing diminution of salmon stocks. A century ago there were an estimated 30 million salmon in Pacific Northwest waters; the total present in the mid-1990s was thought to be only 3 million, of which about half were hatchery-bred. The salmon fishing season has been progressively shortened; in some cases the season is only one weekend long, and in two recent years a complete moratorium was imposed on the salmon fishery of Washington, Oregon, and northern California. The precipitous salmon decline has been ascribed to various factors, most notably the degradation of spawning streams, dams and water diversions, and climatic fluctuations caused by El Niño ocean conditions.

A major pollock fishery has developed in the past decade. Although pollock is a low-value fish, it is caught in enormous quantities. Another finfishery of special note is midwater trawling for herring, which includes the gathering of herring roe to send to the luxury Japanese market. Pacific cod and sablefish round out the major pelagic (surface or mid-level feeders) fisheries of the Northwest Pacific (Fig. 18-24).

Halibut is the largest and most valuable of the bottom fish of the North Pacific. They are long-lived (up to 35 years) and can reach enormous size [up to 500 pounds (225 kg)]. They are caught by longline hooks or by power trawling. After many years of declining catches, an International Pacific Halibut Commission, consisting of representatives of the United States and Canada, was established, and the fishery has been stabilized at a productive level with strict quotas enforced.

Although oysters, crabs, and other shellfish have been taken commercially for a long time in regional waters, only in recent years has an outstanding shellfishery developed. This situation was stimulated primarily by greatly increased catches of king crabs in the Gulf of Alaska, beginning about 1960. Since then the king crab fishery has almost vanished, as has the shrimp trawl fishery. Other species of crab, notably snow crabs and tanner crabs, have emerged to take up most of the slack.

In terms of both quantity and value, Alaska is by far the leading fishing state, and the towns of Dutch Harbor-Unalaska and Kodiak consistently are the leading fishing ports of the nation. British Columbia is Canada's largest fishing province, with Prince Rupert as its leading fishing port.

POWER GENERATION

Because of mountainous terrain and heavy rainfall, the region has one of the greatest hydroelectric potentials of any part of North America. Some of the potential was developed early; Victoria, for example, had electric street lights in 1882, just 1 year after Edison's developments in New York, and a commercial hydroelectric power plant in 1895, 2 years after the first such plant at Niagara Falls. Some large facilities were established to provide power to the Trail smelter shortly after the turn of the century, but Grand Coulee and Bonneville

FIGURE 18-24 Purse seiners at work in Puget Sound (TLM photo).

A CLOSER LOOK   *The Saga of the Salmon*

The Pacific salmon is the leading commercial fish in North America in terms of value of catch. It is easily the most important fish caught in Alaska, the leading fishing state, and British Columbia, the leading fishing province.

Because of its remarkable life cycle, vast numbers, and susceptibility to entrapment and depletion, it is both a major resource of the region and a focal point of controversy and dispute. There are five species of Pacific salmon: chinook (*Oncorhynchus tshawytscha*), coho (*O. kisutch*), sockeye (*O. nerka*), pink (*O. gorbuscha*), and chum (*O. keta*). All five species are anadromous—that is, they spend most of their life in the ocean but migrate up freshwater streams to spawn. Only a few weeks or months after being hatched, salmon fingerlings swim downstream to the sea. They spend from one to five years in saltwater before returning to their place of birth (up the same river, the same tributary, and the same creek) to spawn. After the female lays eggs in the stream gravel and the male fertilizes them, the adults die.

When in the ocean, salmon must face the normal hazards of the sea and increasing pressure from oceangoing fishermen with their incredibly long and effective driftnets. However, it is after the spawning instinct begins to govern their behavior that they cluster in immense numbers and become liable to almost total entrapment. As salmon congregate in estuaries and bays at the mouths of rivers and particularly as they begin to move upstream in singleminded response to the urge to propagate, it is relatively easy to capture entire runs by using nets, seines, and fishtraps.

In addition to the potential for overfishing, the concentration in rivers of migrating fish (both adults moving upstream and fingerlings going downstream) produces the risk of severe depletion owing to damming. Although mature salmon are incredibly persistent and tenacious in swimming up rapids and jumping over small obstructions in rushing streams, only the smallest of dams is needed to prohibit their upstream progress totally. In most cases where dams were built on salmon rivers, fish ladders have been constructed to ease upstream migration, and fish boats have been designed to transport fingerlings downstream, thus bypassing the dams. These are costly schemes that are sometimes very successful but not always.

Damming on the Columbia River, the principal salmon river in the conterminous states, has a continuing adverse effect. Fish ladders permit salmon to pass around low dams like the Bonneville, but high dams, such as Grand Coulee, are apparently impassable. Even more serious is the staggering mortality rate among fingerlings, which must either cascade over the spillways or be sucked through the turbines as they move downstream to the sea. There are now 11 major dams on the Columbia River; it has only a single 50-mile (80-km) stretch of free-flowing water in its entire 750-mile (1 200-km) length between the Canadian border and the Pacific Ocean. More than half the natural spawning area of the Columbia Basin has been denied to anadromous fish by dams on the Columbia and its tributaries (Fig. 18-A).

The conflict of interests between advocates of hydroelectricity generation and salmon fishermen is a major long-term cause of disharmony in the region. It is increasingly clear that the two resources are virtually incompatible and that one can be developed only at the expense of the other. In the United States, the dam builders have the upper hand. In Canada, however, provincial regulations prohibit dam-building on major salmon rivers, such as the Fraser (which has the largest natural salmon run in the world), the Stikine, and the Skeena.

Regardless of varied controversies however, the saga of the salmon represents one of the most blatant examples of fishery mismanagement imaginable. Here is one of the most valuable fishery resources in the world, a resource that can be exploited at relatively low cost, that is capable of sustained yield management, and that has a strong market demand that can be translated into high prices for the product. In rational economic terms, this situation should lead to a stable and prosperous fishing industry. What has happened, however, is the opposite: the history of the salmon industry consists of alternations of sporadic bursts of expansive prosperity and long, dragging periods of economic hardship.

The resource is not constrained by property rights, and almost anyone can become a commercial salmon fisherman with a moderate investment in equipment and licenses. As a result, the fishery has consisted mostly of a frantic scramble to take fish quickly before someone else takes them. Although increasingly complex fishing regulations were introduced in both Canada and the United States, they were planned primarily to ensure sufficient escape-

Dams were the first major projects on the U.S. side of the border.

Most major power-generating facilities are associated with dams situated to the east of the North Pacific Coast Region, particularly on the Columbia River and some of its tributaries. In addition, the continent's largest atomic-generating plant is adjacent to the Columbia at Hanford, Washington. The availability of "firm" power was greatly enhanced by the Columbia River Treaty, whereby water stored behind Canadian dams can be released at low water periods and thus diminish seasonal fluctuations in stream flow and power generation.

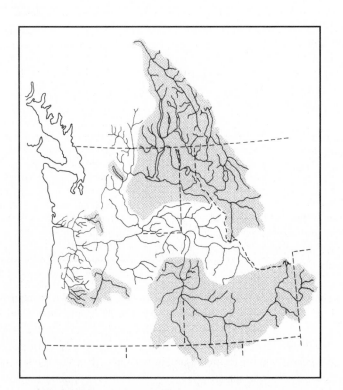

FIGURE 18-A Many parts of the watershed of the Columbia River and its tributaries are now inaccessible to salmon because of dams. The shaded areas represent those portions of the drainage basin from which salmon have been excluded (after Ed Chaney, "Too Much for the Columbia River Salmon," *National Wildlife*, 8 (April–May 1970), 20).

example, the Columbia watershed, by far the most important salmon area in the conterminous states, has experienced a decline of more than 75 percent from its peak several decades ago. In the early 1990s the annual salmon run in the Columbia consisted of only about 2.5 million fish. Moreover, less than one-fourth of these were "wild" salmon; the remainder were hatchery stock, which consistently were weaker and more vulnerable.

More enlightened approaches to management have been followed in recent years, and indications are that it may be possible to have both biological and economic stability in the salmon fishery. Starting in 1968, Canada led the way with an innovative change in management strategy that phased in a program of both limiting and reducing the number of licensed fishermen and boats in the industry. Alaska, with a more complex situation, followed suit in 1973. Later Washington adopted a similar, but less comprehensive, program. There are still many burdensome regulations that restrict the efficiency of the fishermen, for maintenance of a sustained yield resource must still be the keystone to any management program. Nevertheless, at least a start has been made at balancing harvesting capabilities with resource productivity. Finally, after nearly three decades of negotiations, the United States and Canada signed a bilateral agreement about joint and separate harvesting of salmon coming into the Strait of Georgia area. This had been a matter of serious dispute for some time.

ment so that the salmon stocks could be maintained. To accomplish this objective, most regulations were specifically designed to reduce the efficiency of vessels and gear, which had some biological merit but was economically disastrous. Consequently, wasteful duplication of capital and labor occurred in an industry that was already overdeveloped.

There has been a long-term downward trend in salmon numbers particularly in the southern part of their range (California, Oregon, Washington). For

TLM

The principal markets for Columbia Basin power have been the large cities of Washington and Oregon and major aluminum factories, most of which are in the North Pacific Coast Region. The rapid recent increase in generation capacity has led to a search for new markets and the establishment of intertie facilities that link the electric systems of 11 western states in the largest electrical transmission program ever undertaken in the United States. Thus the circular cycle continues: build more dams to supply more power and then seek new markets requiring more power that call for more dams, and so on.

MINERAL INDUSTRIES

Various discoveries of gold, silver, lead, iron ore, coal, and copper have led to moderate flurries of mining activity in southern Alaska and on Vancouver Island in the past, but almost all the operations are of historical interest only. Even the huge gold mine at Juneau, with its nearly 100 miles (160 km) of tunnels and shafts (said to be the world's largest low-grade gold mine), has been shut down for half a century.

Petroleum is another matter. Alaska's dynamic petroleum industry has been a focal point of excitement and controversy for nearly two decades. Until recently, attention was focused primarily on the North Slope–producing area and the pipeline corridor, which are largely outside the North Pacific Coast Region. However, in 1989 an oil tanker collided with an underwater reef in Prince William Sound, near the southern terminus (Valdez) of the pipeline, producing the most massive oil spill in American history. This accident resulted in immense damage to the fragile environment, an unprecedented cleanup effort, hundreds of lawsuits, and notable political and economic repercussions that will continue for years.

Almost lost in the excitement of North Slope petroleum is the realization that major oil and gas extraction have been occurring beneath the waters of Cook Inlet and on the nearby Kenai Peninsula (both areas just southwest of Anchorage) since 1957. Although production has declined in recent years, it continues at a relatively steady pace.

Offshore areas of Alaska are thought to be the most promising prospects for future petroleum development in the United States. Various offshore districts will eventually be prospected, but the first tangible development is taking place in the nearshore waters of the Gulf of Alaska, southwest of the St. Elias mountain knot. The first exploratory wells were completed in the early 1980s, but exploitation has proved to be both slow and difficult. The waters of the Gulf of Alaska are tumultuous at best and supremely hazardous during frequent winter storms. Moreover, these coastal waters are some of the most biologically prolific in the world, and the potential environmental damage from oil spills or leaks is a major point of contention, particularly since the 1989 catastrophe in Prince William Sound.

URBANISM: MAJOR NODES AND SCATTERED POCKETS

Most residents of the North Pacific Coast Region—even the Alaskan portion—are urban dwellers (Table 18-1). There are

TABLE 18-1

Largest Urban Places of the North Pacific Coast Region, 1995

Name	Population of Principal City	Population of Metropolitan Area
Anchorage, AK	276,000	282,000
Bellingham, WA	51,000	156,000
Bremerton, WA	46,000	225,000
Corvallis, OR	43,000	
Eugene, OR	121,000	314,000
Medford, OR	48,000	159,000
Nanaimo, BC	61,000	
Olympia, WA	38,000	189,000
Portland, OR	457,000	1,710,000
Salem, OR	118,000	312,000
Seattle, WA	527,000	2,205,000
Tacoma, WA	195,000	647,000
Vancouver, BC	622,000	1,825,000
Victoria, BC	86,000	312,000

five prominent metropolitan nodes and a number of smaller cities and towns, many of which exist in relative isolation (Fig. 18-25). Four nodes and several smaller centers serve conspicuously as coastal gateways to river or pass openings through the mountain barrier(s) that blockades access to the interior.

Burgeoning Metropolitan Centers

Portland (gateway: Columbia River barge route with paralleling railways and highways) is the dominant commercial center of the lower Columbia and Willamette valleys. The city's dual harbor facilities, part on the Willamette River and part on the Columbia, are among the most modern in the nation. Bolstered by the transshipment of grain and ores that are barged down the Columbia, Portland's bulk ocean freight business is approximately equal to that of Seattle. Wood products and food processing dominate the industrial structure.

Seattle (gateway: Cascade Mountains passes with rail and highway routes to the interior) is the focus of a highly urbanized zone that fronts the eastern shore of Puget Sound from Bellingham and Everett on the north to Tacoma and Olympia on the south (Fig. 18-26). Seattle itself is located on a hilly isthmus between the sound and Lake Washington, which gives it a constricted central area in contrast to the sprawling suburbs to the north and south. The deep and well-protected harbor requires no dredging, and Seattle has long been a major Pacific-oriented port as well as the principal gateway to Alaska. Seattle has outstanding container-han-

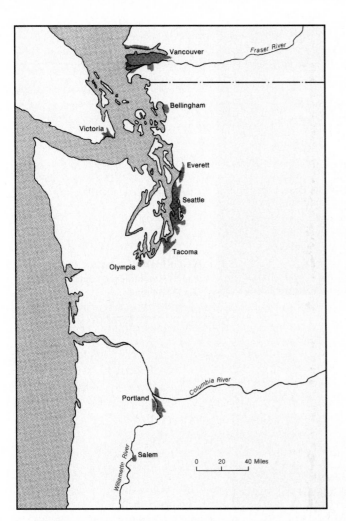

FIGURE 18-25 Metropolitan complexes of the Pacific Northwest.

Vancouver is clearly the primate city of western Canada as well as the nation's fourth largest industrial center (Fig. 18-28). The development at Roberts Bank, 25 miles (40 km) to the south, of a bulk products superport that is primarily for exporting coal and ores to Japan, has added to the city's urban primacy.

Victoria, British Columbia's capital, is the second largest city in the province. It is perhaps the continent's most attractive city and experiences Canada's mildest climate, which makes it a major goal for tourists and retired people.

Anchorage (gateway: Alaska Railroad and two highways to the interior) is the largest city in Alaska, with nearly half the state's population in its metropolitan area. It has been the dominant growth center for both population and economic activity in the state. Despite oil to the south, farming to the north, and its role as an international air-transport hub, the city's economy is primarily dependent on government, especially military, activities.

FIGURE 18-26 A high-altitude view of the Seattle isthmus. Puget Sound is on the left, Lake Washington is on the right, and the waterway connecting the two is near the top of the picture (TLM photo).

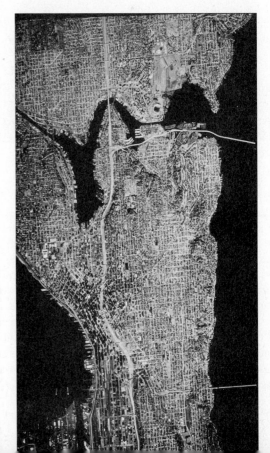

dling facilities and is one of the 10 largest container ports in the world; two-thirds of its port traffic is containers.

The Boeing Company has been the mainstay of the city's economy since the 1920s, although its employment has fluctuated wildly with defense contracts, ranging between 50,000 and 100,000 over the past decade. Wood products and shipbuilding are other major industries in the metropolis (Fig. 18-27). The population of the extended metropolitan area approximates 2 million.

Vancouver (gateway: Fraser and Thompson River canyons, Trans-Canada Highway, and both transcontinental railway lines) overcame its early rivalry with Victoria for economic dominance of British Columbia and has grown to become Canada's second-ranking port. As the principal western terminus of Canada's transcontinental transport routes,

FIGURE 18-27 Seattle is one of the leading shipbuilding centers in the nation (TLM photo).

FIGURE 18-28 Vancouver's skyline, as seen from the south. The Coast Mountains rise abruptly in the distance (TLM photo).

Urbanization in Isolation: The Case of the Alaska Panhandle

In the long stretch of coastland between Vancouver and Anchorage most urban places are isolated and remote. Only the dual towns of *Prince Rupert* and *Kitimat*, on the north coast of British Columbia, have useful surface transport connections with the rest of the continent; the former is a railway terminus and fishing port and the latter is one of the world's largest aluminum-refining localities.

In the Alaska panhandle are seven small urban centers that exist in remarkable isolation in an area of magnificent scenery and persistent rain (Fig. 18-29). Only the two northernmost, *Haines* and *Skagway,* have land transport connections with the rest of the world; the other five, *Ketchikan,*

Wrangell, Petersburg, Juneau (Fig. 18-30), and *Sitka,* are all situated either on islands or on mountain-girt peninsulas; yet the narrow streets of these hilly towns are crowded with autos. All have highly specialized economies, mostly oriented toward commercial fishing or forestry.

Juneau's principal claim to fame, the fact that it is the state capital, is presumably a temporary distinction. The voters of Alaska, in statewide referendums in 1974 and 1976, decreed that the capital would be shifted to a much more accessible site 70 miles (112 km) northeast of Anchorage, where a new planned city would be built. The cost of such an endeavor, however, will be astronomical and the voters have twice (in 1978 and 1982) rejected approval of a bond issue to finance the operation.

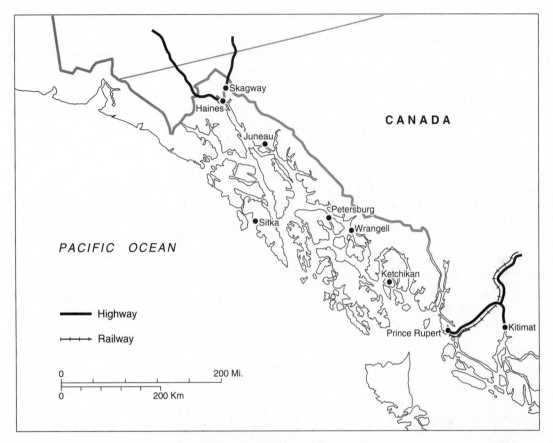

FIGURE 18-29   Alaska panhandle urbanism.

FIGURE 18-30   Juneau is nestled at the foot of the mountain front on the shore of Gastineau Channel (TLM photo).

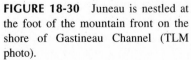

A CLOSER LOOK   North American Grain Transport

Grain is the only North American product that is exported from all four of the continent's principal seacoasts: Atlantic, Pacific, Gulf of Mexico, and Hudson Bay. Canada's northern port of Churchill on Hudson Bay was built during the 1930s as a shortcut route to Europe for wheat grown in the Prairie Provinces. Wheat movements from central Saskatchewan to Liverpool require roughly 1000 miles less travel via Churchill compared with the Great Lakes route. Grain shipping on Hudson Bay has remained marginal, however, because of the short navigation season and the high costs associated with operating a single-purpose port. Despite being closest to Europe, Churchill handles only a small fraction of the total outbound grain destined for European markets.

The traditional export flow of Canadian and American wheat begins at numerous country elevators in the Prairie Provinces, the Dakotas, and eastern Montana, where it is received from the farmers who grow it. From these country shipping points grain is sent eastward by rail, either to Thunder Bay, for Canadian shipments, or Duluth–Superior, if the grain is coming from the northern spring wheat region of the United States. Grain shipping on the Great Lakes was stimulated by construction of the St. Lawrence Seaway, which permitted larger oceangoing vessels to reach Lake Superior. But most grain still is handled in smaller ships that shuttle between the Great Lakes and export terminals in the Gulf of St. Lawrence, where all-year ocean shipping is possbile. Corn and soybeans raised in the Middle West also move via the Great Lakes–St. Lawrence route, especially from ports at Toledo and near Chicago (Fig. 18-B).

Land transportation is more expensive than water transport, but some grain moves more than a 1000 miles by rail to Atlantic Coast ports where it can be loaded on a year-round basis. Canada's major Atlantic grain port is Halifax, which is closer to Europe than any other port except Churchill. Corn and soybeans raised in the Ohio Valley and soybeans from the mid-Atlantic Coastal Plain move to ports such as Baltimore and Norfolk in trainloads of 50 to 100 cars. Grain exports from the Atlantic Coast of the United States are increasingly split between European and African destinations. Although they can be aggressive merchants of wheat in the international marketplace, Europe's wheat-raising countries, notably France, now are more inclined to supply Europe's needs rather than compete for the world's trade.

The Gulf of Mexico is the most important grain coast in North America. A portion of all the subcontinent's various grains and oilseeds are exported from the Gulf. On the Mississippi River, corn and soybeans are loaded into barges as far north as St. Paul; wheat and corn are poured into Missouri River barges at Sioux City; and wheat and sorghum fill Arkansas River barges moving south from Tulsa. All the export-destined grain eventually reaches the Baton Rouge–New Orleans section of the river, where it is reloaded for export. Flow down the Mississippi represents the largest single component of the grain export business of the United States, accounting for many billions of bushels each year and representing hundreds of millions of dollars in foreign trade.

Just as corn and soybeans dominate the downriver traffic at New Orleans, wheat and sorghum raised in the Great Plains account for much of the flow of grain through the port of Galveston, Texas, where it arrives mainly by rail. Winter wheat is produced in the southern Great Plains in such quantity that the crop typically far exceeds the U.S. domestic need. The surplus is exported, especially to African, Asian, and Latin American countries, which also purchase quantities of southern Plains sorghum and other feed grains. Galveston and New Orleans are well positioned to serve these newer markets.

Within the maze of exports are many shorter-distance shipments from grain growers to domestic processors and millers. In both Canada and the United States these include shipments of wheat to Eastern flour millers. Shipments of corn and soybeans to poultry feed processors, especially in the Southeast, account for substantial flows within the United States. Most corn, soybeans, and canola crushed for oil are shipped relatively short distances before they are processed, whereas wheat is shipped unprocessed over relatively long distances before it is milled into flour.

Rapidly expanding East Asian markets for U.S. and Canadian wheat, canola, barley, corn, soybeans, and sorghum have reoriented grain flow patterns throughout North America in recent years. Corn raised in Iowa moves in 100-car trains to Pacific Coast ports such as Tacoma and Portland. The continent's newest grain export terminal is located at Prince Rupert, British Columbia, the North American port closest to Japan. Partnerships between Canadian-prairie grain-marketing organizations have constructed massive new terminals for handling export wheat at both Vancouver and Prince Rupert in the past decade. Navigation improvements on the Snake River have allowed barges carrying export grain to be loaded as far inland as the Washington–Idaho border, thus creating a new outlet for Palouse wheat producers.

The Pacific Rim now constitutes the world's largest grain market. Together, Vancouver and Prince Rupert account for more than half of Canada's wheat

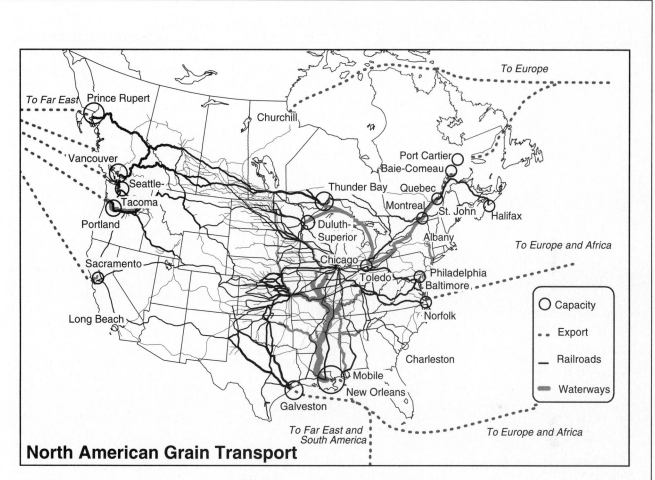

North American Grain Transport

FIGURE 18-B   Cartography by Ryan Baxter.

exports. Within the United States, sharp reductions in railway rates have made it economical to ship much of the export-destined grain raised west the Mississippi River—and nearly all that raised west of the Missouri—to Pacific Coast ocean ports. In terms of the grain trade, North America now looks more to the west and south than it does to the east.

The North American Free Trade agreement has stimulated grain sales from both Canada and the United States to Mexico. More recently, free trade has increased Canadian wheat sales to the United States as American flour millers have found it cheaper to import wheat from north of the border than to purchase it on the domestic market. Changing logistics in the Canadian and U.S. grain trade reflect a variety of world-scale trends, such as rising standards of living in many African and Asian countries, a breakdown of traditional export strategies focused entirely on the European market, and the impacts of new agreements that remove centuries-old tariff barriers between nations. Grain-laden ships departing North America's "four seacoasts" are part of a world pattern of trade flows originating in the grain-growing heartlands of the United States and Canada. They radiate as components of economic transactions encompassing the entire world.

Professor John C. Hudson
Northwestern University
Evanston, Illinois

A CLOSER LOOK   Vancouver, B.C.—Changing Land Use and Functions

The site of Vancouver was occupied by Europeans in the 1870s as one of many single-industry resource-based towns that were to become characteristic of the hinterland of Canada later in the twentieth century. Sawmills were built on both sides of the excellent, sheltered harbor of Burrard Inlet. The tall, straight trees of a luxuriant and untouched West Coast coniferous forest were cut on the gentle mountain slopes around this harbor and northward along the island-fringed, indented and unoccupied West Coast. Huge logs were felled at temporary logging camps along the coastal hinterland and towed through sheltered coastal waterways to large, "central place" sawmills near the delta mouth of the Fraser River. Sailing ships, and later steamships, came to the deep harbor for long, high-grade lumber and took it to growing urban settlements in western United States and East Asia.

This was one of the beginnings of a resource-based economy that was to dominate the development of British Columbia and Canada in general for the next century. Other tiny villages of British settlers dotted the southwestern corner of British Columbia, clustered around Georgia Strait which links Vancouver Island to the mainland of Canada. These "outposts of Empire" were dependent on fish canneries, mines, fur-trading posts, and sawmills—all using the untouched resources of the natural environment and sending their partially processed products to urban markets elsewhere.

These Georgia Strait settlements preceded the arrival of the Canadian Pacific transcontinental railway to Vancouver in 1886. That little sawmill town was to become a major Canadian and world port. But these changes did not happen immediately. Large foreign markets were far away, in time and in distance, until the opening of the Panama Canal during World War I. Vancouver became an industrial city in the first half of the twentieth century, primarily supplying consumer goods to its own rapidly increasing population and to the

million people concentrating in southwestern British Columbia. In one sense, local industries were protected by long distances and high transport costs from the industrial producers of eastern Canada and northeastern United States.

Industrial land use spread along three belts in Vancouver. Land around the Burrard Inlet harbor was occupied by industries served by both the Canadian Pacific railway and ocean or coastal shipping. These included lumber mills, a sugar refinery, shipbuilding, fish canneries, and later grain elevators and oil refineries. This industrial land use spread eastward into adjoining Burnaby, Port Moody and to North Vancouver on the north side of the harbor. The primary markets of these industries were Pacific centers; much of the raw materials came from the Prairie Provinces to the east or from northward along the coast.

A second industrial belt, consisting of wood-based industries such as sawmills, shingle and furniture plants, and shipbuilding, clustered around the shallow water of False Creek south of the city's original commercial core (Fig. 18-C). After the Canadian National Railway (a second transcontinental railway) and the Great Northern Railway (from the United States) came during and after World War I to the filled-in, reclaimed land at the east end of False Creek, more secondary industries spread eastward along rail and highway routes into Burnaby. Much of this production was for local Vancouver and British Columbia markets.

A third industrial zone, along the distributary arms of the Fraser River on the south side of Vancouver, spread eastward into New Westminster. The industries in this zone included sawmills and fish canneries, and like the Burrard Inlet harbor, were served by railways and shallow-draft ocean transport. Modest small wooden houses on narrow lots were built nearby by industrial workers, creating a working-class "village," known as Marpole, along the Fraser River, far from the Vancouver city center.

Industrial workers also built small

wooden homes in the northeastern part of Vancouver and adjoining North Burnaby, near the areas of industrial land use. These eastern parts of the city were "working man's" Vancouver and included many immigrants from central and southern Europe and some from China and Japan. The latter were low-income commercial and service workers. In social and economic contrast, business, management, and commercial workers built larger and nicer homes on the west side of the city, near the growing commercial core south of the harbor and southwest of the original railway station. Most of these residents were of British origin, who flooded into the west side of Vancouver between 1920 and 1940 mostly from eastern Canada and the Prairie Provinces. Vancouver was becoming a city of ethnic, economic, and occupation contrasts between its east and west sides.

The industrial characteristics of Vancouver began to change after about 1950. When the population of the metropolitan area exceeded 1 million persons in 1971, it became a large internal market for a wide range of secondary manufacturing and small consumer-product plants. Primary industries such as big sawmills, and major employers such as shipbuilding, moved from the Burrard Inlet harbor and from around False Creek. The old homes of industrial workers around False Creek were torn down and replaced by attractive apartments and condos. This dirty, polluted industrial area of the 1930s had become an attractive residential area for commercial, professional, and service workers by the 1980s.

Industrial land use around the Burrard Inlet has also changed from manufacturing to transfer and transport uses. Raw materials such as coal, sulfur, chemicals, potash, and wheat, mainly from the western Prairie Provinces, come by rail to storage and container facilities around the harbor, plus lumber by water transport from coastal British Columbia. These products are nearly all for export to Pacific and world markets

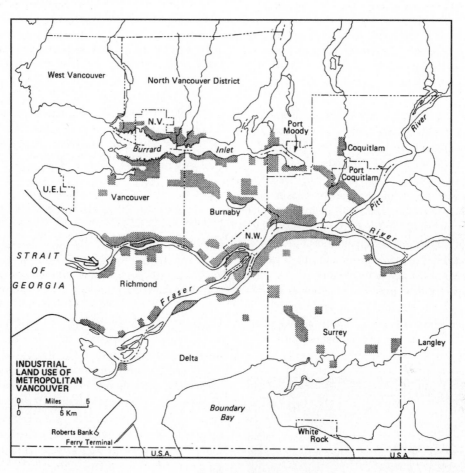

FIGURE 18-C

and need little additional processing in Vancouver. The dockworkers there, operating modern machinery, are not low-income employees.

Manufacturing and assembly for the growing internal consumer market of Vancouver and British Columbia has spread to cheaper land in the suburbs around Vancouver (see Fig. 18-C). These industrial areas are now dispersed, related mainly to truck transport, and industrial workers have spread over the western Fraser delta. Vancouver itself has become a commercial, business, management, service, enter-tainment city. The industrial areas within the city have decreased and changed to other functions.

Vancouver is an example of changing urban land use and functions, characteristic of many North American cities. It started as a single-industry resource-based town and then developed dirty wood-processing industries for local and Pacific markets. It became Canada's main West Coast port, through which funneled many raw materials and natural resources of Western Canada for Pacific and world markets. New, secondary industries for local markets oc-cupied land in the suburbs. People in Vancouver are now employed more in commercial, management, and service occupations, many working in high-rise office towers in the downtown core face the same congestion, transport, and parking problems as their counterparts in other large commercial city centers of North America.

Professor J. Lewis Robinson
University of British Columbia
Vancouver

SPECTACULAR SCENERY

The North Pacific Coast is perhaps the continent's most scenic region and is hailed as one of its most desirable outdoor recreation areas.[6] Remoteness from large population centers and the inaccessibility of many scenic spots, however, have retarded the development of tourism. Improved transport by road, air, and ferry have significantly stimulated the tourist business.

The section south of the international border is best developed, with six magnificent national parks (Fig. 18-31); one of the finest combined scenic-and-sandy coastlines in the world, the Oregon coast (Fig. 18-32); splendid forests; and a plenitude of accessible winter sports areas. The Canadian section and the Inside Passage of southeastern Alaska rank among the world's most scenic areas, with their spectacular mountains, glaciers, and fiords. In south coastal Alaska, however, is found the subcontinent's most magnificent landscape, thousands of square miles of ice and rock and forest, culminating in the grandeur of Denali National Park.

It is to be expected that land-use controversies are frequent and ecological confrontations numerous in a region of such scenic splendor. The conflicts vary but focus particularly on commercial timber cutting and potential oil spills.

FIGURE 18-31   Mount Rainier on a rare cloudless day (TLM photo).

[6] The scenery, however, is frequently shrouded in fog, mist, or rain. The famous Seattle weather forecast still pertains: "If you can see Mount Rainier, it's going to rain; if you can't, it is raining."

One of the most notable developments in the tourist industry of the North Pacific Coast Region has been the rapid proliferation of cruise ship visitation. This activity is focused on the "Inside Passage" to Alaska, normally involving a 7- or 10-day voyage that begins and ends at either Seattle or Vancouver and visits several ports or scenic spots en route (always Juneau and Glacier Bay; sometimes Prince Rupert, Ketchikan, Misty Fiords, Sitka, Skagway). Each ship carries several hundred passengers, and more than 200 cruises ply the Inside Passage each summer. Indeed, the visitation has become so heavy that advance reservations are needed for a cruise ship to enter the waters of Glacier Bay so as to avoid too much disturbance to the marine fauna.

THE VITAL ROLE OF FERRIES IN THE REGION

Highways and railways, the more prosaic forms of transportation, are well established from the Fraser Valley southward despite topographic hindrances. In the northern two-thirds of the region, however, both road and rail routes are chiefly limited to lines extending inland from the ports of Prince Rupert, Skagway, Haines, Cordova, Valdez, Seward, and Homer. Air transport thus becomes very important throughout the region.

As in no other part of North America, ferries play a specialized and vital role in the transportation of this region. There are three large government ferry networks and a few small private systems (Table 18-2). The emphasis in each is on passenger traffic, with roll-on/roll-off facilities for automobiles and trucks. (Most freight is carried by regular oceangoing vessels or tug-propelled barges.)

The principal systems are as follows:

1. The Washington State Ferry system operates throughout the Puget Sound area. Most of its service radiates from Seattle, but there is also interisland

TABLE 18-2

Selected Statistics of Major Ferry Systems, 1995

	Washington State Ferry System	British Columbia Ferry System	Alaska Marine Highway System
Vessels in service	25	40	8
Ports served	20	44	31
Passengers carried	20,864,000	20,665,000	402,700
Vehicles carried	8,380,000	8,150,000	121,000

FIGURE 18-32   The Oregon coast is very scenic from one end to the other. This view is near Port Orford in the south (TLM photo).

service in the San Juan Islands, plus several individual connections between island and peninsula ports.

2. The British Columbia provincial government maintains an excellent ferry network in the Strait of Georgia, interconnecting mainland and Vancouver Island ports, with other service extending northward along the mainland coast and into the Queen Charlotte Islands.

3. The Alaska Marine Highway System has two discrete route networks. The largest serves the 15 principal settlements of the panhandle, with external connections to Prince Rupert and Seattle (Fig. 18-33). The other connects about a dozen ports in south-central Alaska (Prince William Sound, Kenai Peninsula, Anchorage, and Kodiak Island). In the near future there are plans to connect these two networks with service across the turbulent waters of the Gulf of Alaska.

THE OUTLOOK

The characteristics and relative significance of the North Pacific Coast Region are likely to change little in the near future. It will continue to be a region of specialized economy, Pacific Ocean orientation, magnificent scenery, and conservation controversy.

The forest industries, long outstanding in the economy, are caught up in both international economic price squeezes and conservation controversies. These factors are likely to cause a diminution of cut, even from the present depressed levels. It is clear that the prosperity of the wood products industries will continue to fluctuate, and the long-term outlook is not too favorable. An increased housing demand would be a great boon, but this scenario is not likely. Meanwhile, the

FIGURE 18-33   A British Columbia government ferry in the Grenville Channel portion of the Inside Passage (British Columbia government photo).

A CLOSER LOOK   Communities in Transition: Tourism and Economic Diversification in British Columbia

Traditionally the economy of British Columbia, like that of other provinces beyond the industrial heartland of southern Ontario and southern Quebec, has been dependent on the harvesting and extraction of natural resources, primarily for export. Indeed, beyond the main urban concentrations of Vancouver and Victoria, the majority of British Columbia's small towns are dependent on a single industry, usually forestry or mining. Global economic restructuring has resulted in significant job loss in the traditional resource-based sectors as mines and mills have downsized or closed, and high-tech machines have replaced workers. Small communities throughout the province have been struggling to develop alternative economic activities to sustain their communities. Tourism is an option that many communities are considering as an element of diversification.

Tourism in British Columbia has shown steady growth for the past decade. The provincial tourism slogan is "Super, Natural British Columbia," a theme echoed by Tourism Vancouver, which promotes Vancouver as "Spectacular by Nature." Undoubtedly it is the striking mountain and coastal scenery that draws both North American and international tourists to the province. The majority of tourists, however, remain in the southwest of the province around the main urban centers of Vancouver and Victoria as well as the

FIGURE 18-D   One of the many murals depicting the town's history adorns the side of the Visitor Information Center in Chemainus, a forest industry–turned–tourist town on Vancouver Island (photo by Alison M. Gill).

resort of Whistler, 72 miles (120 km) to the north of Vancouver. This planned "resort municipality" has been a great success story, drawing attention the potential contribution of tourism to the provincial economy. With the assistance of both federal and provincial funding during the late 1970s and early 1980s, private developers constructed an internationally renowned ski destination. For several years Whistler has been voted by North American skiers as the

best North American ski resort, and it also ranks as the number one destination for Japanese skiers, who constitute 20 percent of the 1.5 million skier-days annually. Recently, the resort has expanded into a "four-season" destination with the development of many summer recreation opportunities, especially golf.

But not all communities have the locational or resource advantages of

---

higher-quality and more accessible timber supplies grow scarcer, which means that the costs of logging continue to rise. Communities heavily dependent on the forest industries are, with a few exceptions, unlikely to flourish.

Agriculture should become even more specialized than it is today, although dairying should be unrivaled as the principal farm activity throughout the region. There is little likelihood of expanding farm output.

The commercial fishing industry will probably be characterized by considerable fluctuation in the annual catch of

the various species. Year-to-year variations in the availability of fish reflect both natural factors and the erratic results of overexploitation. Fishermen have demonstrated remarkable versatility and resiliency in the past, shifting with great rapidity from an overfished species to an underutilized one. As fish processing and marketing facilities develop a similar measure of versatility, the entire industry will become more stabilized. Problems of overexploitation and conflicts of interest, however, will continue to cloud the scene, particularly in regard to salmon.

Whistler, and each community must consider its own resource potential and competitive position. A few communities have successfully made the transition from a resource-based economy to tourism, and these are often held up as examples of what is possible. Chemainus, as suggested by its slogan—"The little town that did"—is one community that has been successful in effecting the turnaround. The community is located on the southeastern coast of Vancouver Island, north of Victoria. In 1983 the major employer, MacMillan Bloedel (Canada's largest forest products company), underwent corporate restructuring, closed its sawmill, and laid off 650 workers. When the new, highly computerized mill reopened in 1985, there were only 140 employees. In anticipation of this downsizing, the town, with assistance from the provincial government, set about downtown revitalization to attract tourists. Based on the ideas of one energetic community member, artists were commissioned to paint murals depicting the evolution of the local area and especially the forest industry on downtown buildings. Over a period of several years 32 murals were developed (Fig. 18-d). As tourist numbers have grown, so too has the local tourism economy, with art, craft and souvenir shops, tea shops, and a local theater. The community has "survived" and attracts around 300,000 tourists every year—most passing through on tours of the island. Tourism and the new mill coexist, and the town continues to attract new residents, mainly retirees. It is not the same type of town it used to be, and there is no record of how many unemployed mill workers remained in the town or became employed in the tourism sector. Here as elsewhere, those people traditionally employed in the resource sector often disdain jobs in the tourism industry as low paid and seasonal.

Although the murals in Chemainus are a created attraction, the fundamental tourism resource in British Columbia is still the natural environment and the opportunities it provides for a wide array of outdoor recreational activities. Ecotourism (which is environmentally and socially responsible tourism) and adventure tourism (which involves some challenging form of outdoor recreation such as heli-skiing, rafting, or ocean kayaking) are the fastest growing segments of the British Columbia tourism product. They generally sustain lower numbers of tourists than concentrated urban attractions but do offer business and employment opportunities for people in less accessible areas. On the west coast of Vancouver Island, for example, Tofino has developed as a center for viewing the migration of the gray whales as well serving as the gateway community to Pacific Rim National Park and the coastal wilderness area of Clayquot Sound. Tourists from around the world have been attracted here due to media attention surrounding the fight to preserve the old-growth forest in this area from logging.

The degree to which tourism development is a realistic goal for communities facing the loss of traditional resource industries is dependent on many factors. In the examples given the communities have either unique natural assets (Whistler, Tofino), strong government support (Whistler), locational advantage (Chemainus, Whistler), and/or a strong vision and leadership (Chemainus, Whistler). The provincial government until recently operated a program (the Community Tourism Action Program) to assist communities in understanding and realistically assessing their potential for tourism development. For some this led to the realization that tourism is not the economic panacea that can replace jobs lost in the resource sector. Other communities, especially those with active citizen involvement, have begun planning to ensure that the resources on which tourism is dependent such as attractive landscapes, clean water, and recreational access are valued and protected in resource decision-making. Such resources serve not only to attract tourists but also enhance the quality of life for residents.

Professor Alison M. Gill
Simon Fraser University
Burnaby, British Columbia

The impact of the oil spill in Prince William Sound is likely to have continuing and far-reaching reverberations. Oil-happy developers in Alaska will find diminishing enthusiasm among the citizenry, and environmental safeguards probably will be expanded and extended in a multiplicity of arenas.

Commercial ties with Asia in general and with Japan in particular will undoubtedly grow rapidly. The bustling Japanese economy should provide a significant market for the quantity products of the North Pacific Coast's primary industries and the products from the continental interior that are shipped from the region's ports. A continuation of the high level of Japanese capital investments in the region is to be anticipated.

The established urban areas of the southern portion of the region will probably remain the major centers of population and economic growth. Most of the region, however, will remain largely a wilderness, dominated by a few extractive industries and continually beckoning North Americans for all forms of outdoor recreation.

SELECTED BIBLIOGRAPHY

"Admiralty Island: Fortress of the Bears," *Alaska Geographic,* 18 (1991), entire issue.

"The Alaska Peninsula," *Alaska Geographic,* 21 (1994), entire issue.

"Alaska's Salmon Fisheries," *Alaska Geographic,* 10 (1983), entire issue.

"Anchorage," *Alaska Geographic,* 23 (1996), entire issue.

"Anchorage and the Cook Inlet Basin," *Alaska Geographic,* 10 (1983), entire issue.

ASHBAUGH, JAMES G., ED., *The Pacific Northwest: Geographical Perspectives.* Dubuque, IA: Kendall/Hunt Publishing Co., 1994.

"British Columbia's Coast," *Alaska Geographic,* 13 (1986), entire issue.

BROWN, ROBERT C., AND IAN T. JOYCE, "The Roe Herring Fishery on Canada's West Coast," *Yearbook,* Association of Pacific Coast Geographers, 56 (1994), 75–88.

BROWNING, R. J., "Fisheries of the North Pacific," *Alaska Geographic,* 1, (1974), entire issue.

DICKEN, SAMUEL N., AND EMILY F. DICKEN, *Oregon Divided: A Regional Geography.* Portland: Oregon Historical Society, 1982.

DOWNIE, BRUCE K., "Kluane, One of our Most Exciting National Parks," *Canadian Geographic,* 100 (April–May 1980), 32–38.

DREW, LISA, "Here's Your Land: Now Make Money," *National Wildlife,* 30 (December 1991–January 1992), 26–45.

ELLIS, DEREK V., ED., *Pacific Salmon: Management for People,* Western Geographical Series, Vol. 13. Victoria: University of Victoria Department of Geography, 1977.

ERICKSON, KENNETH A., "The Tillamook Burn," *Yearbook,* Association of Pacific Coast Geographers, 49 (1987), 117–138.

EVENDEN, L., ED., *Vancouver: Western Metropolis.* Victoria: University of Victoria, Department of Geography, 1979.

FARLEY, A. L., *Atlas of British Columbia: People, Environment, and Resource Use.* Vancouver: University of British Columbia Press, 1979.

FORWARD, CHARLES N., ED., *British Columbia: Its Resources and People.* Victoria: Department of Geography, University of Victoria, 1987.

———, *The Geography of Vancouver Island.* Victoria: University of Victoria Department of Geography, 1979.

FRANKLIN, JERRY F., AND C. T. DYRNESS, *Natural Vegetation of Oregon and Washington.* Corvallis: Oregon State University Press, 1988.

"Glacier Bay: Icy Wilderness," *Alaska Geographic,* 15 (1988), entire issue.

HAMILTON, WILLIAM G., AND BRUCE SIMARD, "Victoria's Inner Harbour, 1967–1992: The Trasformation of a Deindustrialized Waterfront," *Canadian Geographer,* 37 (Winter 1993), 365–371.

HARRIS, STEPHEN L., *Fire Mountains of the West: The Cascade and Mono Lake Volcanoes.* Missoula, MT: Mountain Press Publishing Company, 1991.

HAYES, DEREK W., "Fog and Cloud in British Columbia," *Canadian Geographical Journal,* 83 (December 1971), 200–203.

JACKSON, PHILIP L., AND A. JON KIMERLING, EDS., *Atlas of the Pacific Northwest.* 8th ed. Corvallis: Oregon State University Press, 1993.

JOHANNESSEN, CARL L., ET AL., "The Vegetation of the Willamette Valley," *Annals,* Association of American Geographers, 61 (1971), 286–302.

"Juneau," *Alaska Geographic,* 17 (1990), entire issue.

KENNEDY, DES, "Fraser Delta in Jeopardy," *Canadian Geographic,* 106 (August–September 1986), 34–43.

———, "Whither the Gulf Islands? Nature's Largesse Spawns a B.C. Dilemma," *Canadian Geographic,* 105 (October–November 1985), 40–49.

"Kodiak," *Alaska Geographic,* 19 (1992), entire issue.

LACASSE, DOMINIQUE, "The Adams Run," *Canadian Geographic,* 115 (March–April 1995), 60–69.

LEEPER, JOSEPH, "Humboldt County: Its Role in the Emerald Triangle," *Canadian Geographer,* 30 (1990), 93–110.

OKE, TIM, AND JOHN HAY, *The Climate of Vancouver.* British Columbia Geographical Series, no. 50, 2d ed. Vancouver: University of British Columbia, Department of Geography, 1994.

O'NEIL, ERNEST, "Science Doubles Number of Spawning Salmon," *Canadian Geographic,* 99 (December 1979–January 1980), 62–65.

POWERS, RICHARD L., ET AL., "Yakutat: The Turbulent Crescent," *Alaska Geographic,* 2 (1975), entire issue.

"Prince William Sound," *Alaska Geographic,* 20, (1993), entire issue.

ROBINSON, J. LEWIS, ED., *British Columbia.* Toronto: University of Toronto Press, 1972.

———, "Sorting Out All the Mountains in British Columbia," *Canadian Geographic,* 107 (February–March 1987), 42–53.

SCHREINER, JOHN, "The Port of Vancouver: Its Health Is Vital to Western Canada's Economy," *Canadian Geographic,* 107 (August–September 1987), 10–21.

SCOTT, JAMES W., ET AL., *Washington: A Centennial Atlas.* Bellingham: Western Washington University, 1981.

"Southeast Alaska," *Alaska Geographic,* 20 (1933), entire issue.

THEBERGE, J. B., "Kluane: A National Park Two-thirds Under Ice," *Canadian Geographical Journal,* 91 (September 1975), 32–37.

———, ED., *Kluane: Pinnacle of the Yukon.* Toronto: Doubleday Canada Ltd., 1980.

———, AND J. A. KRANLIS, "The Saint Elias: Our Highest, Youngest, and Iciest Mountains," *Canadian Geographic,* 105 (December 1985–January 1986), 36–45.

"The Kenai Peninsula," *Alaska Geographic,* 21 (1994), entire issue.

TOWLE, JERRY C., "Man and the Pacific Salmon: A New Era?" *California Geographer,* 22 (1982), 13–32.

WAHRHAFTIG, CLYDE, *Physiographic Divisions of Alaska,* Geographical Survey Professional Paper 482, Washington, DC: Government Printing Office, 1965.

"Wrangell–Saint Elias: International Mountain Wilderness," *Alaska Geographic,* 8 (1981), entire issue.

WYNN, GRAEME, AND TIMOTHY OKE, EDS., *Vancouver and Its Region.* Vancouver: University of British Columbia Press, 1992.

YOUNG, CAMERON, "The Last Stand: Timber Running Low as Confrontations Continue in B.C.," *Canadian Geographic,* 106 (February–March 1986), 8–19.

The Boreal Forest

19

The largest region of North America sprawls almost from ocean to ocean across the breadth of the subcontinent at its widest point, in subarctic latitudes. The Boreal Forest Region is primarily a Canadian region, occupying almost half of that nation's areal extent, although it also encompasses the Upper Lakes States and central Alaska in the United States.

It is a region of rock, water, and ice, but particularly of forest—interminable, inescapable forest (Fig. 19-1). The forest extends for hundreds of miles over flattish terrain with relatively little variation in either appearance or composition. Its very endlessness arouses a feeling of monotony in some people. Edward McCourt, in describing a traverse in western Ontario, emphasized this point:

> In Canada there is too much of everything. Too much rock, too much prairie, too much tundra, too much mountain, too much forest. Above all, too much forest. Even the man who passionately believes that he shall never see a poem as lovely as a tree will be disposed to give poetry another try after he has driven the Trans-Canada Highway.[1]

This is a region in which nature's dominance has been but lightly challenged by humankind. The landscape is largely a natural one and people are only sporadic intruders. Much of the region consists of the Canadian Shield, a land of Precambrian crystalline rock, rounded hills almost devoid of soil, fast-flowing rivers, and innumerable lakes, swamps, and muskegs. Long, cold winters characterize most of the region; its short growing season makes agriculture only locally important. From boundary to boundary most of the inhabitants are engaged overwhelmingly in *extractive* pursuits.

Most North Americans have never seen this region and the vast majority will never set foot in it. Yet it is a region that has enriched the autochthonous literature and folk music of both countries; the sagas of Paul Bunyan, the stories of Jack London, the poetry of Robert Service, and many other literary contributions have brought the "North Woods" into public consciousness. This is particularly true in Canada, where the forest syndrome has played a major role in the concept of the Canadian character. Subarctic expert William Wonders pointed out this emotional attachment:

> Forest, rock, water—these are the elements which exert a near-irresistible call for most Canadians. Even if most of the year is spent in a very different setting, the

[1] Edward McCourt, *The Road Across Canada* (Toronto: Macmillan of Canada, 1965), p. 110.

FIGURE 19-1   The landscape of the Boreal Forest Region is dominated by trees and water. This is the Rushing River near Kenora in western Ontario (TLM photo).

traditional northern ingredients are sought out for rest and relaxation. Though they may live to the south of it and rail against its harshness, many Canadians probably find in the Subarctic the "emotional heartland" of their nation.[2]

The boundaries of the Boreal Forest Region are nowhere clear-cut but can easily be conceptualized (Fig. 19-2). The northern margin is the tree line, which separates the Subarctic (Boreal Forest Region) from the tundra of the Arctic Region. As with most such vegetation boundaries, the idea of a line is misleading; the interfingering of forest and tundra is complex, and the northern margin of the region is a transition zone rather than a line. This northern floristic boundary is further reinforced by its approximate coincidence with the 50°F (10°C) July isotherm and its almost absolute separation of Dene (to the south) and Inuit (to the north) settlements.

The region's southern boundary is more complicated and should be visualized as follows. In the east it is approximately coincidental with the limit of close agricultural settlement north of the St. Lawrence and Ottawa Rivers. Across southeastern Ontario it follows the southern margin of the Canadian Shield. It crosses northern Michigan and central Wisconsin and Minnesota as a transition zone separating es-

sentially agricultural land on the south from essentially forest land on the north.

From western Minnesota to western Alberta the boundary coincides with the change from prairie landscape to forest landscape. In western Canada and Alaska the boundary is largely determined by topography; the mountainous terrain of the Rockies and the North Pacific Coast Region ranges mark the southern limits of the Boreal Forest Region in this part of the continent.

The eastern (Labrador) and western (Alaska) margins of the region are generalized as separating the zones of coastal settlement from the almost unpopulated interior, with the added distinction of a forest-tundra ecotone in western Alaska.

A HARSH SUBARCTIC ENVIRONMENT

From a human point of view the natural environment of the Boreal Forest Region is a harsh one. It is a land dominated by winter, and winter temperatures are the most severe to be found on the subcontinent. The surface is frozen for many months; during the summer poor drainage inhibits land transportation. The rocky forested landscape, with its filigree of lakes and rivers, was so inhospitable that Canadian settlement expansion was delayed and deflected. The Canadian Shield stood as a barrier between eastern and western Canada, denying the nation the strength and vigor that come with geo-

[2] W. C. Wonders, "The Forest Frontier and Subarctic," in *Canada: A Geographical Interpretation,* ed. John Warkentin (Toronto: Methuen Publications, 1968), p. 477.

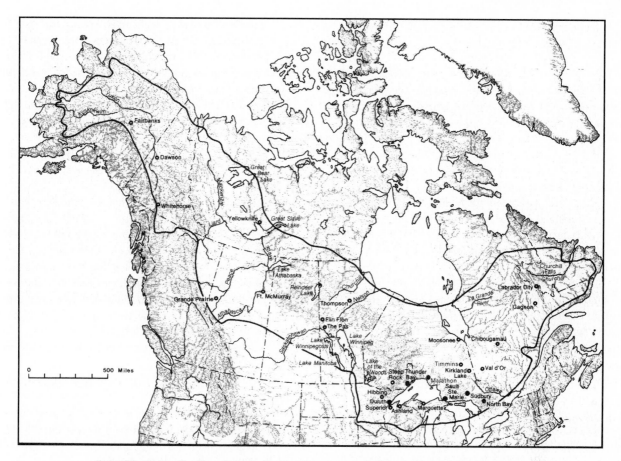

FIGURE 19-2   The Boreal Forest Region (base map copyright A. K. Lobeck; reprinted by permission of Hammond, Inc.).

graphical and political unity. Thin soils, poor drainage, hostile climate, and pestiferous insects made penetration and settlement anywhere in the region a difficult and expensive undertaking.

Terrain

The eastern two-thirds of the region is underlain largely by the Canadian Shield, a vast, gently rolling surface of ancient crystalline rocks that has been scraped and shaped by the multiple glaciations of the Pleistocene epoch. In Labrador and Quebec are several sprawling ranges of mountains and hills that reach elevations exceeding 4000 feet (1200 m). The southern edge of the Shield in Quebec is marked by the spectacular and complex Laurentide Escarpment. Elsewhere on the Shield the topography is more gently undulating, and elevations are generally well under 2000 feet (600 m). The remarkable sameness of terrain over vast expanses of the

Shield can be explained by the constancy of rock types; more than 80 percent of the surface consists of predominantly gneissic granitic rocks.[3]

Scouring by Pleistocene ice sheets remolded the surface and removed most of the preexisting soil. Drainage was totally disarranged, bare rock was left exposed on most of the upland surfaces, and glacial and glaciofluvial debris was deposited in countless scoured valleys. In some places the accumulation of glacial debris and deposits on the floors of old glacial lakes mantled the underlying Precambrian rocks to considerable depth, most notably in the large lowland surrounding James Bay and the western side of Hudson Bay; such areas have exceedingly flat surfaces.

To the west of the hardrock Shield is a broad lowland of vast extent that is underlain by softer sedimentary materi-

[3] J. Brian Bird, *The Natural Landscapes of Canada* (Toronto: Wiley Publishers of Canada Ltd., 1972), p. 136.

als. It has been build up by deposition of sands and silts washed off the Rockies and the Shield. Consisting for the most part of a plain with scattered hilly districts, it is almost as poorly drained as the Shield, with the result that water features (rivers, lakes, muskegs) are commonplace. Some of the largest rivers (Mackenzie, Slave, Athabaska, Saskatchewan) and lakes (Great Bear, Great Slave) in North America are found here. Most of the sedimentary plain, sometimes referred to as the forested northern Great Plains, drains northward to the Arctic Ocean via a complicated hydrographic system dominated by the Mackenzie River.

In the Yukon Territory the region encompasses an area of more complicated geology, greater relief, and greater variety of landscape. There are several rugged mountain ranges and deeply incised plateau surfaces as well as two lengthy structural trenches.

In central Alaska the drainage basin of the Yukon River occupies the broad expanse of land between the Alaska Range to the south and the Brooks Range to the north, widening ever more broadly westward. The lower basin of the Yukon and the equally extensive basin of the Kuskokwim River to the south consist of broad flat lowlands that are quite marshy in early summer.

Hydrography

Water is abundant in most of the region during the summer, chiefly as a result of glacial derangement of drainage patterns and the fact that the subsoil is at least partly frozen throughout much of the region, thus preventing downward percolation of surface moisture (Fig. 19-3). There are water bodies of every conceivable shape and size, from small ponds to some of the largest lakes in the world. [Water] spills from one to the next in swift, dashing streams and sweeps along in major rivers which reach such proportions that in the case of the Mackenzie, the width of its lower reaches is measured in miles. Rapids and falls are common features on almost all rivers.[4]

Lakes are so numerous that parts of the Shield might almost be described as water with occasional land (Fig. 19-4).

Most rivers originate within the region and flow outward from it: some to the Atlantic via the Great Lakes–St. Lawrence system; some in a centripetal pattern into Hudson Bay; some directly from Quebec or Labrador into the Atlantic; and some to join the Mackenzie, Yukon, or Kuskokwim systems in their path to the Arctic Ocean or the Bering Sea. Only in the so-called Nelson Trough area of northern Manitoba are there rivers that flow *across* the region; the Nelson and Churchill systems originate in the Rocky Mountains and Prairies to the west before flowing eastward into Hudson Bay.

Swamps and marshes are also widespread, as is that peculiar northern feature, muskeg (poorly drained flat land covered with a thick growth of mosses and sedges). These areas severely inhibit overland transportation and provide an extensive habitat for the myriad insects that swarm over the region during the brief summer.

[4] Wonders, "The Forest Frontier and Subarctic." p. 474.

FIGURE 19-3 Moving water is commonplace throughout the region in summer. This is the Kenebec River near Elliot Lake, Ontario (TLM photo).

Climate

In so enormous an area important differences of climate from north to south or east to west might be expected, but actually the differences are relatively slight. Everywhere the climate is continental (i.e., there is little oceanic influence); this is mainly the result of interior location, great distance from oceans, and the barrier effect of the western cordilleras blocking eastward moving air masses from the Pacific.

This continentality is the dominating feature of the region's natural environment. Winters are long, dark, and bitterly cold. Temperatures occasionally drop to 60°F below zero (−51°C) over most of the region, and in the northwest readings of less than −70°F (−57°C) have been recorded.

The region is saved from recurrent glaciation only by the warmth of summer; summer temperatures are generally mild, but the occasional incursion of warm air masses sometimes produces decidedly warm weather. The transition seasons are short but stimulating. Spring is characterized by considerable muddiness, which makes it the most difficult season for overland transportation.

Precipitation is relatively light except in the east (Quebec and Labrador) but is quite effective because evaporation is scanty. Summer is the period of precipitation maximum; virtually the entire landscape is covered with snow throughout the winter (Fig. 19-5).

FIGURE 19-4   The region contains hundreds of thousands of lakes. This is the Sioux Narrows section of Lake of the Woods in western Ontario (TLM photo).

Soils

Soils of the region are characteristically acidic, severely leached, and poorly drained, and low in fertility. Spodosols, Inceptisols, and Histosols predominate. They are mostly shallow, permeable soils, developed on coarse material exposed

FIGURE 19-5   Snow does not accumulate to great depths over most of the region, but it stays on the ground for many months. This March scene is near Lake Astotin, a few miles northeast of Edmonton, Alberta (TLM photo).

by Pleistocene ice sheets. Where farming is carried on, it is invariably in an area of more productive soil, generally a clay-based soil that was derived from glacial lake sediments.

Permafrost

Roughly half the surface area of Canada and about three-fourths of that of Alaska are underlain with permafrost, or permanently frozen subsoil (Fig. 19-6). In the zone of continuous permafrost the *active layer* (that part that freezes in winter and thaws in summer) of the soil is only 1.5 to 3 feet (0.45 to 0.9 m) in depth.[5] Beneath it is the permafrost layer, which is usually several hundred feet thick and has been measured to depths of 1300 feet (390 m) in both Canada and Alaska. In the zone of discontinuous permafrost various factors contribute to the presence of sporadic unfrozen conditions; where permafrost occurs, the layer is much thinner and the overlying active layer may be more than 10 feet (3 m) in depth.

[5] R. J. E. Brown, "Permafrost Map of Canada," *Canadian Geographical Journal,* 76 (February 1968), 57.

The presence of permafrost creates many engineering difficulties. Although providing a solid base for supporting structures (buildings, roadbeds, railways) when frozen, permafrost readily thaws when its insulating cover is removed and then loses its strength to such a degree that it will not support even a light weight; this condition results in buckling and displacement of any structure built on it. Various engineering techniques have been developed to overcome these difficulties, the most important being to remove the insulating cover (vegetation, surface soil) several months before actual construction begins, thus allowing permafrost conditions to stabilize to a new equilibrium.

Natural Vegetation

The distinctive natural vegetation of the region is coniferous forest with a deciduous admixture. From south to north the trees become smaller, those in the extreme north being small and scraggly. In this region of slow growth the trees never reach great heights, although they are normally close growing except where interrupted by bedrock or poor drainage.

Two great east–west bands of approximately equal width can be distinguished:

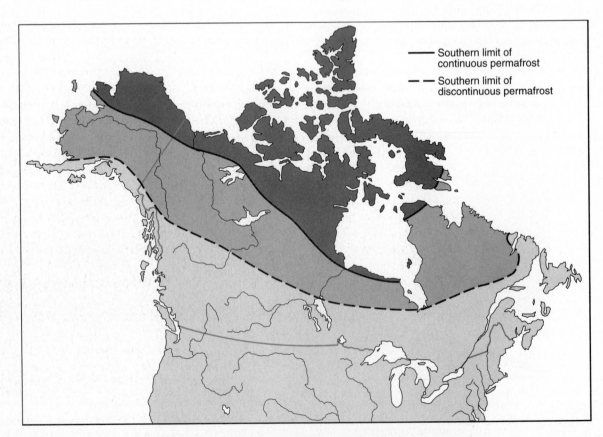

FIGURE 19-6 The southern limit of permafrost in North America.

1. The southern band is a continuous forest that forms canopies that permit moss, herbs, and shrubs to thrive at ground level.

2. The northern band is a more open lichen forest that is little known because it contains few human settlements or roads but contains a plague of blackflies.

The boreal forest is often referred to as *taiga*; a similar expanse of northern continental forest extends across Eurasia in similar latitudes. Conifers, growing in relatively pure stands, dominate the plant associations of the taiga. The principal species are white spruce, black spruce, tamarack, balsam fir, and jack pine. Associated with the dominant coniferous species is a small group of hardwoods that are quite widespread; most notable are white birch, balsam poplar, aspen, black ash, and alder.

Although relatively pure stands of conifers are very extensive, hardwoods are very common in the southern part of the region. A mixed forest (coniferous and deciduous), with deciduous species predominating, is prominent in the southern part of the region in eastern Canada and the Lake States, whereas across the central part of the Prairie Provinces the forest is almost completely deciduous.

In the northern part of the boreal forest, there is much natural hybridization of tree species (cf., lodgepole pine and jack pine; black cottonwood and balsam poplar; Alaska birch and paper birch), so that identification of actual species often is difficult.

Fire is a prominent ecological force in the Boreal Forest Region, often burning thousands of square miles before it is finally quenched; moreover, the incidence of fire in the taiga has greatly increased in recent years. Burned areas regenerate slowly in these subarctic latitudes. Hardwood species, especially birch and aspen, attain their greatest areal extent as an initial replacement for the conifers after a forest fire; they are eventually superseded by conifers as the climax association becomes reestablished.

About 9000 forest fires are recorded annually in Canada, the vast majority of them in the Boreal Forest Region (Fig. 19-7). Most are small and burn only a few acres, but some are huge, raging for weeks and consuming 250,000 acres (100,000 ha) or more. An average of 7 million acres (2.8 million ha) are burned avery year, which is about three and a half as much timber as is cut annually. About 85 percent of the fires are started by lightning.

Although a wildfire is a catastrophic event, it doesn't necessarily cause a catastrophe. Fire means renewal in the boreal forest and is a central part of the life cycle. From the charred ground, new life emerges. Viewed from the air, the boreal landscape is a patchwork of even-aged stands that developed following fires or insect outbreaks.

Native Animal Life

Native animal life was originally both varied and abundant; the variety is still great, although abundance has considerably diminished. Furbearing animals have had the greatest economic influence because of the trapping carried on by natives and outsiders alike. The beaver, distributed throughout the region (Fig. 19-8), has been the most important species, but also notable are the muskrat, various mustelids (ermine, mink,

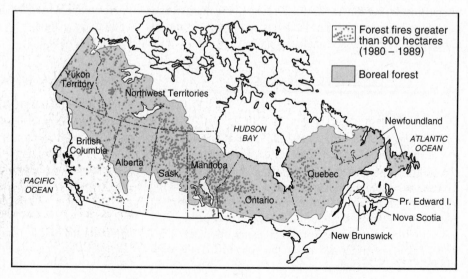

FIGURE 19-7 The distribution of large forest fires in Canada during the 1980s (after *Canadian Geographic*, 1996).

FIGURE 19-8 Beaver are still numerous in many parts of the region. This beaver dam and lodge are in Mount Tremblant Park in southern Quebec (TLM photo).

marten, otter, fisher, and wolverine), canids (wolf and fox), the black bear, and the lynx. Ungulates are another significant group. Woodland caribou occupy most of the region, and vast herds of barren-ground caribou spend the winter in the taiga. Moose and deer are also found, as well as forest-dwelling bison.

Bird life is abundant and varied in summer when vast hordes of migratory birds, especially waterfowl, come to the region to nest. Only a handful of avian species, however, winter in the Boreal Forest Region.

Insects are superabundant in summer. For two or three months of the year mosquitoes, blackflies, no-see-ums (biting midges), and other tiny tormentors rise out of the muskeg and the sphagnum moss in veritable clouds and make life almost unbearable for people and animals alike.

THE OCCUPANCE

Native Peoples

Before the arrival of Caucasians, the Boreal Forest Region was occupied by widely scattered bands of seminomadic people. Algonkian speakers, particularly Crees and Ojibwas, were in the eastern portion, and various Athabaskan-speaking tribes lived in the west. The material culture and economy of these widely dispersed tribes were quite uniform, presumably reflecting the homogeneity of the taiga environment. They were hunters and fishermen, depending primarily on the wandering caribou for their principal food supply. The relative

scarcity of inhabitants was due to a warlike social tradition and to the fact that hunting and trapping always preclude a dense population, for they entail the great disadvantage of uncertainty. Life was precarious, poverty extreme, and starvation not uncommon.

With the coming of whites, fur trapping offered significant trading opportunities. Many First Nations women, particularly Ojibwas and Crees whose hunting territory was where most of the fur trade was carried on, intermarried with whites, especially in the earlier years of contact. The offspring of such unions are called *Métis*. The Métis component has grown rapidly through the years so that it is probable that there are more Métis than full-blooded First Nations in the region today, although no accurate statistics on the Métis population are available.

White Penetration and Settlement

Over most of the Boreal Forest Region the evolution of white settlement has been associated with exploitive activities: trapping, mining, and forestry, or with governmental functions. Only in relatively limited areas and under often marginal conditions has settlement on a broader scale or greater diversity been attempted.

In the eighteenth and early nineteenth centuries furs were almost the only stimulus to white penetration of the region. By the middle of the nineteenth century, however, lumbermen were actively at work on the southern margin of the taiga, especially in the Upper Lakes States. At about the beginning of the present century exploitive activities were accelerated, with the initiation of pulping, hydroelectricity generation, and large-scale mining. Although Lake Superior iron ores were being mined as early as the 1850s and there were other isolated mineral discoveries in the region before 1900, mining in the region is largely a twentieth-century activity.

The agricultural frontier advanced into the Boreal Forest Region only slowly and hesitantly. Initial farming settlement consisted mostly of a northward overflow from the lower St. Lawrence Valley. Pioneer farmers pushed into various sections of northern Michigan, Wisconsin, and Minnesota, more or less concurrently with logging in those areas. More intensive agricultural settlement came later in the Clay Belt of Ontario and Quebec. Still later, the farmlands of Alaska's Tanana Valley were settled. Finally, about the time of World War I, the Peace River block in Alberta began to be occupied with some intensity.

White settlement of the Boreal Forest can thus be seen as a push from the east and the south, one primarily motivated by a search for furs, timber, and ores. Agrarian colonies came later and often persited on the economic margin. Urban settlement was sporadic and, for the most part, urban prosperity depended on either the stability of mineral output or some specialized transportation function.

Present Population

Most of the population of the Boreal Forest Region is located on its southern fringe and nowhere is there a significant density. The few nodes of moderate density are associated with some sort of economic opportunity, such as farming in the Peace River district or mining in northern Minnesota.

The population of the region is a little over 4 million, amounting to less than 2 percent of the two-country total. Although occupying half of Canada's areal extent, the Canadian portion of the region contains only about 10 percent of that nation's populace.

Native people constitute a significant proportion of the total; they are less numerous on the southern margin of the region, although even Wisconsin and Minnesota have a considerable number of Native Americans; in Canada the First Nations and Métis are the fastest-growing ethnic group in the nation, increasing at a rate of 3 percent annually.

The Boreal Forest Region has an economy that is primarily resource oriented, which means that it is sensitive to national and international economic trends. Until the last few years there was a rapid rate of population increase, based on a high birthrate among native people and a prominent influx of outsiders who were lured north by the availability of high-paying jobs for which local people lacked requisite skills. Since the mid-1970s, however, the regional economy has stagnated, resulting in a net out-migration (mostly whites who had moved North temporarily). This factor, coupled with a downturn in the birthrate, produced a zero population growth rate for the region in recent years.

THE ECONOMY

Most of the Boreal Forest Region remains economically underdeveloped, but that portion lying immediately north of the St. Lawrence River and around Lake Superior has been settled for many decades, and small but increasing numbers of people have found their way into other parts of the region, especially the Mackenzie Valley. In most areas the economy is based on a single exploitive activity, but in three widely separated portions of this vast region a broader development pattern has unfolded, resulting in a more diversified economic base, a higher density of settlement, and a more stabilized, although not necessarily prosperous, economy.

Areas of Broader Development

The Upper Lakes Area   Surrounding lakes Superior and Huron are parts of northern Michigan and Wisconsin, northeastern Minnesota, and southern Ontario that are functionally focused on the Great Lakes waterway, along which are transported vast quantities of the area's ore output (Fig. 19-9). The long-established mineral and transportation industries of the Upper Lakes provide a base for the most diversified subregional economy in the Boreal Forest Region.

Mining   The Upper Lakes area has a limited variety of mineral resources, but they occur in tremendous quantities. Southwest and northwest of Lake Superior lie the continent's largest and most favorably located iron ores. On Michigan's Keweenaw Peninsula are long-mined but diminishing copper deposits. On the west shore of Lake Huron are valuable beds of metallurgical limestone.

Iron mining has long been the outstanding economic enterprise in the subregion, hundreds of millions of tons of red hematite ore having been extracted and shipped down the Lakes. The area, which accounts for more than nine-tenths of U.S. production, is spread over 11 counties (3 in Minnesota, 5 in Michigan, and 3 in Wisconsin). Minnesota alone contributes two-thirds of the total, chiefly from its world-famous Mesabi Range (Fig. 19-10). Moreover, significant iron-ore deposits were long exploited at Canada's Steep Rock mine, 140 miles (225 km) northwest of the head of Lake Superior.

Present (and recent) production is generally of low-grade taconite and jasper ores, which must undergo an expensive pelletizing process to yield a higher-grade

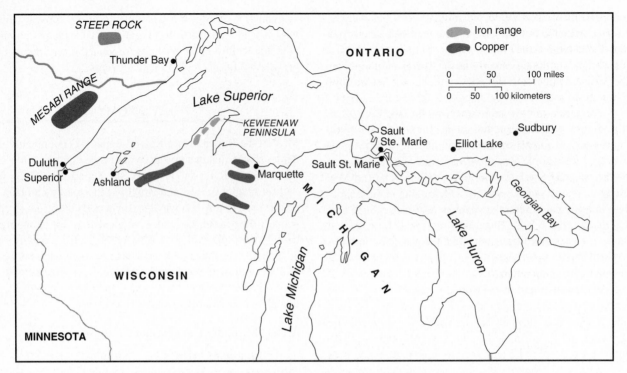

FIGURE 19-9   The Upper Lakes area.

FIGURE 19-10   An open-pit iron mine on the Mesabi Range in Minnesota (TLM photo).

concentrate before shipping. The pellets are loaded into rail cars, which are assembled into trains and taken to the various Lake Superior ore ports (particularly Duluth, Thunder Bay, and Marquette). The trains move onto the loading docks that jut out into the lake like huge peninsulas and dump their ore into "pockets," from which it can be dropped through hatches into the holds of lake vessels (Fig. 19-11). The lake carriers operate a busy one-way traffic, carrying a vast tonnage of ore across Lake Superior, through the Soo Canals, and down to the Lower Lakes ports. The return trip is often without cargo despite lower rates offered for upbound freight (Fig. 19-12). In most years the ore-shipping season on the lakes is about 10 months long, beginning in April and being closed by ice in January.

Although mining and concentrating facilities of the iron ranges are modern and efficient, the beneficiation process adds a notable cost factor. Higher-grade direct shipping ores from some foreign sources are available at much lower costs. Moreover, the North American steel industry has faltered conspicuously because of foreign competition; so it requires only about half as much ore as it did a decade ago. Consequently, the Lake Superior ore producers have closed their more expensive operations and increased the operating levels of their more efficient facilities. The net result in this area is a stifling unemployment rate, with production at less than one-fourth its previous level.

Logging   The Upper Lakes subregion was a major source of lumber during the latter half of the nineteenth century. Heavy exploitation so denuded the forests that they diminished to relative insignificance for several decades. Second- and even third-growth timber is now being exploited. There is a considerable, although scattered, forestry industry today in the three states and one province that constitute the subregion. Pulpwood production is more important than logging for lumber.

Fishing   Fishing was a flourishing enterprise in lakes Huron, Michigan, and Superior, antedating lumbering and

FIGURE 19-11   A typical ore dock (at Marquette, Michigan) with an ore train moving slowly into position to dump (TLM photo).

mining. Whitefish and lake trout in 1880 made up 70 percent of the catch. Since then the fishing industry has steadily declined as a result of overfishing, depredation by the destructive sea lampreys, destruction of immature fish, fouling of waters by city sewage and industrial waste, and changing physical conditions in the lakes.

An unusual addition to the recreational and commercial fishery of the Great Lakes is the coho salmon. In the late 1960s young salmon were deliberately released in Michigan streams and soon established their anadromous life cycle in

FIGURE 19-12   An ore boat returning to the harbor at Duluth to pick up another load of iron ore. The return journey usually is made without cargo, but in this case the deck of the boat is loaded with new automobiles picked up in Detroit (TLM photo).

the Upper Lakes. A large niche in the lakes ecosystem was unoccupied, and the coho filled this vacuum and occupied a place in the food chain without much competition to native species. Such exotic introductions must be approached carefully, but early results of the coho experiment seem positive.

Farming   Although a great deal of land in the northern part of the Upper Lakes States was cleared for farming, most of it was only of marginal value. Short growing season, lack of sunshine, poor drainage, and mediocre soils combine to produce very limited areas for prosperous crop growing or animal husbandry. Dairying is the principal farm activity in the area, and grains and root crops are cultivated.

Farming is somewhat more prosperous on the Canadian side, where pockets of good soil are used and the poor areas have not been settled.

Urban Activities   This portion of the Boreal Forest Region has several important urban nodes, each of which is a lake port of significance. Duluth–Superior is the funnel at the western end of the Great Lakes waterway on the U.S. side of the border; Thunder Bay performs an analogous function on the Canadian side (Fig. 19-13). The twin cities of Sault Ste. Marie, Ontario, and Sault St. Marie, Michigan, are at the crossroads of the east–west land and water route and the north–south land route of the Upper Lakes. The former city is also one of Canada's leading iron and steel manufacturing centers.

The Clay Belt   The so-called Clay Belt of the Quebec–Ontario borderland actually consists of two areas: the Great Clay Belt, which extends about 100 miles (160 km) both east and west from Lake Abitibi; and the Little Clay Belt, which occupies a more restricted area northeast and

northwest of Lake Timiskaming (Fig. 19-14). Although some of the more productive soils are clay derivatives from glacial lake sediments, much of the soil development is from glacial and lacustrine deposits that contain only minor clay elements.[6]

Agricultural settlement in the area is a product of this century. Many immigrants from overseas came into the Ontario portions, whereas most settlers on the Quebec side of the border were French Canadians who moved from the more closely settled portions of Quebec.

Agriculture in the area was fostered from early days by mineral discoveries. Silver and cobalt ores were discovered in 1903 and the town of Cobalt became the mining center. A major gold discovery in 1909 soon evolved into the Timmins-Porcupine complex. A second locale of high-grade gold deposits, around Kirkland Lake, began to be developed 2 years later. Several dozen mines were opened in the next two decades, mostly producing gold. The mines provided a local market for farm products, which enabled farmers to achieve some stability of production despite the high cost of their operations and the limited cropping possibilities.

Some mines closed down during World War II, and farming became an increasingly profitless occupation. However, a mining rejuvenation occurred in the 1960s, with the development of zinc, copper, and silver ores near Timmins and the opening of a large iron ore mine, near Cobalt. In the late 1970s higher prices for gold stimulated the reopening of several old gold mines and the development of several new ones. Although farming continues to be marginal in the area, a semblance of subregional prosperity has returned.

[6] For more details on this topic and the resulting confusion, see J. Lewis Robinson, *Resources of the Canadian Shield* (Toronto: Methuen Publications, 1969), pp. 99–106.

FIGURE 19-13   Canada's lakehead port, Thunder Bay, is marked by an array of huge shoreline grain elevators (courtesy Ontario Ministry of Industry and Tourism).

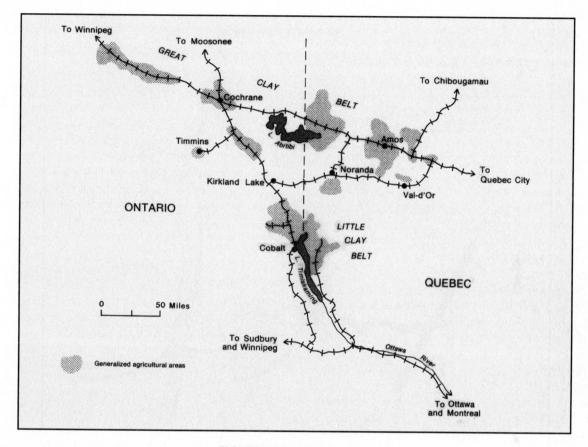

FIGURE 19-14   The Clay Belt.

The Peace River District   In the northwestern part of Alberta and adjacent British Columbia is an enclave of black soil and grassy parkland that has some 25,000 square miles (65,000 km^2) of potentially cultivable land (Fig. 19-15). This is Canada's northernmost area of satisfactory commercial agriculture. The Peace River district was first entered by settlers as long ago as 1879, but it was only sparsely occupied until well into the twentieth century, particularly after a rail line was laid from Edmonton to Grande Prairie in 1916.

Mixed farming is characteristic, with an emphasis on cash grains (especially barley) in the older areas, and livestock (particularly beef cattle and hogs) in the newer ones. Lately there has been a phenomenal increase in canola production, a continuing increase in barley, and an accompanying decrease in both wheat and oats.

Completion of a Vancouver railway connection in 1958 gave the Peace River farmers access to the markets of southern British Columbia for the first time. This improved transportation has accounted for a considerable increase in production of feed grain, which is railed to Vancouver for export or for use in the Lower Fraser Valley.

The district is no longer an area of frontier farming. Even though the pioneer fringe of agriculture is being pushed slowly northward toward Great Slave Lake, commercial agriculture is now an accomplished fact. A lot of land clearing is associated with the improvements of already existing farms.

Lumber and pulp are being produced in increasing volumes, primarily from white spruce and lodgepole pine. Natural gas and, to a lesser extent, petroleum production is increasingly significant and bolsters the economy of the entire area. Improved transportation by road and rail have stimulated growth, and the larger towns of the district are rapidly expanding.

Faunal Exploitation: Hunting and Trapping

Both First Nations and Métis led seminomadic lives in the past, depending on wildlife and fish for subsistence and sometimes trapping as a sideline for trading. The taiga today

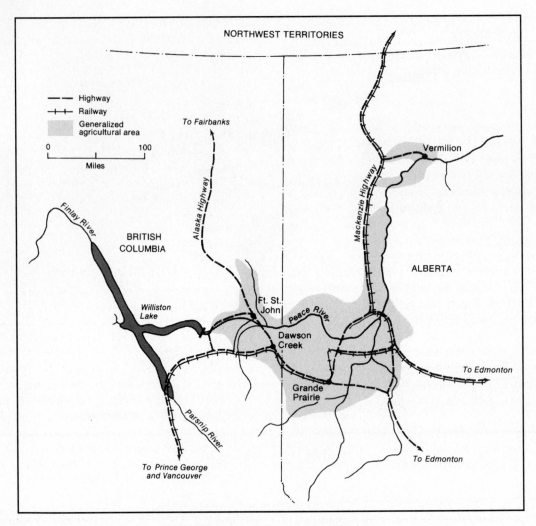

FIGURE 19-15 The Peace River District.

still has a moderate number of these activities, but nomadism is a thing of the past. The introduction of modern arms and ammunition had the predictable effect of making it easier to kill game at first, but the deer, moose, and caribou that once supplied most of the meat are greatly depleted.

Until recent years trapping was a major source of income for many native inhabitants of the region. Within the last three decades, however, its importance has declined, and in many areas trapping has almost completely disappeared. The availability of wage labor and government social service payments has removed much of the incentive for the trapper to spend long, hard weeks in the wilderness, and antitrapping campaigns organized by various humane and animal-rights organizations have greatly diminished the market for furs.

Federal and provincial government programs to bolster the wild-fur industry have been promulgated, especially in Manitoba. The establishment of a Registered Trapline System brought order to a previously chaotic endeavor. Controlled harvesting of muskrats on "fur rehabilitation blocks" has brought both stability and increased productivity in various floodplain and delta areas. Other programs are aimed at trapper training, trapline development grants, and fur marketing improvements.

Fur Farming

In recent years the fur-farming trend in Canada has been markedly upward. The proportion of pelt value from fur farms of the total pelt value (wild and farmed) has steadily increased from about one-third a half century ago to two-thirds today.

Foxes, mainstay of the early fur farmers, are rarely raised today. Mink farms make up more than 90 percent of the total farms and chinchillas most of the remainder. Fur farms are found in every region and every province of Canada, but they are characteristic of the Boreal Forest Region, with Ontario as the leading producer.

Forestry

Logging and lumbering started early in the accessible edges of the Boreal Forest Region, although only in the Upper Lakes area was there intensified exploitation. Once lumbering got underway in any locale it was usually only two or three decades before the entire stand had been cut-over, the loggers had moved on, and settlements were abandoned or stagnated.

The pulp and paper industry came later and has continued as a prominent activity in many parts of the region, although it, too, is concentrated on the southern margin. This is Canada's leading industry; the nation is second only to Sweden as an exporter of pulp and produces nearly half the world's newsprint. Much of the industry is controlled by U.S. capital.

In the last decade or so there has been a significant increase in forestry activity in the previously untapped forests of northwestern Alberta (northwest of Edmonton). The pulpwood is utilized in local new pulp mills or shipped to British Columbia mills.

The geographical distribution of pulp and paper mills in Canada has been remarkably stable since the early 1920s. Cheap power is an important factor in the location of pulp plants, as is shown by their distribution (Fig. 19-16). A string of mills (all water driven) lines the southern edge of the forest from the mouth of the St. Lawrence to Lake Winnipeg. Farther west in the region, sawmills and pulp mills are usually large, but they are fewer in number and much more scattered in distribution.

Logging depends on snow in this region of long, cold winters. The cutters begin their work in autumn because the logs must be moved out while the ground is frozen. Tractors pull the sleds laden with logs to rivers, where the cut is piled to await the spring thaw and the freshets, which transport the logs by the hundreds of thousands to downriver mills. River driving is an economical method of transporting logs.

Commercial Fishing

In the last three decades commercial fisheries have developed in several large lakes of the western taiga. The industry is on a small scale compared with oceanic fishing but adds measurably to the local economy. It is a year-round activity, but emphasis is on winter ice-fishing and gill-netting. Nets are set in winter with the aid of an ice jigger, a simple machine that walks along the undersurface of the ice. Transportation is significantly simplified during the cold months when tractor trains can reach the remoter lakes over frozen ground that would be soddenly impassable in summer.

Great Bear Lake is too deep, cold, and barren to provide a commercial fishery, but Great Slave Lake has an expanding industry centered at Hay River. Whitefish and lake

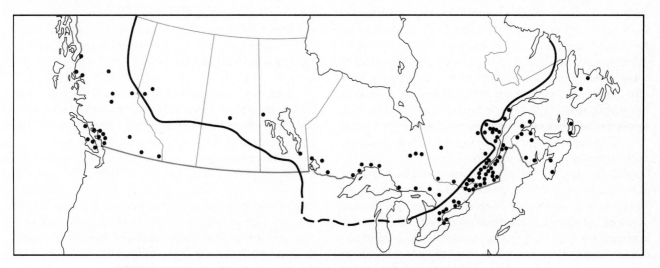

FIGURE 19-16 The location of pulp mills in Canada. Although most of the pulp logs are cut in the Boreal Forest Region, most of the mills are located south of the region. (The southern boundary of the Boreal Forest Region is shown by a black line.)

trout are caught, primarily for export to the central United States. Commercial fishing also is sporadically important in Lake Winnipeg.

Several other large lakes in the region support commercial fishing on a regular basis, and a recent summer innovation, pulse fishing, has made it economically feasible to conduct concentrated fishing operations on small lakes without overusing the resource. Pulse fishing involves rotational use of small lakes—that is, heavy fishing for one season followed by no fishing at all for 5 to 7 years; this practice also allows for economical use of portable ice-making and refrigerated fish-holding facilities.

Mining

Mineral industries have been the mainstay of the regional economy throughout most of its history. Mining has formed the basis for most of the urban settlements. Today the outstanding significance of mining continues despite the fluctuating fortunes that accompany variations in local supply and worldwide demand. The hardrock nature of the Shield provides an unfavorable basis for agriculture but abundant opportunities for mineralization in economic quality and quantity. Also, the sedimentary formations of the Northwest are favorable for hydrocarbon accumulation.

It is difficult to stay current with the opening and closing of mines, because changes are so frequent. Ore bodies are erratically distributed over the region, but several mining districts can be recognized on the basis of relative location.

Labrador Trough   Ore bodies had been known in the Labrador Trough—which extends from near the estuary of the St. Lawrence northward to Ungava Bay—for decades before exploitation was finally initiated in the mid-1950s. Transportation was the big problem until completion of the initial railway. So difficult is the terrain that during construction bulldozers and other heavy equipment, food supplies, and people had to be flown into the area. The 360-mile (575-km)-long railway follows the winding Moisie River for part of the way from the mines at Schefferville to Sept-Iles on the shore of the Gulf of St. Lawrence.

Four principal mining centers were developed. *Schefferville,* at the end of the long railway from the major ore port of Sept-Iles, was the first and largest. *Labrador City–Wabush* is connected by spur to the same railway line. *Gagnon,* farther southwest in Quebec, is connected by a 200-mile (320-km) rail line to Port Cartier on the St. Lawrence. Farther east are the massive ilmenite (titanium) deposits of *Lac Allard,* which are sent by rail 25 miles (40 km) to Havre-St.-Pierre for shipment (Fig. 19-17).

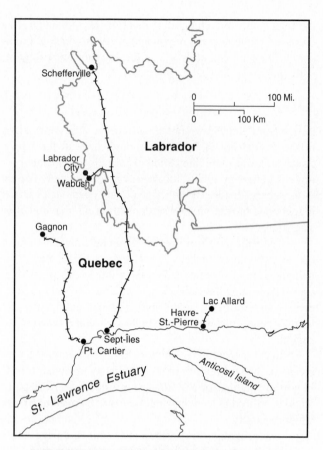

FIGURE 19-17   The Labrador Trough area, showing mining towns, rail lines, and air ports.

As with other iron ore producers in America, the Labrador Trough mines have been plagued by falling demand and high costs in the last few years. As a result, the mines at Schefferville were closed, presumably forever, and output decreased at the other three locales.

Sudbury–Clay Belt District   Considering the magnitude of the Shield, this is a district of rather closely spaced mining activity. Many metals are recovered, but mostly nickel, copper, and gold.

Modern mining began in Canada after the discovery of the Sudbury nickel–copper ores in 1883. It is the outstanding nonferrous metal mining center in the nation, producing about one-fifth of the world's supply of nickel and considerable amounts of copper, silver, cobalt, and platinum. Both open-pit and shaft mining are used, and most of the smelting is done locally. Fumes from the smelter smokestacks have denuded the nearby countryside of trees, giving a moonscape appearance that is only slowly being alleviated by intensive regreening efforts.

In the Clay Belt country straddling the Ontario–Quebec boundary is a major gold–copper zone that extends from Timmins on the west 200 miles (320 km) to Val d'Or on the east. Various mines open and close, but mining activity has been continuous for many decades. At the time of this writing, more than a dozen gold mines are in operation, and base metal production is expanding.

Two hundred miles (320 km) northeast of Val d'Or is the still-developing mining complex of Chibougamau. Lead, zinc, copper, silver, and gold are produced from nearly a dozen mines, the area being served by rail connections with both the Clay Belt and the Lake St. John lowland.

Northern Manitoba District   Although copper–zinc ore was discovered in 1915 at Flin Flon, exploitation did not begin until a rail line was built from The Pas in 1928. Flin Flon, a well-established center, is a major copper producer and also yields considerable zinc, gold, and silver.

Other producing centers are located northeast of Flin Flon and north of the railway line to Churchill. Snow Lake produces gold and copper. But the outstanding development is at the planned town of Thompson, a "Second Sudbury" 200 miles (320 km) northeast of Flin Flon, which began producing nickel in 1961. It is the world's second-largest nickel producer.

Isolated Mining Centers   At several other localities in the region are more isolated mines. However, they open and close with such frequency that any list or map is out of date within a few weeks. At the present writing there is much uranium activity in northern Saskatchewan, and three of Canada's newest large gold mines are in production near Marathon on the central north shore of Lake Superior.

Hydrocarbons   In the northwestern part of the region there are vast possibilities for production of mineral fuels. Coal of inferior quality is widespread in central Alaska, but the small population, lack of manufacturing, and limited railroad mileage keep production low. It is, however, important locally because transportation costs are high and the winters bitterly cold. Most coal mined in Alaska is from the Healy River Field not far from Fairbanks. Ladd and Eielson Air Force bases are the principal consumers of the coal.

Significant production of oil and gas has thus far come only from the Peace River district, where the Rainbow and Zama Lake oil fields are being intensively developed.

For many years it has been known that within the boreal forest of Alberta a fabulous petroleum potential is locked up in tar sands (Fig. 19-18). Recoverable reserves are estimated to be half as much as the total reserves of convention-

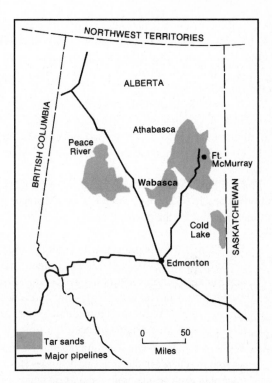

FIGURE 19-18   The major tar sands deposits of Alberta.

al crude oil in the entire world. There are four major tar sands accumulations, generally 200 to 250 miles (320 to 400 km) northeast and northwest of Edmonton.

Complex technological requirements and extraordinary capital investment costs have deterred development. It is both difficult and expensive to separate the oil from its matrix of imprisoning sand. Two small refineries are now in operation and others are under construction. The liquid petroleum is transported by pipeline to Edmonton. Fort McMurray is the center of activity; it is now a city of more than 50,000 people, making it the largest urban place in the western part of the region. Despite the high cost of production and various environmental objections, the ultimate economic success of these ventures seems ensured. Production is increasing and will become more significant as Alberta's conventional oil fields are depleted.

Productive Water Usage

Hydroelectricity is the principal source of electric power in Canada, accounting for about 70 percent of the total. More than half the installed hydroelectric-generating capacity of the nation is within the Boreal Forest Region. Hydroelectric power facilities have been critical to the success of most of the pulp milling, mining, smelting, and refining that occur within the region. In addition, a great deal of power has been transmitted southward for use in southern Quebec, southern Ontario, the prairies, southwestern British Columbia, and even the United States.

The continually increasing demand for power led to the near maximal development of hydroelectric power sources on the tributaries to the St. Lawrence in Quebec and Ontario, with the greatest development on the Ottawa, St. Maurice, and Saguenay Rivers.

Presently there are four major development schemes in operation, or nearly so, each of which is a great distance from the ultimate market for the power:

1. The Churchill Falls project in Labrador, which delivered its first power in 1972, was at that time the largest construction undertaking in Canadian history and is now the largest single-site hydroelectric scheme in the Western Hemisphere. Most of its power is used in southern Quebec.

2. The government of Quebec initiated construction of a massive scheme in the area east of James Bay. The total plan involves damming and/or diversion of six rivers, building two new towns, constructing several hundred miles of roads, building several wilderness airports, and stringing 10,000 miles (16,000 km) of transmission lines (Fig. 19-19). The first stage, in the watershed of the La Grande River, went into full operation in 1985. Native people (Cree and Inuit) in the area signed a contract approving the James Bay project in 1975 but now claim that the contract applies only to the first stage and are now objecting strenuously to the proposed second and third stages of the project.

3. In northern Manitoba a complex scheme is under construction on the Nelson River, supplemented by a large diversion from the Churchill River. Seven dams have been built and another 11 generating stations are proposed. Presumably the project will supply all the province's power needs until the end of the century and provide a surplus for export, but at a massive disruption of the environment.

4. The Peace River project in British Columbia, sends most of its power to the Vancouver area, nearly 600 miles (960 km) away.

Agriculture

Despite the unattractiveness of the Boreal Forest Region for agriculture, farmers were settling along the edge of the Shield in Quebec and Ontario by the middle of the nineteenth century. Farming settlements were attempted in many places, generally reaching a peak of expansion in the 1930s. Frontier farming on the edge of the Shield declined for the next four decades, although there was some agricultural expansion in the western portion of the Boreal Forest Region, notably in the Peace River district.

The Clay Belt and Peace River district, the only real agricultural areas, were discussed earlier. Otherwise farming is scattered and marginal, with a few minor concentrations in such places as the Delta Junction area southeast of Fairbanks and the Tanana Valley near Fairbanks. Most frontier farms emphasize beef production, and most of their cultivated acreage is in hay; other typical products are barley, oats, hogs, potatoes, and cool-season vegetables, such as cabbages and cauliflowers.

In the last few years there has been some expansion of fringe agriculture along the southern margin of the Boreal Forest Region from Manitoba to Alaska. Most of this activity does not involve the establishment of "new" farms; rather, it is the enlargement of already existing farming enterprises. As such, it does not imply pioneering in the traditional sense. A long-time researcher on the agricultural fringe has noted:

> The . . . fringe has lost much of its frontierlike appearance; increasingly it resembles established farming districts. . . . Many log cabins and crude barns [have been] replaced by more elaborate and substantial structures or sometimes with modern mobile homes. . . . The con-

FIGURE 19-19   The LG 3 reservoir spillway in the massive James Bay hydroelectric project in Quebec (courtesy Hydro Quebec).

version of raw land to cultivation . . . still involves a degree of experimentation. . . . Abandoned farms here and there attest that success is not guaranteed.[7]

SUBARCTIC URBANISM: ADMINISTRATIVE CENTERS AND UNIFUNCTIONAL TOWNS

Although cities are scarce in the Boreal Forest Region, intimations of increased urbanism can be recognized (see Table 19-1 for a listing of the region's largest urban places). The population continues to cluster as bush dwellers settle in small settlements, and residents of small settlements move to larger centers. The availability of wage-labor opportunities and the advantage of having a stable mailing address for government social service checks provide the major attractions, but the well-recognized amenities of town life are inducements in the Subarctic just as in Megalopolis.

TABLE 19-1

Largest Urban Places of the Boreal Forest Region, 1995

Name	Population of Principal City	Population of Metropolitan Area
Bemidji, MN	10,000	
Chibougamau, QU	11,000	
Dawson Creek, BC	13,000	
Duluth, MN	82,000	243,000
Escanaba, MI	12,000	
Fairbanks, AK	31,000	
Fort McMurray, AL	48,000	
Grande Prairie, AL	29,000	
Hibbing, MN	17,000	
Marquette, MI	21,000	
North Bay, ON	62,000	
Owen Sound, ON	24,000	
Prince Albert, SK	36,000	
Rouyn, QU	21,000	
Sault St. Marie, MI	15,000	
Sault Ste. Marie, ON	85,000	
Sudbury, ON	91,000	168,000
Superior, WI	28,000	
Thompson, MN	14,000	
Thunder Bay, ON	116,000	129,000
Timmins, ON	46,000	
Traverse City, MI	16,000	
Val d'Or, QU	21,000	
Wausau, WI	37,000	120,000
Whitehorse, YT	18,000	20,000
Yellowknife, NWT	15,000	

[7] Burke G. Vanderhill, "The Passing of the Pioneer Fringe in Western Canada," *Geographical Review*, 72 (April 1982), 217.

Most urban places in the region are essentially unifunctional, depending largely on a single type of economic activity for their livelihood. In some cases, they were heterogeneous bush towns that grew haphazardly around a mine, a mill, or a transportation crossroads. The modern mining towns, however, are planned communities, designed for a specific size.

Two of the three largest urban centers in the region, Duluth–Superior and Thunder Bay, are Lake Superior ports whose well-being is intimately associated with the shipping of bulk products on the Great Lakes (Fig. 19-20). The twin cities of Sault Ste. Marie, Ontario, and Sault St. Marie, Michigan, have a crossroads function, although the big steel mill in the former center provides another dimension to its economy. Sudbury is an example of a mining–smelting town that grew into a subregional commercial center. Other places that serve significantly as somewhat diversified subregional centers include Timmins and Val d'Or in the Clay Belt, North Bay (Ont.), Fort McMurray (Alta.), and Fairbanks. Smaller mining or mining–smelting centers, such as Thompson and Hibbing, are more clearly unifunctional.

It should be noted that income from government sources, in the form of both wages and social service payments, also significantly contributes to the local economy for many urban places in the region. Whitehorse and Yellowknife have become territorial capitals, whereby their economy has profited. Many other towns in the region also serve as modified administrative centers, with district or sub-

FIGURE 19-20 The prime function of Superior, Wisconsin, is indicated by this huge grain elevator, where grain from the northern Great Plains is stored prior to shipping down the lakes (TLM photo).

district offices of various government agencies headquartered there. Often the economic base for small, subarctic settlements can be summarized as "government-supported economy supplemented by trapping." The wisdom of diffusing the administrative infrastructure into communities that no longer have a viable economic base, thereby perpetuating stagnant communities that often have serious economic and social problems, continues to be questioned.

TRANSPORTATION: DECREASING REMOTENESS AND INCREASING ACCESSIBILITY

Transportation difficulties have often been cited as the principal deterrent to economic prosperity in the region. With increasing technological sophistication, however, it is now clear that mere remoteness is no longer a major handicap. If the economic prize is sufficiently promising, transportation can be provided.

Early transport in the region was by canoe in summer and dog team and sledge in winter. The maze of waterways with only short portages enabled the canoe to go unbelievable distances, and the accumulation of snow, because of the absence of thaws in winter, made sledging relatively easy.

Water transport became notable from the earliest days of European penetration of the region, utilizing the long rivers, large lakes, and intricate network of streams. The Quetico area (that part of Ontario and Manitoba between Lake Superior and the edge of the prairie) was traversed by a varied flotilla of watercraft, for instance, and "during certain periods, the volume of traffic made this one of the busiest regions of interior North America."[8]

The Yukon and Mackenzie Rivers were heavily used waterways, and the latter is still used by barges. Many other rivers and lakes had, and continue to have, considerable usage, but the Great Lakes are the major waterway of commercial importance. Their traffic consists almost entirely of bulk products—iron ore, grain, and limestone downbound, and coal upbound—although considerable St. Lawrence Seaway shipping traverses the Huron–Michigan route destined for Chicago.

Canada's two transcontinental railway lines cross the Shield through Ontario, providing critical transport links for the nation and encouraging mineral and forestry exploitation. Two rail lines were built across the region to Hudson Bay, reaching Churchill in 1929 and Moosonee in 1932. Most other railway construction in the region was designed as feeder

lines to connect mining localities with the outside world: the Labrador Trough lines, Chibougamau, the Clay Belt, the various Manitoba mining centers, and the Great Slave Lake Railway to Pine Point and Hay River.

In the far northwest are two railway routes of different origin. The *White Pass and Yukon Railway* was built in 1898 during the feverish boom days of the Klondike. It is 111 miles (178 km) long and extends from Skagway over White Pass to Whitehorse. The *Alaska Railroad,* extending 470 miles (750 km) from Seward to Fairbanks, was built by the U.S. government to help develop and settle interior Alaska. It was completed in 1923.

Although vital to the economy of their local areas, neither route is profitable. Indeed, the former ceased operating in the early 1980s. The Alaska Railroad is still owned by the federal government, but negotiations are underway to transfer its ownership to the state of Alaska.

The relatively few miles of roads and highways in the region are being rapidly expanded. The Upper Lakes area and Clay Belt have been fairly well served for years, but only recently has roadway construction been accelerated elsewhere. The *Alaska Highway,* extending 1500 miles (2400 km) from the Peace River district to Fairbanks, has been an important connection ever since its construction in 1941–1942; it is an all-weather road. The *Mackenzie Highway* is a 650-mile (1040-km) link from the Peace River district to Great Slave Lake, extending as far north as Yellowknife. The *Dempster Highway* (Fig. 19-21) has been built from the central Yukon

FIGURE 19-21   A typical scene along the Dempster Highway in the Yukon Territory near Dawson City (TLM photo).

[8] Bruce M. Littlejohn, "Quetico Country: Part I," *Canadian Geographical Journal*, 71 (August 1965), 41.

northward to the Mackenzie delta, but a planned roadway down the full length of the Mackenzie River in the Northwest Territories has been suspended, presumably awaiting settlement of native land claims in the area. In Alaska an all-weather road, the *Dalton Highway,* has been built along the route of the Trans-Alaska Pipeline from Fairbanks to the Arctic coast; in the mid-1990s it was opened to travel for the general public. In addition, roads usable only in winter have been bulldozed to provide seasonal access in many remoter localities.

Perhaps the most important development in the history of the region was the introduction of the airplane in about 1920. It revolutionized communications and accelerated the pattern of economic and social progress. There are increasing networks of scheduled airline service, particularly in the Mackenzie Valley and central Alaska, and nonscheduled bush-pilot flying is significant. Light aircraft can be equipped with pontoons, skis, nosewarmers, and other devices to permit operation into almost any area in both summer and winter. Short TakeOff–and-Landing (STOL) modifications make it possible to use mere puddles or meadows as an airstrip, and helicopters do not even require that much room.

"Personal" transportation in the bush was greatly enhanced by the proliferation of motorized ATVs (all-terrain vehicles) and snow machines. Although expensive to purchase and operate, these machines provide speed and flexibility never before available.

TOURISM

The recreational activities of the region are of two principal types: brief winter and summer visits from nearby population centers and more extensive expeditions involving considerable travel.

A number of major population centers lie just beyond the southern margin of the region, and short-term visitors from these areas constitute the major component of tourism in the Boreal Forest Region. The Upper Lakes area is within driving distance from many midwestern cities whose summer weather is sufficiently uncomfortable to urge "North Woods" vacations on its populace. Thus many Detroiters visit northern Michigan; people from Chicago and Milwaukee go to northern Wisconsin; and Twin Cities residents travel to the Minnesota north country.

A similar situation prevails for urbanites in southern Canada. Principal areas of attraction include the Laurentides Park area, particularly for the people of Quebec City; the Central Laurentians, favored for skiing as well as summer activities, especially by Montrealers; Algonquin Park near

Toronto, notable for fishing and canoeing; the lake country of southern Manitoba and Riding Mountain National Park, for Winnipeg; Prince Albert National Park near Saskatoon; and Elk Island National Park, a wildlife preserve within picnicking distance of Edmonton.

The abundant fish and wildlife resources of the taiga are major attractions for hunters on both sides of the international boundary. They make long trips, frequently by air, to fish in remote lakes for trout, whitefish, pike, perch, and muskie, and to hunt moose, caribou, bear, and deer.

NATIVE LAND CLAIMS IN THE NORTH

The native inhabitants—First Nations, Inuit, Aleut, and Métis—of the North have emerged from decades of relatively unobtrusive docility to assert, with increasing stridency, claims for much greater control of their own social, economic, and political destiny. These claims focus particularly on land rights and have been unusually effective in attracting attention in both Ottawa and Washington because they have arisen at least partially in direct opposition to previously discussed energy developments.

Judge Berger stated with reference to the Mackenzie Valley:

[Our] society has refused to take native culture seriously. European institutions, values and use of land were seen as the basis of culture. Native institutions, values and language were rejected, ignored or misunderstood and ... [we] had no difficulty in supposing that native people possessed no real culture at all. Education was perceived as the most effective instrument of cultural change: so, educational systems were introduced that were intended to provide the native people with a useful and meaningful cultural inheritance, since their own ancestors had left them none.[9]

Native people constitute the bulk of the permanent population in the North. Although a white population, in general, is permanent, a large proportion of individual whites are only temporary. Thus it seems logical that the future of the North should be determined to some degree by its permanent residents. Such is the underlying thought behind native claims.

The first "settlements" of claims were made in the early 1970s. Since then, however, the nature of the claims has

[9] *Northern Frontier, Northern Homeland: The Report of the Mackenzie Valley Pipeline Inquiry,* Vol. 1 (Ottawa: Minister of Supply and Services Canada, 1977), p. xviii.

become much more complex, and clearly native people are now seeking a fundamental reordering of the relationship between native and non-native components of contemporary society.

After seemingly endless negotiations, the seven "comprehensive" (as opposed to "specific" claims that deal with relatively small areas claimed by individual tribes) claims in the North have either been settled or are very close to settlement at this writing. (It should be noted that several of the "specific" claims, especially in British Columbia, may become of major significance.)

Alaska Native Claims Settlement Act (ANCSA), 1971

Representatives of the Alaskan native populace (Native American, Inuit, and Aleut) formed the Alaskan Federation of Natives in 1966, which instituted lawsuits claiming native ownership of three-fourths of the state's 375 million acres (150 million ha) of land (Fig. 19-22). Congress, goaded by the fact that the litigation prevented construction of the Trans-Alaska Pipeline, passed ANCSA in 1971, remanding to Alaska's estimated 50,000 native people the title to 44 million acres (18 million ha) of land (equivalent in area to the state of Washington) and a cash settlement of nearly $1 billion.

This legacy is managed by an elaborate system of native corporations that hold title to the lands and receive the financial benefits. The state was divided into 12 regions on the basis of common native cultural heritage and interests and a regional corporation was established in each (a thirteenth corporation was added later for natives who had moved away from Alaska). In addition, some 250 village corporations were formed. Each native residing in a village and a region in 1971 automatically became a stockholder in that village corporation and that regional corporation.

Village corporations were entitled to select a stipulated acreage (prorated on the basis of population) of land relatively near the village; regional corporations were allowed to choose vast acreages within their regional boundaries, and they also own the subsurface mineral rights to all village land within their regions. The ultimate goal of ANCSA is assimilation of native people into the general society and capitalistic economy of Alaska. After the first two decades of effectiveness (1971–1991) there were no special rights or guarantees for the natives or their land. Meanwhile, regional corporations are using their patrimony in various ways in attempts to better the economic and social conditions of their stockholders. Some corporations have already lost great sums of money, whereas others have made remarkable profits. Throughout Alaska, however, the native people now have greatly expanded opportunities—as well as temptations.

James Bay and Northern Quebec Agreement, 1975

Some 10,000 natives (mostly Cree and some Inuit) who inhabit northern Quebec agreed to relinquish all claim to lands affected by the James Bay hydroelectric project in exchange for $225 million, exclusive right to 5000 square miles (12,950 km^2) of land, and hunting-fishing-trapping rights to another 60,000 square miles (155,000 km^2). In 1990 the

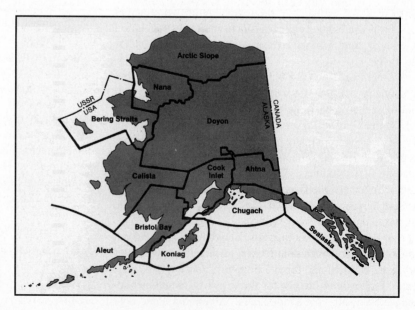

FIGURE 19-22 The areas served by the Alaska Native Regional Corporations.

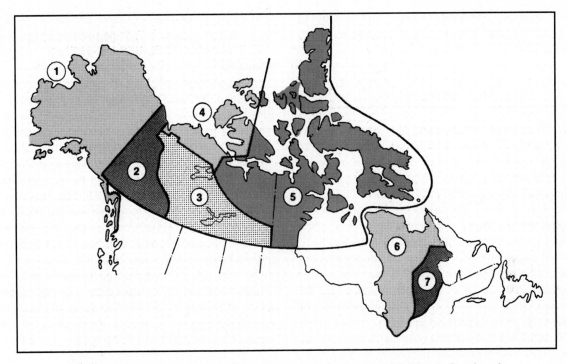

FIGURE 19-23   The areas encompassed by major northern native land claims in Canada and Alaska: (1) Alaska Native Claims Settlement Act (ANCSA), (2) Council for Yukon Indians, (3) Dene Nation/Métis, (4) Committee for Original People's Entitlement (COPE), (5) Tungavik Federation of Nunavut (TFN), (6) James Bay/Northern Quebec, (7) Northeastern Quebec.

Crees entered into litigation to invalidate the agreement, claiming that the governments of Quebec and Canada had not fulfilled all legal obligations (Fig. 19-23).

Northeastern Quebec Agreement, 1978

This is supplementary to the James Bay agreement and deals with the Naskapi tribe of the Schefferville area.

Western Arctic (Inuvialuit) Claims Settlement Act, 1984

This agreement was between the federal government and the Committee for Original Peoples' Entitlement (COPE), representing the Inuvialuit (native people) of the Western Arctic. The Inuvialuit received land, cash ($45 million), a $10 million Economic Enhancement Fund, wildlife harvesting and management rights, and Inuvialuit participation on advisory boards dealing with land-use planning and environmental management. About 4500 individuals are involved.

Dene Nation/Métis Agreement, 1990

This agreement provides the 8000 status First Nations and 5000 non-status First Nations and Métis of the Mackenzie Valley with surface rights and exclusive control over development on 70,000 square miles (180,000 km^2) and shared management of an area five times as large.

Council for Yukon Indians, 1990

The 7000 Yukon First Nations will receive entitlements similar to those of the Dene/Métis agreement.

Tungavik Federation of Nunavut, 1990

This federation represents the 15,000 Inuit of the central and eastern Arctic. An agreement-in-principle was signed in 1990, and final negotiations continue at the time of this writing.

It seems clear that extensive portions of subarctic and arctic North America are, or soon will be, governed at the regional and local level by public bodies controlled by native peoples. The implications of these developments are tremendous.

THE OUTLOOK

The Boreal Forest Region is primarily an empty area as far as human occupancy is concerned and logically will remain so. It represents, however, the northern frontier and as long as there are frontiers there will be people trying to push them back. Yet today many parts of the region are emptier than they were 50 or 100 years ago, as bushdwellers have increasingly congregated in larger, permanent settlements.

The region's population is youthful. The Northwest Territories and Yukon Territory have a smaller proportion of old people than any Canadian province, and Alaska has a smaller proportion than any of the other 49 states. This preponderance of young adults results in a high birthrate; the two territories have the highest in Canada and Alaska's is the highest in the United States. But the population is not a stable one. There is considerable seasonal fluctuation, with a summer peak and a winter ebb.

Many people enter the region for temporary summer jobs and then leave at the end of the season. Even many year-round residents, attracted by high wages in "hardship" posts, have come to the region on a temporary basis; they are sojourners, not settlers. Although improved transportation makes it easier for them to "go south" for a holiday, it also makes it easier to go south permanently.

For the region as a whole, mineral industries have been and will continue to be the keystone to the economy. More major ore deposits will undoubtedly be found and developed.

It is unlikely that there will ever be a second Sudbury, but more Scheffervilles and Thompsons can be anticipated, which only emphasizes the boom-and-bust syndrome that is characteristic of such activities and communities.

Large-scale energy developments—oil and gas exploration and production, oil sand development, pipeline projects, and dam building—will dominate the scene in many areas. The ramifications of these major projects are far-reaching and long-lasting. Now that native land claims in the Mackenzie Valley appear to be settled it seems likely that major transportation ventures will be initiated, perhaps including a highway, a gas pipeline, and an oil pipeline.

The role of native people should become increasingly significant in most parts of the region. Their political power is increasing and financial settlements of native claims are providing them with some economic leverage for the first time. Still, a high proportion of the native population will continue to be socially and economically disadvantaged. Elements of "white" material culture and values have been introduced throughout the region. In many cases, the native villages have neither the physical nor the organizational foundations to support these introductions, and the results are likely to be severe sociopsychological maladaptations. Many young-adult natives of the region suffer from an identity crisis that is often manifested as a generational conflict within the village or tribe.

As with so many "wilderness" regions, there will be an increasing emphasis on tourism and recreation. Genuine tourists will begin to appear where now only fishermen and hunters go. There is only a short time lag in North America between the completion of any fairly negotiable road and the appearance of motels, trailer camps, picnic tables, roadside litter, and new money.

SELECTED BIBLIOGRAPHY

"Alaska's Great Interior," *Alaska Geographic*, 8 (1980), entire issue.

BONE, ROBERT M., *The Geography of the Canadian North: Issues and Challenges*. Don Mills, Ont.: Oxford University Press Canada, 1992.

BRADBURY, JOHN H., AND ISABELLE ST. MARTIN, "Winding Down in a Quebec Mining Town: A Case Study of Schefferville," *The Canadian Geographer,* 27 (Summer 1983), 128–144.

BROSNAHAN, MAUREEN, "After the Inferno: Northern Manitoba Slowly Comes Back to Life One Year after the Worst Forest Fires on Record," *Canadian Geographic,* 110 (December 1990–January 1991), 68–75.

BROWN, R. J. E., *Permafrost in Canada*. Toronto: University of Toronto Press, 1970.

———, "Permafrost Map of Canada," *Canadian Geographical Journal,* 76 (1968), 56–63.

DARRAGH, IAN, "Gatineau Park," *Canadian Geographic,* 107 (December 1987–January 1988), 20–29.

ELLIOTT-FISK, DEBORAH L., "The Stability of the Northern Canadian Tree Limit," *Annals,* Association of American Geographers, 73 (December 1983), 560–576.

FAHLGREN, J. E. J., AND GEOFFREY MATHEWS, *North of 50°: An Atlas of Far Northern Ontario*. Toronto: University of Toronto Press, 1985.

FRENCH, HUGH M., AND OLAV SLAYMAKER, EDS., *Canada's Cold Enviroments*. Montreal: McGill-Queen's University Press, 1993.

FULLER, W. A. "Canada's Largest National Park," *Canadian Geographical Journal,* 91 (December 1975), 14–21.

GORRIE, PETER, "The Bruce: Our Newest National Park," *Canadian Geographic,* 107 (October–November 1987), 62–71.

HAMLEY, WILL, "Tourism in the Northwest Territories," *Geographical Review,* 81 (October 1991), 389–413.

HARE, F. KENNETH, AND J. C. RITCHIE, "The Boreal Bioclimates," *Geographical Review,* 62 (1972), 333–365.

KAKELA, PETER J., "The Superiority of Low-Grade Iron Ore," *Professional Geographer,* 33 (February 1981), 95–102.

KROTZ, LARRY, "Dammed and Diverted: Hydro Projects in Northern Manitoba Have Disrupted the Land and a Way of Life," *Canadian Geographic,* 111 (February–March 1991), 36–45.

LANTIS, DAVID W., DONALD MEARES, AND VALENE SMITH, "The Alaskan Bush in Transition: An Overview," *Yearbook,* Association of Pacific Coast Geographers, 51 (1989), 125–134.

LINTON, JAMIE, " 'The Geese Have Lost Their Way': The James Bay Hydroelectric Project Has Turned the Lives of Northern Quebec's Natives Upside Down," *Nature Canada,* 20 (Spring 1991), 27–33.

LIPSKE, MICHAEL, "Playing for Power in Quebec's North," *International Wildlife,* 21 (May–June 1991), 10–17.

McKAY, DONALD, *Heritage Lost: The Crisis in Canada's Forests.* Toronto: Cross Canada Books, 1985.

MEREDITH, THOMAS C., "The George River Caribou Herd," *Canadian Geographer,* 29 (Winter 1985), 364–366.

"The Middle Yukon River," *Alaska Geographic,* 17 (1990), entire issue.

OLSON, ROD, FRANK GEDDES, AND ROSS HASTINGS, EDS., *Northern Ecology and Resource Management.* Edmonton: University of Alberta Press, 1984.

ROBERGE, ROGER A., "Resource Towns: The Pulp and Paper Communities," *Canadian Geographical Journal,* 94 (February–March 1977), 28–35.

ROBINSON, J. LEWIS, *Resources of the Canadian Shield.* Toronto: Methuen Publications, 1969.

SALISBURY, RICHARD E., *A Homeland for the Cree: Regional Development James Bay, 1971–1981.* Montreal: McGill-Queen's University Press, 1986.

SEABORNE, ADRIAN A., AND PATRICIO N. LARRAIN, "Changing Patterns of Trade Through the Port of Thunder Bay," *Canadian Geographer,* 27 (Fall 1983), 285–290.

STRUZIK, ED, "Yellowknife and Whitehorse: Sister Cities North of Sixty," *Canadian Geographic,* 106 (June–July 1986), 24–33.

"The Tanana River," *Alaska Geographic,* 16 (1989), entire issue.

TYNER, GERALD E., AND JUDITH A. TYNER, "Tourism in Canada's Northwest Territories: Aspects and Trends," *The California Geographer,* 18 (1978), 137–149.

USHER, PETER J., "Unfinished Business on the Frontier," *The Canadian Geographer,* 26 (Fall 1982), 187–190.

VANDERHILL, BURKE G., "The Passing of the Pioneer Fringe in Western Canada," *Geographical Review,* 72 (April 1982), 200–217.

WELSTED, JOHN, JOHN EVERITT, AND CHRISTOPH STADEL, EDS., *The Geography of Manitoba: Its Land and Its People.* Winnipeg: University of Manitoba Press, 1996.

WILKINS, CHARLES, "East Meets West at Thunder Bay," *Canadian Geographic,* 111 (February–March 1991), 16–29.

WINTERHALDER, KEITH, "The Re-greening of Sudbury," *Canadian Geographic,* 103 (June–July 1983), 23–29.

WRIGHT, ALLEN A., "Yukon Hails Opening of the Dempster Highway," *Canadian Geographic,* 98 (June–July 1979), 16–21.

The Arctic

20

Enormous in size but sparse in population, the Arctic Region sprawls across the vastness of the northern edge of North America. In few parts of the earth is nature more niggardly, more unyielding, or more unforgiving, and nowhere else are people's ways of living more closely attuned to the physical environment.

This is primarily a region of the Inuit and the Aleut; yet they occur only in small numbers and in scattered settlements. Although parts of this region have been known to nonnatives for six centuries, only a few "outsiders" have come to the Arctic, and only a tiny fraction of that few have been more than visitors or sojourners for a relatively short period of time. Conversely, only a small number of Inuit or Aleuts have departed from the region on anything other than a temporary basis.

Such urban places as those in the midlatitudes are virtually nonexistent. In the more remote areas a settlement nucleus may include only a handful of families. Wherever they are and whatever their size, the settlements of the region are always dominated by the immensity of the environment. The inhabited places are separated by great distances of trackless and treeless land or by equally barren water or ice. Within a settlement, the buildings and artifacts of people take on a peculiarly aggressive significance. There are no trees or shrubs

to cover mistakes, provide transitions, or ease the exposed rawness. The sparse, slow-growing, unobtrusive vegetation survives with difficulty, and where it has been ripped away by human endeavor, the nakedness persists for a long time. Settlements are inevitably scars on the fragile landscape.

This, then, is a region in which nature thoroughly dominates humankind. Such items as ice thickness, windchill factor, permafrost depth, caribou migration route, hours of daylight, blizzard frequency, abundance of harp seal, and formation of fast ice are critical to human existence.

Conversely, what people do in this region has remarkably long-lasting effects on the environment. Although the surface of the land is rock-hard through the long winter months, it is extremely susceptible to the impress of human activities during the brief summer period. The structural fragility of the ground-hugging tundra plants and the spongy soil beneath them is such that any type of compression leaves a mark that is not soon erased. The scrape of a bulldozer blade will leave a scar for generations, the track of a wheeled vehicle will be visible for years, and even a single footprint may be obvious for months.

The Arctic Region includes the part of North America that extends from the Bering Sea on the west to the Atlantic Ocean on the east and from the Boreal Forest on the south to

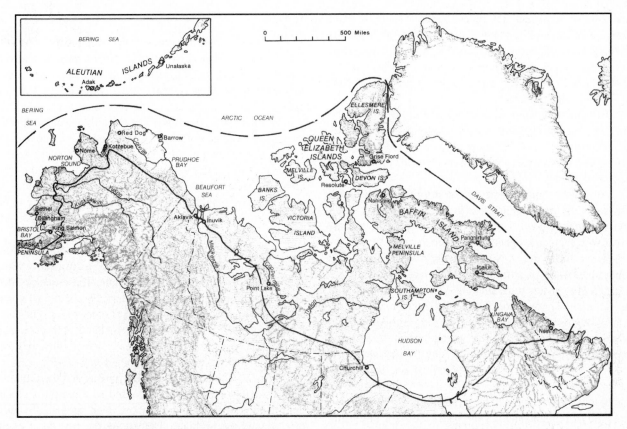

FIGURE 20-1 The Arctic Region (base map copyright A. K. Lobeck; reprinted by permission of Hammond, Inc.).

the Arctic Sea on the north; it also includes the vast Arctic Archipelago north of the Canadian mainland and the far-flung chain of Aleutian islands that arcs westward from the southwestern corner of mainland Alaska (Fig. 20-1).

Although some basic characteristics of the region are broadly uniform, there are many aspects of heterogeneity. It is possible to generalize validly about the region, but, as in any extensive region, many exceptions and variations must be considered. Particularly notable in this respect is the fact that the eastern and western extremities—coastal Labrador in the east and the Bering seacoast–Aleutian Island area in the west—have a pronounced orientation toward commercial fishing that is quite unlike the situation over most of the region. This is a function of the availability of exploitable marine resources, which is governed particularly by climatic differences.

As a remote and unpopulous region, the Arctic has generally received little attention from most citizens of Canada and the United States. Lately, however, it intrudes into our consciousness more often. Vast energy resources are being discovered; environmental and ecological concerns are being raised; political, social, and economic demands of the native people are being voiced more stridently. Public interest in the Arctic Region is rising, and both Washington and Ottawa have taken note.

THE PHYSICAL SETTING

Climate

The Arctic is not, as novelists would have it, a land of perpetual ice and snow. Winter temperatures are low, but they are higher than in the taiga to the south. Point Barrow has yet to record winter temperatures as low as those characterizing certain stations in North Dakota and Montana. Nevertheless, most of the time over most of its area, the Arctic is cold. It is a global heat sink—a low-energy environment with mean annual temperatures around or below freezing and winters that are spectacularly frigid and long.

Temperature extremes become greater south of the coast, for the country increases in altitude and is more remote from the ameliorating effects of the ocean. Thus the temperature range at Allakaket, 350 miles (560 km) south of Point Barrow, is much greater; whereas the lowest and highest temperatures at Point Barrow are $-56°F$ ($-49°C$) and $78°F$ ($26°C$), respectively, those at Allakaket are $-79°F$ ($-67°C$) and $90°F$ ($32°C$). But everywhere in the Arctic Region winters are long and summers short.

Although the temperature range at Point Barrow is the more limited, the growing season is only 17 days; at Allakaket, however, it is 54. Snow may be absent for 2 to 4 months. The growing season along much of the Arctic coast is less than 40 days.

Air at low temperatures cannot absorb or retain much water vapor, so the precipitation is light and varies over much of the region from 5 to 15 inches (12 to 38 cm). In the far east and far west there are higher totals, but part of the High Arctic is the most arid area of North America; most of Ellesmere Island, for example, is a frigid desert that receives less than 2 inches (5 cm) of moisture annually. The precipitation that does fall in the region is mostly fine dry snow or sleet.

Winds, especially in winter, are very strong and frequently howl day after day. They greatly affect the sensible temperature; thus on a quiet day a temperature as low as $-30°F$ ($-34°C$) is not at all unpleasant if one is suitably clothed, but on a windy day a temperature of zero ($-18°C$) may be quite unbearable. The wind sweeps unobstructed across the frozen land and sea and packs the snow into drifts so hard that they often take no footprints, and no snowshoes are required for human locomotion. Winter windchill not only discourages people and animals from moving about but also significantly contributes to the slow growth of plants. But it is not always cold. The long daylight hours of summer combine with continual reflection off water surfaces to produce heat that can occasionally become intolerable.

The coastal areas of Labrador, the Bering Sea, and the Aleutian Islands experience widespread overcast conditions, considerable fogginess, and more storminess than other parts of the region. The relative warmth of the Aleutians contributes to much heavier precipitation there; persistent mist and rain are characteristic. On the average, the Aleutians experience only 2 clear days per month.[1]

Perhaps the climatic phenomenon of greatest significance to humans is the seasonal fluctuation in length of days and nights. Continual daylight in summer and continual darkness in winter persist for lingering weeks, and even months, in the northerly latitudes.

[1] The region as a whole is noted for its inhospitable climate, but the Aleutian Islands area is particularly famous for its wild weather. Survivors of a visit to the Aleutians have been known to report that "the islands have two seasons: fog and storm." Native inhabitants of the Aleutians, however, claim that a more valid statement is "the Aleutians have two seasons: this winter and next winter."

Terrain

The gross topographic features of the Arctic Region are similar to those of other parts of the continent; only in relatively superficial details does the unique stamp of the arctic environment appear. Most typical are flat and featureless coastal plains, which occupy much of the Canadian central Arctic, as well as most of the northern and western coasts of Alaska. Prominent mountains occur in several localities. The massive Brooks Range separates the Yukon and Arctic watersheds in northern Alaska. The eastern fringe of arctic Canada, from Labrador to northern Ellesmere Island, is mountain-girt with numerous peaks over 6000 feet (1800 m), rising to the 8544-foot (2563-m) level of Barbeau Peak in the far north (Fig. 20-2).

Of special interest is the 1000-mile (1600-km) long chain of the Aleutian Islands, extending westward in a broad arc from the tip of the Alaska Peninsula. These treeless, desolate, fog-shrouded islands are essentially a series of volcanoes built on a prominent platform of older rocks. The chain contains about 280 islands, mostly small, but is more notable in that it has more than twice as many active volcanoes (46) as the rest of North America combined (Fig. 20-3).

The coastline adjacent to Baffin Bay and the Labrador Sea is notably embayed and fiorded as a result of glacial modification of the numerous short, deep, preglacial valleys that crossed the highland rim. Along the east coast of the three islands (Ellesmere, Devon, and Baffin) and Labrador are innumerable fiords, some of which penetrate inland for more than 50 miles (80 km). Offshore is a fringe of rounded, rocky islets called skerries.

FIGURE 20-2   Tundra flowers blooming during the brief summer in northernmost Manitoba. Hudson Bay is in the background (E. C. Williams/Visuals Unlimited photo).

FIGURE 20-3   Smoke wafts from the snow-covered crater of Mount Shishaldin on Unimak Island in the Aleutians (photo by Aeromap US, Inc.).

A number of large rivers flow into the Arctic Ocean and Bering Sea from the continental mainland, but the Arctic islands have no streams of importance. Most notable of the rivers are the Mackenzie and the Yukon, both of which form extensive deltas; that of the former river contains literally thousands of miles of distributary channels and as many as 20,000 small lakes.[2] Other major rivers of the region are the Kuskokwim and Colville in Alaska and the Coppermine and Thelon in the Northwest Territories.

Lakes, large and small, abound in the region, including the Arctic islands. For example, Lake Hazen (included in Ellesmere Island National Park Reserve), at latitude 82°, is 45 miles (72 km) long and 900 feet (270 m) deep.

Distinctive Topographic Features of the Arctic

Three types of distinctive landform features in the Arctic Region are limited to this harsh environment.

Icecaps   Icecaps and glaciers constitute less than 5 percent of the ground cover of arctic Canada, but many are quite large in size.[3] Baffin Island contains two icecaps that are larger in area than the province of Prince Edward Island, and the islands north of Baffin (Bylot, Devon, Axel Heiberg, and Ellesmere) each contain icecaps that are still more extensive in size (Fig. 20-4). Almost all these ice features have been diminishing in area and thinning in depth during the twentieth century, and the evidence of glacial recession is conspicuous.

[2] J. Ross Mackay, "The Mackenzie Delta," *Canadian Geographical Journal*, 78 (May 1969), 148.

[3] J. Brian Bird, *The Physiography of Arctic Canada* (Baltimore: Johns Hopkins Press, 1967), p. 23.

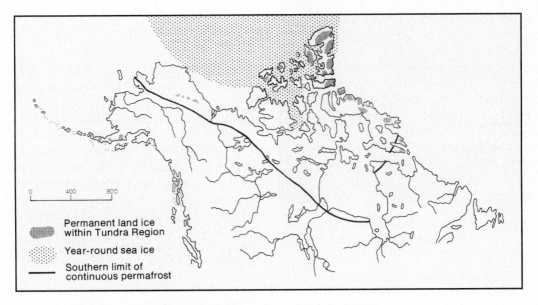

FIGURE 20-4   Ice conditions in the Arctic Region.

Ground newly uncovered by the retreating ice front is raw, light-coloured, unvegetated, in contrast to the ground beyond the trimline which was the position reached by the glacier at its greatest recent extent. Many small glaciers have completely disappeared since the turn of the century, larger ones have become 300 to 500 feet thinner and in some instances their snouts have retreated several miles.[4]

Raised Gravel Beaches　Adjacent to many portions of the present coastline are relics of previous sea levels. Characteristically, they appear as gravelly beaches that may extend inland for several hundred feet above the contemporary coastline. These subaerial beaches indicate emergence of the land after the great weight of Pleistocene ice sheets was removed by melting. The postglacial recovery of the land from ice depression varies from 100 (30 m) to as much as 900 vertical feet (270 m) in some places.

Patterned Ground　The most unique and eye-catching of Actic terrain is patterned ground; the generic name applied to various geometric patterns that repeatedly appear over large areas in the Arctic. The patterns, consisting of circles, ovals, polygons, and stripes, are of apparently varied but still unknown origin, although it is generally accepted that frost action is instrumental in their formation (Fig. 20-5).

FIGURE 20-5　Polygons of patterned ground on the tundra near Prudhoe Bay, Alaska (Steve McCutcheon/Visuals Unlimited photo).

A very distinctive type of patterned ground is the tundra polygon ... [which resemble] enormous mud cracks, such as those of a dried-up muddy pool, but with diameters of from 50 to 100 feet. The tundra polygons may be nearly as regularly shaped as the squares on the checkerboard, but most are irregular, somewhat like the markings on turtle shells. The boundary between two adjacent polygons is a ditch. Beneath the ditch there is an ice wedge of whitish bubbly ice which tapers downwards, like the blade of an axe driven into the ground. Some ice wedges are more than ten feet wide at the top and are tens of feet deep. ... On a smaller scale, the ground observer may see stones arranged in circles or garlands a few feet across, like stone necklaces; or the ground may have stripes trending downhill. ... Of particular interest to people in the western Arctic are the conical ice-cored hills called pingos, an Eskimo word for hill [Fig. 20-6]. The pingos are most numerous near the Mackenzie Delta, where there are nearly 1500 of them. The pingos may reach a height of 150 feet and so are prominent features in the landscape. They are found typically in shallow or drained lakes and are believed to have grown as the result of the penetration of permafrost into a thawed lake basin. Each pingo has an ice core of clear ice. If the ice core should melt, a depression with a doughnut-shaped ring enclosing a lake is left behind.[5]

The principal significance of patterned ground is that it demonstrates the mobility of tundra terrain, emphasizing the role of soil ice in producing geomorphic processes that are largely unknown farther south.[6]

Permafrost

Most, but not all, of the Arctic Region is underlain with permafrost. In the Aleutian Islands and part of the Alaska Peninsula it is unknown. In most of the region, however, permafrost is both continuous and thick. It has been measured to a depth of 1600 feet (480 m) in some places.

This means that water cannot flow at depth at any time of the year; thus surface water either must run off entirely or saturate the active layer above the permafrost. Such conditions make both wells and sewage lines impossible. Moreover, melting permafrost subsides unevenly, posing special

[4] J. D. Ives, "Glaciers," *Canadian Geographical Journal,* 74 (April 1967), 115.

[5] J. Ross Mackay, "Arctic Landforms," in *The Unbelievable Land,* ed. I. Norman Smith (Ottawa: Department of Indian Affairs and Northern Development, n.d.), p. 62.

[6] J Brian Bird, *The Natural Landscapes of Canada: A Study in Regional Earth Science* (Toronto: Wiley Publishers of Canada, Ltd., 1972), p. 160.

FIGURE 20-6   A pair of pingos near the northern edge of the Mackenzie delta. These ice-cored hills are conspicious in an otherwise featureless landscape (courtesy Canadian Government Travel Bureau).

problems for construction of buildings, roads, and other features.

Sea Ice

A distinctive feature of the Arctic environment is the presence of heavy sea ice for much or all of the year. Some areas—the southern Beaufort Sea, Hudson Bay, Hudson Strait, and Davis Strait—usually become ice-free for about 3 months in summer whereas the ice pack of most of the waters of the Central and High Arctic often fails to break up during the summer, and northwest of the Arctic Islands there is a permanent ice cover.

There are areas, called *polynyas* (after a Russian word for "clearing"), where open water recurs at the same place every winter. Their origin is not clearly understood; apparently they are formed by various combinations of currents, tides, upwellings, and winds. Polynyas are relatively rare and occupy only a small part of the entire Arctic waters, but they attract a wealth of sea mammals, and their presence has profoundly influenced human exploration and settlement patterns in the Arctic. The largest and best known polynya is

North Water, which is an area about the size of Lake Superior between southern Ellesmere Island and Greenland.

Natural Vegetation

Most of the Arctic as considered here refers to that part of North America lying north of the tree line, the great coniferous forest belt. The line on the map separating the tundra (*tundra* is a Finnish word meaning "barren land") from the forest symbolizes a zone within which trees gradually become smaller and more scattered until they disappear altogether. It coincides rather closely with the 50° (10°C) isotherm for the warmest month. This zone, in most instances, lies south of the Arctic Circle, even reaching as far south as the 55th parallel on the west side of James Bay (Fig. 20-7).

From area to area, however, the boundaries between taiga and tundra differ, and the extent to which the taiga penetrates the tundra seems to depend on a combination of low temperatures, wind velocity, and availability of soil moisture. The forest boundary extends farthest north in the valley of the Mackenzie River, where a forest of white spruce reaches into the southern part of the delta, at about the 68th parallel. There

FIGURE 20-7 An Arctic tundra scene in the Northwest Territories of Canada, with the Richardson Mountains in the background (William J. Weber/Visuals Unlimited photo).

are also significant, although not large, forest outliers in the valleys of the Thelon and lower Coppermine Rivers.

The characteristic vegetation association of this region is tundra. The tundra consists of a great variety of low-growing and inconspicuous plants that belong to seven principal groups:

1. Lichens, which grow either on rocks or in mats on the ground where they form the principal food supply of the migrating caribou herds
2. Mosses
3. Grasses and grasslike herbs
4. Cushion plants
5. Low shrubs
6. Dwarf trees
7. Flowering annuals

This is one of the world's harshest floristic environments, and the growing season is so short that there is simply not time during the brief summer for the life processes of annual plants to be completed. Instead, the plant cover consists of hardy perennials, which can remain dormant for as long as 10 months and then spring to life for an accelerated annual cycle during the abbreviated summer period. This means that arctic plants have a very slow weight increase. Studies of arctic willow *(Salix artica)* on Cornwallis Island, for example, show an annual increment of only one-third of total plant weight; in the midlatitudes such an increase can take place in a week.[7]

In the drier parts of the region [and some areas have an annual precipitation of less than 2 inches (5 cm)] conditions for plant growth are even more trying. Vegetation "is restricted largely to poorly drained areas, such as small depressions, or to areas where remnant snow patches provide local seepage during much of the melt season."[8]

However limited the tundra vegetation may be, it is of critical importance to animal life. It nourishes the entire terrestrial food chain, from mosquitoes and muskoxen to carnivores.

Native Animal Life

The biological regime is characterized by low species diversity, relatively simple food and energy chains, instability of populations, and slow growth rates. Many species have population cycles of dramatic amplitude. Although they must be hardy to survive the rigorous environment, at the same time they are vulnerable. The larger mammals, both marine and terrestrial, tend to travel great distances; either they migrate seasonally in herds or they wander as individuals over a large territory.

In no other region are the fauna so important to people. There has long been an intimate association between the abundance or scarcity of animal life and the welfare of the natives of the region. This close relationship is diminishing but is still pronounced.

Aquatic mammals have long been the mainstay of Inuit livelihood. Several varieties of *seals* range throughout the region and they are the most common quarry of Inuit hunters. The ringed seal is the most successful and widespread of the marine mammals because it can use breathing holes to live under fast ice (solid surface ice) all winter.

The *walrus is* a ponderous, slow-breeding creature (the female does not reproduce until the sixth year and then has only one pup every other year). It must live in an area of strong currents that keep the sea ice moving all winter, for unlike the seal it does not gnaw a breathing hole through the ice. The Atlantic walrus inhabits most of the eastern Arctic Ocean, with particular concentration around the Melville Peninsula and Southampton Island. The Pacific walrus inhabits the Bering Sea; there is a 1000-mile (1600-km) gap between the ranges of the two.

Whales of various species are much sought by the Inuit, but by far the most common is the white whale, or *beluga.*

[7] Patrick O. Baird, *The Polar World* (New York: John Wiley & Sons, 1964), p. 112.

[8] John England, "Ellesmere Island Needs Special Attention," *Canadian Geographic,* 103 (June–July 1983), 14.

They occur in considerable numbers throughout the Arctic, chiefly in saltwater but often going up the larger rivers (they have been seen as far up the St. Lawrence as Quebec City). They gather in remarkable abundance upon occasion; for example, up to 5000 cluster off the Mackenzie delta in summer. The *narwhal is* much less common but is highly prized because of the ivorylike horn of the male. Varieties of the large whales, such as the *bowhead,* are limited and sporadic in distribution and, with limited exceptions, are now fully protected.

The *polar bear* ranges widely in the Arctic, roaming primarily on sea ice except for land denning to give birth. They are sometimes found several tens of miles from the nearest land. There is an unusual denning concentration along the Hudson Bay lowland between James Bay and Churchill, where the Ontario government has established a large provincial park for their protection. Unlike other bears, which are omnivorous, polar bears are almost wholly carnivorous.

The outstanding land animal is the *barren-ground caribou,* which numbered in the millions a few decades ago (Fig. 20-8). There are several large herds, each of which makes an annual migration from taiga in winter to tundra in summer. Thousands of First Nations and Inuit have depended—some still do—significantly on caribou meat as a dietary staple. During the 1980s and 1990s there has been a significant increase in caribou numbers in much of Arctic North America, particularly in northeastern Quebec, where the George River herd is considered to be the world's largest.

The Arctic islands have a scattering of barren-ground caribou, particularly on Baffin Island. Several northern islands also contain a limited population of the smaller *Peary caribou.*

The only other native ungulate of the Arctic Region is the *muskox.*[9] In general, they are protected from hunters, and their numbers have been rebuilding from low points reached in the 1930s and 1940s. Prominent concentrations have long been found in the Thelon Valley of the Keewatin district and in the Lake Hazen area of northern Ellesmere Island. In recent years there has been considerable increase in numbers on several Arctic islands, particularly Banks Island. Muskox were exterminated in Alaska long ago but were reintroduced in the 1930s to Nunivak Island in the Bering Sea, where they have flourished enough to provide seed stock for reintroduction to the Alaskan mainland.

Furbearers of note include the *arctic fox,* a prolific breeder whose population seems to run in cycles; the *lemming,* a queer nocturnal burrowing rodent noted for its seemingly pointless migrations; the *arctic wolf,* principal large predator of the region; and the *arctic hare,* whose population pattern follows wildly fluctuating cycles.

Of the birds, the snow owl, ptarmigan, gyrfalcon, raven, and snow bunting are year-round inhabitants. Others, summer residents, arrive by the millions to breed in the seclusion and security of the tundra. They are also attracted by the prolific insect life.

Because of the abundance of poorly drained land, insects find this region a paradise during the short summer season. More than 1000 insect species occur north of the tree line, and about half of them belong to the order *Diptera* (two-winged flies), which includes mosquitoes, blackflies, and midges. According to one observer, "On a warm, cloudy,

[9] Moose and Dall sheep sometimes venture into the southern edge of this region, particularly on the north side of the Brooks Range.

FIGURE 20-8 A group of caribou using a wildlife crossing embankment over the Trans-Alaska Pipeline on the north slope of Alaska (photo by BP America, Inc.).

A CLOSER LOOK   Good-Bye Northwest Territories

The map of Canada is scheduled to lose one of its most historic names, with the disappearance of the "Northwest Territories" (NWT). Originally, the name was applied to the vast area acquired in 1870 from the Hudson's Bay Company and Great Britain (Rupert's Land and the North-Western Territory), which lay northwest of central Canada, to which the Arctic Islands were added by Great Britain in 1880. It thus made up most of Canada's national area. Since then large portions of the NWT have been carved off from time to time, to be added to individual provinces and to create new provinces and the Yukon Territory (Fig. 20-a) Nevertheless, the Northwest Territories of today still are the largest political subdivision within Canada (1.3 million square miles, or 3.4 million km^2), 34 percent of the national area.

The scheduled disappearance of this historic name reflects a unique situation among Canadian primary political subdivisions: Only in the NWT do aboriginal peoples form the majority of the population (some 55 percent of the total 57,649 in 1991) and also the majority of the elected members in the legislature. With devolution of government from federal to territorial control, aboriginal voters and aboriginal legislators in the Territories now are able to ensure that political and economic decisions reflect their values and their views rather than those of the dominantly white Canadian population at large. Their legal position has been strengthened further by the recent inclusion of "aboriginal rights" in the Canadian Constitution, even though definition of those rights remains to be established. Finally, after almost two decades of negotiations, comprehensive land claims of almost all the aboriginal peoples in the NWT have been settled with the Government of Canada, resulting in major acquisitions of financial and natural resources.

In this situation of newfound political strength, the aboriginal peoples are confronted with difficult decisions. They seek to preserve and strengthen their traditional culture, yet in many ways they have become integrated into the wider Canadian society at least as far as social services are concerned—educational and health services, transportation and communication links, modern housing, and the like. The only realistic way to pay for these services would seem to be exploitation of natural resources, essentially mineral, of the NWT, a process that is environmentally destructive to some degree at least, and thereby threatening to the traditional way of life of the local people. The fur trade, which for centuries was the major economic support of the aboriginal people, has been severely reduced by the animal-rights movement of the outside world, further compounding the problem.

The population of the NWT is the most youthful in Canada, with a median age of 25 in 1991, as compared with 34 for the national average. There is great urgency to provide employment for the young and increasingly better educated people. Although administration and

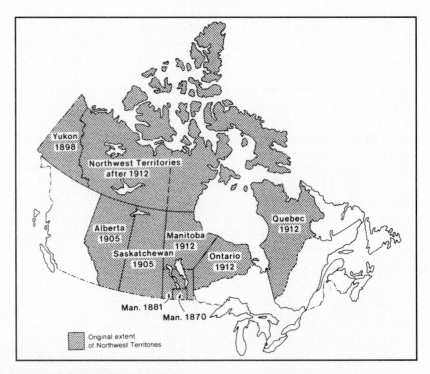

FIGURE 20-A   The changing boundaries of the Northwest Territories through the years.

windless day, the insect life on the Barrens defies description."[10] Happily such days are rare; most days are windy and cause the insect hordes to lie low.

[10] Eric W. Morse, "Summer Travel in the Canadian Barren Lands," *Canadian Geographical Journal*, 74 (1967), 162.

THE PEOPLE

This vast region has fewer than 90,000 inhabitants, two-thirds of whom are Inuit. Perhaps 6 percent are Aleut, and most of the rest are Caucasians. These people reside chiefly in widely scattered small settlements, of which there are

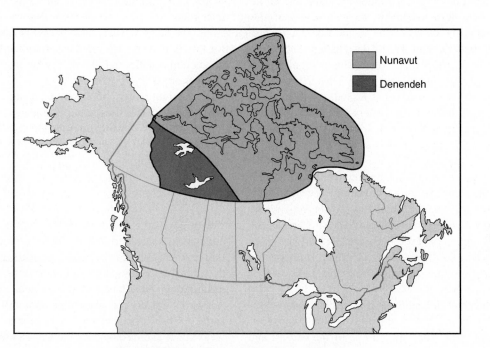

FIGURE 20-B The proposed territory of Nunavut.

service jobs exist, only resource development can increase employment opportunities sufficiently.

The Inuit of the Arctic believe that the only way they can achieve an acceptable balance between the realities of the modern world and their traditional cultural values is through creation of their own territory (Nunavut—"Our Land") but remaining a part of Canada. In 1991 the Inuit and the Government of Canada reached a lands claim agreement, the largest such in Canadian history, involving an area of over 772,000 square miles (2 million km²) (approximately one-fifth of the area of Canada, or twice the area of British Columbia, or larger than the total area of the three Maritime Provinces).

Under the terms of the agreement the Inuit will receive title to 135,000 square miles (350,000 km²) of this area, including 13,900 square miles (36,000 km²) of subsurface mineral rights, as well as hunting, fishing, and trapping rights over all Nunavut, and $1.15 billion over 14 years. Division of the Northwest Territories was narrowly approved by the territorial residents in 1992; in 1993 the Canadian Parliament approved the division and the official creation of the new Territory of Nunavut by 1999 (Fig. 20-b). (The Western Arctic areas were covered by a 1984 agreement with the Inuvialuit people and are not included in Nunavut.) In 1995 the Inuit voted to make Iqaluit (formerly Frobisher Bay), the largest community in Nunavut, the capital of the new territory. In 1991 its population was 3552.

While the 20,000 people (four-fifths of them Inuit) deal with their immense new territory the 37,700 people in the "other part" of the NWT include a much more diverse population of both aboriginal people and whites. Earlier cooperation between First Nations and Métis people splintered, with some parties reaching land claim agreements and others still not finalized. It remains to be seen how this more heterogeneous western area and the former NWT capital of Yellowknife (1991 population 15,179) deal with the complexities following territorial division.

Professor William C. Wonders
University of Alberta
Edmonton

about 50 in Canada and 130 in the Alaskan portion of the region. No settlement has a population of more than 4000, and only half a dozen exceed 2500.

The Inuit

The Inuit are one of those rare races readily identifiable by all three basic anthropological criteria: physical characteristics, culture, and language. They have well-marked physical homogeneity, a distinctive culture, and a language spoken by themselves and no one else.

The term "Eskimo" (or a translation thereof) has traditionally been used in all major languages of the world to designate the tundra dwellers of the North American Arctic and Greenland. The name, originally applied by Algonguian Indians to their northern neighbors, means "eaters of raw meat." In recent years the "Eskimo" people have increasingly sought a new generic term for self-designation. The choice was not easily reached, for there were different ethnic designations in common use in different areas. Thus in Greenland the "Eskimos" refer to themselves as Kalaallit; in the eastern Canadian Arctic, as Inuit; in the western Canadian Arctic, as Inuvialuit; in northern Alaska, as Inupiat; and in southwestern Alaska, as Yupik. These names all mean approximately the same thing—"the people." In the late 1970s a pan-Eskimo conference was held in Point Barrow for the purpose of choosing one official name for the "Eskimo" race and its culture. The choice was "Inuit," which has now been generally accepted by all native tundra dwellers of Alaska, Canada, and Greenland as a replacement for "Eskimo." It should be noted that "Inuit" is the plural form; an individual is referred to as "Inuk," whereas the "Eskimo" language is known as "Inuktitut. "

In their aboriginal condition the Inuit have shown remarkable ingenuity in adapting themselves to an almost impossible environment. They live in one of the coldest and darkest parts of the world and in one that is among the poorest in available fuel; yet they have not only survived but have also enjoyed life in self-sufficient family groups. Originally their entire livelihood depended on fishing and hunting, whereas practically all Eurasian tundra people were herders.

Inuit culture history revolves around successive waves of people and cultural innovations spreading eastward from the Bering Sea area. The earliest Inuit, or "proto-Inuit," of North America presumably derived their culture in Siberia and originally migrated from there. This culture stage—called the Arctic Small Tool Tradition or simple Pre-Dorset—had an indefinite tenure in Alaska but lasted in Canada until about 800 B.C., when it was replaced by the eastward-spreading Dorset Culture. After about 20 centuries the Dorset Culture was, in turn, replaced by the Thule Culture, which also spread eastward from Alaska.

Thule Culture had two principal attributes lacking in the Dorset era: domesticated dogs to aid in hunting and in pulling sleds and a full range of gear for hunting the great baleen whales, a major food source not available to Dorset people. Thule Culture evolved into the contemporary Inuit culture in about the eighteenth century, in a transition stage marked by the decline of whale hunting and the almost complete depopulation of the northern Canadian islands.

The coming of whites to Inuit country marked the beginning of the end for their way of life. European diseases, midlatitude foods, liquor, rifles, motorboats, and a new set of mores were introduced; the overall result was more often bad than good.

The last few years have been a period of rapid change for the Inuit. The birth rate has exploded; nearly half of the population is under the age of 15. Material and technological changes are striking. Virtually every home now has electricity, refrigeration, telephone, and television. Daily life in the North has become, by southern standards, much more comfortable and convenient. But jobs are scarce, and indexes of social disruption (alcoholism, domestic violence, suicides, etc.) are climbing.

The Aleuts

The origin of the Aleuts is unclear, although generally they are considered to derive from Inuit, or proto-Inuit, who settled in a maritime environment in southwestern Alaska and developed a livelihood based almost entirely on fishing and sea hunting. Aleuts occupied the Aleutian Islands and the Alaska Peninsula in considerable numbers (perhaps 25,000) at the time of Russian contact in the eighteenth century. They were killed and enslaved by the Russian fur seekers, and in relatively few years their number was reduced by 90 percent.

Today there are some 6000 Aleuts in southwestern Alaska, most of mixed blood. They are primarily commercial fishermen, sealers, and workers in salmon fishing and canning in the Bristol Bay area. In addition, they provide most of the labor for the fur-sealing industry of the Pribilof Islands, where the village of Saint Paul is the largest single Aleut settlement in existence.

The First Nations

There are fewer than 1000 First Nations in the entire Arctic Region and almost all are in Canada, where they are called *Dene*. Generally speaking, the tree line has served as the northern boundary of Dene occupance, just as it has served as the southern border of Inuit settlement. Dene and Inuit live adjacently in any numbers in only four localities: Aklavik and Inuvik in the Mackenzie delta, Churchill, and Poste-de-la-Baleine on the eastern shore of Hudson Bay. The Dene livelihood in this region is based on hunting, trapping, fishing, and temporary construction work.

The Whites

Whites living in the region now number some 35,000,—a total that is growing. Well over half are in Alaska, where Nome is the largest white settlement. Most whites who live in the region today are in control of the defense and weather installations or are government officials, oil company employees, prospectors, fur traders, fishermen, or missionaries.

THE SUBSISTENCE ECONOMY

Throughout history most of the population of the Arctic Region was involved in essentially subsistence activities: fishing, hunting, and trapping. After World War II there was a decline in these pursuits and their partial replacement by a money economy. Despite the downturn in subsistence activities, however, they still contribute significantly to the well-being of the people in providing "country food" (the name given to subsistence food obtained from the land or water), material for clothing and implements, and furs for sale or trade.

Although the exchange component (trapping and commercial fishing) of the native economy declined during the 1950s and 1960s, the domestic sector (hunting and fishing for local consumption) did not experience a long-term retrenchment except locally. Recent studies of Canadian Inuit show that three-fifths of households rely largely on hunting and fishing for their meat, five households out of six are involved in some form of hunting and/or fishing, and one in six is involved in trapping.

Hunting

Hunting for sea and land animals is widespread (Fig. 20-9). In some areas where winter conditions persist almost year-round, such as in the Queen Elizabeth Islands, ice hunting for sea mammals is a specialty. In other areas, such as the Colville River country of Alaska, inland groups specialize in caribou hunting. But most native settlements are on the coast and there are well-marked seasonal changes in hunting patterns. A typical "quarry rotation" might be seals in early fall, walrus in late fall, bear and fox in winter, seals in spring, fish and beluga in early summer, and caribou in later summer. In hunting areas, late winter and spring are the most important seasons for resource use. Midwinter is too dark and cold; ice and fog handicap long trips by small boat in summer; and autumn is stormy and newly forming ice is hazardous.

Some animals, such as caribou, birds, and hares, are hunted on land. Others, such as walrus and narwhal, are taken from boats. In many cases, the most important hunting is done on the sea ice, largely for seals and polar bear.

Modern technology has made it much easier for the Inuit to hunt. Repeating rifles, improved ammunition, snow machines, all-terrain-vehicles, and motorboats have provided greater mobility and killing power. The long-term result has been a decline in the number of potential quarry.

Fishing

Subsistence fishing is carried on wherever possible. Stone fish traps used to be constructed along some streams, but the vast majority of all fishing has always been done in the ocean. Summer fishing is easier, but ice fishing also takes place throughout the winter. Subsistence fishing is less important now than in the past because sled dogs, which are major consumers of fish, are much more scarce.

FIGURE 20-9   Butchering a walrus on St. Lawrence Island in the Bering Sea (Steve McCutcheon/Visuals Unlimited photo).

Trapping

Besides supplying meat and clothing, trapping has long been the major source of money income for natives of the Arctic Region. Various animals are trapped. Most often caught is the muskrat (about 100,000 are trapped annually in the Mackenzie delta alone), but the most important is the white arctic fox. The most reliable fox-trapping areas in the Arctic are in the southwesternmost portion of the Arctic Archipelago, particularly on Banks Island but also on Victoria Island. Other major trapping areas include northern Labrador, the Grise Fiord area of southern Ellesmere Island, and the deltas of the Yukon and Kuskokwim Rivers.

Trappers are licensed and carefully regulated in both countries. Only natives are allowed to trap in the Canadian Arctic, and only a very few white trappers are permitted in Alaska.

Trapping in the Arctic Region has been on a general decline for several decades, but in the 1980s it decreased dramatically. Humane, animal-rights, and antitrapping advocates mounted an international campaign that destroyed virtually the entire seal-fur market and had a lesser but notable effect on trapping of other furbearers (Fig. 20-10).

The ability of Inuit and Dene to rely on trapping as a source of income has been severely compromised, but trapping, hunting, and fishing continue as important subsidiary activities for a considerable proportion of the native population.

THE RISE OF A MONEY ECONOMY AND AGGLOMERATED SETTLEMENTS

The most important trend in the contemporary geography of the Arctic Region is the increasing concentration of the pop-ulation in fewer and larger settlement centers where life is based on a money economy. This pattern is being followed throughout the region. Wage labor is now the preferred means of livelihood, and more and more families are abandoning their seminomadic hunting camps and settling down to a sedentary existence in a settlement node where the younger generation grows up relatively ignorant of the techinques requisite to a subsistence economy.

As a result, large areas of the tundra, such as the Keewatin Barrens, the Colville delta, and much of the northern Labrador coast, are now almost unpopulated. Dozens, and sometimes hundreds, of Inuit have moved into settlements that previously had very small populations. The possibility of a steady job and the lure of stores, medical facilities, and perhaps housing are proving irresistible (Fig. 20-11).

Unfortunately, the rate of natural increase in the population is very high—generally the highest on the continent—and the supply of jobs is inadequate. Construction work has been the principal provider of wage labor in the past. During and after World War II there was a great deal of construction of defense and meteorological installations: radar stations, communications bases, and airfields. Such construction activity inevitably decelerated, but it gave the people of the region a taste of sedentary regularity that they cannot forget.

There are few "steady" jobs for the local people, except in government. Indeed, government is by far the largest enterprise in today's Arctic. Throughout the region there is a very limited resource base for employment, although there may be for food.

Population clusters have clearly been a mixed blessing. They made it possible for the Inuit to enjoy generally improving housing, a well-developed school system, access to medical facilities and social services, rudimentary community government, opportunities for the growth of local cooper-

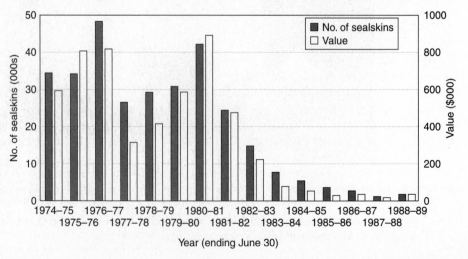

FIGURE 20-10 Sealskin production and value to Arctic communities, Northwest Territories, 1974–1989. The abrupt downturn of the early 1980s reflects the crash of the international market for sealskins (source: Government of the Northwest Territories, 1990).

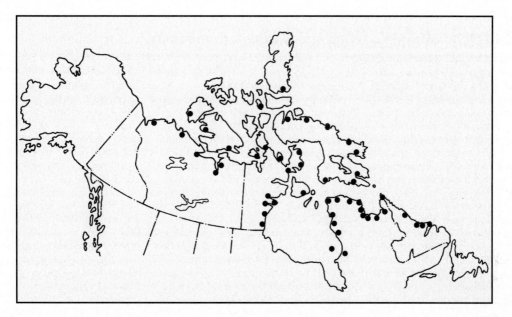

FIGURE 20-11   The distribution of Inuit settlements in Canada.

atives, introduction to "outside" amenities (ranging from A&W Root Beer stands to color television brought by satellite transmission to communities that had never known newspapers or magazines), and an awakening cultural awareness.

Negative effects, however, are also notable. Agglomerated living with limited employment opportunities and the incursion of outside temptations has caused all indexes of social pathology to skyrocket. Intrafamily violence, divorce, mental illness, suicide, and drug abuse are increasing rapidly, and alcoholism has become a ubiquitous menace. Several dozen communities across the region have enacted prohibition regulations, but even most of these settlements have major problems with alcoholism. And despite access to improved medical facilities, health problems have increased dramatically. Many settlements are hygienically as well as socially grim. Providing water and removing waste are problems almost everywhere. Dependence on packaged foods has dramatically increased the incidence of tooth decay and malnutrition.

The agglomeration of population and the consequent dependence on modern technology result in a much higher cost of living. Purchased food is expensive and much of it must be air-freighted in, which adds markedly to the cost. But the greatest expense is for power and fuel. Most Arctic settlements are overwhelmingly dependent on oil—a necessity for fueling home stoves, heating dwellings, powering village

electric generators, and propelling the ubiquitous snowmobiles and motorized boats.[11] In many areas more than half the family income goes for electricity and fuel costs.

Associated with agglomeration has been the growth of cooperatives. Prior to 1960 all Inuit art in Canada was marketed through Hudson's Bay Company or the Department of Indian Affairs and Northern Development. Since then cooperatives have proliferated so that they are now the largest single employer of Inuit labor in the Canadian Arctic. In addition to the production and marketing of arts, crafts, and sculpture, they are often involved in providing municipal services, construction, commercial fishing, retailing, and even mining.

Cooperatives continue to be vital economic and social components for almost all Canadian Inuit communities, but during the 1980s their business, and thus their vitality, went into a serious slump that is cause for great concern in the region. Cooperatives are much less common in the Alaskan tundra, primarily because each village has a corporation of its own under the Alaska Native Claims Settlement Act.

[11] One of the unsavory ironies of life in Alaskan villages is that the oil they use makes a 5000-mile (8000-km) round trip before it is received. Pumped from the North Slope fields on the shore of the Arctic Ocean, the crude oil is taken to California for refining and then the heating oil is returned to Alaska for sale.

A CLOSER LOOK   Making a Living in a Northern Community

For many northern native peoples, earning a living "from the land"—through hunting, fishing and trapping—is still vitally important. Being able to hunt well, supplying your extended family and others with food, is a matter of pride, with social and cultural importance, but also fundamental economic importance. Research in the Northwest Territories has shown that the average harvester produces the equivalent of $10,000–15,000 per year in "country food," fuel, and materials just as *domestic income* (that used for home consumption rather than sale). That is quite a "wage," especially in communities where few other cash-paying jobs are available.

Unemployment rates are very high in northern communities; in the Northwest Territories, for example, unemployment rates for aboriginal peoples can be 30 percent, compared with 6 percent for non-natives. This difference may be related to educational or skill levels but also to remote locations and preferences (Fig. 20-C). It must be noted that unemployment rates measure only participation in the *formal economy* (wage-paying or cash-based). Those who work in the home, in subsistence activities or in a barter network, for instance, are invisible in such data.

The problem is that the traditional harvest brings in little cash income (especially as antisealing or antitrapping campaigns produce their effect), and living in communities in the 1990s requires increasing amounts of cash to pay for utilities, hunting equipment, store purchases of food, laundry detergent, CDs, t-shirts—all the things one is used to buying in southern North America and that come with communi-

ty life and exposure to southern-style consumer value (Fig. 20-D).

The government of Canada encouraged native peoples to move into communities, roughly 40 to 50 years ago (and many Inuit have lived in communities a much shorter time), with the assumption that they would adopt the lifestyle and work style of southern Canada. But the work-style revolution, particularly, has not occurred as planned. As noted above, there are few jobs available in most northern communities, and furthermore, they are often filled by workers from the south.

Even when full-time work is available, experience from the oil booms in Alaska and the Northwest Territories has shown that native peoples may still pre-

fer to work for a period of time only, earn the cash they require, then go back to hunting, fishing, or trapping. Finally, jobs in the oil and gas or mining sectors can be a "boom-and-bust" experience, giving people large cash incomes for a period of time, then disappearing as world prices change or as resources become depleted.

Is there an alternative to this way of life? Rather than relying only on southern-style industry, perhaps development plans should include the traditional lifestyles, resources, skills, and values that have always supported native northerners as the foundation for community economies. If other kinds of industrial or non-renewable resource development come along, that's gravy.

FIGURE 20-C   The village of Pond Inlet at the northern tip of Baffin Island typifies the remote location and difficult environment of most northern communities (photo by Heather Myers).

One of the most important factors in the change from a subsistence to a commercial economy in the Arctic has been the adoption of personal motorized travel vehicles (Fig. 20-12). Generally called "snowmobiles" in Canada and "snow machines" in Alaska, these small, versatile, noisy vehicles have displaced dog sleds and sled dogs throughout most of the region. A snowmobile can cover 100 to 150 miles (160 to 240 km) in a day in contrast to 20 or 30 miles (32 to

48 km) by a dog sled. As the population increases, nearby game becomes scarcer and so the distance needed to travel for even the basic necessities of hunting or trapping becomes much greater. Moreover, the need to find food for hungry dogs all year is obviated, although that of finding cash to buy fuel becomes acute. In some areas snow machines are being replaced by all-terrain vehicles, which provide the flexibility of year-round transport. And in a few localities dog sleds

In the meantime, well-managed renewable resources could provide both domestic income and some cash income.

Some early development projects in the north attempted this, for example through reindeer herding, fisheries, or crafts production. Their success was variable, partly related to their emphasis on non-native business assumptions and practices, rather than native values or preferences.

Increasingly now, native entrepreneurs are trying out their own ideas, for instance, in forestry, fisheries, country food, tourism, sport hunting, and sport fishing. Examples of these northern businesses include small community sawmills, which serve local needs and perhaps a larger nearby mill; native-style tourist camps; fruit and berry preserving; fish smoking or canning; sea-cucumber collecting for the Japanese market; native women's groups tanning moose skins for local and regional craftswomen; boat and snowmobile tours of northern areas; eiderdown collection and duvet production; preparation of processed meats from the country food harvest, such as corned muskox, seal pastrami or caribou jerky.

These enterprises are often small, seasonal, and flexible enough to complement other activities and obligations, such as hunting or child-minding—in fact, these characteristics may enhance the success of such businesses. They often do not generate large incomes—perhaps a few thousand dollars a year—but they can provide some necessary cash that complements their participants' domestic income. They are also the foundation for future economic development, because they provide business experience, investigate market potentials, and generate local pride.

Many people in North America now do not have jobs, or full-time jobs, whether because of choice or availability. Perhaps we need to reevaluate the economic paradigms advocated in the north. There, in many places, a traditional economy still exists; rather than instituting southern patterns and expectations, should not employment and development options be adapted to northern resources, cultures, and existing economic strengths?

This is an exciting opportunity to experiment with alternative economic approaches, but it also raises some questions. Can part-time or seasonal jobs, plus domestic harvesting, continue to satisfy northern natives? Will northern peoples continue to be satisfied with "enough" rather than wanting more and more, like southern consumers? Can existing resource management structures, and the increasing number of comanagement processes that include native harvesters, ensure sustainable resource use? Will the largely southern, urban, antiharvesting campaigns allow native communities to develop economies appropriate to their own cultures, resources, and values? What other alternatives exist in remote, northern communities? How *do* you define sustainable development?

FIGURE 20-D   Animals killed for food sometimes also can furnish useful byproducts. Here is a polar bear skin stretched for drying at Pond Inlet (photo by Heather Myers).

Professor Heather Myers
University of Northern British Columbia
Prince George

are making a comeback, based on the high cost of fuel and the fact that if one becomes stranded, "one can't eat a snowmobile."

NODES OF SETTLEMENT

Because the people of the Arctic Region increasingly tend to cluster, a nodal pattern of population distribution is becoming apparent. Most settlements lack many attributes of towns in other parts of the continent: Their form is often sprawling, amorphous, and unregimented to a street pattern; their urban functions are extremely limited; their buildings are often raised above ground level to keep the permafrost from thawing and buckling the foundations; and provision of utilities is generally primitive or totally absent (sometimes the piped conduits for water, sewage, and gas are in "utilidor" or

FIGURE 20-12   A snowmobile pulling a sled at a hunting camp on offshore ice near Barrow, Alaska (Steve Mc-Cutcheon/Visuals Unlimited photo).

FIGURE 20-13   Modern apartment houses in Inuvik, with a utilidor system in the foreground (courtesy Richard Hart-mier/Northwest Territories Researces, Wildlife and Economic Development, Government of the Northwest Territories).

"servidor" systems raised above the ground in heavily insulated tunnels) (Fig. 20-13).

Despite the agglomerating tendencies, there are only a few settlements of any size (Table 20-1). The larger settlements are listed.

1. *Adak* and *Unalaska* are the largest communities in the Aleutian Islands. The former is primarily a naval base, but dependents are permitted to live there; so it has a relatively normal urban form and function. Unalaska, which has incorporated the across-bay settlement of Dutch Harbor, is now a major commercial fishing port.

2. *Bethel* and *Kotzebue* are Bering Sea towns of primarily Inuit population but some diversity of function. Bethel is a commercial fishing center, and Kotzebue has become an attraction to fly-in tourists who want to see a "real" Inuit village.

TABLE 20-1

Largest Urban Places of the Arctic Region, 1995

Name	Population
Adak, AK	3600
Barrow, AK	3500
Churchill, MN	1200
Iqaluit, NWT	3300
Inuvik, NWT	3500
Kotzebue, AK	2600
Nome, AK	2800
Pangnirtung, NWT	1100

3. Despite its relative isolation, *Nome* has managed to function as a subregional commercial center (Fig. 20-14). Only about one-third of its population is Inuit.

4. *Barrow is* the administrative center of the North Slope Borough, an oil-rich local government unit that has built new schools, sewers, and water lines, but that has run up an incredible $1 billion debt. It is an oil operations center and probably has the largest Inuit population of any settlement in the region.

5. Twin towns of the Mackenzie delta are *Aklavik* (the old settlement) and *Inuvik* (the newer planned town and administrative center for the Western Arctic). Both are situated on the edge of the boreal forest but have strong functional relationships with the Arctic Region.

6. *Iqaluit* (previously known as *Frobisher Bay),* on Baffin Island, is the administrative center for Canada's Eastern Arctic and is distinguished by an impressive town center, a large federal building, a high-rise apartment block, a relatively busy commercial airfield, and a weekly newspaper. It contains the largest Inuit community in Canada.

7. *Churchill* is a grain-shipping port during its 3-month navigation season, for its ocean route for transporting Prairie wheat to Europe is 1000 miles (1600 km) shorter than the route via eastern Canada. Churchill experiences a flurry of tourist visitation in the fall owing to the attraction of polar bears migrating right through the town.

ECONOMIC SPECIALIZATION

The commercial economy of the Arctic Region involves only a few specialized activities on a generally limited scale.

FIGURE 20-14 Massive grain elevators dominate the scene at Churchill, Manitoba (Tim Hauf/Visuals Unlimited photo).

Commercial Fishing

There are two areas where commercial fishing is of considerable significance. In both places it is primarily a seasonal activity, and many participants are outsiders who come into the region for only a few weeks during the fishing season. The Bristol Bay area of southwestern Alaska is one of the state's principal salmon fisheries. Canneries operate at a feverish pitch during late summer and fall at King Salmon, Dillingham, and other localities. There is also a prominent Bering Sea fishery for halibut and crabs.

Along the north coast of Labrador there has long been an important commercial fishing venture, primarily for cod but also supplying char and salmon. The area is almost uninhabited now, and with the recent decline of the cod stocks, the fishery has diminished in importance. Essentially all the commercial fishing in the area is offshore by vessels from the island of Newfoundland.

Small and sporadic commercial fisheries are also in operation at other places in the region, most notably for arctic char, a freshwater species.

Commercial Reindeer Herding

Reindeer were introduced into Alaska from Siberia in the 1890s, with the idea of establishing a viable pastoral enterprise among the Inuit. The Canadian government subsequently brought some Alaskan reindeer to the Mackenzie delta in the late 1920s to serve as a nucleus for eventual dispersal.

In neither case, however, has the reindeer experiment been very successful. Although the Alaskan herds increased to a total of more than 600,000 reindeer in the 1930s, problems developed and the numbers diminished rapidly. There are now about 40,000 reindeer in Alaska most of them in 14 separate herds on the Seward Peninsula. Since the early 1970s the sale of antlers has become the driving force in the industry, with most sales to Korea.

The Mackenzie delta herd reached a peak of about 9000 animals in 1942, but interest waned and the herd has declined to under 3000 today.

Mineral Industries

Exploration for oil and gas has been extensive and is continuing. The only commercially feasible discovery thus far has been the tremendous reserves of Alaska's North Slope, centering on Prudhoe Bay. The North Slope now provides about 20 percent of all U.S. oil production, and Alaska ranks second only to Texas among producing states. Only a relatively few wells are involved (about 1000), but the investment is enormous. Some 6000 people work on the North Slope in summer, and about half that many in winter.

About 125 miles (200 km) east of Prudhoe Bay is another potentially commercial petroleum deposit that is currently the site of the bitterest environmental controversy in Alaska. The oil underlies the coastal plain of the Arctic National Wildlife Refuge. Enormous dispute about the economic and ecological desirability of "development" of this

A CLOSER LOOK   Precious Jewels in the Tundra

The Northwest Territories (NWT) has been a steady supplier of gold for over 60 years, and there are two lead/zinc mines there as well, but now the prospect of diamond mining in the Lac de Gras area is causing excitement (Fig. 20-e). Though the potential for diamonds to exit in the NWT has been known for some time, it was not until the early 1990s that diamond prospecting discoveries led to a "staking rush" in the tundra 180 miles (300 km) north of Yellowknife. At the peak, in March 1993 alone, 2572 applications were made to record diamond mining claims, covering 9030 square miles [November 1991, by comparison, saw 87 applications for 283 square miles (735 km²) and the entire year of 1991 saw 831 claims staked for 2572 square miles

(6,660 km²)]. A total of 102,364 square miles had been staked by May 1996.

Diamonds are found in association with kimberlite pipes, extrusions of magma in earlier geologic times that pushed up through overlying layers of rock. In the NWT, these pipes tend to be softer than the surrounding "country rock," so they have often become the sites of lakes, after glaciation. The way it is proposed to mine these pipes is to carve down into the rock, in open-pit and underground mines, encircling each of the pipes and extracting the ore. Processing the gems is relatively easy, compared to the requirements for other minerals; the ore is crushed, the rough diamonds are extracted, sorted and sent out to be cut or processed further. Other minerals

such as gold or lead/zinc may require several of chemical proccessing, which results in toxic waste by-products that need to be treated, stored or disposed of.

What do you need in order to set up a diamond mine in the tundra? The Lupin gold mine, already operating in the area, annually builds a winter road, consisting of plowed stretches across lake ice (once the ice is strong enough to support heavy vehicles!) and cleared portages across land. This will be used by the new diamond mine to bring in fuel, equipment and materials by truck. This is much less expensive than airfreight which must be used in the summer—it just requires planning to make sure shipments occur during the winter! There will be 17 miles (29 km) of all-

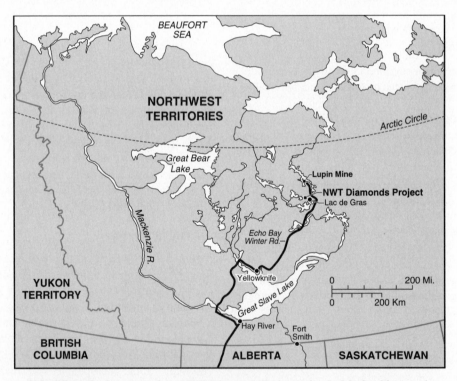

FIGURE 20-E   Location of the NWT Diamonds Project (after L. MacLochlan, et al., 1996, *NWT Diamonds Project; Report of the Environmental Assessment Panel,* June 1966, Canadian Environmental Assessment Agency, Minister of Supply and Services Canada, Ottawa).

weather or year-round, built around the mine sites themselves. An airstrip is required to receive Hercules C 130 planes carrying freight, as well as Boeing 727s and 767s carrying workers, food, perishables, critical material in, and taking diamonds out. There also must be a 400-person permanent camp for workers, a power plant, maintenance shop, office and warehouse space, and a security building in which to keep the diamonds. Systems need to be set up to store and distribute fuel, water, and sewage, and to dispose of wastes.

Tailings, the waste rock and by-products of processing, are a principal concern in mine design and environmental management. Over the life of this project, 146 million tons (133 million tonnes) of ore will be mined, generating thirty-five to forty million tonnes of waste rock each year; this will be contained in tailings ponds or, later in the life of the mined-out pit. The mine's plan includes letting the tailings settle and consolidate in layers; eventually they would be revegetated and converted into a wetland. Water is part of the tailings, and is discharged from these ponds, but it must meet prescribed standards of purity first.

The proposed mine was reviewed by an Environmental Assessment Panel in 1995–1996. Issues raised in the hearing included impacts on wildlife, particularly grizzly bears; quarrying for gravel in eskers, which are critical denning, feeding and travelling habitat for wildlife; impacts on water quality and fish from draining 5 lakes in order to access the kimberlite pipes; access by northerners and local native communities to jobs; social impacts in northern communities created by large incomes, periods of absence, and a probable influx of alcohol and drugs; settlement of native land claims before undertaking further development; access for the territorial government to the royalties generated by the mine, in order to augment territorial government revenues and balance some of the costs of providing more services;

preparing workers and communities for the inevitable closure of the mine (expected in 25 years).

The project review generated some heasted responses from intervenors, but the reality is, the mine proposal has come at a time when employment is desperately needed for northerners, not to mention that royalty revenues are needed by governments. The currently proposed mine involves operations that are limited in geographic scale 28 square miles (73 km^2) and probable impact, though this might change should other diamond mines or developments occur in the area (unfortunately, the review panel was not permitted to deal with such cumulative impacts). The Panel approved the mine subject to some conditions, and made recommendations for monitoring of socio-economic and environmental effects, and requirements for the mine to negotiate impact-benefit agreements with native peoples.

The mine company estimated that the project could employ 830 workers per year, with 60% being NWT residents and 25% being aboriginal; this would generate perhaps $39 M/year in wages to northern residents. Workers at the mine would follow a "2 in/2 out" rotation, working at the site for two weeks, then going home to their communities for two weeks off. Training, community liaison and other programs were proposed to remove barriers to aboriginal peoples' employment at the mine.

The Panel noted that employment from the mine could relieve poverty and instability in the wage economy, but that the needs and aspirations of northern peoples must be recognized, that peoples' close connections to family and community must not be strained, that there must be no damage to the relationship between people and the land, and that other economic/lifestyle options must not be limited. The Panel also commented that there is generally a low level of interest among northern native peoples regarding mine-work, and

that there is, furthermore, skepticism about the long-term prospects for mine employment and thus a desire to protect the capacity of the land to supply a subsistence livelihood should jobs disappear.

The proposal and review have included native peoples' concerns in two interesting ways. This is the first Environmental Assessment in Canada which has explicity called for the use of Traditional Indigenous Knowledge in assessment studies, which means that local indigenous peoples' understanding of environmental and social processes would be incorporated. This raised questions about how such knowledge should be collected and used, and by whom—indigenous peoples, government, or industry? Second, indigenous witnesses from Papua New Guinea were brought to the hearings in Yellowknife by the local Dene intervenors, to talk about their experiences with the mining company's operations in PNG, where a large tailings pond spill has affected their communities and livelihoods.

What are the implications—whether for native communities, the territorial government in the NWT, or the federal Canadian government, of having a lucrative development like this owned and operated by a foreign-multinational company? Will northern concerns and interests be fully understood and considered? Will larger international economic forces begin to dominate the NWT economy and policy regime, if 50% of employment in the mining industry (already a large sector of the territorial economy) is dependent on this diamond mine? Should this be just the first of other diamond mines, will the scale and pace of development swamp the environment and peoples of the North or will they be enriched by it?

Professor Heather Myers
University of Northern British Columbia
Prince George

deposit appeared about to be resolved by governmental permission for drilling when the 1989 *Exxon Valdez* oil spill in Prince William Sound provoked agonizing reappraisal of the enterprise. At the time of this writing, no final decision has been reached, but it seems likely that "development" will win out over "preservation" in this case.

Considerable oil, especially in the Mackenzie delta–Beaufort Sea area, and natural gas, particularly on Melville and King Christian islands, have been discovered in the Canadian Arctic, but the indicated reserves thus far are considerably smaller than those of the North Slope. Moreover, production costs are very high (drilling expense in the Beaufort Sea is about $30 million per hole) and the massive problem of transportation awaits future solution.

Considerable metallic mineral wealth is known to exist in this region, but transportation is so expensive that only a few mines have been developed. Notable among these are the following:

> The Polaris mine on Little Cornwallis Island is the northernmost metal mine in the world. A zinc and lead mine that opened in the early 1980s, it has about 200 employees. There is no town; all employees live in a large accommodation block and work on a rotating basis.
>
> The zinc–lead–silver mine at Nanisivik started production in 1976. The population of Nanisivik is about 320, of whom 200 work in the mine.
>
> Alaska's most important hardrock mine is the Red Dog mine in the DeLong Mountains about 90 miles (150 km) north of Kotzebue. It is already the largest zinc mine in North America and may become the largest zinc–lead mine in the world. It is a joint operation of a Canadian mining corporation and the local native corporation (NANA). Production began in 1989, and more than 350 people work in the mine and the adjacent concentrating mill. The concentrate is trucked on a 54-mile (90-km) road to a new port, where it is stored in the largest building in Alaska.

The most exciting ore discovery of recent years is a nickel–copper–cobalt body discovered a few miles southwest of Nain, Labrador, in 1993. The ore body is close to the surface, close to the coast, and should be relatively inexpensive to exploit.

TRANSPORTATION

Inadequate transportation is the outstanding deterrent to economic progress in the Arctic. All-weather roads have been almost nonexistent, and the only railway line is that connecting Churchill with the Prairie Provinces. In other words, conventional forms of land transportation are almost totally lacking. Tractor trains, temporary winter roads, and snowmobiles provide minimal transport during the cold season, and some useful riverboat service exists during summer. By and large, however, regional transport depends on oceangoing vessels, which have a short navigation season, or on aircraft, which are expensive.

There is now a great deal of scheduled air service in the region. Resolute on Cornwallis Island, for example, receives regular air traffic from both Montreal and Edmonton and is the hub of the High Arctic. A surprising number of small settlements are served at least occasionally by scheduled flights. Nonscheduled flying fills in many gaps. Construction of an airfield has therefore become an essential for most population nodes, and frequently the airfield is the location of the most modern and desirable facilities to be found locally. The light aircraft has become an almost ubiquitous link between the inhabitants of the region and the products and services of the postindustrial civilization on which they are dependent. In the Canadian Arctic the per capita airline boardings amount to nearly five; for the rest of Canada the figure is less than one.

Many settlements, however, still receive their bulk supplies from government patrol boats that often cover thousands of miles on a single summer trip, bringing foodstuffs, fuel, and other materials that must last until the ship returns the following summer. This service is particularly characteristic of the eastern Canadian Arctic and the northern coast of Labrador.

THE OUTLOOK

The region doubtless will continue as a land of great distances and few people where nothing more than a scanty livelihood is obtainable by trapping, hunting, fishing, or grazing. Cities in the true sense will, as now, be nonexistent. Fur trapping will continue to be important to the natives, but it is too dependent on the vagaries of fashion, pressures from the animal-rights movement, fluctuation of prices, and biological cycles to provide a steady means of livelihood for large numbers of people.

Expansion of commercial fishing offers considerable promise in a few localities, but in most of the region its growth possibilities seem quite limited. Expansion of reindeer herding seems logical in theory, but the lesson of seven decades of history is thoroughly negative.

Family and community life continue to deteriorate over most of the Arctic Region, as the population agglomerates in fewer settlement nodes and acquires a taste for an alien way of life but has scant opportunity to make a living. The traditional and the modern are in a state of continual collision; the children apathetically watch an American soap opera on television while their grandmother chews a caribou skin to soften it.

Efforts are being made to retain traditional skills, virtues, and values. Some villages restrict both TV and alcohol use. The government of the Northwest Territories has introduced an "outpost camp programme" to encourage hunters to live at least part-time in outposts in order to pursue a traditional way of life; more than 100 such camps have been established.

Despite such efforts, it is likely that most natives of the region will make the transition to a "civilized" way of life, and the transiton probably will be a difficult and painful one. They love the North and are adjusted to northern living and so could become the backbone of northern development if they could acquire new skills without losing their identities. But with the perishing of the "old" way of life, employment opportunities must be made available or the native is likely to sink into a slough of apathy and degradation, as the monumental problem of alcoholism clearly demonstrates. Is it possible to develop a sound economic base for the native people of the Arctic without endangering their cultural survival?

Mineral exploitation will almost surely provide the major economic development stimulus for the region. The actualities of Prudhoe Bay and the potentialities of the Beaufort Sea area and the Mackenzie Valley corridor will be the principal foci of activity. What will this do for local inhabitants?

Will they receive "early, visible, and lasting benefits" from such developments, as is the Canadian government's stated ambition? Oil drilling and pipeline operation can be either a boon or a burden (or both) to the region.

The regional and village corporations in the Alaskan portion of the region now have considerable capital to work with and it is quite likely that native claim payments will significantly augment the economy in the Canadian tundra as well (as is already the case with the James Bay Cree). The wise use of these windfall monies could do much to alleviate the bleakness of the long-term prospects for the native people.

The establishment of Nunavut in 1999 probably will set in motion a variety of economic and social activities that are yet unforeseen. The reaction of Alaskan Inuit and Canadian Dene and Métis will be watched with interest.

Even in this thinly populated and rarely visited region questions of environmental preservation are being raised and debated at length. The major focus of concern involves oil and gas activities, which have generally been subjected to very restrictive environmental protection regulations. Significant areas were set aside as conservation reserves of one sort or another in the Alaskan tundra. Such actions have thus far been much more limited in the vaster expanses of the Canadian tundra, but an extensive inventory of the land resource has been accomplished, with the result that six major wilderness parks have been proposed and some 136 "significant conservation" areas have been recognized.

Tourism undoubtedly will be an economic growth pole. The "hostile" natural environment becomes a natural resource that will bring in visitors and money and not "use up" the resource. Iqualuit on Baffin Island already has more than 200 hotel rooms, and a surprising amount of tourism has come to Resolute, mostly through wilderness outfitters.

Decision-makers of both nations are faced with the difficult dilemma of dealing with a region that has a fragile resource base, a rapidly changing geopolitical scenario, a marginal economy, and heavy pressure for exploitation of energy minerals. It is in many ways a classic conflict of colonialism.

SELECTED BIBLIOGRAPHY

BIRD, J. BRIAN, *The Physiography of Arctic Canada.* Baltimore: Johns Hopkins Press, 1967.

BRUEMMER, FRED, *The Arctic.* Englewood Cliffs, NJ: Prentice-Hall, 1974.

———, "Churchill: Polar Bear Capital of the World," *Canadian Geographic,* 103 (December 1983–January 1984), 20–27.

———, "Northern Oases: Polynyas, Where Arctic Waters Teem with Wildlife," *Canadian Geographic,* 114, (January–February 1994), 54–63.

ENGLAND, JOHN, "Ellesmere Island Needs Special Attention," *Canadian Geographic,* 103 (June–July 1983), 8–17.

EVERETT, K. R., "Summer Wetlands in the Frozen North," *Geographical Magazine,* 55 (October 1983), 510–515.

HARRY, DAVID G., "Banks Island: Gem of the Western Arctic," *Canadian Geographic,* 102 (October–November 1982), 40–49.

HERRINGTON, CLYDE, *Atlas of the Canadian Arctic Islands.* Vancouver: Shultoncraft Publishing Co., 1969.

"Islands of the Seals: The Pribilofs," *Alaska Geographic,* 9 (1982), entire issue.

LAMONT, JAMES, "Pangnirtung: Gateway to Our Only Arctic Park," *Canadian Geographic,* 100 (August–September 1980), 34–37.

LYNCH, WAYNE, "A New Arctic Refuge," *Canadian Geographic,* 115 (March–April 1995), 24–35.

———, "Our Unknown National Park," *Canadian Geographic,* 112 (July–August 1992), 18–28.

MARSH, P., AND C. S. L. OMMANNEY, EDS., *Mackenzie Delta: Environmental Interactions and Implications of Development.* Saskatoon: National Hydrology Research Institute, 1991.

MATTHIASSON, J. S., *Living on the Land: Change among the Inuit of Baffin Island.* Peterborough, Ont.: Broadview Press, 1992.

MAXWELL, J. B., *The Climate of the Canadian Arctic Islands and Adjacent Waters,* Vols. I and II. Hull, Quebec: English Publishing, 1980.

"Nome, City of the Golden Beaches," *Alaska Geographic,* 11 (1984), entire issue.

"North Slope Now," *Alaska Geographic,* 16 (1989), entire issue.

PELLY, DAVID F., "Dawn of Nunavut," *Canadian Geographic,* 113 (March–April 1993), 20–31,

———, "Barrens and Borders; Passing the Torch to Protect the Thelon's Wild Side," *Canadian Geographic,* 116 (March–April 1996), 60–65.

———, "Pond Inlet," *Canadian Geographic,* 111 (February–March 1991), 46–52.

———, "The Faces of Nunavik," *Canadian Geographic,* 115 (January–February 1995), 24–37.

PYNN, LARRY, "The Dempster," *Canadian Geographic,* 109 (June–July 1989), 32–39.

RIEWE, R., ED., *Nunavut Atlas.* Edmonton: Canadian Circumpolar Institute, 1994.

STIX, JOHN, "National Parks and Inuit Rights in Northern Labrador," *The Canadian Geographer,* 26 (Winter 1982–1983), 349–354.

"The Aleutian Islands," *Alaska Geographic,* 22 (1995), entire issue.

"The Lower Yukon River," *Alaska Geographic,* 17 (1990), entire issue.

"Unalaska/Dutch Harbor," *Alaska Geographic,* 18 (1991), entire issue.

WENZEL, G. W., *Animal Rights, Human Rights: Ecology, Economy, and Ideology in the Canadian Arctic.* Toronto: University of Toronto Press, 1992.

INDEX

X

Y

Z

MAJOR POLITICAL UNITS
OF THE UNITED STATES AND CANADA

	ABBREV.	AREA MI²	AREA KM²	AREA RANK-ORDER	POPULATION TOTAL*	POPULATION RANK-ORDER	CAPITAL
Alabama	AL	51.6	133.6	29	4,253	22	Montgomery
Alaska	AK	586.4	1518.8	1	604	48	Juneau
Arizona	AZ	113.9	295.0	6	4,218	23	Phoenix
Arkansas	AR	53.1	137.5	27	2,484	33	Little Rock
California	CA	158.7	411.0	3	31,589	1	Sacramento
Colorado	CO	104.2	269.9	8	3,747	25	Denver
Connecticut	CT	5.0	13.0	48	3,275	28	Hartford
Delaware	DE	2.1	5.4	49	717	46	Dover
Florida	FL	58.6	151.8	22	14,166	4	Tallahassee
Georgia	GA	58.9	152.6	21	7,201	10	Atlanta
Hawaii	HI	6.5	16.8	47	1,187	40	Honolulu
Idaho	ID	83.6	216.5	13	870	41	Boise
Illinois	IL	56.4	146.1	24	11,830	6	Springfield
Indiana	IN	36.3	94.0	38	5,803	14	Indianapolis
Iowa	IA	56.3	145.8	25	2,842	30	Des Moines
Kansas	KS	82.3	213.2	14	2,565	32	Topeka
Kentucky	KY	40.4	104.6	37	3,860	24	Frankfort
Lousiana	LA	48.5	125.6	31	4,342	21	Baton Rouge
Maine	ME	33.2	86.0	39	1,241	39	Augusta
Maryland	MD	10.6	27.5	42	5,042	19	Annapolis
Massachusetts	MA	8.3	21.5	45	6,074	13	Boston
Michigan	MI	58.2	150.7	23	9,549	8	Lansing
Minnesota	MN	84.1	217.8	12	4,610	20	St. Paul
Mississippi	MS	47.7	123.5	32	2,697	31	Jackson
Missouri	MO	69.7	180.5	19	5,324	16	Jefferson City
Montana	MT	147.1	381.0	4	870	44	Helena
Nebraska	NE	77.2	200.0	15	1,637	37	Lincoln
Nevada	NV	110.5	286.2	7	1,530	38	Carson City
New Hampshire	NH	9.3	24.1	44	1,148	-42	Concord
New Jersey	NJ	7.8	20.2	46	7,945	9	Trenton
New Mexico	NM	121.7	315.2	5	1,685	36	Santa Fe
New York	NY	49.6	128.5	30	18,136	3	Albany
North Carolina	NC	52.6	136.2	28	7,195	11	Raleigh
North Dakota	ND	70.7	183.1	17	641	47	Bismarck
Ohio	OH	41.2	106.7	35	11,151	7	Columbus
Oklahoma	OK	69.9	181.0	18	3,278	27	Oklahoma City